SpringerWienNewYork

CISM COURSES AND LECTURES

The series presents lecture notes, monographs, edited works and proceedings in the field of Mechanics, Engineering, Computer Science and Applied Mathematics.
Purpose of the series is to make known in the international scientific and technical community results obtained in some of the activities organized by CISM, the International Centre for Mechanical Sciences.

INTERNATIONAL CENTRE FOR MECHANICAL SCIENCES

COURSES AND LECTURES – No. 487

ROMANSY 16

ROBOT DESIGN, DYNAMICS, AND CONTROL

EDITED BY

TERESA ZIELIŃSKA, CEZARY ZIELIŃSKI
WARSAW UNIVERSITY OF TECHNOLOGY

SpringerWienNewYork

This volume contains 292 illustrations and 34 tables

SPIN 11791393

In order to make this volume available as economically and as rapidly as possible the authors' type scripts have been reproduced in their original forms. This method unfortunately has its typographical limitations but it is hoped that they in no way distract the reader.

ISBN 3-211-36064-6
ISBN 978-3-211-36064-4

PREFACE

This volume contains the papers presented at the 16th Symposium on Theory and Practice of Robots and Manipulators, Warsaw, June 21–24, 2006. All papers had been reviewed by two independent reviewers before they were accepted for final publication and presentation at the Symposium. The event was organized under the supervision of international Steering Committee consisting of: M.Ceccarelli (IFToMM Secretary General, University of Cassino, Italy), I-Ming Chen (Nanyang Technological University, Singapore), B.Heimann (Chair of Technical Committee Robotics, Hannover University, Germany), E.Martin (Space Agency, Canada), O.Khatib (Stanford University, USA), W.Schiehlen (CISM representative, Technical University Munich, Germany), O.Takanishi (Waseda University, Japan), T.Zielińska (Warsaw University of Technology, Poland). Local Organizing Committee consisted of T.Zielińska, M.Olszewski, C.Zieliński, K.Kędzior from Warsaw University of Technology, K.Kozłowski from Poznań Technical University, and K.Tchoń from Wrocław Technical University. The symposium was held at the Faculty of Mechatronics, Warsaw University of Technology.

Main Building, Warsaw University of Technology

The 1st CISM-IFToMM Symposium on Theory and Practice of Robots and Manipulators was held on Sept. 5–8, 1973, in Udine, Italy, not long after IFToMM had been founded in 1969. The first ROMANSY, or Ro.Man.Sy, as the Symposium used to be referred to, marks the beginning of a long-lasting partnership between two international institutions, CISM, the Centre International des Sciences Mécaniques, and IFToMM, the International Federation for the Promotion of Mechanism and Machine Science. ROMANSY is one of the activities of IFToMM Technical

Committee for Robotics. The Symposium has taken place every even-numbered year with only one exception for the first symposium. It is traditionally a limited gathering of scientists that encourages informal discussions and focuses on recent trends and advances in robotics. The volume is organized into nine chapters with more than 50 papers in all. The Authors from 17 countries discussed the problems grouped in the following thematic parts:

— *Robot Design, Mechanism Performance,*
— *Motion Planning and Synthesis,*
— *Control Methods and Systems,*
— *Humanoids,*
— *Biology and Robotics – Specialized Tools and Methods,*
— *Innovative Technologies in Robotics,*
— *Space Robotics,*
— *Vision and Navigation.*

The key-note presentations dealt with the problems of the coexistence of humans and personal robots providing assistance to people in their housework, or to the elderly and the handicapped, as well as the robots working with or without human help in space missions. It is interesting that the cultural aspects influencing the robotics research also attracted the attention of the Scientists. Modeling and control methods of complex human-like robotic systems are developing very fast with the goal to produce a robot with human motion skills. To effectively work and cooperate with us, robots must exhibit abilities that are comparable to those of humans. The speakers focused on the ongoing efforts to design and develop human-friendly robotic systems that can safely and effectively interact and work with humans. The progress in robotics is also stimulated by human will to explore outer space. This issue creates specific requirements, limitations and targets for the designers. Standardization in space robots is necessary to enable the creation of advanced cooperative systems, where different technologies, requirements, control systems etc. meet. Finally, standardization is a critical element in having large numbers of robots working safely side-by-side their human counterparts.

*The 16th ROMANSY solicited papers providing a vision of the evolution of the robotics disciplines and identifying new directions in which these disciplines are foreseen to develop. The papers are devoted to novel robots, humanoids and bio-robotics problems, challenges in control and motion synthesis, kinematical and dynamical analysis of robotic systems, perception problems, space robots, and to other promising innovative mechanisms and technologies. We hope that the material included in this volume does not limit itself to just reporting the ongoing research, but will also stimulate the Reader to create new ideas and solutions, as: „**Every scientists is an artist and every artist is a scientist in a part**" (from: Summa Technologiae, by Stanisław Lem, 1964, Wydawnictwo Lubelskie).*

Warsaw, 20th May 2006

Teresa Zielińska
Cezary Zieliński

CONTENTS

Keynote Lectures

Humanoid Robotics, Culture and Society of Japan

Atsuo Takanishi

Department of Mechanical Engineering / Humanoid Robotics Institute, Waseda University

3-4-1 Ookubo, Shinjuku-ku, Tokyo 169-8555

Japan

Even though the market size is still small at this moment, applied fields of robots are gradually spreading from the manufacturing industry to the others in recent years. One can now easily expect that applications of robots will expand into the first and the third industrial fields as one of the important components to support our society in the 21st century. There also raises strong anticipations in Japan that robots for the personal use will coexist with humans and provide supports such as the assistance for the housework, care of the aged and the physically handicapped, since Japan is one the fastest aging societies in the world. Consequently, humanoid robots and/or animaloid robots have been treated as subjects of robotics researches in Japan such as a research tool for human/animal science, an entertainment/mental-commit robot or an assistant/agent for humans in the human living environment. Over the last couple of years, some manufactures including famous global companies started to develop prototypes or even to sell mass production robots for the purposes mentioned above, such as TOYOTA, TMSUK, SONY, HONDA, Mitsubishi Heavy, ZMP, etc. Most of those robots have some life-likeness in their appearances and behaviors. Why are so many Japanese companies developing humanoid robots? I believe there is a special reason which strongly relates to the national character of Japan in terms of the history, the religion and the culture of Japan. On the other hand, Waseda University, where I belong to, has been one of the leading research sites on humanoid robot research since the late Prof. Ichiro Kato and his colleagues started the WABOT (WAseda roBOT) Projects and developed the historical humanoid robots that are WABOT-1 and WABOT-2 done in the early 70s and 80s respectively. One of the most important aspects of our research philosophy is as follows: By constructing

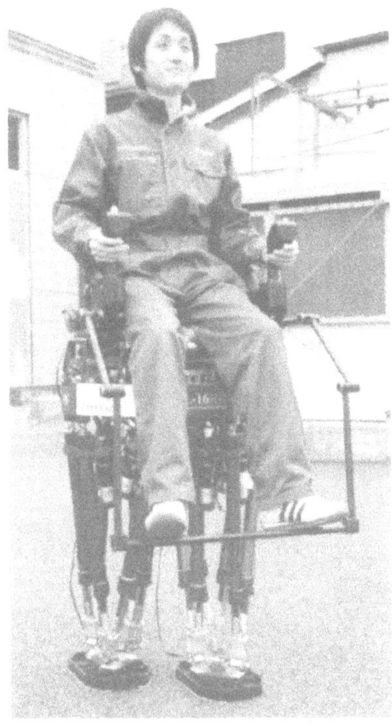

anthropomorphic/humanoid robots that function and behave like a human, we are attempting to develop a design method of a humanoid robot having human friendliness to coexist with humans naturally and symbiotically, as well as to scientifically build not only the physical model of a human but also the mental model of it from the engineering view point. Based upon the philosophy, I and my colleagues have been doing researches on humanoid robots, such as the Biped Walking Robots, Emotion Expression Robots, Mastication Robots, Flute Player Robots, Speech Production Robots, etc. In this plenary speech, I will introduce the national character of Japan in terms of the historical, religious and cultural backgrounds of Japan, as well as the research philosophy of humanoid robotics, the design concept of the humanoid robots and its applications with the robots mentioned above as examples.

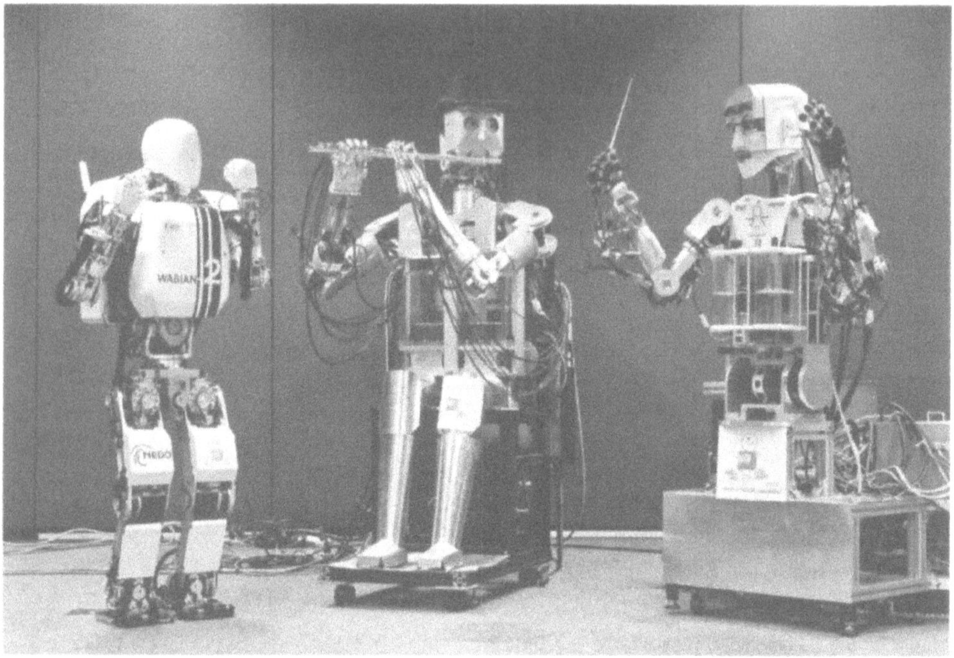

The Human Frontier: Robotics New Quest and Challenge

Oussama Khatib

Department of Computer Science
Stanford University, Stanford, California, USA
khatib@cs.stanford.edu

Summary

Robotics is rapidly expanding into human environments and vigorously engaged in its new emerging challenges. Interacting, exploring, and working with humans, the new generation of robots will increasingly touch people and their lives. The successful introduction of robots in these environments will rely on the development of competent and practical systems that are dependable, safe, and easy to use. To effectively work and cooperate with a person, robots must display abilities and skills that are compatible with those of humans. The discussion focuses on the ongoing effort for the design and development of human-friendly robotic systems that can safely and effectively interact and work with humans.

A major component in these developments is a new framework for the modeling and control of complex human-like robotic systems. In this framework, the various problems associated with (i) the motion coordination of the large number of degrees of such robots; (ii) the effective control of their contacts and interactions with the environment; (iii) the maintenance of their internal and external constraints; (iv) and the strategies for dealing with their underactuation and balance are all treated in a unified fashion within a general whole-body control structure. This is a task-oriented control structure that addresses the whole body dynamics for specifications involving multiple distributed tasks and postures in consistency with the requirements of multiple distributed contacts and constraints.

The second component in this effort is concerned with the synthesis of natural human movements to produce human-like robot behaviors. The objective is to unveil the underlying characteristics of human motion through an elaboration of its physiological basis. The aim is to formulate general strategies for whole-body robot behaviors. This exploration has employed models of human musculoskeletal dynamics and made use of extensive experimental studies of human subjects with motion capture techniques. Our study of human motion has revealed the dominant role physiology plays in shaping human motion. The characteristics of human motion revealed in this study have allowed the development of generic motion creteria that efficiently and effectively encode human motion behaviors.

The third component in our effort is concerned with the critical issue of safety in robot design. Our work in human-friendly robot design has led to the development of a new actuation methodology which has been shown to be well-suited for the emerging generation of robots conceived to operate in human environments. This methodology of distributed macro mini actuation, DM^2, addresses both the safety and performance characteristics of a robot. The approach has led to the design and construction of several prototypes, the last of which is a two-arm on a common torso robotic testbed. This new system represents a unique platform to explore the competing issues of safety and performance in the design of robot mechanisms. The new two-arm torso testbed is being used to validate the promise of safety and performance and to establish meaningful measures for safety and performance. Of particular interest is the analysis of impact forces in a three dimensional collision between a robot and its surroundings. Two safety standard measures are used to quantify the improvement in safety in terms of reduction of impact force, while the robot performance characteristics are evaluated against traditional design.

Other fundamental issues in human-centered robotics will be also examined in this presentation. These include the elastic planning methodology for real-time modifications of existing plans, and various other effective methodologies and efficient algorithms that address the computational challenges associated with human-like robotic structures.

Standardization: A Logical Step in Sustained Space Exploration

Arthur Bradley[1]

[1] NASA Langley Research Center, USA, Arthur.T.Bradley@nasa.gov

Abstract. As NASA pushes ahead with its initiative to create a sustainable campaign of space exploration, there is a growing recognition that a paradigm shift will have to occur in the space robotics industry. Specifically, robotic systems-of-systems must be realized through a renewed focus towards interoperability, modularity, and commonality. The practice of developing unique mission-specific robots operating independently and communicating through custom interfaces will have to give way to standardization.

1 Motivation

In January of 2004, NASA was given a presidential directive to "gain a new foothold on the moon and to prepare for new journeys to the worlds beyond our own." As part of this initiative, NASA is focused on returning to the moon by 2020 to serve as the launching point for missions beyond. Robotic probes are expected to be on the lunar surface by 2008, with a human mission following as early as 2015, "with the goal of living and working there for increasingly extended periods of time." Based on this mandate, NASA published the Vision for Space Exploration (VSE), a roadmap for achieving sustainable exploration. In 2005, NASA's Exploration System Architecture Study (ESAS) subsequently recommended that robotics development focus on three things: (1) human-robot interaction, (2) material handling and transportation, and (3) improved surface mobility.

The focus of lunar and Martian expeditions will migrate from early exploration, to site preparation, to long-duration habitation. Given the hostile environmental conditions, each of these activities will require the extensive use of robotic systems. Such ambitious objectives will introduce tremendous new challenges – requiring system planners to think far beyond simply getting to the remote location safely and making scientific measurements or taking high-resolution photographs. Robots that operate independently of one another, like those seen in the past (e.g. Sojourner, Spirit/Opportunity), will be inadequate to accomplish the complex tasks associated with these challenges. Rather, complex systems-of-systems will be required in which robots work cooperatively by widely exchanging information, planning and dividing complex tasks, sharing common resources, and physically cooperating to manipulate objects. The challenges associated with cooperative robotics are many, including communications, smart mechanisms, control, autonomy, physical compatibility, sensor processing, and operator control.

In this paper, we describe the issues considered by an early investigation (i.e. JTARS) into standardization of space robotics. The benefits, scope, and steps necessary for enabling advanced cooperative systems are all briefly discussed.

2 Joint Technical Architecture for Robotic Systems

In 2005, NASA's Office of Exploration Systems Research & Technology funded the development of the Joint Technical Architecture for Robotic Systems (JTARS). JTARS charter was to identify the interface standards necessary to achieve interoperability among space robots. The working group was made up of recognized leaders in the field of space robotics including representatives from seven NASA centers along with academia and private industry. The working group's early accomplishments included addressing key issues required for interoperability, defining which systems are within the project's scope, and framing the JTARS standards manuals around classes of robotic systems. Unfortunately, a subsequent realignment within NASA left the JTARS effort without the broad participation necessary to complete the task. The working group is currently looking for follow-on opportunities to fund this valuable work.

3 Benefits of Standardization

There are many benefits to standardizing space robots. The most obvious benefits would come in the form of enhanced capabilities, cost savings, and risk reduction.

Enhanced Capabilities
Standardization would enable the creation of integrated robotic systems-of-systems that incorporate numerous individual units working cooperatively. Such complex systems would be capable of meeting the upcoming technical challenges, including off-planet resource handling (extraction, processing, and transport) and habitat development (construction, inspection, and maintenance). Standardization would also prove invaluable in developing systems that are capable of being controlled by universal (i.e. shared) operator control stations. Finally, standardization is a critical element in having large numbers of robots working safely side-by-side their human counterparts.

Cost Savings
Clear and relevant standards would serve as guidelines for acquisitions. Such standards would promote competition by effectively "leveling the playing field." This would allow for smaller companies to enter the market by reducing the high research costs associated with developing mission-unique systems. This increase in competition would ultimately lead to better product selection and cost reduction. It is also expected that standards-driven systems would be more modular in nature, leading to the sharing and reuse of components (both hardware and software). Finally, standards-based systems would be easier and cheaper to validate and verify through the use of common reusable test sets and simulators.

Risk Reduction
Significant risk reduction would result from replacing unique "one of a kind" systems with robots equipped with well-understood interfaces and capabilities. Reliability is also expected to improve as the designs mature and are subsequently reused. By increasing competition, the risks currently seen by being tied to a small set of vendors would also decrease. Finally, standards are expected to be "living documents," useful in distributing lessons learned, and thus helping to ensure that all vendors are aware of important design considerations.

4 Scope

The space community has many systems under development, from satellites to rovers, many of which can be considered "robotic" in nature. An important consideration of any standardization program is therefore to clearly define the scope of the effort, specifically answering the question, "What is a robot?" Two broad approaches have generally been considered. One divides systems based on where and how they are to be used, and the other divides systems based on capabilities.

The metrics considered descriptive of robotic systems by NASA's Joint Technical Architecture for Robotic Systems (JTARS) working group are: processing capability, communications, and the ability to physically interact with the world. Figure 1 illustrates how these three metrics are used to determine which systems fall within the scope of a standardization program.

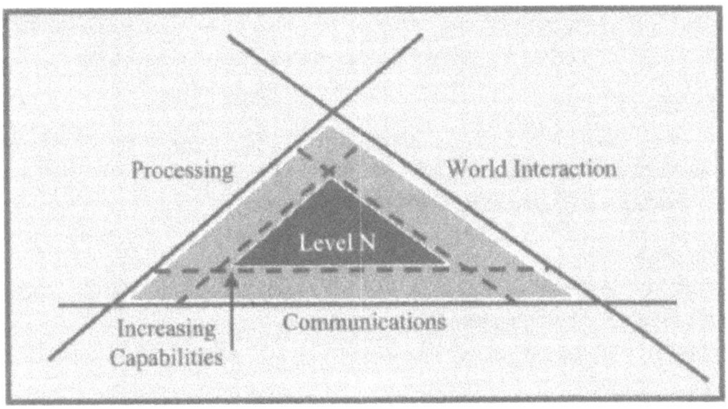

Figure 1. JTARS metrics used to define a "robot."

Systems that meet a minimum level of all three metrics would be considered robotic in nature and thus within scope. Within the core bounded region, sub-regions could also be identified to indicate systems with advanced capabilities. Additional requirements would likely be levied on more capable systems, enabling them to perform advanced cooperative activities. Therefore, it is expected that there would be different levels of compliance depending on system capabilities.

There are several important implications that result from this definition of scope. Specifically, the definition suggests that standards would be applied to robotic systems independent of their:

- Location (c.g. Mars, Earth, moon, in-space, etc.)
- Function (e.g. rovers, station crawlers, air vehicles, etc.)
- Level of autonomy (teleoperated to fully autonomous)

A broad inclusive definition of this nature removes the burden of asking the question "Does this robotic system need to be compliant?" Simply put, if the space system meets the minimum metrics defining a "robot," then it must be compliant to some minimum level, and perhaps a greater level dependent on what cooperative tasks it will be assigned. An important exception however, is that standards requirements should not supersede higher-level heritage system-use requirements. For example, new maintenance robots to be used on the International Space Station (ISS) would have to be compliant with ISS requirements first and foremost, even at the expense of the goals of long-term standardization.

5 Enabling Cooperative Behavior

For robots to cooperate autonomously, as well as work with nearby astronauts, standardization must extend to the systems that interact with the robots. Examples of such systems include: end-effectors/tools, sensor suites, overhead satellites, operator control stations, communication networks, ground support equipment, and other cooperative robots. Complimentary systems of this nature must be capable of interfacing with robotic systems in a uniform way.

Four high-level topics have been identified as necessary for achieving robot standardization: physical interactions, information exchange, command structure, and reference frames.

Physical Interactions

Standards will need to identify physical interactions associated with robots working cooperatively, including (but not limited to) end effectors, tools, mounts and hard points, maximum loads, workspace constraints, and electrical mates.

Information Exchange

Defining the methods of exchanging information between robots and/or supporting systems requires the identification of interconnection standards, communication protocols, and message sets. The International Organization of Standards Open Systems Interconnection (OSI) 7-layer model is a good example of a top-to-bottom interconnection standard around which various protocols have been developed. The Department of Defense's Joint Architecture for Unmanned Systems (JAUS) is a relevant example of a message set designed for communications to/from robotic systems.

OSI Model

The OSI model defines a networking framework for implementing protocols in seven layers. Other variants (e.g. 5-layer) have also been proposed. With either model, control is passed from one layer to the next, starting at the application layer in one station, proceeding to the bottom physical layer, then over the channel to the next station and back up the hierarchy of the receiver. Additional information can be found at ww.iso.org.

JAUS Message Set

The Joint Architecture for Unmanned Systems (JAUS) was developed by the Department of Defense as a common message set between unmanned systems. It is currently being converted to a Society of Automotive Engineering (SAE) standard. JAUS addresses many of the important considerations of communicating between unmanned (robotic) systems, including:

- Commands/Sequences
- Responses
- Queries

- Data
- Events
- Periodic Communications

Command Structure

For robots to cooperative effectively, a command structure will also need to be defined. The command structure will serve as a policy for processing commands, sharing data, and changing the authority structure within individual units or across the larger system. The structure would define which system(s) have authority to direct subordinate systems. Figure 2 is a simple illustration showing how the command structure would influence the handling of incoming and outgoing information.

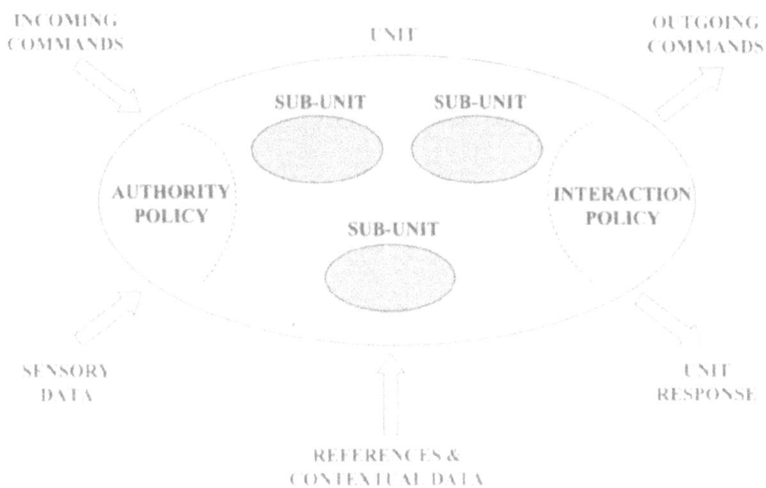

Figure 2. Command structure concept.

Reference Frames

Robotic systems working cooperatively will also require common frames of reference, including time, spatial distance, units of measure, and workspace definitions. Given the global collaboration in space activities, such definitions will have to address conventions that change across international boundaries.

Bibliography

United States Department of Defense Joint Technical Architecture, Vols. I and II, Version 6.0, October 2003.

Human and Robotic Technology Formulation Plan, Version 5.1, 29 July 2004 (Amended 13 September 2004), Development Programs Division, Office of Exploration Systems, National Aeronautics and Space Administration.

Technical Standards Program, information found at www.standards.nasa.gov.

Autonomy Levels for Unmanned Systems (ALFUS) Framework, *NIST Special Publication 1011*, September 2004.

Joint Architecture for Unmanned Systems, *JAUS Domain Model*, V3.0, December 2003.

Bradley, A., et. al., (2005). Enabling Interoperable Space Robots with the Joint Technical Architecture for Robotics Systems (JTARS), *8th International Symposium on Artificial Intelligence, Robotics, and Automation in Space*. Munich, September 2005.

Bradley, A., Holloway, C., *Final Report for Joint Technical Architecture for Robotic Systems*, submitted to NASA ESMD, January 2006.

Chapter I

Robot design

Design and Singularity Criteria of Parallel Manipulators

Victor Glazunov* and Alexander Kraynev*

*Mechanical Engineering Research Institute of RAS, Moscow, Russia,
e-mail: vaglznv@mail.ru

Abstract. This paper addresses the criteria of synthesis and singularity analysis of parallel manipulators. The dynamical decoupling, spatial oscillations as well as kinematic, static and dynamic singularities are considered. The proposed design criteria are intended only for dynamically decoupled manipulators whereas the singularity criteria can be applicable for general parallel manipulators.

1 Introduction

Parallel manipulators offer high stiffness and good position accuracy, they have been studied by numerous researchers. As it is said by J. Angeles (2004) "...the number of novel designs either published in conference proceedings and archival journals or disclosed in patent files is too rich to be recorded exhaustively". Let us mention only some books in which parallel manipulators are considered: Merlet (2000), Ceccarelli (2004), Glazunov, Koliskor, and Kraynev (1991). Most researchers have investigated maximal working volume or dexterous workspace (Chablat, Wenger, and Merlet, 2004, Glazunov and Munitsyna, 1994). In contrast with mentioned approaches other criteria (for example dynamical decoupling Kraynev and Glazunov, 1991, or reduction of the number of actuators in spatial oscillators Borozna, Glazunov, et al., 1992) involve different structures of parallel manipulators. To describe the various types of singularity Gosselin and Angeles (1990) have given the singularity criteria based on Jacobian matrices. Glazunov et al. (1990) proposed other singularity criterion based on the approach developed by Dimentberg (1965). This paper is focused on influences of the criteria mentioned above to corresponding results.

2 Dynamical decoupling

In this section, we represent the criterion of design of parallel manipulators taking into account the main motivation, which is the reduction of the dynamical coupling of the actuators. The basic idea is to represent the kinetic energy as the quadratic polynomial including only the squares of the generalized velocities.

Let us consider the Gough-Stewart platform. This manipulator is composed of a mobile platform connected to a fixed base via six kinematic sub-chains (legs) comprising one prismatic and two spherical pairs. Let the length of the i-th leg (the generalized coordinate) be denoted as l_i $(i=1...6)$.

The kinetic energy can be expressed by means of the 6×6 matrix (E) constituted by Plücker coordinates of the unit screws of the axes of the prismatic pairs:

$$(E)=\begin{pmatrix} x_1 & y_1 & z_1 & x_1^o & y_1^o & z_1^o \\ x_2 & y_2 & z_2 & x_2^o & y_2^o & z_2^o \\ \cdot & \cdot & \cdot & \cdot & \cdot & \cdot \\ \cdot & \cdot & \cdot & \cdot & \cdot & \cdot \\ \cdot & \cdot & \cdot & \cdot & \cdot & \cdot \\ x_6 & y_6 & z_6 & x_6^o & y_6^o & z_6^o \end{pmatrix} \tag{1}$$

Now let $(\dot{L})$ be the vector of the generalized velocities $(\dot{l}_1, \dot{l}_2, \dot{l}_3 ..., \dot{l}_6)^T$ and $\boldsymbol{\Omega}$ be the twist of the platform. The generalized velocities can be written as follows:

$$(E)\boldsymbol{\Omega}=(\dot{L}) \tag{2}$$

where $\boldsymbol{\Omega}$ is expressed as the vector $(v_x, v_y, v_z, \omega_x, \omega_y, \omega_z)^T$. Using the inverse matrix $(E)^{-1}$

$$(E)^{-1}=\begin{pmatrix} p_1^o & p_2^o & \cdot & \cdot & \cdot & p_6^o \\ q_1^o & q_2^o & \cdot & \cdot & \cdot & q_6^o \\ r_1^o & r_2^o & \cdot & \cdot & \cdot & r_6^o \\ p_1 & p_2 & \cdot & \cdot & \cdot & p_6 \\ q_1 & q_2 & \cdot & \cdot & \cdot & q_6 \\ r_1 & r_2 & \cdot & \cdot & \cdot & r_6 \end{pmatrix} \tag{3}$$

the twist $\boldsymbol{\Omega}$ can be obtained as:

$$(E)^{-1}(\dot{L})=\boldsymbol{\Omega} \tag{4}$$

Let m be the mass and J_x, J_y, J_z be the inertia moments of the platform. Then assuming that the mass of the platform is much more than the masses of the legs and using the Eqs. (3),(4), the kinetic energy T cab be expressed as (see Dimentberg, 1965, Kraynev and Glazunov, 1991):

$$T=\frac{m}{2}\left[\left(\sum_{i=1}^{6} p_i^o\dot{l}_i\right)^2+\left(\sum_{i=1}^{6} q_i^o\dot{l}_i\right)^2+\left(\sum_{i=1}^{6} r_i^o\dot{l}_i\right)^2\right]+\frac{J_x}{2}\left(\sum_{i=1}^{6} p_i\dot{l}_i\right)^2+\frac{J_y}{2}\left(\sum_{i=1}^{6} q_i\dot{l}_i\right)^2+\frac{J_z}{2}\left(\sum_{i=1}^{6} r_i\dot{l}_i\right)^2 \tag{5}$$

where $\dot{l}_i$ $(i=1...6)$ are the generalized velocities.
The Lagrange equations can be written as:

$$\frac{d}{dt}\left(\frac{\partial T}{\partial \dot{l}_i}\right)-\frac{\partial T}{\partial l_i}=Q_i \tag{6}$$

where Q_i are the generalized forces $(i=1...6)$.
The dynamical coupling can be determined using the Eqs. (6). The expression of each generalized force comprises all generalized velocities and accelerations.
In order to reduce the dynamical coupling we represent the kinetic energy as follows:

$$\begin{aligned} T=&\frac{m}{2}\left[\sum_{i=1}^{6}\left(p_i^o\dot{l}_i\right)^2+2\sum_{i=1,j=1,i\neq j}^{6} p_i^o p_j^o\dot{l}_i\dot{l}_j+...+\sum_{i=1}^{6}\left(r_i^o\dot{l}_i\right)^2+2\sum_{i=1,j=1,i\neq j}^{6} r_i^o r_j^o\dot{l}_i\dot{l}_j\right]+ \\ &+\frac{J_x}{2}\left[\sum_{i=1}^{6}\left(p_i\dot{l}_i\right)^2+2\sum_{i=1,j=1,i\neq j}^{6} p_i p_j\dot{l}_i\dot{l}_j\right]+...+\frac{J_z}{2}\left[\sum_{i=1}^{6}\left(r_i\dot{l}_i\right)^2+2\sum_{i=1,j=1,i\neq j}^{6} r_i r_j\dot{l}_i\dot{l}_j\right] \end{aligned} \tag{7}$$

According to the Eq. (7) the design criterion is defined so that the columns of the following matrix *(D)* are to be orthogonal:

$$(D)=\begin{pmatrix} p_1^o\sqrt{m} & p_2^o\sqrt{m} & \cdots & p_6^o\sqrt{m} \\ q_1^o\sqrt{m} & q_2^o\sqrt{m} & \cdots & q_6^o\sqrt{m} \\ r_1^o\sqrt{m} & r_2^o\sqrt{m} & \cdots & r_6^o\sqrt{m} \\ p_1\sqrt{J_x} & p_2\sqrt{J_x} & \cdots & p_6\sqrt{J_x} \\ q_1\sqrt{J_y} & q_2\sqrt{J_y} & \cdots & q_6\sqrt{J_y} \\ r_1\sqrt{J_z} & r_2\sqrt{J_z} & \cdots & r_6\sqrt{J_z} \end{pmatrix}=\begin{pmatrix} \sqrt{m} & 0 & 0 & 0 & 0 & 0 \\ 0 & \sqrt{m} & 0 & 0 & 0 & 0 \\ 0 & 0 & \sqrt{m} & 0 & 0 & 0 \\ 0 & 0 & 0 & \sqrt{J_x} & 0 & 0 \\ 0 & 0 & 0 & 0 & \sqrt{J_y} & 0 \\ 0 & 0 & 0 & 0 & 0 & \sqrt{J_z} \end{pmatrix}(E)^{-1}=(M)(E)^{-1} \tag{8}$$

From the theory of matrices it is known that to satisfy the design criterion defined above the rows of the inverse matrix $(D)^{-1}=(E)(M)^{-1}$ are to be orthogonal. From this, the following matrix *(E)* is proposed:

$$(E)=\begin{pmatrix} 0 & 0 & 1 & 0 & -\sqrt{J_y/m} & 0 \\ 0 & 0 & -1 & 0 & -\sqrt{J_y/m} & 0 \\ 1 & 0 & 0 & 0 & 0 & -\sqrt{J_z/m} \\ -1 & 0 & 0 & 0 & 0 & -\sqrt{J_z/m} \\ 0 & 1 & 0 & -\sqrt{J_x/m} & 0 & 0 \\ 0 & -1 & 0 & -\sqrt{J_x/m} & 0 & 0 \end{pmatrix} \tag{9}$$

The determinant of the matrix (9) can be written as:

$$\det(E)=8\sqrt{J_xJ_xJ_x}\Big/m^{3/2} \tag{10}$$

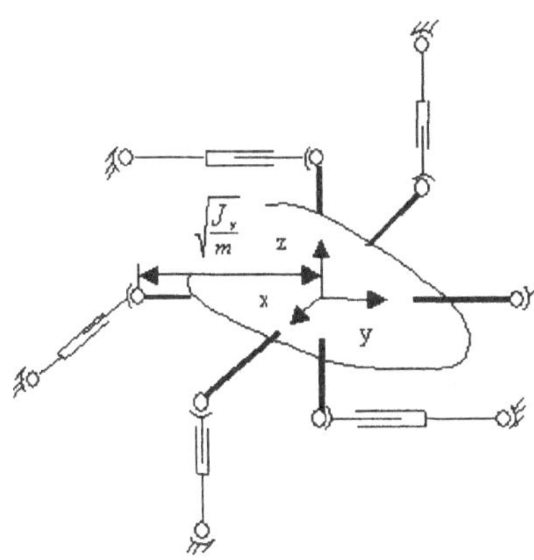

Figure 1. The parallel manipulator with dynamical decoupling.

In this case, the Eq. (5) can be simplified. The kinetic energy can be written as:

$$T=0.25m\sum_{i=1}^{6}\dot{l}_i \tag{11}$$

Obtained matrix (9) corresponds to the Figure 1. Here the center of the mass of the platform coincides with the center of the co-ordinate system *xyz* and the axes of the legs are parallel to the main central inertia axes of the platform. The proposed approach can be applicable for manipulators characterized by small displacements and high speeds.

3 Spatial oscillations

In this section, the parallel manipulator applicable as a spatial oscillator is considered. For instance such mechanism can be applied for testing electronic devices or as a flight simulator. It is clear that the well-known Gough-Stewart platform can be used for mentioned purposes. But this mechanism can cause several disadvantages because it is difficult to control six actuators in the compliance. From this, the criterion of design is introduced which defined as the minimum of the number of actuators. According to this criterion the device is proposed by Borozna, Glazunov, et al. (1992). In the Figure 2, the new mechanism is shown constructed by adding four universal joints situated between kinematic sub-chains.

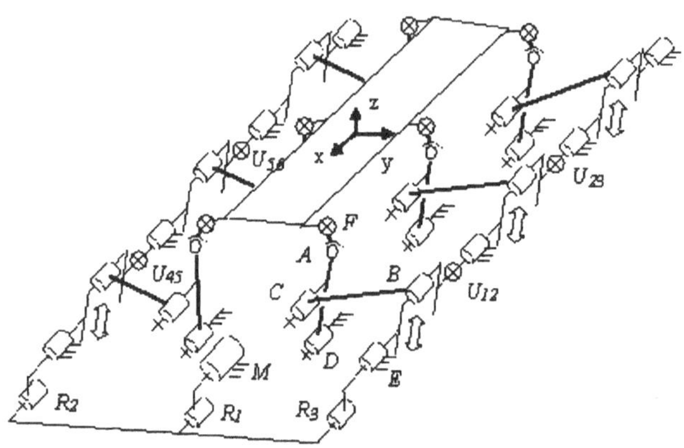

Figure 2. The parallel mechanism for spatial oscillations.

The mechanism consists of a platform connected to a fixed base via six kinematic sub-chains. The motor M situated on the base actuates the device by three rotating kinematic pairs R_1, R_2, R_3 and four universal joints U_{12}, U_{23}, U_{45}, U_{56}. Each kinematic sub-chain comprises a planar four-bar $B_iC_iD_iE_i$, the spherical pair A_i and the universal joint F_i $(i=1...6)$. The form of oscillations (the twist of oscillations) is determined by previous adjusting the lengths of the links B_iE_i. Since displacements of the platform are small, we use the linear equations to find the lengths of these links.

Let $\boldsymbol{\Phi}$ denote the required twist of the platform, $\boldsymbol{\rho}_i=(\rho_{xi}, \rho_{yi}, \rho_{zi})^T$ denote the vector directed from the origin to the point F_i, $(i=1...6)$. $\boldsymbol{\Phi}$ can be expressed by its Plücker coordinates $(\varphi_x, \varphi_y, \varphi_z, r_x, r_y, r_z)^T$. The displacement of the point F_i can be written as: $\boldsymbol{S}_{Fi}=(s_{Fxi}, s_{Fyi}, s_{Fzi})^T$ where

$$s_{Fxi}=r_x+\varphi_y\rho_{zi}-\varphi_z\rho_{yi}, \quad s_{Fyi}=r_y+\varphi_z\rho_{xi}-\varphi_x\rho_{zi}, \quad s_{Fzi}=r_z+\varphi_x\rho_{yi}-\varphi_y\rho_{xi}, \tag{12}$$

Now let $\boldsymbol{L}_i=(L_{xi}, L_{yi}, L_{zi})^T$ denote the vector directed from the point A_i to the point F_i, $\boldsymbol{K}_i=(K_{xi}, K_{yi}, K_{zi})^T$ denote the unit vector along the displacement of the point A_i, and S_{Ai} denote the amplitude of this displacement. The latter can be written as:

$$S_{Ai} = l_{BEi}\, l_{ADi} / l_{CDi} \tag{13}$$

where $l_{BE\,i}$, $l_{AD\,i}$, $l_{CD\,i}$ are the lengths of the links BE_i, AD_i, CD_i, correspondingly $(i=1...6)$.
The projections of the velocities of the points A_i and F_i to the line A_iF_i passing through these points can be expressed as follows:

$$S_{A\,i}\, \boldsymbol{K}_i \boldsymbol{L}_i = \boldsymbol{S}_{Fi} \boldsymbol{L}_i \tag{14}$$

Using the Eqs. (12)-(14) the lengths $l_{BE\,i}$ of the links BE_i can be obtained $(i=1...6)$. If one of these lengths is more than the maximum $l_{BE\,max\,i}$ determined from design then all the lengths $l_{BE\,i}$ are to be multiplied by the coefficient $k = l_{BE\,max\,i} / l_{BE\,i}$.
Thus the parallel mechanism for spatial oscillations is presented that comprises only one motor. In proposed device the system of control is not needed to coordinate the displacements in the sub-chains.

4 Singularity criteria

Gosselin and Angeles (1990) have given the singularity criteria based on Jacobian matrices. Velocities of a parallel manipulator can be written in matrix form

$$(A)\dot{x} = (B)\dot{\theta} \tag{15}$$

where *(A)* and *(B)* are Jacobian matrices, $\dot{x}$ and $\dot{\theta}$ are absolute and generalized velocities correspondingly. According to the proposed criterion singularities exist if the determinants of the matrices *(A)* or *(B)* vanish.
It is known (see Mohammed and Duffy, 1985, Glazunov et al., 1990) that the matrix *(E)* (see the Eq. (1)) can be substituted into the Eq. (15). Each line in the matrix *(A)* can be interpreted as Plücker coordinates of the wrenches reciprocal to the unit vectors of passive joints of the corresponding kinematic sub-chains and the elements of the matrix *(B)* can be expressed as the mutual moments of the mentioned wrenches and the unit vectors of the actuated joints. Any element of the diagonal matrix *(B)* can be equal to zero if the unit vectors of all joints of this sub-chain are dependent. From this, the singularity criterion is proposed according to which singularities exist if the determinant *det(E)* of the matrix *(E)* comprising the Plücker coordinates of the reciprocal wrenches vanishes

$$det(E)=0 \tag{16}$$

or if the determinant consisting of the Plücker coordinates of the unit vectors of joints of any sub-chain vanishes. The condition (16) corresponds to the geometrical interpretation that six wrenches are reciprocal to the twist of possible infinite motion of the output link (see Dimentberg, 1965, Glazunov, Koliskor, and Kraynev, 1991). The criterion (16) brings opportunity to find the twists situated inside singularities.
Let us consider the Gough-Stewart platform, Figure 3. Let $\boldsymbol{\rho}_{Ai}=(x_{Ai}, y_{Ai}, z_{Ai})^T$, $\boldsymbol{\rho}_{Bi}=(x_{Bi}, y_{Bi}, z_{Bi})^T$ denote the vectors directed from the origin to the points A_i and B_i respectively $(i=1...6)$, l_i denote the distance between these points. Furthermore let $\boldsymbol{E}_i$ denote the unit screw situated along the *i*-th leg, $\boldsymbol{d\Phi}$ denote the twist of the platform. $\boldsymbol{E}_i$ consists of the unit vector $\boldsymbol{e}_i$ ant its moment $\boldsymbol{e}^o_i = \boldsymbol{\rho}_{Bi} \times \boldsymbol{e}_i$ and can be expressed by Plücker coordinates $\boldsymbol{E}_i = (x_i, y_i, z_i, x^o_i, y^o_i, z^o_i)$, as well as the twist $\boldsymbol{d\Phi} = (\boldsymbol{d\varphi}, \boldsymbol{dr}) = (d\varphi_x, d\varphi_y, d\varphi_z, dr_x, dr_y, dr_z)^T$.
An infinitesimal displacement of the point A_i can be written as: $\boldsymbol{d\rho}_{Ai}=(dx_{Ai}, dy_{Ai}, dz_{Ai})^T$ where

$$dx_{A\,i} = dr_x + d\varphi_y z_{A\,i} - d\varphi_z y_{Ai},\; dy_{A\,i} = dr_y + d\varphi_z x_{A\,i} - d\varphi_x z_{Ai},\;\; dz_{A\,i} = dr_z + d\varphi_x y_{A\,i} - d\varphi_y x_{Ai} \qquad (17)$$

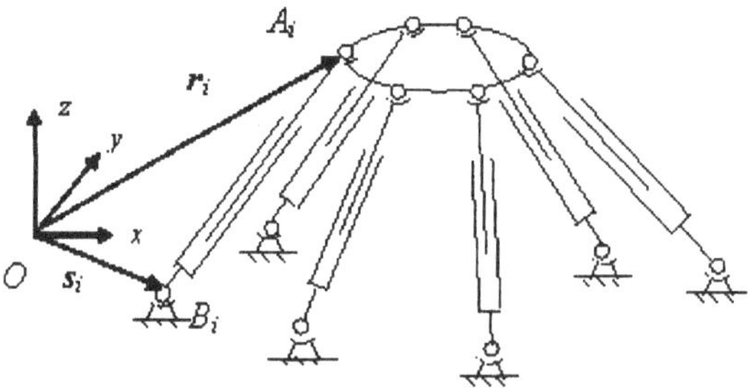

Figure 3. A parallel manipulator.

After this displacement the unit screw $\boldsymbol{E}_i$ can be rewritten (see Glazunov, Koliskor, and Kraynev, 1991, Glazunov et al., 2002) as $\boldsymbol{E}_i+d\boldsymbol{E}_i$ or as $\boldsymbol{e}_i+d\boldsymbol{e}_i$ and $\boldsymbol{e}^o{}_i + d\boldsymbol{e}^o{}_i$ where

$$d\boldsymbol{e}_i = [d\boldsymbol{\rho}_{Ai} - \boldsymbol{e}_i(d\boldsymbol{\rho}_{Ai}\boldsymbol{e}_i)]/l_i, \qquad d\boldsymbol{e}^o{}_i = \boldsymbol{\rho}_{Bi} \times d\boldsymbol{e}_i \qquad (18)$$

Using the Eqs. (17)-(18) the coordinates of the $d\boldsymbol{e}_i$ and $d\boldsymbol{e}^o{}_i$ can be expressed as:

$$dx_i = (\partial x_i/\partial\varphi_x)\, d\varphi_x + (\partial x_i/\partial\varphi_y)\, d\varphi_y + (\partial x_i/\partial\varphi_z)\, d\varphi_z + (\partial x_i/\partial r_x)\, dr_x + (\partial x_i/\partial r_y)\, dr_y + (\partial x_i/\partial r_z)\, dr_z$$
$$\cdots\cdots\cdots\cdots\cdots\cdots\cdots\cdots \qquad (19)$$
$$dz^o{}_i = (\partial z^o{}_i/\partial\varphi_x)\, d\varphi_x + (\partial z^o{}_i/\partial\varphi_y)\, d\varphi_y + (\partial z^o{}_i/\partial\varphi_z)\, d\varphi_z + (\partial z^o{}_i/\partial r_x)\, dr_x + (\partial z^o{}_i/\partial r_y)\, dr_y + (\partial z^o{}_i/\partial r_z)\, dr_z$$

where $\partial x_i/\partial\varphi_x = x_i(z_{Ai}y_i - y_{Ai}z_i)/l_i$, $\partial x_i/\partial\varphi_y = [z_{Ai}(1-x_i^2) + x_{Ai}x_iz_i]/l_i$, $\partial x_i/\partial\varphi_z = [y_{Ai}(x_i^2-1) + x_{Ai}x_iy_i]/l_i$,
$\partial x_i/\partial r_x = (1-x_i^2)/l_i$, $\partial x_i/\partial r_y = -x_i\, y_i/l_i$, $\partial x_i/\partial r_z = -x_i\, z_i/l_i$,
$\partial x^o{}_i/\partial\varphi_x = \{y_{Bi}[y_{Ai} + (z_{Ai}y_i - y_{Ai}\, z_i)z_i] + z_{Bi}[z_{Ai} + (y_{Ai}z_i - z_{Ai}\, y_i)y_i]\}/l_i$
$\partial x^o{}_i/\partial\varphi_y = \{y_{Bi}[(x_{Ai}z_i - z_{Ai}\, x_i)z_i - x_{Ai}] + z_{Bi}(z_{Ai}x_i - x_{Ai}\, z_i)y_i\}/l_i$
$\partial x^o{}_i/\partial\varphi_z = \{y_{Bi}(y_{Ai}x_i - x_{Ai}\, y_i)z_i - z_{Bi}[x_{Ai} - (x_{Ai}y_i - y_{Ai}\, x_i)y_i]\}/l_i$
$\partial x^o{}_i/\partial r_x = x_i(z_{Bi}y_i - y_{Bi}\, z_i)/l_i$, $\partial x^o{}_i/\partial r_y = [z_{Bi}\,(y_i^2-1) - y_{Bi}\, y_i\, z_i]/l_i$, $\partial x^o{}_i/\partial r_z = [y_{Bi}\,(1-z_i^2) - z_{Bi}\, y_i\, z_i]/l_i$, $i=1\ldots6$.

Other partial derivatives can be obtained by index rotation.
By means of the properties of linear decomposition of determinants $d[det(E)]$ can be obtained as the sum of 36 determinants:

$$d[\det(E)] = \begin{vmatrix} \partial x_1/\partial\varphi_x & y_1 & z_1 & x_1^o & y_1^o & z_1^o \\ \partial x_2/\partial\varphi_x & y_2 & z_2 & x_2^o & y_2^o & z_2^o \\ \cdot & \cdot & \cdot & \cdot & \cdot & \cdot \\ \cdot & \cdot & \cdot & \cdot & \cdot & \cdot \\ \cdot & \cdot & \cdot & \cdot & \cdot & \cdot \\ \partial x_6/\partial\varphi_x & y_6 & z_6 & x_6^o & y_6^o & z_6^o \end{vmatrix} d\varphi_x + \ldots + \begin{vmatrix} x_1 & y_1 & z_1 & x_1^o & y_1^o & \partial z_1^o/\partial r_z \\ x_2 & y_2 & z_2 & x_2^o & y_2^o & \partial z_2^o/\partial r_z \\ \cdot & \cdot & \cdot & \cdot & \cdot & \cdot \\ \cdot & \cdot & \cdot & \cdot & \cdot & \cdot \\ \cdot & \cdot & \cdot & \cdot & \cdot & \cdot \\ x_6 & y_6 & z_6 & x_6^o & y_6^o & \partial z_6^o/\partial r_z \end{vmatrix} dr_x \qquad (20)$$

From this, $d[det(E)]$ can be represented as:

$$d[det(E)] = \partial[det(E)]/\partial\varphi_x d\varphi_x + \partial[det(E)]/\partial\varphi_y d\varphi_y + \ldots + \partial[det(E)]/\partial r_y dr_y + \partial[det(E)]/\partial r_z dr_z \qquad (21)$$

Using the Eq. (21) the criterion of determination of the singularity locus can be represented:

$$d[det(E)]=0 \tag{22}$$

This condition imposes only one constraint to motions of the platform therefore the singularity locus is 5-dimensioned for 6-DOF manipulators (and $(w-1)$-dimensioned for w-DOF manipulators). The determinant of the matrix (E) can be imagined as scalar function of six arguments that are the absolute coordinates of the platform. That is why all configurations corresponding to constant values of the determinant $det(E)$ can be considered as hyper-surfaces of the constant level.

The Eq. (21) can be applicable for manipulators with actuation redundancy. It is apparent, that extra actuator must drive the manipulator from singularity as quickly as possible, i.e. along the twist-gradient $d\boldsymbol{\Phi}_g = (d\varphi_g, dr_g) = (d\varphi_{gx}, d\varphi_{gy}, d\varphi_{gz}, dr_{gx}, dr_{gy}, dr_{gz})^T$ where $d\varphi_{gx}=K\partial[det(E)]/\partial\varphi_x$, $d\varphi_{gy}=K\partial[det(E)]/\partial\varphi_y$,... $dr_{gz}=K\partial[det(E)]/\partial r_z$, K is arbitrary coefficient. The condition (16) means that n $(n=2...6)$ of 6 unit screws $\boldsymbol{E}_i$ enter in $(n-1)$-member group. This can be found by exception the row corresponding to the i-th joint $\boldsymbol{E}_i$. If remaining matrix is of the rank 5 then $\boldsymbol{E}_i$ is the member of mentioned group. The extra actuators must be placed only in corresponding sub-chains.

The method to avoid singularity can be based on the Eq. (22). Let ε denote the limit of $det(E)$

$$|det(E)|\geq\varepsilon \tag{23}$$

If a parallel manipulator is situated at a configuration where $|det(E)|=\varepsilon$ and the prescribed path corresponds to the desired twist $d\boldsymbol{\Phi}_d = (d\varphi_d, d\boldsymbol{r}_d)$ leading to the next configuration where $|det(E)|<\varepsilon$ then the corrected twist $d\boldsymbol{\Phi}_c = (d\varphi_c, d\boldsymbol{r}_c)$ can be determined. The components of this twist can be written as: $d\varphi_c = d\varphi_d - d\varphi_g(d\varphi_d\, d\varphi_g)/(d\varphi_g)^2$, $d\boldsymbol{r}_c = d\boldsymbol{r}_d - d\boldsymbol{r}_g(d\boldsymbol{r}_d\, d\boldsymbol{r}_g)/(d\boldsymbol{r}_g)^2$, where $d\varphi_g$, $d\boldsymbol{r}_g$ are the components of the twist-gradient. The corrected twist $d\varphi_c$ satisfies the condition (22) and is the nearest twist to the desired one. The Eqs. (16), (22), (23) express the kinematical criteria of singularity, moreover the Eqs. (16), (23) express the differential condition of singularity loci.

The static criterion of the configurations either singular or nearly singular is based on the pressure angles α_p which are angles between the forces acting along the axes $\boldsymbol{E}_i$ of the legs of parallel manipulators and the vectors of velocities or infinitesimal displacements of the corresponding points A_i (see V. Glazunov et. al., 1998). The six twists $d\boldsymbol{\Phi}_j$ $(j=1...6)$ each of which is reciprocal with respect to five of the six unit screws $\boldsymbol{E}_i$ $(i=1...6)$, can be found to an accuracy of a scalar multiplier. If the twist $d\boldsymbol{\Phi}_i$ is known then the infinitesimal displacement $d\boldsymbol{\rho}_{Ai}$ of the point A_i can be obtained using the Eq. (17). The pressure angle α_{pi} is the angle between the vectors $d\boldsymbol{\rho}_{Ai}$ and $\boldsymbol{e}_i$: $\alpha_{pi}=|arccos(d\boldsymbol{\rho}_{Ai}\boldsymbol{e}_i/|d\boldsymbol{\rho}_{Ai}|)|$. Evidently that if $det(E)=0$ then $\alpha_{pi}=\pi/2$. Let φ_f denote the friction angle: $\varphi_f=arctg\, f$ where f is the coefficient of friction. The static singularity criterion is defined as:

$$\alpha_{p\,max}\leq\pi/2-\varphi_f \tag{24}$$

It is obvious that the working volume restricted by the kinematical criterion (16) is more than the volume restricted by the static criterion (24).

The dynamical criterion of singularity is determined taking into account not only geometrical but also inertial link parameters as well as control algorithm which must minimize the deviations of generalized coordinates from their program meanings (see Glazunov et al., 2004). This criterion is formulated as: the overrun of the generalized moment's marginal tolerance value. It is necessary that the moment surpass the nominal value not more than two times. On reaching such configuration there should be a load transfer with taking extra actuator into account. Note that certain configurations can be detected as singular or not singular by different criteria.

5 Conclusion

Thus, various criteria of design and singularity analysis of parallel manipulators are presented here. According to the criterion of the dynamical decoupling the kinetic energy is expressed as the polynomial including only the squares of the generalized velocities. The criterion of the design of the spatial oscillator is defined as minimum of the quantity of the actuators. The kinematic criterion of singularity corresponds to linear dependence of wrenches supporting the output link, the static criterion corresponds to the limitation of pressure angles and the dynamical criterion corresponds to the limitations of driving moments. Proposed criteria determine various results of synthesis and of detection of singularity of parallel manipulators.

Bibliography

J. Angeles. The Qualitative Synthesis of Parallel Manipulators. In *Journal of Mechanical Design,* 126: 617-624, 2004.

J.-P. Merlet. *Parallel Robots*. Kluwer Academic Publishers, 2000.

M. Ceccarelli. *Fundamentals of Mechanics of Robotic Manipulations.* Kluwer Academic Publishers, 2004.

V. Glazunov, A. Koliskor, and A. Kraynev. *Spatial Parallel Structure Mechanisms.* M.: Nauka, 1991.

D. Chablat, P. Wenger, and J.-P. Merlet. A Comparative Study between Two Three-DOF Parallel Kinematic Machines using Kinetostatic Criteria and Interval Analysis. In *Proceedings of XI World Congress in Mechanism and Machine Science*, Tianjin, China, pages 1209-1213, 2004.

V. Glazunov and N. Munitsyna. Optimal Design of Parallel Structure Manipulators for Extreme Environments of Textile Industry. In *Pr. of High Schools. Technology of Textile Industry.* 5: 85-89, 1994.

A. Kraynev and V. Glazunov. Parallel Structure Mechanisms in Robotics. In *MERO'91, Sympos. Nation. de Roboti Industr.,* Bucuresti, Romania, 1: 104-111, 1991.

Patent of Russia N 1753322 *The Six-Coordinate Vibration Stand* / A. Borozna, V. Glazunov et al. (G 01M 7/06), 1992.

C. Gosselin and J. Angeles. Singularity Analysis of Closed Loop Kinematic Chains. In *IEEE Trans. on Robotics and Automation*, 6(3): 281-290, 1990.

V. Glazunov, A. Koliskor, A. Krainev, and B. Model. Classification Principles and Analysis Methods for Parallel-Structure Spatial Mechanisms. In *Journal of Machinery Manufacture and Reliability,* Allerton Press Inc., 1: 30-37, 1990.

F. Dimentberg. *The Screw Calculus and its Applications in Mechanics.* M.: Nauka, 1965. (English translation: AD680993, Clearinghouse for Federal Technical and Scientific Information, Virginia).

M. Mohammed and J. Duffy. A Direct Determination of the Instantaneous Kinematics of Fully Parallel Robot Manipulators. In *ASME J. Mech., Trans., Autom. Des.,* 107(2): 226-229, 1985.

V. Glazunov, A. Kraynev, G. Rashoyan, R. Bykov, and N. Novikova. Neighboring Special Configurations of Parallel Manipulators. In *Theory and Practice of Robots and Manipulators. (RoManSy), Proceedings of XIV CISM-IFToMM Symposium,* Springer Wien New York, pages 59-66, 2002.

V. Glazunov, A. Kraynev, G. Rashoyan, and A. Trifonova. Design of the Paths of Parallel-Structure Mechanisms and Construction of their Working Zones with Allowance for Singular Positions. In *Journal of Machinery Manufacture and Reliability,* Allerton Press Inc., 5: 40-45, 1998.

V. Glazunov, A. Kraynev, R. Bykov, G. Rashoyan, and N. Novikova. Parallel Manipulator Control while Intersecting Singular Zones. In *Theory and Practice of Robots and Manipulators. (RoManSy), Proceedings of XV CISM-IFToMM Symposium,* Montreal, Canada, 2004.

L-legs for the Design of Mini and Micro Parallel Compliant Mechanisms

Matteo Zoppi and Rezia Molfino*

University of Genova, PMARlab, via all'Opera pia 15A, 16145 Genova, Italy.
{zoppi,molfino}@dimec.unige.it

Abstract. The paper presents a delocalized-compliance mechanism architecture, named L-gimbal, with 3 elastic equivalent degree of freedom and discusses three example applications of it characterized by the same degree of freedom requirements. The L-gimbal is a parallel mechanism realized using modular compliant leg elements named L-legs. It originates from a mechanism in IIM (Increased Instantaneous Mobility) singular configuration, showing that singularities can be exploited in synthesis of compliant mechanisms. The simple geometry of the L-gimbal makes it fit for fabrication with MEMS techniques. Three example applications of this compliant architecture are presented: an elastic joint for the steering-trust module of a worm robot with peristaltic locomotion, a micro wrist for minimally invasive robotic surgery, and an active MEMS mirror support.

1 Introduction

Mechanism and machine design is evolving in a multidisciplinary approach that simultaneously takes into account sensing, actuation and control issues for simultaneous satisfaction of functional and performance requirements. This evolution is mainly driven by the new paradigms of miniaturization of the products; easy, clean and low cost manufacturing; environment sustainability. A guiding concept is *functions fusion* resulting in integration of parts and subsystems each one contributing to make the overall structural, sensing and actuation functionalities merging/emerging in the complete system.

At design level, tools for multiphysics modeling, analysis and simulation are required for the concurrent definition of the parts of the system providing the different functionalities. Integration at physical level includes combination in the systems of multifunctional and new materials with specific characteristics. The fabrication process has a strong influence introducing additional design constraints. MEMS are an example.

Especially at smaller and MEMS scale, biomimetics is an effective approach. An important characteristics of bio mini and micro living creatures is the compliant structure of their invertebrate bodies.

Compliant mechanisms, composed of compliant and stiff elements realizing a desired mobility as the result of the elastic deformation of the whole structure rather than because

*Dr. Luca Bisio is kindly acknowledged for the help in the design of the worm robot module. The Italian Ministry of University and Research is acknowledged for the financial support.

of the relative displacement of the links (1), are an alternative to traditional mechanisms composed of kinematics pairs (i.e. revolute, prismatic, planar, etc.) and rigid (or almost rigid) links (*kinematics mechanisms*).

Some pros of compliant mechanisms compared to kinematics mechanisms are: minimum assembly needs; wider choice of shapes and geometric design parameters and materials for tailored applications; easy integration with modern (e.g. electrostatic, piezo-electric) actuators and smart materials; high reliability and MTBF (under fatigue life constraints); absence of joint friction, backlash and need of lubrication (but presence of a rest configuration and elastic forces); easier design for MEMS fabrication technologies.

We distinguish two types of compliant mechanisms: with localized and delocalized compliance.

Localized-compliance mechanisms are composed of rigid or almost rigid links and *localized-compliance joints*. They can be obtained from a kinematics mechanism (*reference mechanism*) in a certain configuration (*reference configuration*) by replacing the kinematics joints with compliant joints. Generally, in this case the compliant joints are designed to replicate at best the mobility of the kinematics joints they replace (same invariants and suitable distribution of stiffness), from which the name localized-compliance, e.g., for flexure hinges, minimum displacement of the rotation axis with the angle of rotation, stiffness about this axis low and rapidly increasing changing the axis of rotation (2; 3).

Any localized-compliance mechanism can be analyzed using the same methods and algorithms as for the reference mechanism, and the same kinematics models can be used, e.g. in the control system to compute the end-effector pose and displacement. Moreover, the actuators can be the same.

Localized-compliance mechanisms are the most used and studied today but applications exist where deformable structures with distributed (or *delocalized*) compliance (*delocalized-compliance mechanisms*) may result preferable. The paper deals with three such applications from the domain of robotics: • an elastic coupling in the steering-trust module of a worm robot with peristaltic locomotion; • a micro wrist for a new generation of small diameter surgical robots; • an active MEMS mirror support. All three applications are satisfied by the same compliant mechanism architecture, the L-gimbal, described in the following. As for kinematics mechanisms, the selection of a compliant mechanism architecture is mostly based on the desired mobility. In all three example applications considered, the required mobility is 2 rotational freedoms about a center of rotation O fixed to the base and 1 translational freedom (extrusion) in direction fixed to the end-effector.

In delocalized-compliance mechanisms, the mechanism structure as a whole provides the desired elastic constraint (4; 5). Compared to localized-compliance mechanisms, in general pros of delocalized-compliance mechanisms are larger mobility and higher reliability. On the other hand, delocalized compliance mechanisms may result less accurate and their synthesis and design much more complex (6). Because of these difficulties, even considering the recent advances in design methodologies, e.g. (7), an approach based on a-priori analyzed compliant modules and related heuristic design rules is advantageous.

In the selection of the delocalized-compliance mechanism presented in the paper, intelligent design rules including criteria like symmetry to achieve higher accuracy were

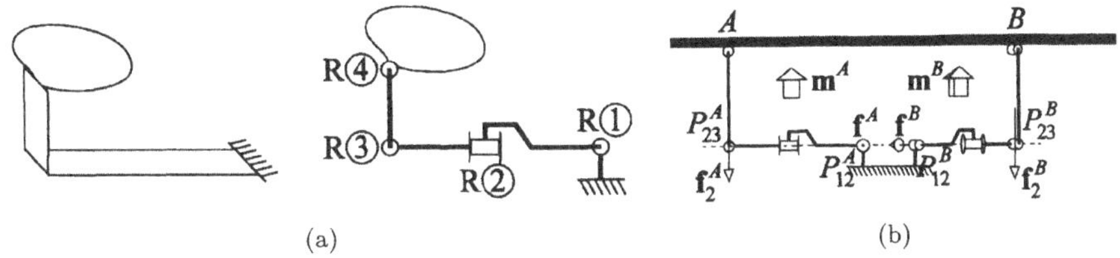

Figure 1. (a) L-leg: schematic (left) and the reference kinematics chain (right); (b) reference mechanism

adopted.

2 L-gimbal delocalized-compliance mechanism with L-legs

This section is devoted to the description of the L-gimbal mechanism and its modular legs; some hints are also given on the synthesis methodology, which starts from a parallel kinematics mechanism in singular configuration.

The L-gimbal is composed of two or more cylindrical elastic elements (compliant structures) with "L" shaped section, called *L-legs*, schematically represented in Fig. 1(a), connecting the end-effector to the base. The reference mechanism for the L-leg is the 4R serial chain in Fig. 1(a) in the shown reference configuration. A brief discussion of its mobility may be useful to understand the functioning of the L-gimbal (8).

The axes of the revolute joints are called $\boldsymbol{\xi}_i$, $i = 1, \ldots, 4$: $\boldsymbol{\xi}_1 \perp \boldsymbol{\xi}_2 \perp \boldsymbol{\xi}_3 \parallel \boldsymbol{\xi}_4$. The intersection point of $\boldsymbol{\xi}_1$ and $\boldsymbol{\xi}_2$ is called P_{12}; P_{23} is the intersection point of $\boldsymbol{\xi}_2$ and $\boldsymbol{\xi}_3$; π_{12} is the plane defined by $\boldsymbol{\xi}_1$ and $\boldsymbol{\xi}_2$. The system of structural constraint (with all leg joints free to move) applied by the leg in the reference configuration is a 2-system spanned by the pure force $\mathbf{f} \parallel \boldsymbol{\xi}_3$ through P_{12} and by the pure moment $\mathbf{m} \perp \pi_{12}$.

The reference mechanism for the L-gimbal is shown in Fig. 1(b) in reference configuration. The minimum number of legs to get the desired 3dof mobility is two (labeled with A and B in the figure), disposed with the axes of the base joints intersecting and not parallel, with the π_{12} coincident. The combined structural constraint applied on the end-effector (sum of the systems of structural constraints applied by the legs) is the 3-system spanned by $\mathbf{f}^A$, $\mathbf{f}^B$ and $\mathbf{m}^A$ (because $\mathbf{m}^A = \mathbf{m}^B$). Instantaneously, the end-effector tilts about axes in π_{12} through the center of rotation O at the intersection between $\boldsymbol{\xi}_1^A$ and $\boldsymbol{\xi}_1^B$, and translates in direction orthogonal to π_{12} (8). This configuration is a IIM (Increased Instantaneous Mobility) singularity (9; 10). Out of singularities the reference mechanism is 2dof; by selecting as reference configuration a IIM singularity it is possible to realize a 3 elastic dof deformable structure with the desired mobility. More details on the selection of the reference mechanism can be found in (8).

The third L-leg is added to realize a symmetric end-effector support: it increases the stiffness of the compliant mechanism without changing the stiffness distribution (low stiffness associated to the desired translations/rotations; high stiffness in the directions that have to be constrained). Moreover, in the symmetric system the displacements of the center of rotation associated with tilt mechanism partially compensate each-other.

Figure 2. L-leg: alternative geometries

Different realizations of the L-leg are for example (Fig. 2): constant thickness elastic strip; layered, using materials with different stiffness characteristics; monolithic, variable thickness with grooves.

The same delocalized-compliance mechanism with purely parallel architecture, the *L-gimbal*, can be used in the three robotics applications presented in the following sections, which represent three realizations of the L-gimbal at meso, milli/micro and micro scale.

3 Worm robot steering-trust module

A modular worm robot with peristaltic locomotion for rescue applications is considered. The worm includes steering-trust modules, sensorial modules and a head module. The different modules are connected each other through modular interfaces allowing fast reconfiguration of the worm depending on the mission. Each module (except the head) is stand alone crossed by a power line and a bus. An umbilical connected to the back end of the robot provides power from a remote station and teleoperation commands. Electronics onboard the modules pilots the actuators and conveys sensor information in the bus.

The steering-trust compliant mechanism designed is shown in Fig. 3. In this module the 2 rotational freedoms are to bend the worm so making it steering, and the extrusion is needed to generate peristaltic trust.

The desired dimensions are: 45 mm diameter and 45 mm rest length, with 10 mm extrusion range and $\pi/18$ tilting angle.

The mechanism is composed of two L-gimbals connected in series (Fig. 3), which realize the desired elastic constraint between the two interfaces ⑤ and ⑥ of the module. The frame of each interface is a ring connected to three L-legs, whose other sides are fixed to the central yoke ④. Each L-leg is composed of two blades ① and ③ of superelastic alloy (for large bending) connected by an elbow vice ②.

Actuation is by SMA (shape memory alloy) wires ⑧ winded around two teflon rings ⑦, with lateral grooves to increase the wire lengths and thus the actuation range. Each wire pulls the corresponding elbow of the opposite leg. Because wires provide one-way actuation (they pull but do not push), the elbow vice angles are higher than $\pi/2$ allowing elastic preload of the module by assigning an initial tension to the actuation wires.

A measure of the effectiveness of the mechanism for this application is the ratio between shear stiffness (required low) and extrusion stiffness (required high). The final compromise value obtained is about 5.3. This is acceptable considering the high actuation forces generated by the SMA wires.

The center of rotation O is in the symmetry plane of the module, at the middle of the yoke ④. Some displacement of O associated with tilting is acceptable. The overall

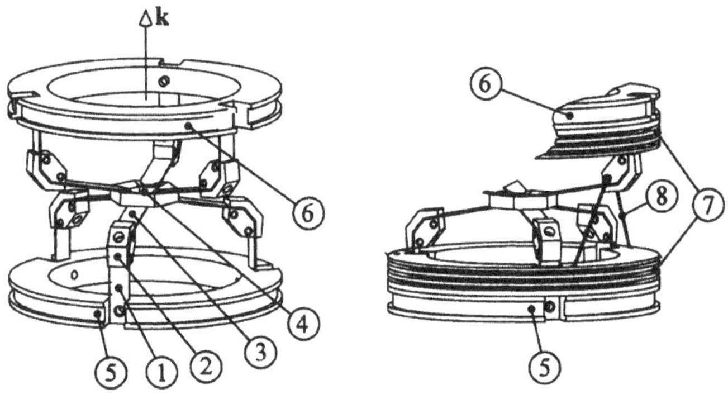

Figure 3. Worm robot steering-trust module: deformable structure of the module (left) and detail view of one actuation wire (right)

result is a robust and cheap mechanism fitting the application requirements. The use of a delocalized compliant mechanism instead of a kinematics or a localized compliant mechanism appears effective, in particular from the points of view of simplicity and large mobility ranges.

4 Micro wrist for minimally-invasive robotic surgery

The second example is at milli/micro scale.

Robotic systems for minimally invasive surgery provide outstanding benefits to the patients. Classes of surgical operations exist that could be carried out with surgical tools smaller than the ones today available in both traditional and robotic surgery. This pulls part of the research in the field of surgical robotics toward miniaturized surgical systems and one research subject is realization of micro wrists for manipulation of micro surgical instruments at the tip of a laparoscope (or other robotic system operating inside the patient).

Compliant mechanisms with parallel architecture, like the L-gimbal, can be particularly effective to get enough agility and stiffness in small dimensions and to keep the fabrication easy (11).

In the minimally-invasive surgical robotic wrist the 2 rotational freedoms are to orient the surgical instrument (e.g. a micro high-frequency electric knife), and the translational freedom is for fine movement of the instrument toward the tissues and backward, during the operation (the surgical robot carrying the wrist is supposed to provide rough positioning of the wrist in the operative region).

Wrist size and mobility considered are: 2 mm external diameter of the links of the surgical robot carrying the wrist (1.3 mm in diameter the wrist end-effector platform); 0.15 mm extrusion range and $\pi/9$ tilting angle.

Figure 4 shows the solution proposed. The L-legs (Fig. 4, right) are fabricated separately and assembled to the cylindric base ⑤ in grooves matching with the square leg

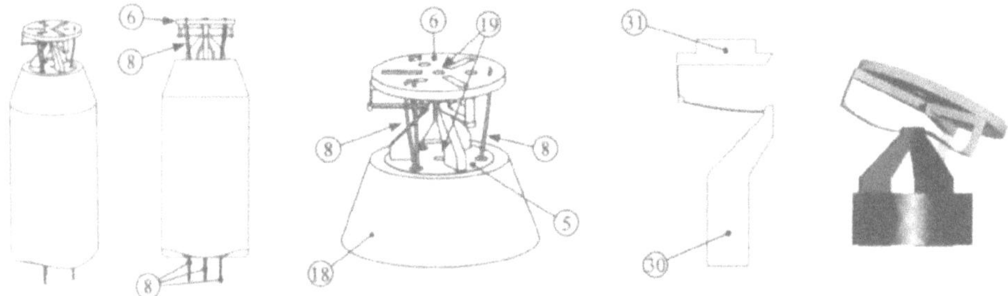

Figure 4. Wrist for surgical robot: full views; detailed view; profile of the L-leg; tilted configuration

ends ㉚. The end-effector platform ⑥ is assembled to the other leg sides ㉛. UV curing glue can be used for fixing the parts. The hole ⑲ in both base and end-effector, is to connect the surgical tool, fixed to the end-effector, to the base of the wrist, e.g. an electrical connection.

All parts are realized with LIGA technology (lithography, electroforming and molding for MEMS fabrication), depending on the aspect ratio.

The actuation proposed is by wires ⑧ pulled by actuators placed inside the link carrying the wrist or in the base of the surgical robot outside the patient. This solution presents several drawbacks such as difficulty to generate accurate movements and the fact that the wires provide one-way actuation, resulting in coupling of tilt and extrusion and in the need of an elastic preload of the legs for outward extrusion (opposite to the pulling direction of the cables). The main advantage of wires is the large actuation range achievable. Other actuations are considered, in particular SMA (Shape Memory Alloy) wires 25 μm diameter applying about 0.017 N pulling force. Figure 4, right, shows the wrist tilted by one wire at maximum force.

The surgical robot is teleoperated. During the operation, the small displacements of the center of rotation of the wrist can be compensated by the surgeon.

5 MEMS active mirror support

The third example, Fig. 5, is an application of the L-gimbal at a scale smaller than the surgical wrist.

The 2 rotational freedoms are required to tilt the mirror about the center of rotation O (reflection of a light beam) and the extrusion (orthogonal to the mirror surface) is for corrections at the wave-length scale of the reflected light such as compensating for nonplanarity in the wavefronts.

The first prototypes that we are realizing, in collaboration with the DXRL beamline Experiments Division of Elettra – Sincrotrone Trieste, to validate the concept have dimensions: 540 μm mirror diameter; 250 μm leg long side by 30 μm short side, 6 μm thickness, 40 μm largeness (aspect ratio about 7). Mirrors with these dimensions can be

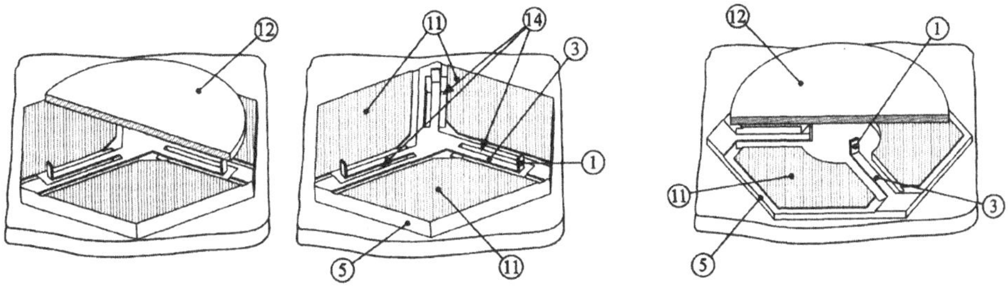

Figure 5. Micromirror active support: outward (a) and inward (b) architectures

easily fabricated using standard LIGA equipment. Then the system will be scaled down to 200 μm mirror diameter.

The main difference with the worm module and the surgical wrist presented in the previous sections is that, due to the size, the fabrication cannot include assembly operations.

The layer-by-layer fabrication starts from the mirror ⑫, Fig. 5(a) (the base layer is numbered ⑤, the legs ⑭). The horizontal ③ and vertical ① leg sides are realized in two steps: ③ in the same layer of the three electrode surfaces ⑪ used to command mirror tilt and extrusion by direct interaction (push/pull) with the mirror bottom surface.

Two leg placements are considered, providing same mobility: *outward* (Fig. 5(a)), with leg vertical sides at the periphery of the mirror and the leg-base connection at the center; *inward* (Figs. 5(b), left), with leg vertical sides at the center and leg-base connection at the periphery.

The fabrication process gives several design constraints, e.g. constant leg section, reducing the number of available design parameters (which can be used, for example, to reduce the displacement of the center of rotation).

The gap between electrodes and mirror can be reduced (in order to increase the electrostatic actuation forces) by fabricating the electrodes on an additional intermediate layer between the base and the mirror.

6 Conclusions

Two types of compliant mechanisms are introduced: localized-compliance mechanisms and delocalized-compliance mechanisms. Looking at applications where compliant mechanisms are preferable compared to kinematics mechanisms, the paper tries to underline that applications exist where delocalized-compliance mechanisms are preferable to localized-compliant mechanisms although their synthesis is more difficult. Reasons are larger workspace and easiness to fabricate, in particular at micro scale. Intelligent design, using modularity, symmetries to improve accuracy and new and smart materials can make things easier and lead to effective solutions. Control can play an important role, in particular for higher accuracy with control loops closed on the pose of the end-effector.

In this framework, a delocalized-compliance mechanism architecture, the L-gimbal, realized with three elastic leg elements called L-legs is presented and discussed. Three example applications are shown: Two already designed in detail and ready to be realized, the worm robot steering-trust module and the micromirror support; one, the surgical robotic wrist, at a development stage.

The surgical wrist and the worm module are presented in this paper for the first time.

In line with the proposed distinction between localized and delocalized compliant mechanisms, another type of smart mechanisms foreseen are deformable mechanisms with localized or delocalized (distributed) plasticity. Materials exist such as Cu–oxygen-free with virtually infinite fatigue life: they can be used to design structures almost without wear and fatigue failure.

Bibliography

[1] S. Smith. *Flexures, Elements of Elastic Mechanisms.* Taylor & Francis, London, England, 2000.

[2] J.M. Paros and L. Weisbord. How to design flexure hinges. *Int. J. Machine Design*, 37(27):151–156, 1965.

[3] N. Labontiu. *Compliant Mechanisms: Design of Flexure Hinges.* CRC Press, Boca Raton, FL, 2002.

[4] B. Trease, Y. Moon, and S. Kota. Design of large-displacement compliant joints. *ASME Int. J. of Mechanical Design*, 127(4):788–798, 2005.

[5] W. Weinstein. How to design flexure hinges. *Int. J. Machine Design*, 37(13-D):150–157, 1965.

[6] L.L. Howell. *Compliant Mechanisms.* Wiley, New York, 2001.

[7] G.K. Lau, H. Du, and M.K. Lim. Use of functional specifications as objective functions in topological optimization of compliant mechanism. *Int. J. Comp. Methods in Applied Mech. and Eng.*, 190(34):4421–4433, 2001.

[8] M. Zoppi and R. Molfino. Micro L-gimbal parallel mechanism for the active support of a micro mirror. In *36th Int. Symposium on Robotics ISR05*, Keidanren Kaikan, Tokyo, Japan, November 29 – December 1 2005.

[9] D. Zlatanov, R.G. Fenton, and B. Benhabib. A unifying framework for classification and interpretation of mechanism singularities. *ASME Journal of Mechanical Design*, 117:566–572, 1995.

[10] D. Zlatanov, I. Bonev, and C. Gosselin. Constraint singularities of parallel mechanisms. In *2002 IEEE International Conference on Robotics and Automation, ICRA 2002*, pages 496–502, Washington, DC, USA, May 11-15 2002.

[11] M. Zoppi, D. Zlatanov, and C.M. Gosselin. Analytical kinematics models and special geometries of a class of 4-dof parallel mechanisms. *IEEE Trans. on Robotics*, 21(6):1046–1055, 2005.

A singularity free parallel robotic mechanism for aiming antennas and cameras

Reg Dunlop[1], Michael Schlotter[2], Peter Hagedorn[2], Tim Jones[1],

[1] University of Canterbury, New Zealand reg.dunlop@canterbury.ac.nz
[2] Darmstadt University of Technology, Germany hagedorn@dyn.tu-darmstadt.de

Abstract. Several different mechanisms have been used to aim narrow beam communications links between earth stations and orbiting satellites. Only two axes are required to aim the beam but for the three conventional serially connected mechanisms currently used, a singularity about the first axis prevents tracking through a key hole region centered on the first axis and the communication link can fail. This key hole or mechanical singularity is caused by dynamic limitations which can never be overcome with the standard two axis serial mechanism. To keep the aim, it is common to add a third axis, or even add a two axis gimbaled base for marine applications. These serial connections reduce the stiffness of the aiming mechanism and increase the cost. A solution to this problem is to use a parallel robot. Previous work on two axis parallel robots showed that while such a mechanism could provide hemispherical coverage, a singularity in the drive linkages prevented full control of the coverage. A different drive system has been developed to avoid all singularities and is presented here for the first time. The results of dynamic simulations of the system in marine applications are presented to show reduced drive system loadings due to motion of the mounting base. This robotic mechanism is also capable of providing singularity free hemispherical aiming capability for aircraft mounted camera systems.

1 Introduction

Aiming systems provide the ability to point a camera, antenna or weapon in a particular direction. The parallel robotic aiming mechanism to be discussed is shown in Figure 1. It is development of the constant velocity joint given by Hunt (1978) who in turn attributes it to Clemens (Steeds 1940). More recently the mechanism has been used as a wrist and as a joint for a bipedal walking machine (Sellaouti and Ouezdou 2005). The discussion is restricted to the problem of aiming an antenna but applies equally well to other situations. The aiming device is required to point the antenna in a direction

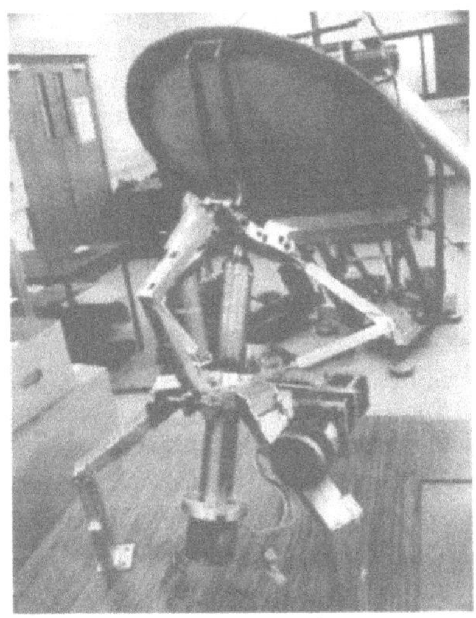

Figure 1. Parallel robot aiming system.

specified by the bearing and elevation or more generally the look angles as shown in Figure 2. Satellite tracking systems were developed after space research programs started in the late fifties. Monopulse tracking systems are widely used at many earth stations throughout the world (c.f. Hawkins et. al. 1988). Typically earth station antenna dish diameters vary from 10m to 30m and they can weigh several hundred tons. As the electromagnetic frequency used increases, the beam width is reduced and the accuracy required to track the satellite is increased.

In marine environments, this is even more difficult because the roll and pitch of the ship changes the direction of an antenna pointed at a communications satellite. The gain of the antenna is usually reduced to widen the beam width so that misalignment does not affect the signal as much. However, the transmission power of both the satellite and the earth station must be increased to compensate for the reduced antenna gain or else the data rates reduced to compensate for the reduced signal to noise. Such power increases are very costly, especially for the satellite end of the link. The degraded signal to noise ratio is not so important for voice communication, as speech contains considerable redundant information and context cues. For data communication, transmission rates are greatly reduced and retransmission is often necessary to correct errors.

In the next section, the standard serially connected mechanisms for mounting an antenna are described. This is followed by a section on parallel mechanisms outlining the capabilities and the limitations of such mountings. A two axis parallel mechanism is then described along with a drive development that provides singularity free hemispherical coverage. Finally simulation results for using such a device on a rolling and pitching ship are presented.

2 Standard Mounting Systems

The azimuth (Az) and altitude (Alt) or elevation angles of the satellite viewed from the earth's surface are shown in Figure 2. The azimuthal plane is rotated about a vertical axis through the earth station so that the satellite is in the plane. The azimuth angle is measured eastward from geographic north, and the elevation angle in the azimuthal plane is measured from the horizon to the satellite. The antenna mounting is used to aim the antenna along the look angles so that the satellite is within the beam of the antenna. There are three standard 2 DOF mounting methods for achieving this. These are the Alt-Az, the X-Y, and the astronomical mountings; all of which were used for the early space object tracking antennas at the NASA Goldstone site in the U.S.

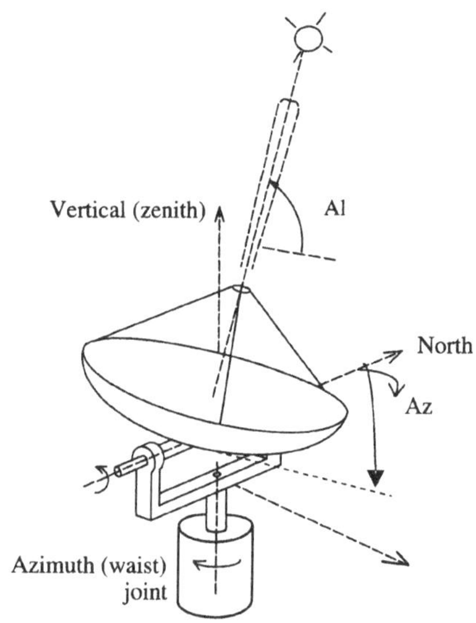

Figure 2. The Alt-Az antenna mount with the Alt-Az *look* angles as shown.

2.1 Alt-Az mountings - the key hole

The standard Alt-Az mount shown in Figure 2 consists of a horizontal axis revolute joint, which is attached to the rear of the antenna and carried on a vertical axis revolute joint. The look angles are set by rotating the vertical base joint through the azimuth angle from North, and then rotating the horizontal joint through the elevation angle from the horizon. If a moving satellite is tracked close to the zenith but not through it, then as the elevation nears 90^0 degrees the azimuth has to rotate through 180^0 degrees. During the time taken for the 180° degree azimuth rotation, the satellite moves out of the beam and the station loses contact. This is known as the key hole problem. While the problem can be anticipated and the azimuthal rotation started early, this only reduces the problem but does not eliminate it (Crawford and Brush 1995).

The problem occurs when the satellite track is through a region close to the zenith as there is a large change in one of the positioning coordinates for a small change in the satellite position i.e. adjacent look angles do not map into adjacent points in the control space. Thus, the system has a singularity about the zenith. The singularity or key hole problem is also known as gimbal lock in weapons systems and slew rate limiting in control system terminology.

For a tracking system mounted on a ship, the problem is particularly severe. The rolling and pitching action action causes the singularity of the alt-azimuth mount to trace out a conical region. Communication within this conical region is unreliable due to the effective increase in size of the key hole. Similar problems can arise with wind loading and vibrations on large narrow beam antennas.

2.2 X-Y and Astronomical mountings

In order to overcome the effect of having a key hole about the vertical axis, the first axis can be set at a different angle. For the X-Y mounting system the axis is horizontal as shown in. Figure 3. For the X-Y mount, each look angle affects both control angles, i.e. the axes are not decoupled and control is more complex. The advantage of the X-Y mount is that it does not have a key hole about the vertical, but it does have two other key holes: one at each end of the first (horizontal) axis. It is effectively the same as the Alt-Az system tipped on its side. When receiving signals from deep space probes, there is only one chance of capturing the data sent by the space craft and a loss of contact is not acceptable. In the case of the NASA deep space exploration antennas, two X-Y mounted antennas (10m and 30m at the Goldstone site) are mounted at each of the three receiving sites. At each site, the horizontal axes of the two antennas are orthogonal so that the key hole of each antenna is covered by the other antenna. This is an effective but expensive solution to the key hole problem.

For astronomical mounts, the first axis is parallel to the earth's N-S axis so when rotated at 1 revolution per day, the stars appear stationary. However, the singularity still remains but it is not a problem for star observations.

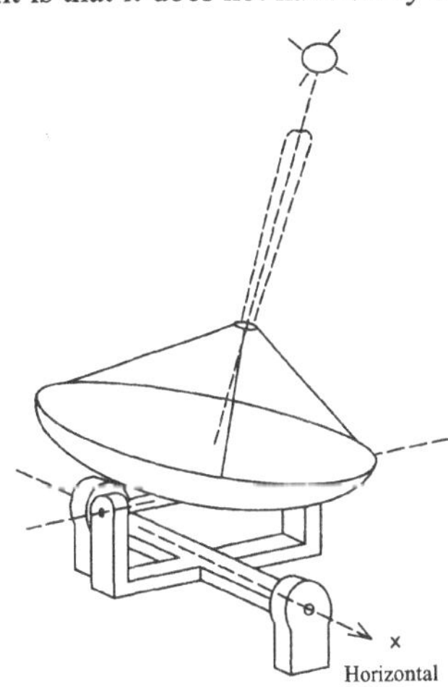

Figure 3. The X-Y mount.

2.3 Marine multi-axis mountings and key holes

The introduction of a third degree of freedom to overcome the limitations of a two axis mount in the vicinity of a key hole is used in marine communication mounting systems, and is quite common in earth resource and weather satellite tracking systems (cf. Johnson 1978). However, the increased mechanical complexity is costly and the axes must be computer controlled so that the mechanical singularities, which cause the key hole, can be avoided. The tracking system key holes are avoided by taking advantage of the extra degree of freedom to steer the mounting mechanism away from these singularities. Also the connecting revolute joints must be very stiff so as to avoid the cumulative errors associated with serial mechanisms.

More commonly, marine systems mount an Alt-Az antenna on a 2 axis gimbal giving a total of four degrees of freedom. The addition of extra serially connected joints to overcome the key hole problem (cf. CCIR report 1978) is expensive. The mass of the extra joint leads to an increase in the mass to be moved. Thus the supporting structure's strength and rigidity must be increased, and the drive motors and gear boxes upgraded further increasing the mass to be controlled. Also the cumulative errors from the joint positioning must be reduced by more precise measurement and control. This requires additional increases in the structural rigidity, which in turn leads to even more mass. The situation is one of diminishing returns for the extra expenditure.

3 Parallel aiming mechanisms

Dunlop, Ellis and Afzulpurkar (1993) devised an extended angular range Gough-Stewart platform mechanism to carry an antenna. Full hemispherical coverage was achieved with this 6 DOF system which used the 4 redundant DOF to maximize the aiming stiffness of the system. However, although the 6 axes were identical, the cost of the extra 4 DOF was considered too great and work continued on other parallel aiming systems. A 3 DOF spherical positioning parallel robot was produced by Dunlop and Jones (1997) based on the constant velocity joint described by Hunt (1978) and is shown in Figure 5. Further work by Dunlop and Jones (1998) resulted in the 2 axis parallel mechanism built to evaluate aiming throughout the hemisphere.

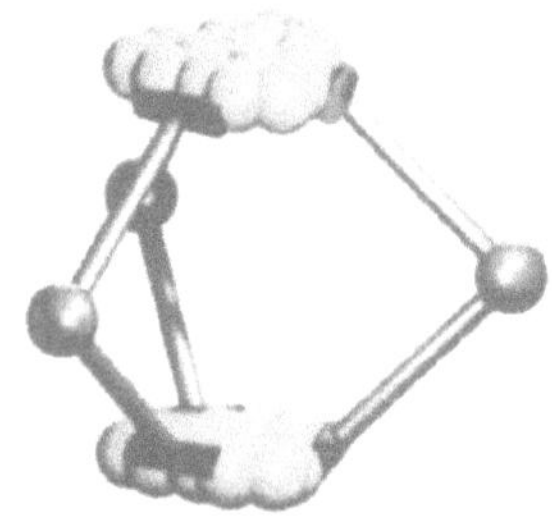

Figure 4. The basic 3 DOF parallel mechanism

4 Two axis parallel aiming mechanisms

The kinematics of the parallel 2 DOF aiming system shown in Figure 5(a) were developed to show the feasibility of the mechanism for a large angular range of movement (Dunlop and Jones 1998). However further investigation of the dynamics of this mechanism (Jones and Dunlop 2003) which included a counter weight (M) showed that a drive singularity occurred when the arms of the mechanism were driven to within 10° of the horizon. Some effect was expected as the symmetry of the mechanism was destroyed when only 2 of the 3 base revolute joints (R_1 and R_2) were driven. One approach was to use a pair of parallel connections to the central strut to overcome the drive singularity problem but these prismatic joints could interfere with the antenna at some look angles. When moving the central strut, the moving platform that carries the antenna is set at the required position and the 3 parallel arms set the angular alignment for that particular position and thus aim the antenna. Using this approach resulted in the 2 DOF robotic mechanism shown in Figure 6(b).

(a) (b)

Figure 5. Kinematics of the parallel 2 DOF aiming mechanism. The lower base triangle carries 3 revolute joints as does the upper platform triangle that carries the antenna. Spherical joints connect the arm pairs, and a central strut fixes the separation of the base and platform. Revolute joints R_1 and R_2 are driven. The spherical joint S_1 in the base of (a) has been replaced by a Cardan joint in (b) where both the Cardan revolute joints are driven as in an X-Y mounting.

Note that the mechanism in (a) has a mobility of 3 while (b) as only the required 2. The central strut in (a) is free to rotate and is a redundant unused DOF. The mechanism shown in Figure 1 is based on the kinematics in (b) and it demonstrated that the range of movement possible is more than that required for hemispherical coverage. The drives are applied via the Cardan joints to move the central strut which, for hemispherical coverage, is < 45° from the vertical and hence well away from the 2 X axis singularities of an X-Y system drive. While the passive mechanism that orientates the antenna is parallel, the positioning system (that determines the operation of the angular alignment) is serially connected. The resulting mechanism is a hybrid mechanism with parallel orientation, and a serial position drive which is operated distant from its R_1 singularities.

5 Marine use of two axis hybrid aiming mechanism

The static counterbalancing requirements for the 2 DOF and 3 DOF systems with driven arms were analysed and while a large reduction in the torque required to move the parallel arms was achieved, complete balancing of the gravitational load was not possible at all positions (Dunlop and Jones 1996). When an aiming system is attached to a moving platform such as for a camera mounted on a drone or an antenna on a ship, the translational movement of the base of the aiming mechanism affects the loading on the drives. A computational study was carried out to determine the dynamic drive requirements for a large antenna mounted on a rolling and pitching ship.

At the bottom of the central strut of the mechanism shown in Figure 1 (based on the kinematic diagram in Figure 5b) there is a single counterbalance weight comprised of the lead counterweights plus the second axis (R_2) drive motor. The counterbalance mass M can be varied and the effective position can be moved along the central strut. Static counterbalancing requires that the mass distance product be set, but dynamic balancing also requires the inertias be adjusted to be equal for the mechanism above and below the lower 2 axis gimbal (Cardan) joint. Because the antenna motion is different from the motion of the upper end of the central strut (i.e. the inertia dyadics differ significantly) perfect counterbalancing is not possible. Thus numerical simulation was used to determine the motor drive requirements for various counterbalance weights and positions.

For the mechanism shown in Figures 1 and 5(b), the object was to minimise the torque demands on the servo motors when the base was undergoing accelerations due to the vehicle movement. The simulation for the mechanism in Figure 1 used the Autolev software program (Kane and Levinson 1985, 2001) and the resulting C code was modified manually and compiled as a mex file for use in Matlab. The results were then output in graphical form. The parameters for all the mechanism components were included, and to indicate the scale, the distance from the driven Cardan joint to the drive motor/counter weight is 0.25m.

Results are presented in Figures 6 7 and 8 which show how the drive torques of the axes vary for the angular and linear accelerations used to simulate the motion of a ship. Figure 6 is a reference plot for no base acceleration. It is for tracking a target from -45° to + 45° about the first drive axis at 10°/s. A distinct minimum torque "valley" can be seen in Figure 6 and a suitable counterweight mass and position can be chosen along this valley. The remaining plots show the effect of base accelerations on the torque requirements for the drive motors.

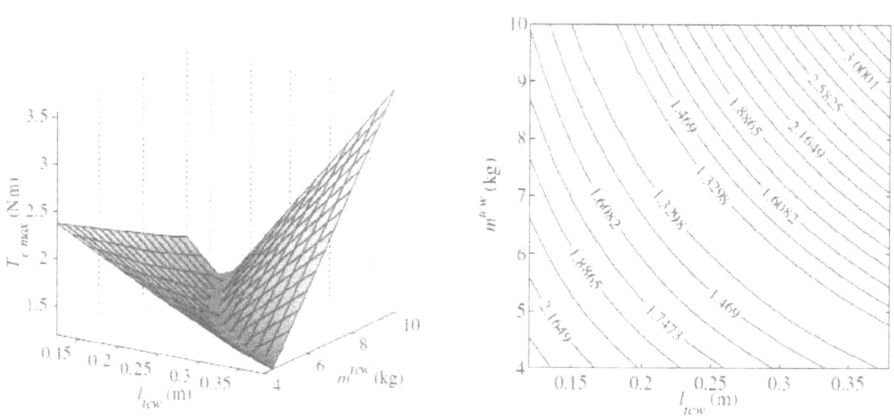

Figure 6. Magnitude of the first drive axis motor torque as a function of counterweight mass and position for tracking about the first axis at 10°/s from -45° to +45°.

The worst counterbalancing situation for the hybrid aiming mechanism occurs when the antenna is aimed horizontally. Figure 7 shows the effect of sinusoidal angular roll, pitch and yaw accelerations of the base over 10°, 8° and 2° with periods of 6.4s, 4.3s and 5.1s, respectively, and a vertical sinusoidal movement of amplitude 1.2m and period 4.3s. The tracker is aimed horizontally and the second axis is changing its position from 5° to 40°. Only the torque for the first motor is shown.

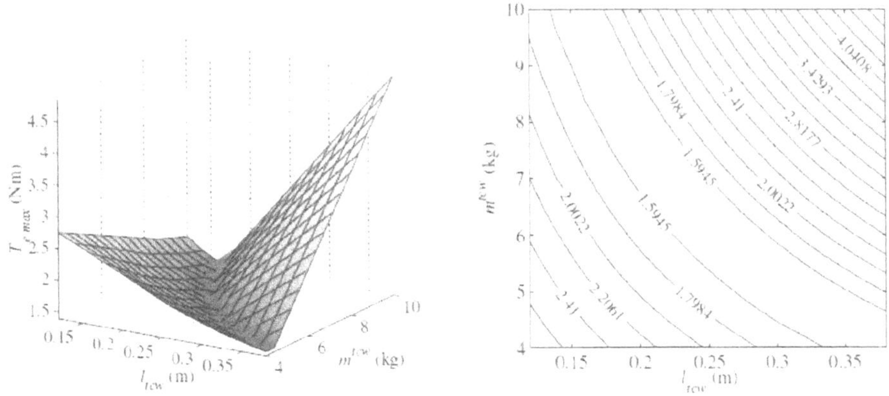

Figure 7. Magnitude of the first drive axis motor torque as a function of counterweight mass and position for tracking about the first axis at 10°/s from -45° to +45° while the base is undergoing angular and linear accelerations.

The effect of a conical scan in the presence of base accelerations is shown in Figure 8. It contains the results for the greatest torque required from either of the two drive motors while conducting a 360° azimuthal scan at an elevation angle of 30°. The base is undergoing the accelerations given for Figure 7, and the azimuthal scan rate is 15°/s.

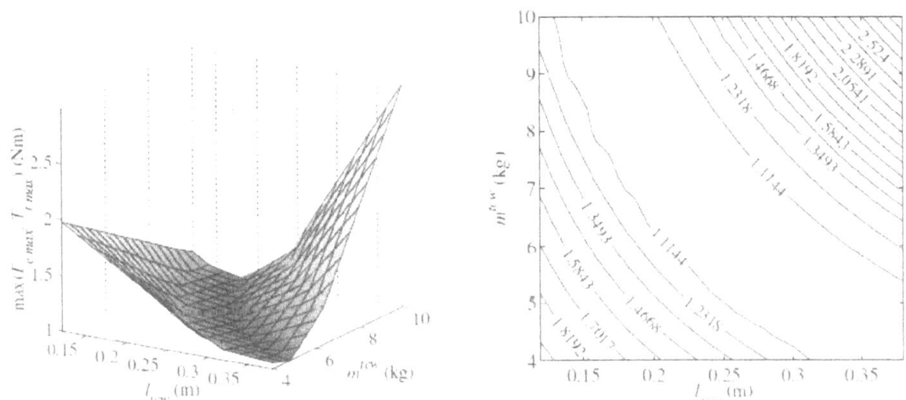

Figure 8. Magnitude of the largest motor torque of either drive axis as a function of counterweight mass and position for 360° azimuthal tracking with an elevation of 30° while the base is undergoing angular and linear accelerations.

Joint friction was also considered in the simulations and while it was found to have some effect on the basic positioning the extra effect when the base is accelerated is negligible as the mechanism is trying to keep the orientation of the antenna fixed i.e. it is not undergoing angular acceleration. Thus positioning is mostly to counter the base movements. In all 3 of the simulations presented, a desirable region for the counterweight mass and position is shown. Combining all 3 results for the aiming mechanism in Figure 1 showed that a counterweight of 6kg positioned 0.3m below the Cardan drive joints gives a satisfactory result.

6 Conclusions

A robotic mechanism combining the singularity free coverage of a 2 DOF parallel mechanism with the singularity free drive characteristics of a limited range X-Y aiming system has been presented. A small demonstration model of the system has been built and a range of simulations calculated for its performance in the presence of base accelerations similar to those experienced in marine applications. It was found that provided good static balancing is achieved, dynamic balancing is also satisfactory.

Bibliography

CCIR 1978 Mobile Services. Recommendations and reports of the CCIR XIVth plenary assembly, Kyoto, Volume VIII, Rep. 594-1, pp 381-399.

Crawford, P. S., and Brush, R. J. H. (1995). Trajectory optimisation to minimise pointing error, IEE Computing & Control Engineering Journal, Vol 6 (2) 61-67.

Dunlop, G. R., Ellis, P. J., and Afzulpurkar, N. V. (1993). The satellite tracking keyhole problem: a parallel mechanism mount solution, IPENZ Trans, Vol 20 (1) EMCh , 1-7.

Dunlop, G. R., and Jones, T. P. (1996). Gravity counterbalancing of a parallel robot for antenna aiming. *Proc. World Automation Congress,* ASME Press Series, ISBN 0-7918-0047-4, Vol 6, 153-158.

Dunlop, G. R., and Jones, T. P. (1997). Position analysis of a 3-DOF parallel manipulator. *Journal of Mechanism and Machine Theory*, Vol 32 (8) 903-920.

Dunlop, G. R., and Jones, T. P. (1998). Position analysis of a 2 DOF parallel manipulator. *Journal of Mechanism and Machine Theory,* Vol 34 (4) 599-614.

Hawkins, G.J., Edwards, D.J., and McGeehan, J.P. (1988). Tracking systems for satellite communications. Proc. IEE, Vol. 135/F (5) 393-407.

Hunt, K. H. (1978) Kinematic geometry of mechanisms. Oxford University Press, ISBN 0-19-856124-5, Fig. 13.8 p397.

Johnson, M. B. (1978) Antenna control system for a ship terminal for marisat. *Proc. Maritime and Aeronautical Satellite Communication and Navigation,* IEE Conference.

Jones, T. P., and Dunlop, G. R. (2003). Analysis of rigid body dynamics for closed-loop mechanisms - its application to a novel satellite tracking device. Proc. IMechE part1 JSCE, Vol 217 (14) 285-298.

Kane, T.R., and Levinson, D.A. (1985) *Dynamics: Theory and Applications,* McGraw-Hill.

Kane, T.R., and Levinson, D.A. (2001) Autolev user's manual. OnLine Dynamics Inc., www.autolev.com

Steeds, W. (1940). Mechanism and the kinematics of machines. Longmans, Green, London.

Sellaouti, R., and Ouezdou, F.B. (2005). Design and control of a 3DOFs parallel actuated mechanism for biped application. Mechanism and Machine Theory, Vol 40 (12) 1367-1393.

Virtual Prototyping of a New Parallel Robot for Milling

Maciej Petko[1], Grzegorz Karpiel[2], Daniel Prusak[3] and Tadeusz Uhl[4]

[1] AGH University of Science and Technology, Kraków, Poland, e-mail: petko@agh.edu.pl
[2] AGH University of Science and Technology, Kraków, Poland, e-mail: gkarpiel@agh.edu.pl
[3] AGH University of Science and Technology, Kraków, Poland, e-mail: daniel.prusak@agh.edu.pl
[4] AGH University of Science and Technology, Kraków, Poland, e-mail: tuhl@agh.edu.pl

Abstract. In the paper, the mechatronic procedure toward parallel robots development with special emphasis placed on the virtual prototyping stage was described on the example of a new 3-DOF parallel manipulator design. For this manipulator, two dynamic models were created and verified: multibody model for simulating the manipulator's dynamic behaviour and the analytical, structural one, for designing of a controller. An initial controller was virtually designed and experimentally tested with physical prototype of the manipulator. The performance of this initial controller is sufficient for conducting identification experiment, which is needed for more advanced nonlinear control strategies.

1 Introduction

One of the development tendencies of the construction of support structures for milling machines is to replace them with parallel manipulators. Milling electrospindles are also attached to parallel manipulators, creating a machine of the HSM (High Speed Machining) class, able to perform complex spatial movements. But kinematic structure of such manipulators is very complex, increasing their cost and restricting applications. Additionally, their stiffness depends on position. A solution to this problem can be a control that allows for shaping of dynamic properties, including stiffness, of the manipulator to ensure vibrostability in a wide range of tool positions. Such a solution can be rated among intelligent constructions, reacting to external influence of a cutting process.

Robots of parallel kinematics possess features that give them certain advantages (Tsai 1999) over serial manipulators: higher rigidity, due to the existence of multiple closed kinematic chains, and, as a result, higher natural frequencies, a high payload capacity, owed to the fact that the load is distributed among the actuators, fast movements, and good accuracy. However, industry has shown more interest in serial manipulators, which has its reasons (Merlet 2000): it is difficult to establish equations for the direct kinematics problem of parallel manipulators, the workspace is limited and very often filled with singularities, at which the manipulator gains or looses a degree of freedom, becoming uncontrollable (Lisowski, 2000).

Development of parallel manipulators appears to be a classical problem of mechatronic design, performed with an integrated, multidisciplinary, iterative approach as a closed kinematic chain complicates control synthesis and kinematic equations. Additionally, actuators usually play also a role of structural members of a manipulator. Therefore, during design of parallel manipulators, all aspects of various physical nature, together with control and its implementation related problems must be taken into account and treated with equal importance from the very beginning of the development process.

2 Design of Parallel Robots

The iterative procedure (Figure 1) starts with the specification, which also serves as a reference for further verification. The construction and the control algorithm are developed simultaneously, and provide models of the manipulator and the controller, which are then simulated together with external interactions. Modifications are introduced till getting a close conformity with requirements. This stage is also called virtual prototyping. Only after successful virtual prototyping, the physical prototype of manipulator is build, and control algorithm is tested in real-time using fast prototyping technique. During the fast prototyping stage, the models used for simulations are automatically transferred into a hardware/software system that is functionally equivalent to the system under test. On this stage, the model of manipulator is improved by identification of manufacturing imperfections, and parameters hard to asses on purely theoretical way, such as friction coefficients. When overall system performance meets specification, control algorithm is implemented on a target hardware platform. The remaining part of the paper concentrates on the virtual prototyping stage of the robot development process.

Virtual prototyping is becoming increasingly important in mechatronic product development (Iserman 2003, Zaeh & Baudisch 2003), as it allows for fast and cheap testing of multiple design variants on a suitable detail level. However it is complicated by the fact that for modelling of features from various physical domains coexisting in the system, and for various purposes, different software tools are deployed. They are usually incompatible and hard to integrate, though recently, a substantial progress can be observed in this area. There exists several approaches toward the realization of a virtual prototyping idea, and three of them seem being useful in practice:

- bond graphs modelling for systems of complex physical nature,
- block diagram approach,
- UML based, for high-level architecture modelling and verification.

Throughout the work, described in this paper, the block diagram approach was used in a form of an integrated software framework built around Simulink environment.

During virtual prototyping, based on the detailed design of mechanical parts, completed in CAD software, multibody model of high accuracy is created. Such a model, due to its size, is inappropriate for control design. Hence, another simpler one is formulated on a standard theoretical way. Parameters of this less accurate model can be tuned to obtain greater conformity with multibody model. Results of simulation of entire system in integrated computational environment provide data for enhancement of the construction and control algorithm. This process is iteratively repeated until obtaining satisfactory performance of the system.

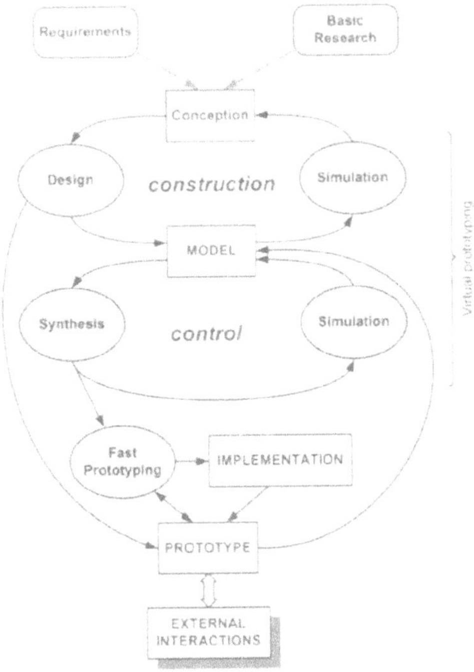

Figure 1. Mechatronics design of parallel robots.

3 Case Study – 3-DOF Spatial Parallel Robot for Milling

The task was to create, using previously presented procedure, a highly robust and versatile, 3-DOF construction for high-speed machining applications, which would exhibit comparatively large working volume (roughly 50 dm^3), high payload capacity to carry an electrospindle and exert sufficient milling forces (several tens of N for aluminium) and good dynamic characteristics (max. velocity ca. 1 m/s, acceleration ca. 1 g) for fast tool return path. An additional requirement was to avoid elements that are expensive and troublesome in manufacturing, as spherical joints or parallelograms.

Several constructions were investigated and virtually prototyped, then one of them was selected for further development. Optimisation of its dimension/workspace was done using heuristic approach based on results of simulations. The final construction combines advantages of Steward Gough platform and Delta based approaches. The developed construction, shown on Figure 2, is a spatial fully-parallel manipulator with three translational degrees of freedom. The manipulator consists of a steady base, which takes the shape of an equilateral triangle, and a moving platform to which an electric spindle is attached. The base is connected with the moving platform by three limbs that consist of two links connected with each other by prismatic joints. These prismatic joints are actuated with LinMot direct drive linear motors, which allow controlling the limb lengths. The limbs are connected to the base platform with the help of universal joints. Each universal joint consists of two revolute joints whose axes of rotation intersect and are orthogonal. The connection of the limbs to the moving platform is facilitated by revolute joints. The moving platform is based on a special construction, which allows the limbs to rotate freely about its centre around an instantaneous axis of rotation that always remains orthogonal to the base platform. This rotation is enabled by a specially designed bearing assembly. The moving platform is provided with a socket for attaching milling spindles equipped with standard taper shank.

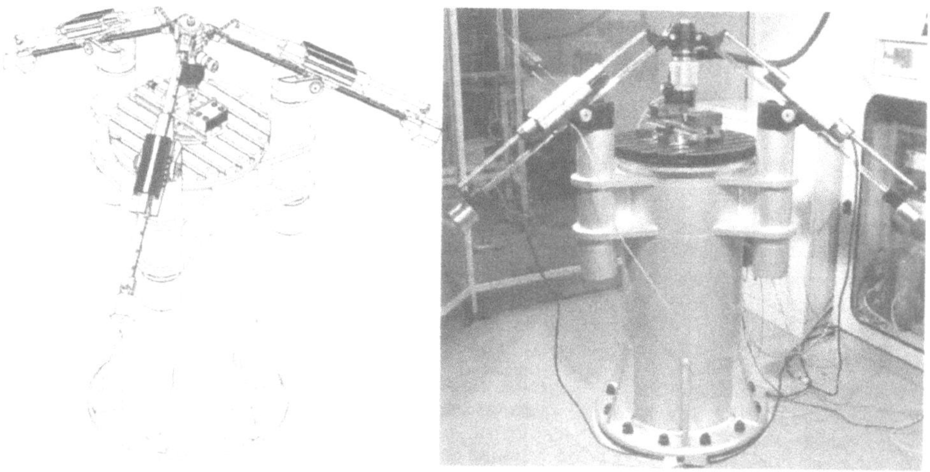

Figure 2. Developed 3-DOF 3-RRPRR manipulator for milling.

The workspace of the manipulator has a shape of cylinder with diameter of 600 mm and height of 300 mm. Inside, with a HFK 95 S 40 spindle from IBAG, maximum attainable speed is ca. 2 m/s and acceleration 2 g. Maximum achievable milling force is over 100 N, which is enough for high speed machining of aluminium.

3.1 The Model of the Manipulator

For the accurate simulation of its dynamics, the multibody model of the manipulator was created in VisualNastran on the basis of the detailed mechanical part drawing, made in CAD software. Some parameters, hard to establish on a theoretical way, as friction coefficients, initially were roughly estimated and corrected later, when experimental data became available. VisualNastran was selected because it allowed for integration with Simulink, where models of the remaining elements of the manipulator, such as sensors or actuators were modelled and all simulations took place.

For control design, another, structural analytical model of the manipulator's dynamic was created based on laws of physics. The dominant source of nonlinearity in the manipulator is its geometry, which entails high variations of inertia and forces caused by gravity inside the working space. Neglecting other nonlinearities, the simplified dynamics takes the general form:

$$\mathbf{M}(\mathbf{x})\ddot{\mathbf{l}} + \mathbf{G}(\mathbf{x}) = \mathbf{F}, \tag{1}$$

where $\mathbf{x}=[x, y, z]^T$ is a position of the moving platform in (x, y, z) Cartesian coordinate system and $\mathbf{l}=[l_1, l_2, l_3]^T$ is a position of the moving platform in the joint coordinate system, as depicted on Figure 3, **M** is an inertia matrix, **G** is a gravity forces (as seen by actuators) matrix, and **F** is a vector of forces exerted by actuators.

Inertia of manipulator can be expressed as function of only prismatic joints lengths, which are measured directly. However, for calculating gravitational part, the position of end effector in Cartesian space is needed. This can be obtained by solving the forward kinematics problem.

Denoting by R the radius of the circle circumscribed about the base triangle, the equations of manipulator's inverse kinematics take the form (Petko et al. 2004):

$$\begin{aligned} l_1 &= \sqrt{x^2 + (y-R)^2 + z^2} \\ l_2 &= \sqrt{\left(x - \frac{1}{2}\sqrt{3}R\right)^2 + \left(y + \frac{1}{2}R\right)^2 + z^2} \\ l_3 &= \sqrt{\left(x + \frac{1}{2}\sqrt{3}R\right)^2 + \left(y + \frac{1}{2}R\right)^2 + z^2} \end{aligned} \tag{2}$$

It is worth noting that these equations are analytically differentiable, so finding the link velocities and accelerations does not pose a problem. It is possible to find close analytical form of forward kinematics:

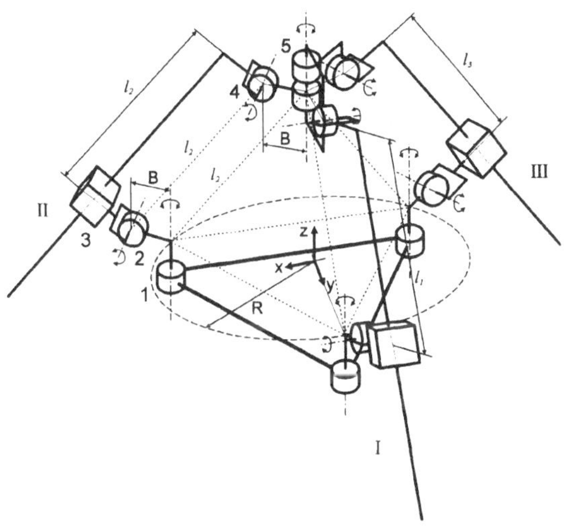

Figure 3. Kinematic structure of the manipulator.

$$x = -\frac{\sqrt{3}\left(l_2^2 - l_3^2\right)}{6R}$$
$$y = -\frac{2l_1^2 - l_3^2 - l_2^2}{6R} \qquad (3)$$
$$z = \sqrt{l_1^2 - \frac{\left(l_2^2 - l_3^2\right)^2}{12R^2} - \left(\frac{2l_1^2 - l_3^2 - l_2^2}{6R} + R\right)^2}$$

Simulations have shown discrepancy of up to 10% between the analytical and the multibody models. It can be explained mainly by too much simplified mass distribution in the analytical model. An accurate description would have to be much more complicated due to the complex limbs geometry and necessity of taking into account other, now neglected, sources of nonlinearities, as friction, backlashes or elasticity. Such further extending of the model would make it useless for control purposes because of too large calculation overheads for real-time execution. Therefore, it is planned to use a black-box model, e.g. neural-network based, for designing of a final, precise controller. Such a model requires an experimental identification of its parameters, and this, in turn, requires an initial, stabilizing controller that is described in the next subsection.

3.2 The Controller

To control the position of the parallel manipulator **l**, to track a desired trajectory $\mathbf{l}_d$, the following feed-forward linearizing (computed torque) control law is utilised (Craig, 1989):

$$\mathbf{F} = \mathbf{M}(\mathbf{l})\left(\ddot{\mathbf{l}}_d + K_P\mathbf{e} + K_V\mathbf{e}_V + K_I\int_0^t \mathbf{e}(\tau)d\tau\right) + \mathbf{G}(\mathbf{x}), \qquad (4)$$

where $\mathbf{e} = \mathbf{l}_d - \mathbf{l}$, $\mathbf{e}_V = \dot{\mathbf{l}}_d - \dot{\mathbf{l}}$ and K_P, K_V, K_I are constants. The structure of the position controller is shown on Figure 4, together with the remaining parts of the robot used during virtual prototyping, performed in Simulink: the trajectory generator, multibody model of the manipulator, and models of power electronics, sensors and sensor interfaces.

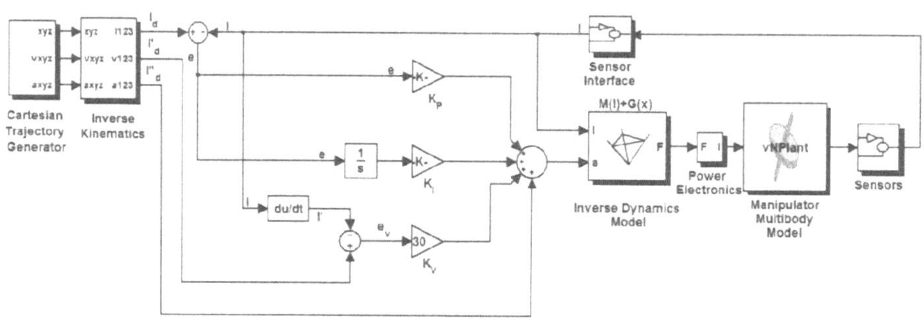

Figure 4. Simulink model used for virtual prototyping of the robot.

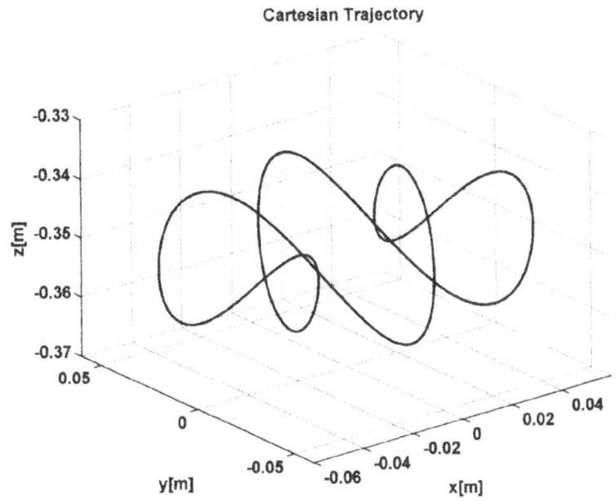

Figure 5. Trajectory used for simulations and experiment.

3.3 Experimental Verification

After controller's parameters tuning, its performance was investigated, using fast prototyping technique on the physical prototype of the manipulator. During experiments, the execution time of the control algorithm, running on a floating-point digital signal processor (DSP), allowed for sampling frequency up to 1 kHz. Having in mind one of the target applications – active control of vibrations emerging during machining, which would require more advanced and complex control strategy – some different approach towards control execution had to be tried. Therefore, the control law was implemented in Altera Stratix family FPGA chip in fixed-point form, using the procedure described in Petko & Karpiel (2003). Such an approach allows, among others, for straightforward bit- and cycle-accurate simulation in Simulink of a hardware-implemented controller, together with the remaining, multidisciplinary part of a mechatronic system.

During experiments, the task was to track a specially prepared trajectory in the Cartesian space. The chosen trajectory was a three-dimensional Lissajoux curve (Figure 5), because it is closed and smooth, both in the Cartesian- and the joint- coordinates, with various speeds and accelerations, effectively exploring a large part of the workspace. Then results (actual platform trajectories) from experiment and virtual prototyping were compared (Figure 6a). Good conformity, obtained between simulation and experiment, indicates high accuracy of multibody model of the manipulator. It is especially visible on the plot of the position errors in the Cartesian coordinate y (Figure 6b). Errors in the two other axes were in the same range (0.5 mm max.). Trajectory tracking performance, though quite fair, can be further improved after experimental identification of more accurate manipulator model used for feedback linearization in control law. Having a good model available, also other approaches to nonlinear control problem can be tried.

4 Conclusions

In the paper, the mechatronic procedure toward parallel robots development was described and justified with special emphasis placed on the virtual prototyping stage. The procedure was successfully applied to the problem of design of a new 3-DOF parallel robot construction. The result fulfils specification and accepted assumptions, has a structure that is simple and inexpensive for manufacturing, comparatively large working space, high payload capacity and fast dynamics with close analytical form of forward kinematics. For this manipulator, two dynamic models were created and verified: multibody model for simulating the manipulator's dynamic behaviour and the analytical, structural one, for designing of a controller. The latter was analysed and conclusions regarding possibilities of its improvements and usefulness for final controller design were presented. An initial controller was designed and experimentally tested with physical prototype of the manipulator. The results confirmed high accuracy of the multibody model. The methodology used, lowered costs and the duration time of the development process and made it possible at all. The performance of this initial controller is sufficient for conducting identification experiment, which is needed for more advanced nonlinear control strategies.

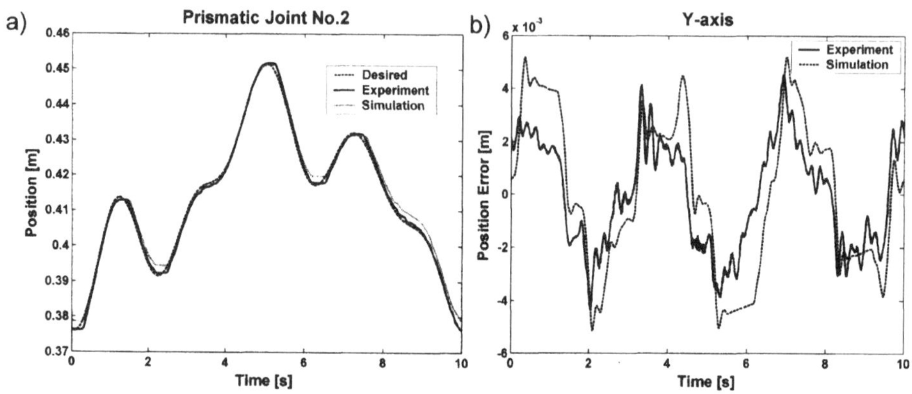

Figure 6. Comparison of manipulator performance during simulation and experiment: a) position in the prismatic joint no. 2; b) trajectory tracking error along y-axis.

Bibliography

Craig, J.J. (1989), *Introduction to Robotics: Mechanics and Control*. Addison-Wesley.

Iserman R. (2003), *Mechatronic Systems: Fundamentals*, Springer.

Lisowski, W. (2000), *Problems of Robots' Manipulators modelling*, Wyd. KRiDM AGH, (in Polish).

Merlet, J.-P. (2000), *Parallel Robots*, Kluwer Academic Publishers.

Petko, M. & Karpiel, G. (2003), *Semi-Automatic Implementation of Control Algorithms in ASIC/FPGA*, 2003 IEEE Conference on Emerging Technologies and Factory Automation: Proceedings, vol. 1, pp. 427-433, IEEE.

Petko, M., Karpiel, G., Prusak, D. & Martowicz, A. (2004), Kinematics of the new type 3-DOF parallel manipulator, *PAK*, 5/2004, pp. 25-28 (in Polish).

Tsai, L.-W. (1999), *Robot Analysis: the Mechanics of Serial and Parallel Manipulators*, John Wiley & Sons.

Zaeh, M.F. & Baudisch, T. (2003) Simulation environment for designing the dynamic motion behaviour of the mechatronic system machine tool, *Proc. Instn Mech. Engrs*, **217**:1031-1035.

A New Approach for the evaluation of kinematic and static performances of a family of 3-UPU translational manipulators

Raffaele Di Gregorio [1] and Vincenzo Parenti-Castelli [2]

[1] Department of Engineering, University of Ferrara, Ferrara, Italy, e-mail: rdigregorio@ing.unife.it
[2] DIEM, University of Bologna, Bologna, Italy, e-mail: vincenzo.parenticastelli@mail.ing.unibo.it

Abstract. This paper presents a new approach for finding kinematic and static performances of a class of 3-UPU manipulators that feature platform and base both planar and similar to each other. Kinematic as well as static issues worth considering for the design of these manipulators are addressed and a novel design procedure is presented.

1 Introduction

Translational parallel manipulators (TPMs) have attracted the attention of researchers and industries since the early appearance of the DELTA robot (Clavel, 1988), and many TPM topologies have been presented and thoroughly studied (see, for instance, Hervé and Sparacino, 1991, Kong and Gosselin, 2002). TPMs with 3-UPU topology are some of these.

3-UPU manipulators have the end effector (platform) connected to the frame (base) by three serial kinematic chains (legs) of type UPU (U and P stand for universal joint and prismatic pair respectively). If in each UPU leg both the two ending revolute pair axes and the two intermediate revolute pair axes are respectively parallel to each other, and the prismatic pair is the actuated joint, a translational manipulator (translational 3-UPU) is obtained (Tsai, 1996). Different architectures of translational 3-UPUs can be obtained by changing the shape of the platform and/or the base (Di Gregorio and Parenti-Castelli, 2002), i.e. by changing the relative position of the three universal joint centers and/or the three revolute pair axes of the base and/or of the platform.

The most popular translational 3-UPU has a symmetric architecture presented by Tsai (1996). In Tsai's 3-UPU, the three centers of the universal joints of the platform (base) are the vertices of an equilateral triangle, and each revolute pair axis of the platform (base) is parallel to the triangle side, opposite to the triangle vertex crossed by the same axis.

The first prototypes exhibited some paradoxical behaviour which could not be justified by advocating either the influence of backlash in the joints or the leg flexibility. The kinematic behaviour of this mechanism was thoroughly investigated and explained (Di Gregorio and Parenti-Castelli, 1998) by the presence of a new type of singularity whose occurrence at some special configuration destroys the translational constraint between platform and base. This type of singularity was later called constraint singularity (Zlatanov et al., 2001).

The result presented in (Di Gregorio and Parenti-Castelli, 1998) highlighted the fact that the design of any translational 3-UPU (and, in general, of any TPM with leg connectivity higher than three) is strictly related to the shape of the singularity loci (Parenti-Castelli and Bubani, 1999), which affects both the workspace and the static behavior of the machine (Parenti-Castelli et al., 2000).

This paper studies the kinematic and static features of a family of 3-UPUs with platform and base whose shapes are planar (i.e. the revolute pair axes of the platform (base) lie on the plane located by the three centers of the platform (base) universal joints) and similar (i.e. the triangle defined by the centers of the platform universal joints is similar to the one defined by the corresponding base universal joints). A brief review of the main results obtained for this family of manipulators will be reported and new issues presented and discussed. A new tool for geometrically identifying the singularity loci, and for establishing a correspondence between configurations of different 3-UPU manipulators will be given. This will allow different architectures of 3-UPUs to be compared from both kinematic and static points of view and a simple design procedure to be provided.

2 Kinematic Aspects Involved in the Design

Figure 1 shows a translational 3-UPU with planar and similar platform and base, and the notations that will be used. With reference to Fig. 1: A_i and P_i (i=1,2,3) are the centers of the universal joints of the base and the platform respectively; R_{ji} (j=1,…,4) is the j-th revolute pair of the i-th leg (j increases from the base to the platform); $\mathbf{w}_{ji}$ is the unit vector of the axis of the revolute pair R_{ji}. Moreover, the straight line passing through A_i and P_i is called the axis of the i-th leg, and $\mathbf{u}_i$ denotes the unit vector of the i-th leg axis (i.e., $\mathbf{u}_i = (\mathbf{P}_i - \mathbf{A}_i)/|\mathbf{P}_i - \mathbf{A}_i|$). Hereafter, it will be assumed that, in each leg, the axes of the two intermediate revolute pairs are orthogonal to the leg axis (i.e. $\mathbf{w}_{2i}$ and $\mathbf{w}_{3i}$ are perpendicular to $\mathbf{u}_i$).

With these notations, the geometric conditions to guarantee the translation of the platform are

$$\mathbf{w}_{1i} = \mathbf{w}_{4i}, \qquad \mathbf{w}_{2i} = \mathbf{w}_{3i} \tag{1}$$

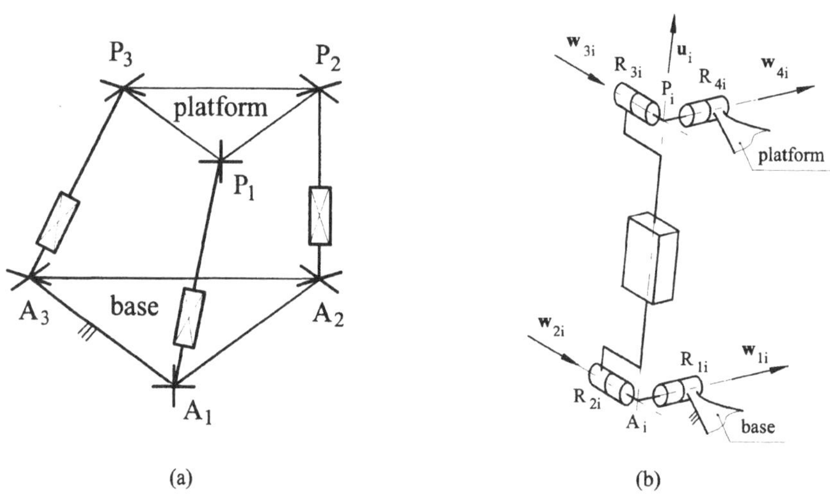

Figure 1. (a) Translational 3-UPU with planar and similar platform and base; (b) i-th leg of UPU type.

The geometric conditions defining the class of translational 3-UPUs that will be analyzed can be given as follows: (a) the unit vectors $\mathbf{w}_{1i}$ ($\mathbf{w}_{4i}$) for i=1,2,3 are coplanar and (b) the triangle $A_1A_2A_3$ is similar to the triangle $P_1P_2P_3$. Condition (b) involves the three leg axes always intersecting each other at one point. Since the base triangle can be obtained by projecting the platform triangle from this intersection point, this point will be called the projection center. The projection center moves when the platform moves.

The singularity conditions of these translational 3-UPUs are (Di Gregorio and Parenti-Castelli, 2002)

$$\mathbf{u}_1 \cdot \mathbf{u}_2 \times \mathbf{u}_3 = 0 \tag{2}$$

$$\mathbf{n}_1 \cdot \mathbf{n}_2 \times \mathbf{n}_3 = 0 \tag{3}$$

where $\mathbf{n}_i$ (i=1,2,3) is the unit vector perpendicular to the revolute pair axes of the i-th leg defined by $\mathbf{n}_i = \mathbf{w}_{2i} \times \mathbf{w}_{1i}$.

Conditions (2) and (3) depend only on the relative directions of the unit vectors $\mathbf{u}_i$ (i=1,2,3). This is clear for condition (2), whereas, for condition (3), it becomes clear when considering that, since vectors $\mathbf{w}_{1i}$, i=1,2,3, are fixed in the base, the i-th vector $\mathbf{n}_i$, i=1,2,3, changes its orientation according to the orientation changes of vector $\mathbf{w}_{2i}$, which is perpendicular to both $\mathbf{w}_{1i}$ and $\mathbf{u}_i$.

Taking the same base shape, the relative orientation of the unit vectors $\mathbf{u}_i$ (i=1,2,3) only depends on the position of the projection center with respect to the base plane. In particular, the values of the triple products $\mathbf{u}_1 \cdot \mathbf{u}_2 \times \mathbf{u}_3$ and $\mathbf{n}_1 \cdot \mathbf{n}_2 \times \mathbf{n}_3$ are uniquely determined when the position of the projection center is given with respect to the base. Thus, all the singularity analyses of this class of translational 3-UPUs can be implemented just by considering the position of the projection center.

Condition (2) identifies configurations where the platform can perform infinitesimal or finite translations even though the actuators are locked (translation singularities). This condition occurs when either the platform triangle $P_1P_2P_3$ lies on the plane of the base triangle $A_1A_2A_3$ (i.e. when the projection center lies on the plane of the base triangle), or the three leg axes are all parallel (i.e. when the projection center goes to infinity).

Condition (3) identifies configurations where the platform can perform infinitesimal rotations (constraint singularities), which may become finite rotations if the actuators are not locked. This condition occurs when the three $\mathbf{n}_i$ vectors are parallel to a unique plane. In particular, it is easy to see that when the projection center lies on the base plane all the $\mathbf{n}_i$ vectors lie on the same plane too; hence this geometric condition identifies singularities that are both translational and constraint. The remaining constraint singularities of the translational 3-UPUs with planar platform and base (Di Gregorio and Parenti-Castelli, 2002) occur when the platform reference point, defined as the platform point used to identify the manipulator configuration and to describe the manipulator workspace, lies on a right cylinder whose intersection with a plane parallel to the base plane is a circular directrix (singularity circle). The singularity circle lying on the base plane can be easily drawn once three points belonging to it are found.

If the orthogonal projection on the base plane of point P_i falls on the axis of the revolute pair R_{1i} (see Fig. 1), vector $\mathbf{n}_i$ will be perpendicular to the base plane. Thus, when at least two such conditions occur for different points P_i, at least two vectors $\mathbf{n}_i$ are parallel to each other and Eq. (3) is verified (i.e. a constraint singularity occurs). By exploiting this geometric condition, three points belonging to the singularity circle can be found as shown in Fig. 2(a). It is easy to realize through this geometric construction that the singularity circle can degenerate either into a straight line, when two revolute pair axes of the base are parallel to each other (Di Gregorio and Parenti-Castelli, 2002), or into a point, when the three revolute pair axes of the base intersect which each other at a single point (Wolf et al., 2002).

If the platform size shrinks to a point, the platform reference point will coincide with the projection center (the leg-axes intersection denoted by H in Fig. 2(b)). Hence the singularity circle drawn by looking at the position of the projection center is the one passing through the intersections of the axes of the base's revolute pairs (see Fig. 2).

2.1 Projection Center as a Design Tool

The quality of a configuration in terms of "distance" from singularities can be measured by the values of the two indices: $k_1 = \mathbf{u}_1 \cdot \mathbf{u}_2 \times \mathbf{u}_3$ and $k_2 = \mathbf{n}_1 \cdot \mathbf{n}_2 \times \mathbf{n}_3$. k_1 and k_2 range from zero to one, and the farther a configuration is from singularities, the greater the values of k_1 and k_2. The position of the projection center with respect to the base plane univocally determines the values of these indices.

The set of translational 3-UPUs with planar and similar platform and base can be separated into disjoint subsets characterized by having a similar shape of the base (similar base subsets). All the manipulators of each subset have the property that, if their sizes are scaled by using as scaling factor the reciprocal of the length of a side of the base triangle, the motion of the projection center with respect to the base plane is the same. This happens because all the scaled platform triangles can be stacked into the same projection cone (see Fig. 3(a)).

Since the motion of the projection center is the same for all the scaled manipulators of a similar base subset (SBS), the singularity circle referring to the projection center is the same for all these manipulators. Moreover, since, for a given position of the projection center, the projec-

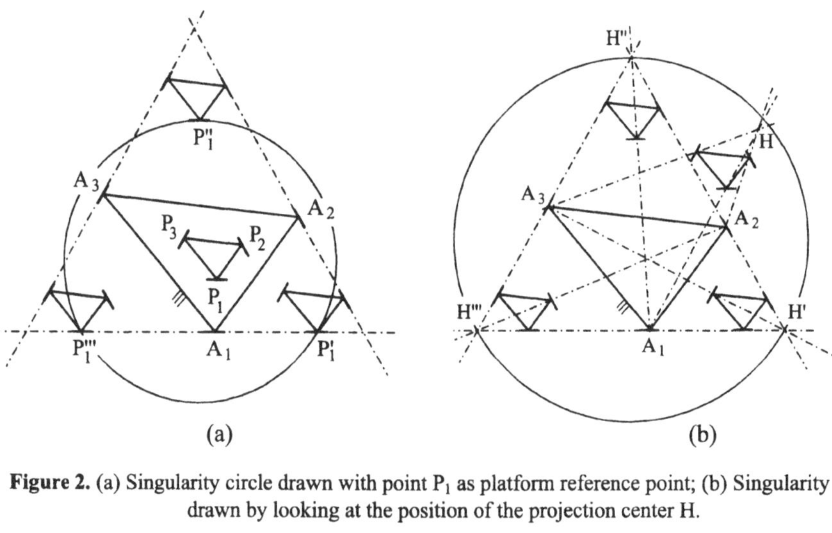

Figure 2. (a) Singularity circle drawn with point P_1 as platform reference point; (b) Singularity circle drawn by looking at the position of the projection center H.

tion center, a one-to-one correspondence between the configurations of two manipulators of the same SBS can be geometrically stated by taking two homologous points as platform reference points. Such a geometric correspondence has the property that the corresponding configurations have the same values as k_1 and k_2 (i.e. they have the same kinematic quality). In fact, two corresponding configurations share the same position of the projection center.

If the workspaces of the manipulators of a SBS are determined by taking homologous points as reference platform points, and by requiring k_1 and k_2 to be greater than some given minimum values, the computed workspaces will collect only configurations corresponding to each other according to the above-defined geometric correspondence. This result will be exploited in the following paragraphs to deduce an analytic relationship between the sizes of two manipulator workspaces.

Figure 3(a) shows platforms belonging to the same SBS. According to Fig. 3(a), the points belonging to different platforms that are homologous to a generic point P of the lowest platform will be denoted by P^z with z=I, II, III,... that indicates the order number of the platform. With this notations, the alignment of the points homologous to the vertices P_i (i=1,2,3) of the first-platform triangle can be expressed as follows:

$$\mathbf{P}_i^z = (\mathbf{P}_i - \mathbf{A}_i)\frac{1-c_z}{1-c} + \mathbf{A}_i \tag{4}$$

where

$$c = \frac{(\mathbf{P}_2 - \mathbf{P}_1)\cdot(\mathbf{A}_2 - \mathbf{A}_1)}{|\mathbf{A}_2 - \mathbf{A}_1|^2}, \quad c_z = \frac{(\mathbf{P}_2^z - \mathbf{P}_1^z)\cdot(\mathbf{A}_2 - \mathbf{A}_1)}{|\mathbf{A}_2 - \mathbf{A}_1|^2} \tag{5}$$

The variable c_z is a signed scaling factor that ranges from c (< 1) to 0 for the platforms below the projection center and becomes negative for the platforms above the projection center (reversed platforms). Without losing generality, point P_1 and its homologous points, P_1^z, will be chosen as reference platform points. Figure 3(b) shows how the position of the projection center changes when point P_1 together with its platform performs a translation equal to **s**. As an effect of the translation of P_1 the homologous point P_1^z performs a translation equal to $\mathbf{s}_z$, and the relationship between homologous points in the new configuration is

$$\mathbf{P}_i^z + \mathbf{s}_z = (\mathbf{P}_i + \mathbf{s} - \mathbf{A}_i)\frac{1-c_z}{1-c} + \mathbf{A}_i \tag{6}$$

The subtraction of Eq. (4) from Eq. (6) yields the following relationship:

$$\mathbf{s}_z = \mathbf{m}\,(1 - c_z) \tag{7}$$

where

$$\mathbf{m} = \frac{\mathbf{s}}{1-c} \tag{8}$$

is a constant vector.

By interpreting the magnitude of $\mathbf{s}_z$ as the maximum travelling distance of P_1^z in the workspace of the z-th platform (e.g. the diameter of a spherical workspace), relationship (7) gives us how the characteristic dimensions of the workspace change for the manipulators of the same

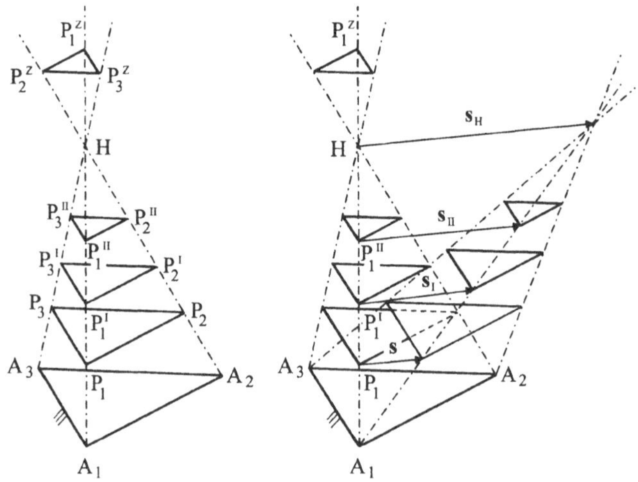

Figure 3. (a) The scaled platform triangles can be stacked into the same projection cone; (b) Variation of the projection cone after a platform translation **s**.

SBS. Relationship (7) justifies the results numerically obtained by Parenti-Castelli and Bubani (1999) for the Tsai's 3-UPUs.

Inspection of relationship (7) reveals that the workspace size linearly increases with the decrease of c_z. For platforms under the projection center, this means that the increase of the workspace size is linearly related to the decrease of the platform size. On the contrary, for the reversed platforms, the negative values assumed by c_z make the workspace size linearly increase with the platform size, and, theoretically, it could become infinite still keeping the values of k_1 and k_2 greater than the minimum values fixed for all the SBS. This characteristic of the reversed platforms makes them appealing. Reversed platforms have been proposed by Bubani (2000) and further analyzed by Di Gregorio and Parenti-Castelli (2000). In Bubani (2000), the problem of manufacturing legs with intersecting axes was also addressed, and it was shown that, if an offset is introduced between the leg axis and the geometric axis of the leg hardware, the collision of the legs never occurs during the platform motion. This fact can be easily understood by considering that, during the platform motion, the leg axes always intersect with each other in the projection center; the introduced offsets are thus always kept during the platform motion.

From the design point of view, the properties of the projection center and of the homologous points of a SBS can be exploited to design a translational 3-UPU that matches an assigned workspace size, and keeps k_1 and k_2 greater than the assigned minimum values inside the workspace. In fact, by fixing the attention on a platform with an arbitrarily chosen size, the size of the workspace that fits the minimum values of k_1 and k_2 and the shape (spherical, cubic, etc.) of the given workspace can be computed for this particular platform. Then, the size of the platform that also fits the given workspace size can be determined by simply using relationship (7), where **s** and c are the values computed for the platform with arbitrarily chosen size.

3 Static Aspects Involved in the Design

With reference to the notations reported in Fig. 4, it was demonstrated in Parenti-Castelli et al. (2000) that the internal loads of the i-th leg consist of a normal load $\mathbf{N}_i$, a torque $\mathbf{M}_{ti}$ and a bending moment $\mathbf{M}_{fi}$ given by the following expressions:

$$\mathbf{N}_i = f_i\, \mathbf{u}_i; \quad \mathbf{M}_{ti} = -M_i\, (\mathbf{u}_i \cdot \mathbf{n}_i)\, \mathbf{u}_i; \quad \mathbf{M}_{fi} = -M_i\, [\mathbf{n}_i - (\mathbf{u}_i \cdot \mathbf{n}_i)\, \mathbf{u}_i] \tag{9}$$

where, with the same external load applied to the platform, the coefficient f_i is inversely proportional to k_1; and the coefficient M_i is inversely proportional to k_2 through a factor that linearly increases with all the f_i (i=1,2,3) and the platform size.

Since relationships (9) show that, with the same external load applied to the platform, the internal loads mainly depend on the relative directions of the vectors $\mathbf{u}_i$ and $\mathbf{n}_i$ (only the dependence of M_i on the platform size is disregarded), it can be concluded that the internal loads practically depend on the position of the projection center.

Nevertheless, the leg lengths (defined as the lengths of the segments $A_iP_i^z$, i =1,2,3) have a crucial role in leg deformation due to the fact that the torque and the bending moment generate a leg deformation that is proportional to the leg length with the same leg cross section. Inspection of relationship (4) reveals that the dependence of the leg lengths on c_z is

$$A_iP_i^z = g_i\, (1 - c_z) \tag{10}$$

with $g_i = A_iP_i/(1-c)$. Since Eq. (10) has the same dependence on c_z of the workspace sizes (Eq. (7)), it can be concluded that the increase of the workspace has as its counterpart the increase of the flexibility of the manipulator. This negative effect can be contrasted by increasing the leg stiffness through the increase of the transversal sizes of the leg. Finally, when large platforms are compared with small platforms the dependence of torque $\mathbf{M}_{ti}$ and bending moment $\mathbf{M}_{fi}$ on the platform size through M_i (Eq. (9)) must also be considered (Parenti-Castelli et al. ,2000).

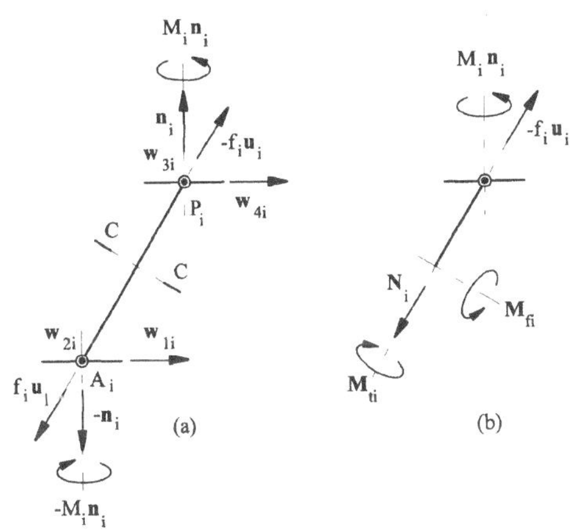

Figure 4. (a) Free-body diagram of the i-th leg; (b) Internal loads in the i-th leg.

4 Conclusions

The kinematic and static features of the family of translational 3-UPU manipulators which have a base and platform planar and similar to each other were studied. After a brief review of the results reported in the literature for this type of manipulators, new issues are presented and discussed. In particular, a new approach for geometrically identifying the singularity loci, and for establishing a correspondence between configurations of different 3-UPU manipulators is presented. The new approach is based on the concept of projection center. The approach makes it possible to compare different architectures of the same class of 3-UPUs from both kinematic and static points of view, and allows the formulation of a simple design procedure.

Acknowledgments

The funding of the Italian MIUR is gratefully acknowledged.

Bibliography

Bubani, F. (1999). Studio delle caratteristiche funzionali di un manipolatore parallelo di pura traslazione e progetto del prototipo. *Master Thesis*, Bologna (Italy): DIEM-Univ. of Bologna.

Clavel, R. (1988). DELTA: a fast robot with parallel geometry. In *Proceedings of the 18th International Symposium on Industrial Robots*, 91-100

Di Gregorio, R., and Parenti-Castelli, V. (1998). A translational 3-dof parallel manipulator. In Lenarcic, J., and Husty, M.L., eds., *Advances in Robot Kinematics: Analysis and Control.* Dordrecht (The Netherlands): Kluwer Academic Publishers, 48-58.

Di Gregorio, R., and Parenti-Castelli, V. (2000). Benefits of twisting the legs in the 3-UPU Tsai mechanism. In *Proceedings of the Year 2000 Parallel Kinematic Machines International Conference*, 201-211.

Di Gregorio, R., and Parenti-Castelli, V. (2002). Mobility analysis of the 3-UPU parallel mechanism assembled for a pure translational motion. *ASME J. of Mechanical Design*, 124:259-264.

Hervé, J.-M., and Sparacino, F. (1991). Structural synthesis of parallel robots generating spatial translation. In *Proceedings of the 5th International Conference on Advanced Robotics*, 808-813.

Kong, X., and Gosselin, M.-C. (2002). A class of 3-DOF translational parallel manipulators with linear input-output equations. In *Proceedings of the Workshop on Fundamental Issues and Future Research Directions for Parallel Mechanisms and Manipulators*, 25-32.

Parenti-Castelli, V., and Bubani, F. (1999). Singularity loci and dimensional design of a translational 3-dof fully-parallel manipulator. In *Proceedings of the Workshop on Advances in Multibody Systems and Mechatronics*, 319-331.

Parenti-Castelli, V., Di Gregorio, R., and Bubani, F. (2000). Workspace and optimal design of a pure translational parallel manipulator. *Meccanica*, 35:203-214.

Tsai, L.-W. (1996). Kinematics of a three-dof platform with three extensible limbs. In Lenarcic, J., and Parenti-Castelli, V., eds., *Recent Advances in Robot Kinematics*. Dordrecht (The Netherlands): Kluwer Academic Publishers, 401-410.

Wolf, A., Shoham, M., and Park, F. C., (2002). Investigation of singularities and self-motions of the 3-UPU robot. In Lenarcic, J., and Thomas, F., eds., *Advances in Robot Kinematics: Theory and Applications*. Dordrecht (The Netherlands): Kluwer Academic Publishers, 165-174.

Zlatanov, D., Bonev, I., and Gosselin, C.-M. (2001). Constraint singularities as configuration space singularities. In *www.parallelmic.org/Reviews/Review008.html*, online paper.

Geometric Configuration in Robot Kinematic Design

Dr Joe Rooney[1]

[1]The Open University, Faculty of Technology, Milton Keynes, UK, j.rooney@open.ac.uk

Abstract. A lattice of geometries is presented and compared for representing some geometrical aspects of the kinematic design of robot systems and subsystems Three geometries (set theory, topology and projective geometry) are briefly explored in more detail in the context of three geometric configurations in robotics (robot groupings, robot connectivities and robot motion sensor patterns).

1 Introduction

Kinematics is naturally focussed on changes, with respect to time, in the relative distances and angles between mechanical systems and/or sub-systems. But distance and angle are not the only concepts involved in kinematics and a range of other entities and constructs arise in a more general treatment. Many of these are essentially geometrical, in that they can be expressed in terms of some type of geometry, although it may not be Euclidean. Characterising a particular geometry usually gives rise to one or more types of geometric configuration. In this paper the emphasis is on geometric configuration associated with robot components, spatial relationships and patterns of motion. It is assumed that a group of robot systems (or subsystems) operates, co-operates or competes kinematically within one or more geometrical regimes, and geometric configuration may be identified whenever robot systems move, change shape or evolve.

2 A Lattice of Geometries

Felix Klein, in his *Erlanger Programm*, organised the (19th century) known geometries into a very elegant hierarchical scheme (Meserve, 1955; Klein, 1939). Since then many more types of geometry have been constructed, including a major class of finite or discrete geometries. Klein's hierarchy (a tree) is now a larger and more general lattice of inter-related geometries.

Klein made sense of the diversity by treating a geometry as a set of points, a set of relations between the points, and a group of transformations on the points. For robots a 'point' might represent: a location in workspace or jointspace; a vision sensor pixel; a complete manipulator system; a robot state; a pattern of motion, etc. A relation might represent: an association of two or more pixels; distances between pairs of robots, or states; the 'connectivities between robot components or sub-systems. A transformation might represent: a rotation of a manipulator link or of a set of pixels forming a vision sensor; a permutation of robot 'names' or identities; a mapping between a robot's jointspace and workspace.

If a transformation is applied to a set of points, then certain attributes of the points and their inter-relationships do not change – they remain *invariant*. Each type of transformation and its invariants characterise the type of geometry. Typically, concepts such as magnitude, direction, dimension, orientation, connectedness etc. can be given formal geometrical meanings (as invariants under particular types of transformation).

By considering a lattice of geometries in robotics it is hoped that something akin to the Brooks hierarchical subsumption architecture (Brooks, 1999) might be achieved. The idea is that each geometry on one level of the lattice subsumes all aspects of the geometries on the adjacent lower levels. Obversely, each geometry is subsumed by the geometries on the adjacent higher levels. A corresponding situation for robot motion behaviour might be that, for example, initially the robot is being controlled to behave within the scope of a particular geometry for a period of time. Then in response to some event, fault or failure it switches to a different level in the hierarchy where it continues to function, but now by being controlled within the scope of the new geometry on the new level. The resulting new behaviour pattern replaces the original behaviour pattern in operation before the fault/failure occurred. The capabilities of the robot under the new geometrical regime might be enhanced, but probably they will be diminished. However, some aspects of the original robot motion could continue to be controlled under the new regime. Moreover, another possible scenario is that a group of robots might be controlled concurrently on a number of different (geometrical) levels in order to produce more complex behaviour and achieve multiple goals, improve efficiency or optimise performance.

3 Examples of Geometries

The methods employed by Klein to describe and analyse geometries used co-ordinates. The most general view is that a co-ordinate is merely a *label* attached to (i.e. co-ordinated with) a point. Notions or measures of distance (or of any other potential invariant) are defined in terms of co-ordinates by specifying some function of the latter. Klein introduced co-ordinate systems to label the points and then defined each geometry by a set of relationships amongst the co-ordinates that remained invariant under some group of co-ordinate transformations. The relationship $(x_2 - x_1)^2 + (y_2 - y_1)^2$ between the Cartesian co-ordinates (x_1, y_1) and (x_2, y_2) of two points in a plane, defines the square of the distance between the points, and both this Pythagorean relationship and the geometrical distance remain invariant under rotations, etc. More general types of labels (co-ordinates) than Cartesian ones, and more abstract relationships than the Pythagorean one are used in the Klein scheme (see Tables 1 and 2). However, the overall principle is the same, namely: a geometry is defined by the set of co-ordinate relations that are invariant under the action of a particular group of co-ordinate transformations, and different types of co-ordinate transformations lead to different geometries.

Klein found that the larger the group of allowed (coordinate) transformations for a geometry, the fewer the number of invariants. So, paraphrasing Klein, the most general types of transformation leave hardly anything unchanged, whereas the least general types of transformation leave many things unchanged. An example of a general transformation is a one-one mapping (essentially a permutation), and this changes everything except the *number* (cardinality) of points or objects. Conversely a Euclidean rotation (a much less general transformation) maintains *distances* between points, *angles* between lines, *areas*, *shapes*, *parallelism*, etc. Table 1 lists six different geometries and their characteristic coordinate transformations. The list includes Set Theory as a geometry as well as the more familiar types. Table 2 shows typical invariants (expressed in terms of co-ordinates) for some of these geometries and for two extra geometries - the symplectic geometry of the phase plane (Weyl, 1939), used in dynamics, and the pseudo-Euclidean geometry of relativistic kinematics (Rindler, 1966).

Table 1. Six geometries and their co-ordinate transformations.

Geometry (reference)	Characteristic Co-ordinate Transformation	Type of Transformation / Types of Invariant
Set Theory (Birkhoff and MacLane, 1965)	$x' = f_1(x,y,z)$ $y' = f_2(x,y,z)$ $z' = f_3(x,y,z)$	One-One
		Cardinality
Topology (Porteous, 1969; Alexandroff, 1961)	$x' = f_1(x,y,z)$ $y' = f_2(x,y,z)$ $z' = f_3(x,y,z)$	One-One and Continuous
		Dimension, Connectivity, Fixed Point
Differential (Kreyszig, 1959)	$dx' = a_1(x,y,z)dx + b_1(x,y,z)dy + c_1(x,y,z)dz$ $dy' = a_2(x,y,z)dx + b_2(x,y,z)dy + c_2(x,y,z)dz$ $dz' = a_3(x,y,z)dx + b_3(x,y,z)dy + c_3(x,y,z)dz$	Continuous and Smooth
		Metric, Nearness
Projective (Klein, 1939; Modenov and Parkhomenko, 1965b)	$x' = \dfrac{a_1x + b_1y + c_1z + d_1}{a_4x + b_4y + c_4z + d_4}$ $y' = \dfrac{a_2x + b_2y + c_2z + d_2}{a_4x + b_4y + c_4z + d_4}$ $z' = \dfrac{a_3x + b_3y + c_3z + d_3}{a_4x + b_4y + c_4z + d_4}$	Rational Linear
		Incidence, Separation, Cross-ratio
Affine (Modenov and Parkhomenko, 1965a)	$x' = a_1x + b_1y + c_1z + d_1$ $y' = a_2x + b_2y + c_2z + d_2$ $z' = a_3x + b_3y + c_3z + d_3$	Linear
		Parallelism, Finiteness, Between-ness
Euclidean (Yale, 1968)	$x' = (l^2(1-\cos\theta) + \cos\theta)x + (ml(1-\cos\theta) + n\sin\theta)y$ $+(nl(1-\cos\theta) - m\sin\theta)z + d_1$ $y' = (ml(1-\cos\theta) - n\sin\theta)x + (m^2(1-\cos\theta) + \cos\theta)y$ $+(mn(1-\cos\theta) + l\sin\theta)z + d_2$ $z' = (nl(1-\cos\theta) + m\sin\theta)x + (mn(1-\cos\theta) - l\sin\theta)y$ $+(n^2(1-\cos\theta) + \cos\theta)z + d_3$	Orthogonal
		Distance, Angle, Shape, Size

Table 2. Five geometries and some of their invariants.

Geometry (reference)	Typical Invariants and their Context
Differential (Kreyszig, 1959)	*Metric*: $ds^2 = g_{11}dxdx + g_{12}dxdy + g_{13}dxdz + g_{21}dydx + g_{22}dydy + g_{23}dydz + g_{31}dzdx + g_{32}dzdy + g_{33}dzdz$ where (x, y, z) and $(x+dx, y+dy, z+dz)$ are two neighbouring points in 3D space
Projective (Klein, 1939; Modenov and Parkhomenko, 1965b)	*Cross-ratio*: $R_c = \dfrac{(x_1 - x_2)(x_3 - x_4)}{(x_1 - x_4)(x_3 - x_2)}$ where x_1, x_2, x_3 and x_4 are four points in 1D space
Symplectic (Weyl, 1939)	*Product*: $\delta^2 = (q_2 - q_1)(p_3 - p_1) - (q_3 - q_1)(p_2 - p_1)$ where (q_1, p_1), (q_2, p_2) and (q_3, p_3) are three points in 2D phase space
Pseudo-Euclidean (Rindler, 1966).	*Interval*: $\tau^2 = (t_2 - t_1)^2 - (x_2 - x_1)^2 - (y_2 - y_1)^2 - (z_2 - z_1)^2$ where (t_1, x_1, y_1, z_1) and (t_2, x_2, y_2, z_2) are two points in 4D spacetime, with $c = 1$
Euclidean (Yale, 1968)	*Product*: $\delta^2 = (x_2 - x_1)(x_3 - x_1) + (y_2 - y_1)(y_3 - y_1) + (z_2 - z_1)(z_3 - z_1)$ *Distance*: $d^2 = (x_2 - x_1)^2 + (y_2 - y_1)^2 + (z_2 - z_1)^2$ *Angle*: $\cos\theta = \dfrac{(x_2 - x_1)(x_3 - x_1) + (y_2 - y_1)(y_3 - y_1) + (z_2 - z_1)(z_3 - z_1)}{[(x_2 - x_1)^2 + (y_2 - y_1)^2 + (z_2 - z_1)^2][(x_3 - x_1)^2 + (y_3 - y_1)^2 + (z_3 - z_1)^2]}$ where (x_1, y_1, z_1), (x_2, y_2, z_2) and (x_3, y_3, z_3) are three points in 3D space

4 Robot Configuration

In the following three sub-sections three different geometries from Table 1, are briefly explored in the context of geometrical configurations in robotics.

4.1 Robot Groupings (Set Theory)

Consider a set of n robots and treat it as a finite set of n points. There are a total of 2^n possible subsets into which the n robots can group themselves. Specifically there are ${}_nC_r$ different possible subsets containing r robots, where ${}_nC_r$ is the binomial coefficient Since there are also ${}_nC_{n-r} = {}_nC_r$ different possible subsets containing $n-r$ robots, this enumerates the complete range of possible ways that up to n robots can partition themselves into two 'teams'.

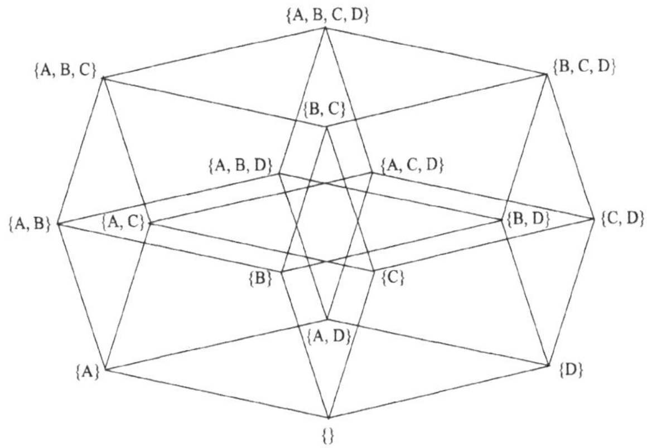

Figure 1. The lattice of subsets of a set {A, B, C, D} of four robots.

One way to represent the interrelationships amongst the subsets or 'teams' of robots is to display them in a lattice with the robots labelled, say, alphabetically. For a set {A, B, C, D} of four robots, the lattice of subsets is shown in Figure 1. The lattice may be used in various ways to characterise robot groupings and to follow changes in robot groupings. For instance this lattice can be viewed essentially as a hypercuboid with the subsets or 'teams' of robots located at its vertices. Subsets at opposite ends of a hyperspace diagonal have no robots in common and so the eight possible hyperspace diagonals represent the eight ways to partition the set of four robots into two non-overlapping 'teams'. This 'geometrical' arrangement can form the basis of a 4-bit coding scheme to label the subsets with a binary codeword (co-ordinate) rather than just using letters. The advantage gained from this code is that the codewords of adjacent subsets in the lattice differ by one bit, and those on opposite ends of a hyperspace diagonal differ by four bits, hence a simple measure is established for tracking changes in the membership of 'teams'.

The paths from one node to another in this lattice can be used as a basis for enumerating fault and failure modes. Thus suppose the top node represents the state in which all four robots are functioning, and the next level down represents states in which one robot is not functioning, and so on. Then the *number* of paths down through the hierarchy from the top to any particular node represents the number of ways (and hence modes of failure) to arrive in the state represented by that node. This leads to a crude measure of probability of ending up in that state. It may also offer modes of recovery by identifying and suggesting strategies for moving to nearby nodes.

The transformations within set theory are one-one transformations, and for a finite set of n points these transformations are permutations (a re-ordering of the points of the set). A collection of n robots might be ordered in some sequence determined by their positions, by their communication protocols, or by their 'seniority', etc. Each possible ordering may be specified by labelling the robots with a sequence of symbols. Then any re-ordering is essentially equivalent to a permutation of the robot symbols. For a set of n robots there are a total of $n!$ different permutations of the labels. They fall into two equally sized classes, namely $\frac{n!}{2}$ even permutations and $\frac{n!}{2}$ odd permutations. Whatever permutation is applied to a collection of robots it does not change the number (cardinality) of robots. But the effects of even and odd permutations are different. Two types of orderings emerge. An even permutation maintains the 'orientation' of the original ordering whereas an odd permutation produces an 'opposite' orientation. The situation is akin to the difference between right-handed and left-handed co-ordinate systems. From the set theory viewpoint there is really only one invariant associated with a collection of robots and that is the number of robots in the collection. However, although the robots may be identical (i.e. indistinguishable) in construction, they might nonetheless at each instant be distinguished by labels, names, or codewords, or by an indication of their places in a chain of command. In this case the different possible labellings will fall into two equal size classes - those related by even permutations, and the remainder. Thus a second attribute of the collection of robots is the 'orientation' of the labelling scheme, but note that this is only an invariant under even permutations.

4.2 Robot Neighbours (Topology)

Given a set of n robots it is possible to introduce more structure into the set by defining a notion of 'neighbourliness' between pairs of robots. This establishes a topology for the set. In general the resulting topology can be very complicated and is often non-intuitive, but in the present context of a simple discrete set of robots the 'neighbourliness' can be specified in terms of 'connections' between pairs of robots. Each connection may be a mechanical or electrical connection, or more generally it can just be a representation of a communication channel or relation between 'neighbouring' robots. The connectedness of a discrete configuration of n robots may be described and explored in terms of graph theory (Wilson, 1996).

Within graph theory each robot is represented by a point (a vertex) and each connection or relation is represented by a line segment (an edge). In general the graph representing the connections amongst a set of n robots may be described in terms of one or more *spanning trees*. Figure 2 shows two different graphs for a set of six robots. Each graph represents one possible way for the robots to be interconnected. In both cases one possible spanning tree is shown by bold edges. A spanning tree may be used to determine two important invariants of the graph, namely its *fundamental cycles* and its *fundamental cutsets*.

In the graph on the left of Figure 2, there are three cycles: $\{e_1,e_2,e_3,e_4\}$, $\{e_3,e_7,e_6,e_5\}$ and $\{e_1,e_2,e_7,e_6,e_5,e_4\}$. But just two are *fundamental* cycles, since the third may be formed as a 'sum' of the first two (where the rule for adding two cycles is to include in the sum all edges except for those common to both of the original cycles). Given a spanning tree, the fundamental cycles are immediately identified since every cycle has an edge in common with the complement of the tree, and so the number of fundamental cycles is the same as the number of edges in the complement of the tree. If there are n vertices and m edges in the graph there are $m-n+1$ edges in the complement and hence $m-n+1$ fundamental cycles. The importance of fundamental cycles stems from the fact that they characterise the circuits and thus the fundamentally distinct pathways through the graph. Hence for a set of robots they therefore characterise how the robots are linked together as a group.

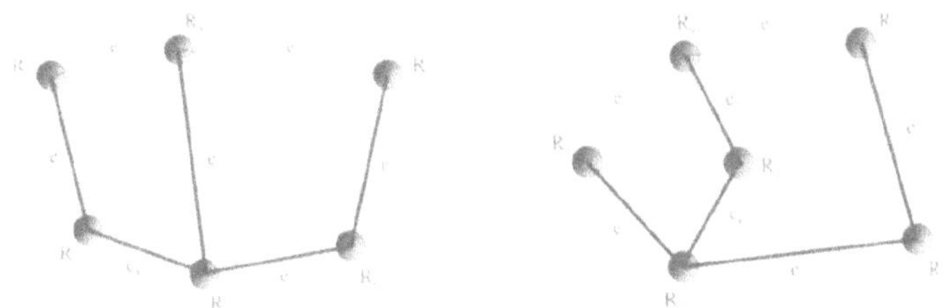

Figure 2. Two different graphs for six robots, showing spanning trees (bold edges).

In the graph on the right of Figure 2, there are sixteen cutsets: $\{e_1,e_2\}$, $\{e_3,e_2,e_7\}$, $\{e_4,e_2,e_7\}$, $\{e_5,e_7\}$, $\{e_6,e_7\}$, $\{e_6,e_5\}$, $\{e_1,e_3,e_7\}$, $\{e_2,e_4,e_5\}$, $\{e_1,e_3,e_5\}$, $\{e_2,e_3,e_6\}$, $\{e_2,e_3,e_5\}$, $\{e_1,e_4,e_6\}$, $\{e_1,e_4,e_5\}$, $\{e_1,e_4,e_7\}$, $\{e_2,e_4,e_6\}$ and $\{e_1,e_3,e_6\}$. But just five are *fundamental* cutsets, since the last eleven can be formed from sums of the first five (where the rule for adding two cutsets is to include in the sum all edges except for those common to both of the original cutsets). Given a spanning tree the fundamental cutsets are immediately identified since every cutset has one edge in common with the tree and so the number of fundamental cutsets is the same as the number of edges in the tree. If there are n vertices and m edges in the graph there are $n-1$ edges in the spanning tree and hence $n-1$ fundamental cutsets. The importance of fundamental cutsets stems from the fact that they characterise the fundamentally distinct divisions of the graph, and hence for a set of robots they therefore characterise how the robots may be divided into co-operating or competing subgroups.

4.3 Robots in Perspective (Projective Geometry)

Pixel configurations in a robot motion sensing system, such as those used in the visual morphology research at the University of Sussex (Harvey I, et al., 1996) often have their outputs grouped into special 2D patterns. The approach has proved to be successful in the interpretation of perspective views acquired by mobile robots. Their performance may be further improved by considering the projective geometry of the situation.

One particular special pattern combines the outputs from four pixels. Any non-collinear group of four pixels in the plane defines a (projective geometry) configuration of seven lines and nine points forming a *complete quadrilateral* (Coxeter, 1993). The extra five points and the seven lines on which they lie have significance in defining auxiliary pixel locations (see Figure 3(a)). An analogous dual configuration of seven points and nine lines, known as a complete quadrangle, defines another potentially useful projective construct for the vision system (see Figure 3(b)).

In Figure 3, if the four pixel positions are replaced by a group of four robots in four positions, the complete quadrilateral and complete quadrangle configurations have significance for the robot grouping. The diagonal points and the intersections of diagonal lines represent significant locations for targets, for goals or for where other associated robots may be placed.

5 Discussion and Conclusions

The notion of a hierarchy of geometries, offers a novel approach to understanding robot kinematic behaviour. A group of robots may behave in a huge range of different ways, co-operating or competing with one another, but whatever their behaviour it almost always has an external kinematic manifestation as movement or change of the spatio-temporal configuration of the group. The geometrical approach may be used to define, characterise and classify various kinematic behaviours. Each different geometry focusses on specific (often subtle) aspects of the kinematics, and may be used either to *analyse* a possible range of kinematic behaviour or to *produce* a required kinematic behaviour. By operating solely within the scope of one level of the hierarchy, or, alternatively, by combining several levels simultaneously, various degrees of complexity of robot kinematic behaviour may be explored.

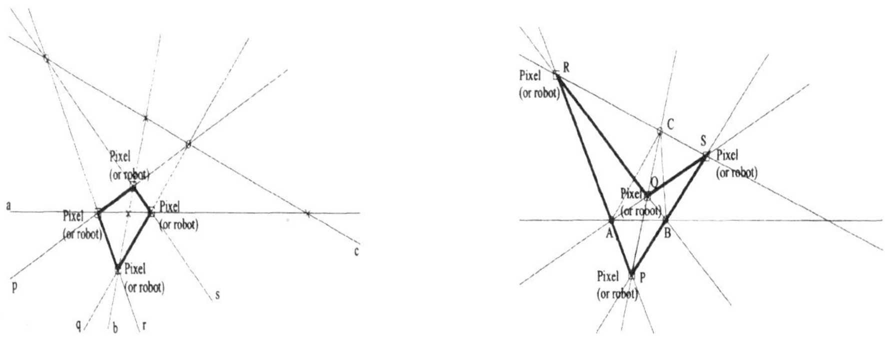

Figure 3(a) Four lines *p, q, r, s* form a complete quadrilateral, with diagonal lines *a, b, c* forming the diagonal triangle)(left); (b) Four points *P, Q, R, S* form a complete quadrangle, with diagonal points *A, B, C* forming the diagonal triangle (right).

Bibliography

Alexandroff P, (1961), *Elementary Concepts of Topology* (Dover Publications, New York)

Birkhoff G, MacLane S, (1965), *A Survey of Modern Algebra* (MacMillan, New York)

Brooks R A, (1999), *Cambrian Intelligence* (MIT Press)

Coxeter H S M, (1993), *The Real Projective Plane* (Springer-Verlag, Berlin)

Harvey I, Husbands P, Cliff D, Thomson A and Jakobi N, (1996), "Evolutionary Robotics at Sussex", in Proc. ISRAM 96: Int. Symp. on Robotics & Manufacturing, Montpelier, France

Hausner M, (1965), *A Vector Space Approach to Geometry* (Dover Publications, New York)

Klein F, (1939), *Elementary Mathematics from an Advanced Standpoint* (two volumes) (Noble) (Dover, New York edition, 1948)

Kreyszig E, (1959), *Differential Geometry*, (University of Toronto Press, Toronto) (Dover, New York edition, 1991)

Meserve B E, (1955), *Fundamental Concepts of Geometry* (Dover, New York edition, 1983)

Modenov P S and Parkhomenko A S, (1965a), *Geometric Transformations, Volume 1: Euclidean and Affine Transformations* (Academic Press, New York)

Modenov P S and Parkhomenko A S, (1965b), *Geometric Transformations, Volume 2: Projective Transformations* (Academic Press, New York)

Porteous I R, (1969), *Topological Geometry* (Van Nostrand Reinhold, London)

Rindler W, (1966), *Special Relativity* (Oliver and Boyd)

Weyl H, (1939), *The Classical Groups*, (Princeton, New Jersey)

Wilson R J, (1996), *Introduction to Graph Theory* (Longman)

Yale P B, (1968), *Geometry and Symmetry* (Holden-Day, San Francisco) (Dover, New York edition, 1988).

Chapter II

Mechanism Performance

Improvement of Positioning Accuracy of PAMINSA (Parallel Manipulator of the I.N.S.A.)

Vigen Arakelian and Sebastien Briot

Département de Génie Mécanique et Automatique
Institut National des Sciences Appliquées (I.N.S.A.) de Rennes
20 av. des Buttes de Coesmes, CS 14315
F-35043 Rennes, France
e-mails : vigen.arakelyan@insa-rennes.fr
sebastien.briot@ens.insa-rennes.fr

Abstract. In this paper we present new design approaches for the positional accuracy improvement of the 4-DOF parallel manipulator PAMINSA. The particularity of this manipulator is the decoupling of displacements in the horizontal plane from its translation along the vertical axis. The positioning errors caused by the inclination of the platform relating to the base are cancelled by means of two different approaches. The first solution is realized by means of two correcting systems mounted on the drive system of the vertical displacements. The second solution is carried out by a correcting mass mounted on the platform. The suggested solutions are illustrated by the numerical examples and the performances of such a design are shown.

1 Introduction

It is well known that the parallel manipulators show a better stiffness than the serial manipulators. Their actuators and gearing are fixed, resulting in a low moving mass. Thus, the payloads capacity of parallel manipulators with lightweight links is more important than the serial manipulators. However they have nonlinear coupled kinematics and dynamics, which is a drawback for the industrial applications. It is obvious that the parallel structure with linear input-output equations is more appealing than the nonlinear one. In recent years, the decoupling of motions of parallel manipulators has attracted researchers' attention and different structures have been proposed. Previous works on this problem may be arranged into two principal groups: (i) fully decoupled parallel structures in which the architecture of the manipulator is such that its input/output equations are linear and (ii) position/orientation decoupled manipulators in which the end-effector's position is independent of the orientation.

One of the first studies devoted to the design of fully decoupled parallel structures was developed by Kong and Gosselin (2002). In this work, a class of 3-legged 3-DOF translational parallel manipulators with linear input-output equations was proposed. Another simple fully decoupled 3-DOF translational parallel mechanism named Tripteron was discussed in the paper Gosselin et al. (2004). Carricato and Parenti-Castelli (2004) proposed a novel parallel mechanism with fully decoupled degrees of freedom. This mechanism consists of two interconnected slider crank linkages, each of which independently actuates one of the Euler angles of the input link.

Several types of translational parallel mechanisms with decoupled motions were discussed in the next work of the same authors (2004). The parallel robot, called T3R1, with four degrees of freedom and decoupled motions was developed at the French Institute of Advanced Mechanics (Bouzgarrou et al., 2004).

In spite of rather encouraging results, it should be noted that the fully decoupling of the motions also brings some disadvantages, such as for example, the increase of the number of joints and a loss of rigidity. It is obvious that it is difficult to solve the problem of the fully decoupling of the motions by keeping the many advantages of the parallel mechanisms. This is why we tried to find a compromise between the decoupling of the motions and the architectural characteristics of the parallel structures.

Another trend of the kinematic decoupling is proposed in the design of the PAMINSA, a new 4-DOF parallel manipulator that has been designed and developed at the I.N.S.A.[1] in Rennes. It consists of decoupling the motions in the horizontal plane and the translation along the vertical axis.

1.1 Description of the PAMINSA

Fig. 1 shows a 3D model of the PAMINSA with three legs. Each leg of the manipulator is realized by a pantograph mechanism (Fig. 2) with two input points A_i and B_i, and an output point C_i (i = 1, 2, 3). Each input point A_i is connected with the rotating drive M_i by means of the prismatic guide mounted on the rotating link.

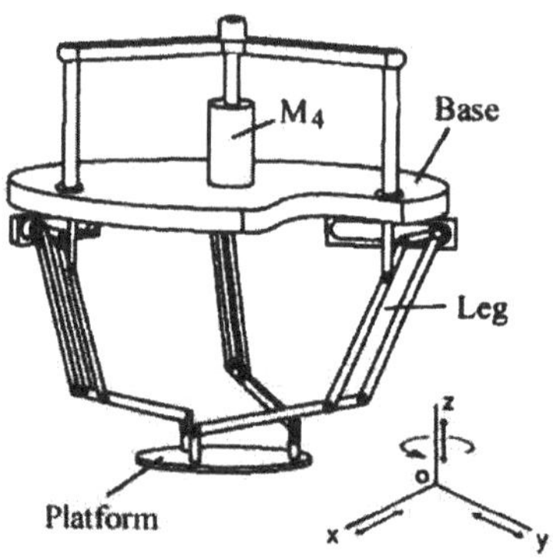

Figure 1. 3D view of the PAMINSA.

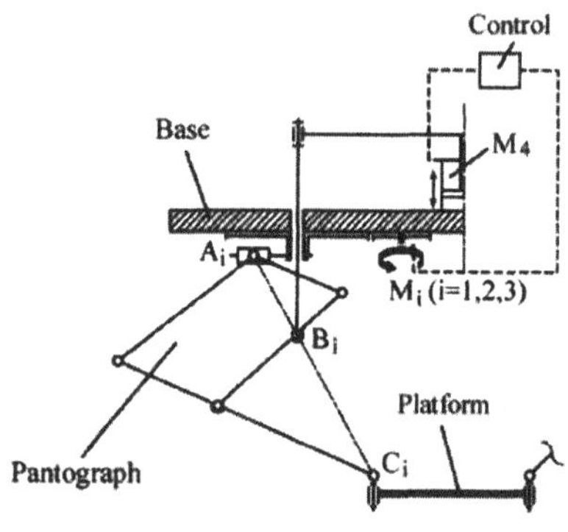

Figure 2. Schematics of each leg.

This type of manipulator architecture allows the generation of motion in the horizontal plane by the rotating actuators M_1, M_2, M_3 and the vertical displacements by the linear actuator M_4. Thus the displacements (x, y, θ) of the platform in the horizontal plane (xOy) are independent of the vertical displacements z.

In the concept of the PAMINSA, the following properties of the hand operated manipulators (see Arakelian, 1998) are used: the work of the gravitational forces of the manipulated object displaced in the horizontal plane is zero because the gravitational forces are always perpendicular

[1] Institut National des Sciences Appliquées (I.N.S.A.) – National Institute of Applied Sciences.

to the displacements. However, the work of the same forces in the case of the vertical displacements is different from zero (the gravitational forces are parallel to the displacements).

Thus, the rotating actuators move the platform in the horizontal plane and the work due to the gravity of the manipulated object is equal to zero. Static properties of PAMINSA were discussed in the study Arakelian et al. (2005), in which was shown that by using an optimal redistribution of masses of the pantograph linkages, the input torques of the rotating actuators of the PAMINSA can be cancelled. In the same study the analytical dynamic modeling of the suggested manipulator by means of Lagrange equations was achieved.

Figure 3a. The prototype of PAMINSA developed at the I.N.S.A. of Rennes.

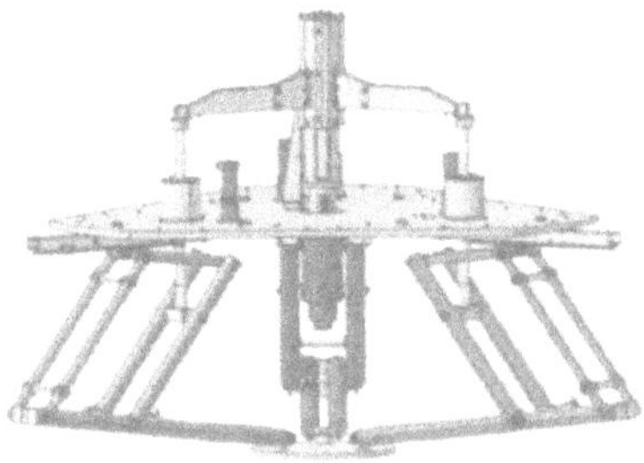

Figure 3b. The CAD model of the prototype.

Fig. 3 shows the prototype of the PAMINSA, which has been developed at the I.N.S.A. of Rennes. It should be noted that the proposed parallel structure has been patented (Arakelian et al., 2005).

1.2 The obvious advantages of PAMINSA

This manipulator has many advantages and it appears to us that the developed manipulator could be used in industrial applications for the manipulation of heavy equipments with great positioning accuracy. We would like to note briefly these advantages: a) a simplification of the vertical control based on the linear input/output equation; b) the improvement of the control in the horizontal plane because the loads of the gravitational forces of the platform on the rotating actuators have been cancelled; c) the improvement of positioning accuracy along the vertical axis because the mechanical locking of the structure does not allow the altitude variations during the displacements in the horizontal plane.

The letter property was discussed in the recent study Arakelian et al. (2005), in which the positioning errors along vertical axis and errors caused by the inclination of the moving platform taking into account the elasticity of links was analyzed. It was shown that the suggested manipulator allows the moving of the platform on the horizontal plane with great accuracy (about 20 μm without charge and 140 μm with a payload of 20 kg applied on the platform). However, PAMINSA has a very good load-carrying capacity and it can support much more.

This contribution deals with the methods and new approaches for the improvement of positioning accuracy of PAMINSA.

2 Improvement of positioning accuracy taking into account the elasticity of links

Many industrial applications of parallel manipulators, such as the assembly of electronic, optical units, or several medical applications require high accuracy. It should be noted that most of parallel manipulators used today are much better at repeatability than at accuracy. For improvement of position accuracy of parallel manipulators, it is possible to use the calibration methods, to increase the rigidity of links or lack of backlashes in drive systems. A new approach called Geometric and Elastic Error Compensation (GEC) was proposed in the study of Meggiolaro and Dubowsky (2001). It was shown that the two techniques can be effectively combined to achieve high absolute positioning accuracy.

In this paper, for improvement of positioning accuracy of PAMINSA, we propose new compensation schemes, which consist of the introduction of complementary units into the initial system making it possible to cancel the positioning errors. Two different approaches are proposed and the performances of such a design are shown.

2.1 Improvement of positioning accuracy of PAMINSA by means of correcting systems mounted on the drive system

Most research papers devoted to the study of parallel manipulators deal with the mechanical structures with rigid links and without clearances in the joints. So in this case the position of the platform is considered perfectly parallel to the base. But in reality, the errors due to the elastic deformations of the mechanical structure of the manipulator change the position of the platform (attitude and inclination).

Static rigidity is defined as the 6×6 symmetrical matrix **K** that maps generalized infinitesimal displacements $\boldsymbol{\Delta X} = [\delta x \quad \delta y \quad \delta z \quad \delta\phi_x \quad \delta\phi_y \quad \delta\phi_z]^T$ of the platform to generalized external loads $\boldsymbol{F} = [F_x \quad F_y \quad F_z \quad M_x \quad M_y \quad M_z]^T$.

Thus, we have

$$\boldsymbol{F} = \boldsymbol{K} \cdot \Delta \boldsymbol{X} . \tag{1}$$

In order to characterize PAMINSA static rigidity, we map the infinitesimal displacements for the whole workspace. The obtained results show that the errors of the elastic deformations of the structure are relatively small. With a payload of 20 kg applied on the platform, the errors, caused by the rotation of the platform about the x and y axes are shown in Fig. 4. The analysis of the obtained results shows that the position in which the structure is less deformed is the central position. When the platform moves away from this position, the manipulator becomes less rigid and loses its accuracy.

These errors are less important if the output point P (Fig.5), i.e. the end of a surgical device or a sensor, is located on the horizontal plane of the platform. But the error becomes more important if this output point is moved from the horizontal plan of the platform. For example, if the output point is located on the plane xOz and it is moved away 200mm from the horizontal plane of the platform (with 100mm radius), the error 0.14 mm of the platform along the z-axis increases at the end of the output point until 0.57 mm (see Fig.5).

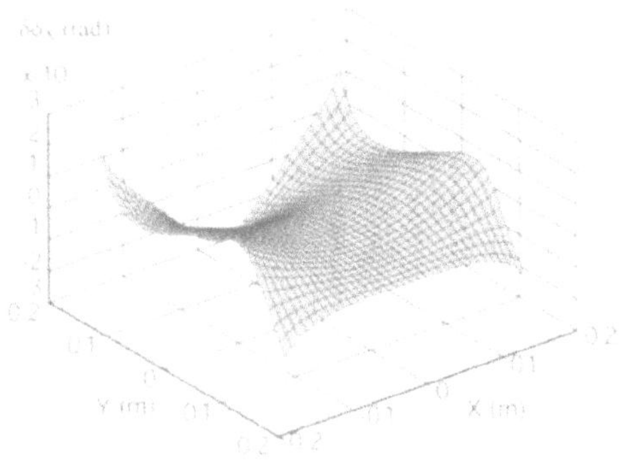

Figure 4a. Errors caused by the rotation of the platform about the x axis.

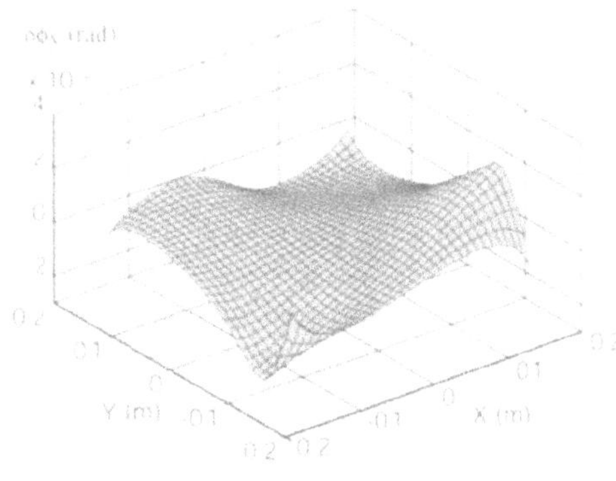

Figure 4b. Errors caused by the rotation of the platform about the y axis.

It is obvious that the improvement of positioning accuracy can be achieved by the increase of the rigidity of links. However it is also promising to develop the design methods for improvement of positioning accuracy by additional correcting systems.

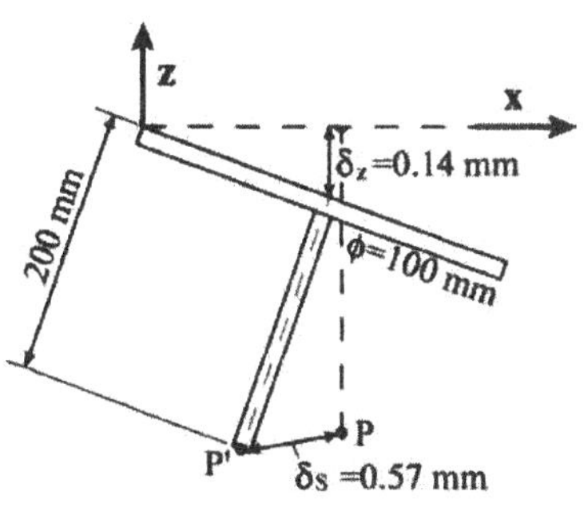

Figure 5. Absolute positioning errors of the output point, which is moved away 200mm from the horizontal plane of the platform.

Fig. 6 shows PAMINSA with two compensation systems, which are presented in Fig.7. It should be noted that in the modified design of the manipulator the joints on the platform are also changed: the universal joints used in the initial version are replaced by the spherical pairs. The compensation systems, which cancel the errors due to the elasticity of links are provided with two complementary actuators M_{ci}. These actuators allow the displacements of the pantograph's points B_i it making possible to eliminate the inclination error of the platform. These modifications allow the correction of the vertical positions of two spherical pairs of the platform, which is sufficient for the cancellation of the positioning error of the inclination of the platform.

The vertical positions of such a spherical pair located on the platform can be determined analytically for the whole workspace (or given altitude) of the manipulator on the base of equation (1) or by using three sensors mounted on the platform. The measuring of the spherical joint motion errors for the manipulators with three and six prismatic joints (for Tripod and Hexapod types), as well as several installation examples of the sensors were discussed in the study of Oiwa (2002).

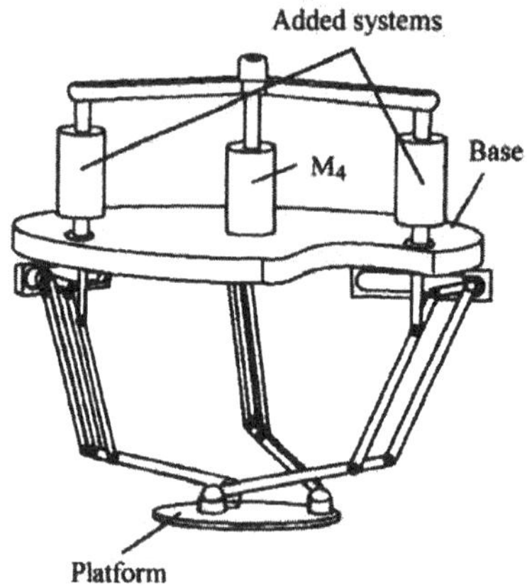

Figure 6. PAMINSA with added compensation systems.

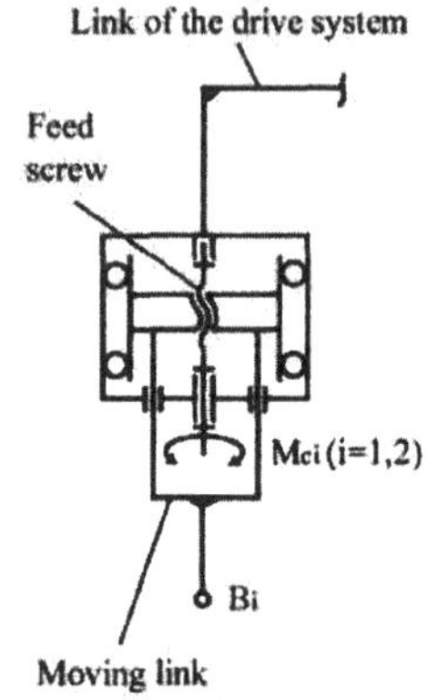

Figure 7. Kinematic schema of the added compensation systems for the correction of the positioning errors of the platform along the vertical axis.

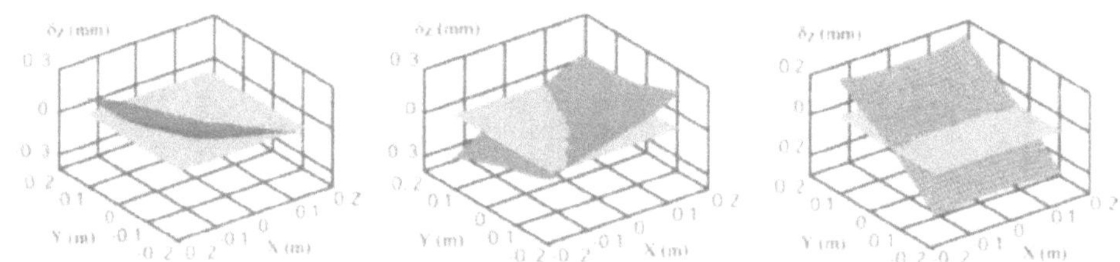

Figure 8. The vertical positioning errors of the platform's joints C_1, C_2 and C_3 of the initial and modified manipulators (the examined case correspond to the platform with orientation ϕ=0° at the altitude z=-0.6 m).

Note that, compensation scheme developed for PAMINSA is constructively more efficient because it is mounted on the drive system of the vertical displacements. As a result, the variable length of the compensation device is always vertical. It should be also noted that it is more simple for computation because the translational displacement in the added system can be found directly from positioning errors of the platform's joint, taking into account the magnification factor of the pantograph linkage. In this manner the significant reduction of errors can be achieved and the obtained results are shown in Fig.8. It is seen that after compensation of the elasticity errors of links the vertical positions of the platform's joints have constant values and the inclination error is eliminated.

2.2 Improvement of positioning accuracy of PAMINSA by means of correcting systems mounted on the platform

Let us consider another correcting system mounted on the platform of PAMINSA. Such a system can be added on any parallel structure for the correction of positioning error of the inclination of the platform. The suggested system consists of a correcting mass, which has the possibility to turn about the vertical axis of the platform and to carry out the translational displacements on the horizontal plane (Fig.9).

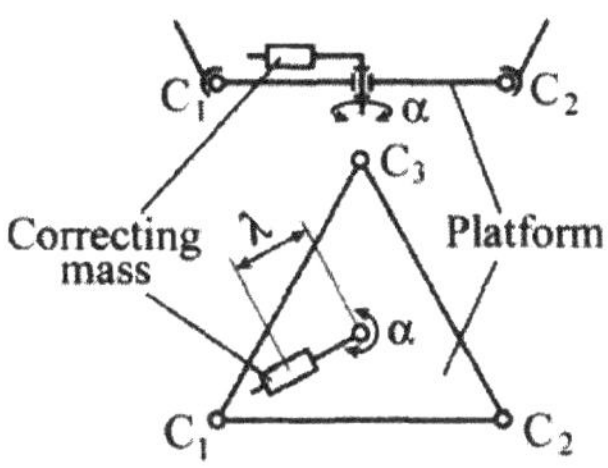

Figure 9. Kinematic schema of the correcting system mounted on the platform.

Introducing the correcting conditions into equation (1), we obtain

$$\Delta X = K_{st}^{-1} \cdot F + K_{st}^{-1} \cdot \begin{bmatrix} 0 & 0 & -mg & M_{st(x)} & M_{st(y)} & 0 \end{bmatrix}^T \quad (2)$$

From which taking into account that after correction

$$\Delta X = \begin{bmatrix} \delta x & \delta y & \delta z & 0 & 0 & \delta\phi_z \end{bmatrix}^T \quad (3)$$

we determine the static moments $M_{st(x)}, M_{st(y)}$ and then the position λ and orientation α of the correcting mass m.

In other words the correcting mass m should be located on the platform in such a manner that its gravity eliminated the inclination error of the platform.

For PAMINSA with parameters of the prototype (see Arakelian et al., 2005) the values of the position λ and orientation α for the correcting mass $m = 3kg$ are shown in Fig. 10.

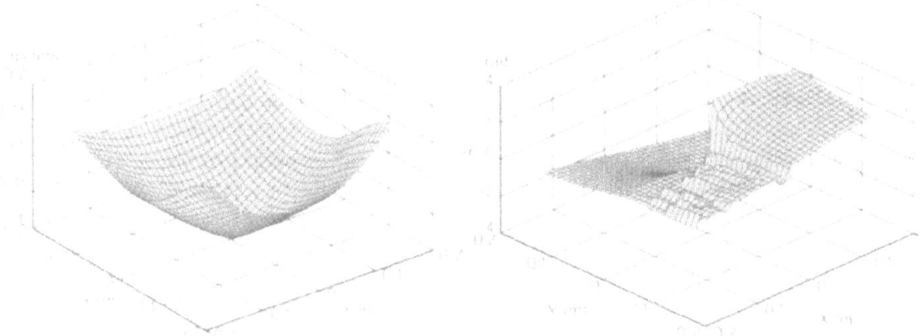

Figure 10. Position λ and orientation α of the correcting mass m.

These values are obtained for the platform having the constant orientation ϕ=0° at the altitude z=-0.6 m. The obtained results are the same as the previous case (Fig.8), i.e. after compensation of the elasticity errors of links, the vertical positions of the platform's joints have constant values and the platform becomes perfectly parallel to the base.

3 Conclusions

In this paper, new design approaches for the improvement of positioning accuracy of 4-DOF parallel manipulator PAMINSA have been discussed. In general, the studies devoted to the parallel manipulators deal with the mechanical structures on the base of rigid body mechanics and consider that the platform is perfectly parallel to the base. Our previous works have shown that the elasticity of links has an influence on the positioning accuracy of the developed parallel manipulator. For cancellation of these positioning errors caused by the elasticity of links, two approaches have been presented. The first solution is realized by means of two correcting systems mounted on the drive system of the vertical displacements. The second solution is carried out by a correcting mass mounted on the platform. The obtained results show that after compensation of the errors caused by the elasticity of links, the vertical positions of the platform's joints have constant values and the inclination of the platform, relating to the base, is cancelled.

Bibliography

Kong, X. and Gosselin, C. M. (2002). A class of 3-dof translational parallel manipulator with linear input/output equations. *Proceesings of the Workshop on Fundamental Issues and Future Research for Parallel Mechanisms and Manipulators*, Québec City, Québec, Canada, 25–32.

Gosselin, C. M., Kong, X., Foucault, S. and Bonev, I. A. (2004). A fully decoupled 3-dof translational parallel mechanism. *Parallel Kinematic Machines International Conference*, Chemnitz, Germany, 595–610.

Carricato, M. and Parenti-Castelli, V. (2004). A novel fully decoupled 2-dof parallel wrist. *The International Journal of Robotics Research*, 23(6), 661–667.

Carricato, M. and Parenti-Castelli, V. (2004). On the topological and geometrical synthesis and classification of translational parallel mechanisms", *Proceedings of the 11th World Congress in Mechanism and Machine Science*, Tianjin, China, 1624–1628.

Bouzgarrou, B. C., Fauroux, J. C., Gogu, G., and Heerah, Y. (2004). Rigidity analysis of T3R1 parallel robot with uncoupled kinematics. *Proceedings of the 35th International Symposium on Robotics*, Paris, France.

Arakelian, V. (1998). Equilibrage des manipulateurs manuels, *Mechanism and Machine Theory*, 33(4), 437-442.

Arakelian, V., Guegan, S., Briot, S. (2005). Static and dynamic analysis of the PAMINSA, *Proceedings of the ASME 2005 IDETC/CIE Conference*, September 24-28, Long Beach, California USA.

Arakelian, V., Maurine, P., Briot, S., Pion, E. (2005) Parallel robot comprising means for setting in motion a mobile element split in two separate subassemblies, WO2006021629, 2006-03-02.

Arakelian, V., Guegan, S., Briot, S., J. Le Flecher (2005). Design and prototyping of new 4, 5 and 6 degrees of freedom parallel manipulators based on the copying properties of the pantograph linkage. *Proceedings of the 36th International Symposium on Robotics,* November 29 – December 1, Tokyo, Japan.

Meggiolaro M. and Dubowsky S. Improving the positioning accuracy of powerful manipulators with application in nuclear maintenance (2001). Anais do 16 Congresso Brasileiro de Engenharia Mecânica (COBEM) , Robotics and Control, Vol. 15, pp. 210-219, ABCM, Uberlândia.

Oiwa T. Accuracy improvement of parallel kinematic machine (2002). *Proceedings of the 6th International Conference on Mechatronics Technology,* September 29 - October 3, Kitakyushu, JAPAN.

The Impact of Friction on the Dynamics of Parallel Robotic Manipulators

Houssem Abdellatif * and Bodo Heimann * and Martin Grotjahn ‡

* Institute of Robotics, University of Hannover, Germany, e-mail: abdellatif,heimann@ifr.uni-hannover.de

‡ IAV-GmbH, Gifhorn, Germany, e-mail: martin.grotjahn@iav.de

Abstract. In this paper an overview on joint-friction for parallel kinematic manipulators (PKM) is proposed. An efficient modeling approach is given, where dissipation in all passive and active joints are uniformly formulated and transformed in the actuation space in parameter-linear form. Furthermore, it is shown, how friction characteristics can be identified independently from any other dynamic influences. Two sections are considered to the practical issues of friction in terms of feedforward and time-optimal control.

1 Introduction

Friction exists in all mechanical and multi-body systems and is the principal cause of energy dissipation. Its consideration is crucial for the improvement of accuracy and control quality. Accurate mathematical models and the identification of their parameters are required Armstrong-Hélouvry et al. (1994). Considering robotics systems, research focussed mostly on serial manipulators or industrial robots. Adaptive friction compensation was proposed more than a decade ago, like in de Wit and Lischinsky (1997). The compensation of dynamic friction is still a current issue and is supposed to be more accurate than classic compensation of static friction Moreno et al. (2002); Zhu and Pagilla (2002). The practicability of such approaches is however not definitively approved, especially for redundant systems and manipulators with high number of degrees of freedom (DOF). Knowledge about Friction can be useful for the nonlinear training of feedforward motion controllers Grotjahn and Heimann (2002). In contrast to serial manipulators, the systematic analysis and investigation of friction and its impact on the dynamics of parallel kinematic manipulators (PKM) are still rare. In fact, major differences between the two mechanism types complicate the task of uniform friction modeling. Parallel manipulator are characterized with a high number of passive joints Angeles (2003); Merlet (2000) and therefore a higher number of energy dissipation spots. Many authors neglect friction in passive joints to avoid the transformation of all friction forces into the actuation space. This simplification is not admissible for every parallel mechanism or system. This paper tries to give an overview on friction modeling and its influences in control issues for parallel manipulators. The discussed topics are substantiated with practical notes and recent experimental results. The latter were carried out on the parallel robot

Figure 1. PKM PaLiDA - left: Test bed at the Hannover Industrial Fair 2001, right: CAD-model.

PaLiDA (Fig. 1). Section 2 gives a general methodology on integrating friction losses in passive and in active joints within a unified actuation model. The approach is presented in a parameter-linear form to allow for the use of computationally efficient linear techniques to estimate the friction coefficients Grotjahn et al. (2004). This topic is detailed in section 3, where a novel method is presented, that allows estimating the friction model parameters independently from rigid-body dynamics. The impact of model-based friction compensation on the improvement of control accuracy is discussed in section 4. Finally time-optimal trajectory planning and control is raised in section 5.

2 Friction Modeling

The modeling approach proposed in this section is discussed for place reasons for friction characteristics. Due to their simplicity and efficiency nonlinear characteristics are well reputed in robotics and give the joint losses in relationship to the joint velocities. Most common approach is the combination of coulomb and viscous damping according to:

$$Q_{\mathrm{f}_j} = \alpha_{1j}\mathrm{sign}(\dot{q}_j) + \alpha_{2j}\dot{q}_j, \tag{2.1}$$

where Q_{f_j} is the friction force or torque of the joint j, $\dot{q}_j$ is the corresponding velocity and α_{kj} are the friction coefficients. Generally, friction characteristics of robotic joints can be described as a linear combination of different elementar nonlinear functions ϕ_k:

$$Q_{\mathrm{f}_j} = \sum_{k=1}^{m} \alpha_{kj}\phi_k(\dot{q}_j). \tag{2.2}$$

The degree of complexity of the characteristics depends on many factors, such the joint type, the used actuation principle, etc.. and has to be experimentally investigated and motivated. Equation (2.1) can be presented in a linear form in respect to the friction coefficients

$$Q_{\mathrm{f}_j} = [\phi_1(\dot{q}_j)\dots\phi_m(\dot{q}_j)]\,[\alpha_{1j}\dots\alpha_{mj}]^{\mathrm{T}}. \tag{2.3}$$

Regrouping friction losses in all n joints yields

$$\boldsymbol{Q}_{\mathrm{f}} = \underbrace{[\boldsymbol{D}_1(\dot{\boldsymbol{q}}), \ldots, \boldsymbol{D}_m(\dot{\boldsymbol{q}})]}_{\boldsymbol{D}_{\mathrm{f}}} \underbrace{\left[\boldsymbol{\alpha}_1^{\mathrm{T}}, \ldots, \boldsymbol{\alpha}_m^{\mathrm{T}}\right]^{\mathrm{T}}}_{\boldsymbol{p}_{\mathrm{f}}}, \tag{2.4}$$

with

$$\boldsymbol{\alpha}_k^{\mathrm{T}} = [\alpha_{kj}, \ldots, \alpha_{kn}], \tag{2.5}$$

and

$$\boldsymbol{D}_k(\dot{\boldsymbol{q}}) = \operatorname{diag}\{\phi_k(\dot{q}_1), \phi_k(\dot{q}_2), \ldots, \phi_k(\dot{q}_n)\}. \tag{2.6}$$

The transformation of all joint losses can be achieved by considering the JOURDAIN's principle of virtual power given in the following power balance equation:

$$\delta\dot{\boldsymbol{q}}^{\mathrm{T}}\boldsymbol{Q}_{\mathrm{a},f} = \delta\dot{\boldsymbol{q}}_{\mathrm{a}}^{\mathrm{T}}\boldsymbol{Q}_{\mathrm{a}} \Rightarrow \boldsymbol{Q}_{\mathrm{a},f} = \left(\frac{\partial\dot{\boldsymbol{q}}}{\partial\dot{\boldsymbol{q}}_{\mathrm{a}}}\right)^{\mathrm{T}}\boldsymbol{Q}_{\mathrm{f}}, \tag{2.7}$$

which means that the total power of all joint losses causes an equal dissipated actuation power. This is nothing but the transformation of all friction forces into the actuation space. It is interesting to notice, that the Jacobian $\boldsymbol{J}_{\mathrm{f}} = \partial\dot{\boldsymbol{q}}/\partial\dot{\boldsymbol{q}}_{\mathrm{a}}$ determines the complexity of the identification task (see also section 3). For non redundant serial manipulators $\boldsymbol{J}_{\mathrm{f}}$ is equal to the identity matrix $\boldsymbol{I}$, which allows single-joint consideration with decoupled models for each joint. However, such a consideration is improper for complex parallel or hybrid manipulators, since $\boldsymbol{J}_{\mathrm{f}} \neq \boldsymbol{I}$. For a more efficient formulation of friction dynamics, it is recommended to use the Jacobian of the parallel Manipulator $\boldsymbol{J} = \partial\dot{\boldsymbol{\theta}}/\partial\dot{\boldsymbol{q}}_{\mathrm{a}}$ and the all-joint jacobian $\boldsymbol{J}_{\mathrm{B}} = \partial\dot{\boldsymbol{q}}/\partial\dot{\boldsymbol{\theta}}$ with respect to the generalized velocities $\dot{\boldsymbol{\theta}}$. Equation (2.7) yields

$$\boldsymbol{Q}_{\mathrm{a},f} = \boldsymbol{J}^{\mathrm{T}}\left(\frac{\partial\dot{\boldsymbol{q}}}{\partial\dot{\boldsymbol{\theta}}}\right)^{\mathrm{T}}\boldsymbol{Q}_{\mathrm{f}}. \tag{2.8}$$

Both Jacobians are simple to obtain after performing the kinematic analysis Grotjahn et al. (2004). Combining equations (2.4) and (2.8) yields the parameter-linear form of the friction model

$$\boldsymbol{Q}_{\mathrm{a},f} = \left[\boldsymbol{J}^{\mathrm{T}}\left(\frac{\partial\dot{\boldsymbol{q}}}{\partial\dot{\boldsymbol{\theta}}}\right)^{\mathrm{T}}\boldsymbol{D}_{\mathrm{f}}\right]\boldsymbol{p}_{\mathrm{f}} = \boldsymbol{A}_{\mathrm{f}}\boldsymbol{p}_{\mathrm{f}}. \tag{2.9}$$

3 Independent Friction Identification

For an independent and robust friction identification, two evaluation steps are required. In the first step local models are identified at single configurations using simple PTP-motions. The second step eliminates configuration dependency and provides the estimate of global friction coefficients. This is necessary, since it is impossible to directly separate friction influences from rigid-body dynamics in the measurements of actuator torques or forces. A detailed description of the experiments can be found in Grotjahn et al. (2004).

At first, different single pose configurations $\boldsymbol{X}_l$ can be arbitrarily chosen within the singularity-free workspace. Simple PTP-motions with trapezoidal velocity profiles are carried out. Each one has to traverse a correspondent configuration. The relevant measurements are collected in the neighborhood of the configuration. To reduce the inertial effects, the velocity of the end effector has to be kept constant, such that acceleration terms are eliminated. Every measured actuation force $\check{Q}_{\mathrm{a}_i}$ contains friction forces, coriolis and gravitational terms:

$$\check{Q}^l_{\mathrm{a}_i}(\dot{q}_{\mathrm{a}_i}) = \left[1\ \dot{q}^2_{\mathrm{a}_i}\ \phi_1(\dot{q}_{\mathrm{a}_i})\ \ldots\ \phi_m(\dot{q}_{\mathrm{a}_i})\right]\left[g_i(\boldsymbol{X}_l)\ c_i(\boldsymbol{X}_l)\ r^l_{1i}(\boldsymbol{X}_l)\ \ldots\ r^l_{mi}(\boldsymbol{X}_l)\right]^{\mathrm{T}}, \quad (3.1)$$

where q_{a_i} is the velocity of the actuator i, c_i is the coriolis term and g_i is the gravitation force affecting the same actuator. The coefficients r^l_{ki} are the local and configuration dependent friction coefficients. This can be explained by the fact that the friction part of $\check{Q}_{\mathrm{a}_i}$ results not only from the actuator dissipation but is also caused by all friction forces in all joints. Like shown by equation (2.7) the global coefficients α^l_{ki} are projected into the actuation space by $\boldsymbol{J}_{\mathrm{f}}$ that is configuration-dependent. It is also important to notice that equation (3.1) can be considered as a local friction characteristic. By carrying different measurements with varying constant velocities at a single configuration $\boldsymbol{X}_l$, local characteristics can be collected for each actuator. Each velocity generates one measurement point of the characteristics (see examples in Fig. 2). The model can be fitted to the measured characteristic by an LS-estimator. The choice of the configurations must excite the parameters of all joints. Therefore, the configurations can be symmetrically distributed within the workspace, e.g. on circles. Figure 2 shows experimental identified local characteristics for the same actuator at different configurations that are distributed on the depicted circle. In configuration nr. 3 and nr. 7 typical forms of viscous damping and dry friction are detected. Their quantitative influences are though unequal. Moving toward configuration nr. 10, the impact of dry friction decreases, whereas the inertial influences become more important. At configuration nr. 1, negative dry friction was detected. This means physically, that dry friction in passive joints is here dominant and that these joints move in opposite direction to the considered actuator ($\dot{q}_{\mathrm{a}_i}\dot{q}_j < 0$). At the next configuration dry frictions in the different joints neutralize each other, so only viscous damping as joint loss is detected. The cyclic form evolution of the identified characteristics can be observed for all actuators, in fact for different configurations but in exactly the same order.

The determination of the physical friction coefficients α_{ij} from the local resulting models is the second step of the identification approach. It is achieved by a second application of an LS-Estimator by combining all local parameters. The over-all model consisting of the actuator friction models as well as the friction models of the passive joints is fitted to the resulting global friction characteristics. In Fig. 3 the resulting fit is depicted for four actuators of the hexapod PaLiDA. The configuration dependency is clearly recognizable. Taking the bad repeatability of friction measurement into account, the measurements and the identified models show a very good agreement. The figure shows the results of two different models: a complex one with different dry friction parameters for each passive joint and a simpler one with one identical parameter for all passive-joints. Figure 3 reveals that the complex model yields only minor improvements. A methodology for a systematic choice of model-complexity by using statistical properties of the measurement and uncertainty of the estimation was published in Abdellatif et al. (2005).

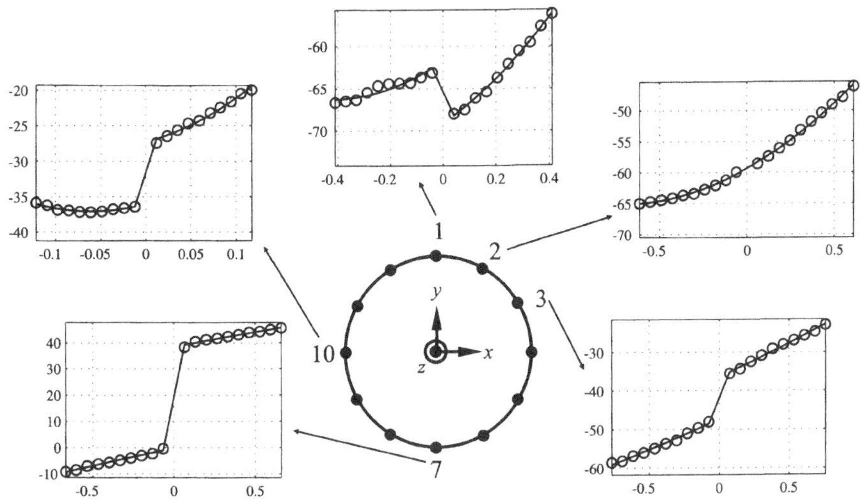

Figure 2. Configuration dependent force-velocity characteristics for single actuator. (○): measured forces, (—): fitted identified local model. x-axis of all figures refer to the actuator velocity $\dot{q}_{a_4}$ [m/s], y-axis correspond to the actuator force Q_{a_4} [N].

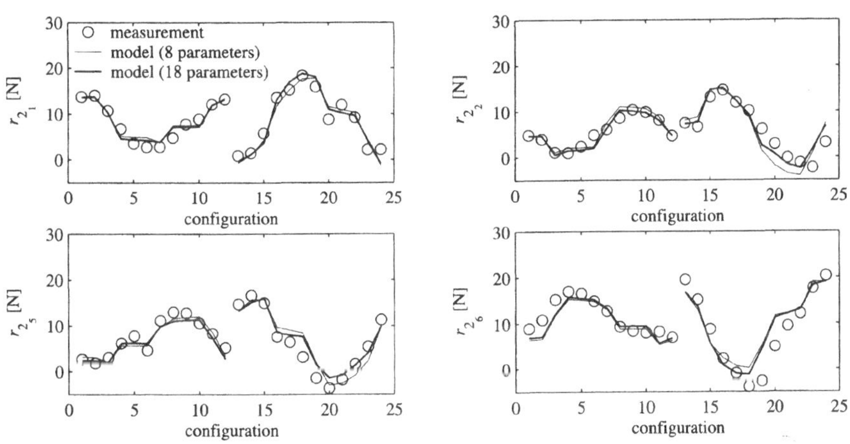

Figure 3. Comparison of calculated and measured dry friction parameters of the robot PaLiDA.

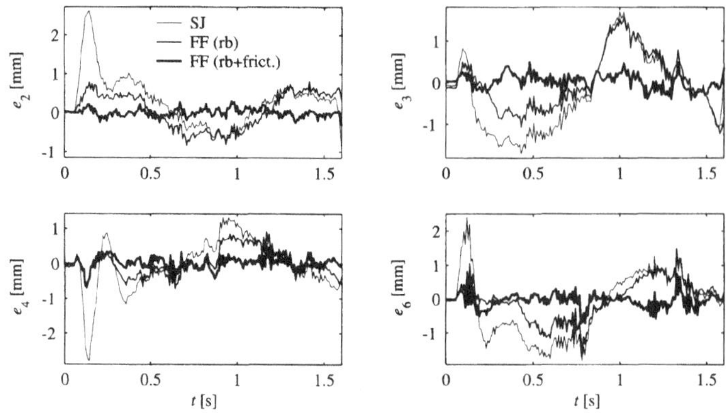

Figure 4. Control errors of actuators while a circular motion. Comparison between single-joint control (SJ), Feedforward control (FF) using only rigid-body dynamics (rb) and with additional friction compensation (rb+frict.).

4 Friction in Control

One of the major aims of modelling friction is its accurate compensation within control to approve accuracy of robotic systems Armstrong-Hélouvry et al. (1994); Grotjahn and Heimann (2002). This can be achieved by adaptive or by feedforward control. The importance of friction compensation is illustrated by the following experiment achieved on the parallel robot PaLiDA. Figure 4 demonstrates the principal results. It shows the tracking errors of four arbitrarily chosen actuators during a circular trajectory with a diameter of $0.4m$ and end-effector path velocity of $1m/s$. The circle is inclined by -45° and is located in the middle of the workspace at a height of $z_0 = -0.7m$. The largest tracking errors are already reduced by the rigid-body model without consideration of friction. These deviations occur at the beginning and the end during acceleration and deceleration phase. However, the deviations during constant velocity phase still remain important. These errors are reduced after additional feedforward of the friction forces. Thus, the implementation of the identified friction model yields excellent reduction of tracking errors, which also leads to significant improvements of path accuracy. It has to be mentioned, that in the illustrated example, coulomb-friction and viscous damping were considered. These effects are the dominant dissipation terms in the considered hexapod, especially in the high dynamics range. For the range of low velocities, pre-sliding and stiction effects may also be compensated like proposed by Moreno et al. (2002) and de Wit and Lischinsky (1997).

5 Friction in Time-Optimal Planning

For time-optimal trajectory planning a geometric path $\mathcal{P}(s)$ with respect to the path parameter s is supposed to be given. In the same sense, the dynamics equations are

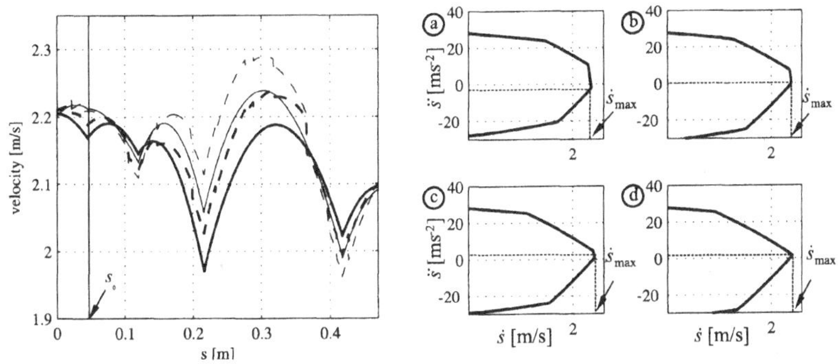

Figure 5. Left: maximum velocity curve with different models: a-(thick line) only rigid-body dynamics, b-(thick dashed line): rigid-body and coulomb friction, c-(thin line): rigid-body and viscous damping, d-(thin dashed line): full dynamics, Right: correspondent admissible motion diagrams at $s = s_0$

transformed to

$$\boldsymbol{Q}_{\mathrm{a}}(s) = \boldsymbol{m}(s)\ddot{s} + \boldsymbol{c}(s)\dot{s}^2 + \boldsymbol{f}(s)\dot{s} + \bar{\boldsymbol{g}}(s), \tag{5.1}$$

where $\boldsymbol{Q}_{\mathrm{a}}$, $\boldsymbol{m}$, $\boldsymbol{c}$ and $\boldsymbol{f}$ are actuator forces, inertia, coriolis and viscous damping terms, respectively. Gravitation and dry coulomb friction are grouped in $\bar{\boldsymbol{g}}$ Abdellatif and Heimann (2005); Pfeifer and Johanni (1987). The optimal trajectory can be obtained by the basic algorithm given early by Shin and McKay (1985), where for each s a region of admissible motion is computed in order to chose the maximal possible velocity and acceleration. The practical implementation differs though, if friction terms are neglected. In Pfeifer and Johanni (1987) viscous damping is neglected to reduce the task of calculating maximal velocities to a linear problem in the $(\ddot{s}, \dot{s}^2)$-plane. In literature, mostly coulomb-friction is neglected to avoid numerical difficulties due to discontinuities. Such simplifications are questionable, especially for parallel mechanisms with high number of joints. This can be proved exemplarily like shown in Fig. 5. For the PKM PaLiDA four different maximum velocity curves were calculated for a half-circle motion in the middle of the workspace and with respect to four different dynamics models. For each model, the admissible region of motion at $s = s_0 = 0,05m$ is illustrated with the respective maximal possible path velocity $\dot{s}_{\mathrm{max}}$ and with the associated path acceleration $\ddot{s}_{\mathrm{max}}$. It is obvious, that the four curves are different, which means that the estimated dynamics capabilities are dependent from the considered model, which is reasonable. Moreover and it is more important that for the same path point s_0, different accelerations were determined. By taking account of the viscous damping (neglected by Pfeifer and Johanni (1987)) positive accelerations are possible. If only dry friction is considered, a zero-acceleration is obtained, whereas by neglecting all friction effects, a deceleration has to be planned at $s = s_0$. These contradictory results demonstrate, that neglecting any dynamics would yield important deviations in the planned trajectories. It is therefore recommended to opt for the full dynamics model Abdellatif and Heimann (2005), to avoid control and path errors caused by wrong estimation of the manipulator capabilities.

6 Conclusions

Due to the presence of numerous passive joints, friction impact in parallel robots is quite different from that in serial manipulators. This paper has shown some related important aspects. The discussion has been substantiated with recent experimental results. It has been shown that the compensation of friction improves the control accuracy. Furthermore, it enhances the model precision for dynamic and time-optimal trajectory planning.

Bibliography

Houssem Abdellatif and Bodo Heimann. Adapted time-optimal trajectory planning for parallel manipulators with full dynamics modelling. In *Proc. of the 2005 IEEE International Conference on Robotics and Automation*, pages 413–418, Barcelona, Spain, 2005.

Houssem Abdellatif, Bodo Heimann, Oliver Hornung, and Martin Grotjahn. Identification and appropriate parametrization of parallel robot dynamic models by using estimation statistical properties. In *Proc. of the 2005 IEEE/RSJ International Conference on Intelligent Robots and Systems, IROS2005*, pages 444–449, Edmonton, Canada, 2005.

Jorge Angeles. *Fundamentals of Robotic Mechanical System.* Mechanical Engineering Series. Springer-Verlag, New York, Berlin, Heidelberg, 2003.

B. Armstrong-Hélouvry, P. Dupont, and C. Canudas de Wit. A survey of models, analysis tools and compensation methods for the control of machines with friction. *Automatica*, 30(7):1083–1138, 1994.

C. Canudas de Wit and P. Lischinsky. Adaptive friction compensation with partially known dynamic friction model. *International Journal of Adaptive Control and Signal Processing*, 11:65–80, 1997.

M. Grotjahn and B. Heimann. Model-based feedforward control in industrial robotics. *International Journal of Robotics research*, 21(1):99–114, January 2002.

M. Grotjahn, B. Heimann, and H. Abdellatif. Identification of friction and rigid-body dynamics of parallel kinematic structures for model-based control. *Multibody System Dynamics*, 11(3):273–294, 2004.

J.-P. Merlet. *Parallel Robots.* Solid Mechanics and its Applications. Kluwer Academic Publishers, Dordrecht, 2000.

Javier Moreno, Rafael Kelly, and Ricardo Campa. On manipulator velocity control using friction compensation. In *Proc. of the 2002 IEEE International Conference on Robotics and Automation*, pages 3417–3422, Washington DC, USA, 2002.

F. Pfeifer and R. Johanni. A concept for manipulator trajectory planning. *IEEE Journal of Robotics and Automation*, RA-3(2):115–123, 1987.

K. G. Shin and N. McKay. Minimum-time control of robotic manipulators with geometric path constraints. *IEEE Transactions on Automatic Control*, AC-30(6):531–541, 1985.

Yongliang Zhu and Prabhakar R. Pagilla. Static and dynamic friction compensation in trajectory tracking control of robots. In *Proc. of the 2002 IEEE International Conference on Robotics and Automation*, pages 2644–2649, Washington DC, USA, 2002.

Dynamics Aspects of Parallel Anthropomorphic Robots

Cornel Brisan [1] and Manfred Hiller[2]

[1] Universiy of Cluj-Napoca, Faculty of Mechanics, Romania, e-mail: cornelbrisan@yahoo.com
[2] Department of Mechatronics, University of Duisburg – Essen, Duisburg, Germany, hiller@mehatronic.uni-duisburg.de

Abstract. This paper deals with the kinematic and dynamics of complex parallel mechanism, such as hybrid parallel mechanisms. Thus, first PARTNER Robotic System is introduced and than main equations for modular development of kinematics and dynamics is presented. Virtual models and also numerical results were used in order to verify theoretical model.

1 Introduction

Anthropomorphic robots are defined as robots which model superior human limb. This definition was adopted in its last form in 2000 by IFToMM.

First robots which were developed according to this definition have serial topology, rotational actuators and from biomechanic point of view are models of the bone – joint system. They are used for industrial tasks (manipulation, assembly etc) or for other non industrial tasks such as: service robots in medical area, accurate manipulation in aero spatial environment etc. These robots are characterized by all advantages and disadvantages of classic serial robots. More than, they are not properly "anthropomorphic" and this is because rotational actuators are not specific for the biological systems.

The second type of anthropomorphic robots takes into consideration not only first two components of the human locomotive system (bones and joints) but also human muscles as "active joints" of this system. It results very complex mechanisms with parallel topology. Of course, these robots have also advantages and disadvantages, many of them are common with those of the parallel robots.

Besides advantages and / or disadvantages of parallel anthropomorphic robots, few particularities of these robots can be emphases:

a) every active joints are prismatic joints; thus, these are surrogate of the contractile component of the muscles;

b) their mechanical architecture is in fact constitute by hybrid mechanisms, because on the one side there are loops which connect the platforms and have not active joints in their structure and on he other side there are usually more then one mobile platform;

c) human locomotive system is working under a very sensitive force equilibrium. This makes from parallel anthropomorphic robots a competitive solution, compared with anthropomorphic robots with serial topology, in modeling phenomena which are characteristic for the movements of the main human segments (superior and inferior limbs, spine etc);

This paper presents results concerning mathematical modeling of topology, kinematics and dynamics of parallel anthropomorphic robots. Virtual model and numerical results are also presented.

2 Topology

Anthropomorphic robots may have an open loop topology (named **AROLT** – **A**nthropomorphic **R**obots with **O**pen **L**oop **T**opology) if only the bone – joint system is modeled or parallel topology if the muscles are also considered (named **ARoPaT** - **A**nthropomorphic **Ro**bots with **Pa**rallel **T**opology). Depending on the type of dof distribution, two variants of **AROLT** (Figure1) can be developed.

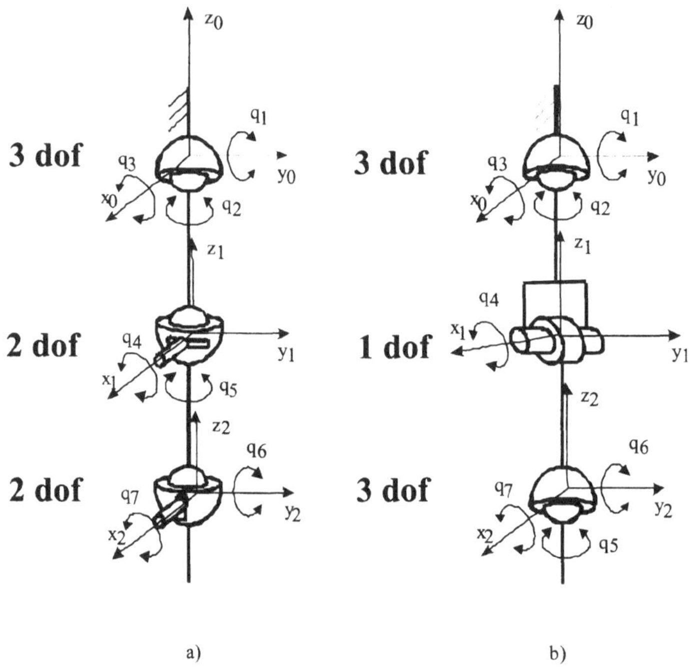

Figure 1. Anthropomorphic robots with open loop topology

Topologic synthesis of ARoPaT can be done according to different criteria, such as:

a) each loop which connect the platforms of the mechanism has a specific topology (for example UPS : universal-prismatic-spherical);

b) each loop has only one actuator (for example in the case of UPS this is a linear actuator);

c) distribution of the number of degrees of freedom (dof) is similar to those of the Superior Human Limb (HSL);

d) topology of the whole mechanism will be developed into a modular and reconfigurable manner.

According to these criteria, special class of parallel anthropomorphic robots with parallel and reconfigurable topology can be developed.

This class of ARoPaT it will be named PARTNER. For the PARTNER components, the following notation is proposed:

PARTNER P(R) abc xyz efgh

where: PARTNER is the name, P(R) – type of actuators (P for prismatic and R for rotational), abc – types of the loops which connect the platforms of the PARTNER (for example UPS-universal, prismatic, spherical joints), xyz – dof distribution for the joints, which directly connect the platforms of the PARTNER, efgh – number of the loops which are connected to the platforms of the PARTNER (e for the first platform, f for the second platform etc).

The number of dof for **PARTNER** is given by:

$$M = 6\,(n + N) - \sum_{K=1}^{5} K(C_K + D_K) - M_P \qquad (1)$$

where M is the number of dof, n is the number of mobile elements which connect the platforms of **PARTNER**, N is the number of mobile platforms of **PARTNER**, C_K is the number of the joints with (6-K) constraints, D_K is the number of the joints with (6-K) constraints and which connect directly the platforms of **PARTNER**, M_P is the number of the passive dof.

The equation (1) becomes:

$$M - 6\,N + 5D_5 + 4D_4 + 3D_3 = 0 \qquad (2)$$

Integer solutions of (2) give all variants of **PARTNER** characterised by assumed hypothesis. It will be useful to find out those solutions of equation (2), which have an appropriate dof distribution with SHL.

3 Kinematics

Kinematics of **PARTNER** robotic system can be characterised by two important properties:

a) forward kinematics has only numeric solutions;

b) inverse kinematics has analytic solutions **only** for two kind of situations: PARTNER variant has only one mobile platform **or** it has more then one mobile platform but there is also a loop which connects the platforms of the mechanism (named **cop** - **c**orresponding **o**pen loop) and which has the same number of dof with the whole mechanism.

General algorithms used to solve direct kinematics in the case of parallel mechanisms consider that for each independent loop of the mechanism one vector equation can be write (Hiller, 1995). Thus, a nonlinear system of scalar equations is obtained. Usually, this system of equations can be solved only with numerical methods and for that an accurate initial value of the solution is required. Of course, this initial value of the solution is strongly related to many parameters such: geometric parameters of the mechanism, boundary conditions for the mobile platforms etc. When these parameters changed also the initial solution must be changed.

According to that and in order to ensure an analytical value for the initial solution, the kinematics of the parallel mechanism has been developed in a modular manner, based on kinematics of the legs which connect the platforms. Thus, each leg is in fact the right (or left) side of one independent closed loop and can be described by two coordinate systems: one attached to the frame and the other one attached to the mobile platform.

For example **PARTNER P UPS 32 352** (Figure 2) has 5 independent closed loops and of course five independent closed equations.

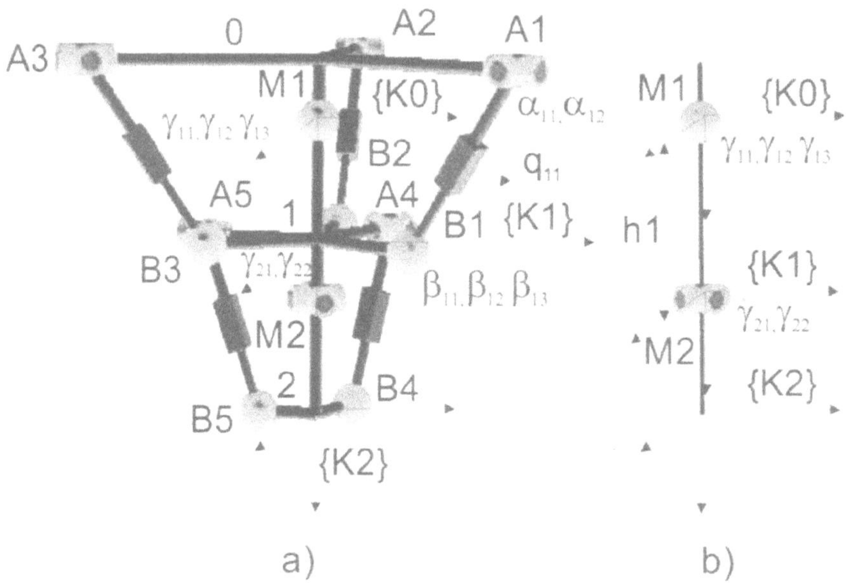

Figure 2. Kinematics of **Partner P UPS 32 352**

For the first mobile platform, it can be written:

$$\begin{aligned} H_{iml} &= A_{i1}(\gamma_{i1})\, A_{i2}(\gamma_{i2})\, A_{i3}(\gamma_{i3}) \\ H_{imr} &= A_{i1}(\alpha_{i1})\, A_{i2}(\alpha_{i2})\, T(q_{i1})\, B_{i1}(\beta_{i1})\, B_{i2}(\beta_{i2})\, B_{i3}(\beta_{i3}) \end{aligned} \quad (3)$$

The equations for the whole mechanism are similar and in this case the system of equations has 30 independent scalar equations.

In the case of forward kinematics, the generalised coordinates q_{i1} are known and α_{i1}, α_{i2} (the rotation angles of the joints from Ai), β_{i1}, β_{i2}, β_{i3} (the rotation angles of the joints from Bi), γ_{11}, γ_{12}, γ_{13} (the rotation angles of the joint from M1), γ_{21}, γ_{22} (the rotation angles of the joint from M2) are unknowns.

In the case of inverse kinematics, five new conditions are added to the previous system of equations (those corresponding to five parameters which are imposed to the movement of the second mobile platform). The generalised coordinates q_{i1} can be found solving this new system of equations.

Inverse kinematics of **PARTNER** system can be solved also into analytical manner if the corresponding open loop is considered (Figure 2b). Thus, the absolute transformation matrix $\mathbf{H}_{20}$, corresponding to the second platform, is known and is given by:

$$\mathbf{H}_{20} = \mathbf{A1}(\gamma_{11})\,\mathbf{A2}(\gamma_{12})\,\mathbf{A3}(\gamma_{13})\,\mathbf{A4}(\gamma_{21})\,\mathbf{A5}(\gamma_{21}) \tag{4}$$

The matrices A1,..,A5 are the transformation matrices corresponding to the relative movements.

Classic algorithms used to solve inverse kinematics for open loops, can be used to find out the angles γ_{11}, γ_{12}, γ_{13}, γ_{21}, γ_{22}. Using these angles, the coordinates of the points B1, ..., B5, A4, A5 can be determinate. For example, the absolute coordinates of the point B1 are given by:

$$\mathbf{P}^{K0}{}_{B1} = \mathbf{A1}(\gamma_{11})\,\mathbf{A2}(\gamma_{12})\,\mathbf{A3}(\gamma_{13})\,\mathbf{P}^{K1}{}_{B1} \tag{5}$$

where $\mathbf{P}^{K1}{}_{B1}$ is the matrix of coordinates of the point B1 related to the frame {K1}, attached to the platform "1".

Finally, the generalised coordinates are given by:

$$q_{i1} = \sqrt{(x_{Bi} - x_{Ai})^2 + (y_{Bi} - y_{Ai})^2 + (z_{Bi} - z_{Ai})^2} \tag{6}$$

The inverse Jacobian matrix, corresponding to the first mobile platform is given by:

$$\mathbf{J}_{1i}{}^{-1} = (\mathbf{M1Bi} \times \mathbf{s}_i) \tag{7}$$

and thus:

$$\dot{\mathbf{q}}_{1i} = \mathbf{J}_1{}^{-1}\,\boldsymbol{\Omega}_1 \tag{8}$$

where $\mathbf{s}_i$ is the unit vector of direction **AiBi** and $\boldsymbol{\Omega}_1$ is the absolute velocity of the first mobile platform.

Similarly, the velocities of the second mobile platform, related to the first mobile platform, are given by:

$$\dot{\mathbf{q}}_{2i} = \mathbf{J}_2{}^{-1}\,\boldsymbol{\Omega}_2 \tag{9}$$

Using (8) and (9) the absolute velocities of the second mobile platform may be determinate:

$$v_{20} = \sum_{i=1}^{2} v_i + \sum_{i=1}^{2} \Omega_i \times r_i \tag{10}$$

where "i" denotes relative movements of the frame $\{\mathbf{K}_i\}$ and $\{\mathbf{K}_{i-1}\}$.

4 Dynamics

Dynamics analysis of Partner robotic system was developed based on d'Alambert principle. Without lose the generality of the presentation, the **PARTNER P UPS 32 352** (Figure 2) will be take as an example.

Thus, let consider platforms "2" (Figure 3a) and "1" (Figure 3b) of the **PARTNER P UPS 32 352** and also the forces, which acting on these platforms.

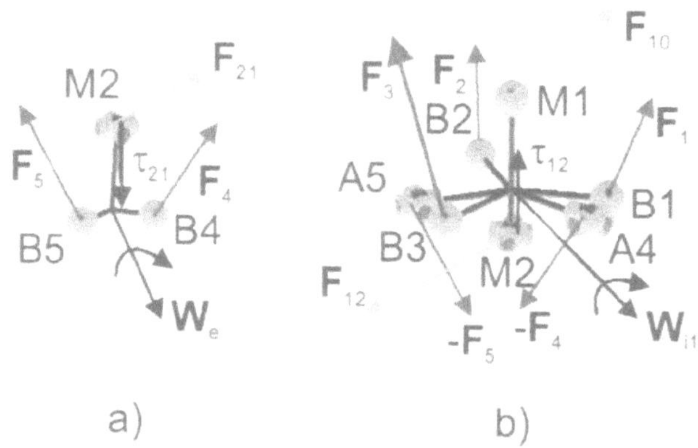

Figure 3. Forces acting on platforms „2" and "1"

We note with: $\mathbf{F}_1, \ldots, \mathbf{F}_5$ – actuator's forces; $\mathbf{F}_{21}, \mathbf{F}_{12}, \mathbf{F}_{10}$, - reaction forces; τ_{21}, τ_{12} - reaction torques. $\mathbf{G}_2$ - weight of platform "2"; $\mathbf{F}_t$ – technological forces; τ_t – technological torques; τ_{24} – torque of $\mathbf{F}_4$ reduced at platform "2"; τ_{25} – torque of $\mathbf{F}_5$ reduced at platform "2"; $\mathbf{F}_{i2} = m_2\, \mathbf{a}_2$ – inertial force, corresponding to the platform "2". $\tau_{i2} = J_2\, \varepsilon_2$ – inertial torque, corresponding to the platform "2". $\mathbf{W}_i\,(\mathbf{F}_i, \tau_i)$ - wrench of inertial forces, which act on the platform "1"; $\mathbf{W}_e\,(\mathbf{F}_e, \tau_e)$ – wrench of inertial generalised forces , weights, and other external forces, such as technological forces;

$$\begin{aligned} \mathbf{F}_e &= \mathbf{F}_{i2} + \mathbf{G}_2 + \mathbf{F}_t \\ \tau_e &= \tau_{i2} + \tau_t \end{aligned} \tag{11}$$

The equilibrium equations are:

a) for the platform "2":

$$\begin{aligned} &F_4 + F_5 + F_{21} + F_e = 0 \\ &\tau_{24} + \tau_{25} + \tau_{21} + \tau_e = 0 \end{aligned} \tag{12}$$

b) for the platform "1":

$$\begin{aligned} &F_1 + F_2 + F_3 - F_4 - F_5 + F_{12} + F_{10} + F_{11} = 0 \\ &\tau_{11} + \tau_{12} + \tau_{13} + \tau_{14} + \tau_{15} + \tau^*_{21} + \tau_{10} + \tau_{i1} = 0 \end{aligned} \tag{13}$$

Equations (12) and (13) can be used for solving both direct and inverse dynamics problems.

5 Virtual models. Numerical results

Virtual models of the PARTNER Robotic System were developed using MOBILE software package (Kecskemethy, 1999). These virtual models were developed into modular and reconfigurable manner, based on the parametric description of the PARTNER P 1 11. Thus, many variants of PARTNER were developed, different kinds of movements were simulated and also numerical results were obtained. Figure 4b presents time variation of the angles of the passive joints of the **PARTNER P 322 3522** (Figure 2) due to the movement presumed and presented by figure 4a.

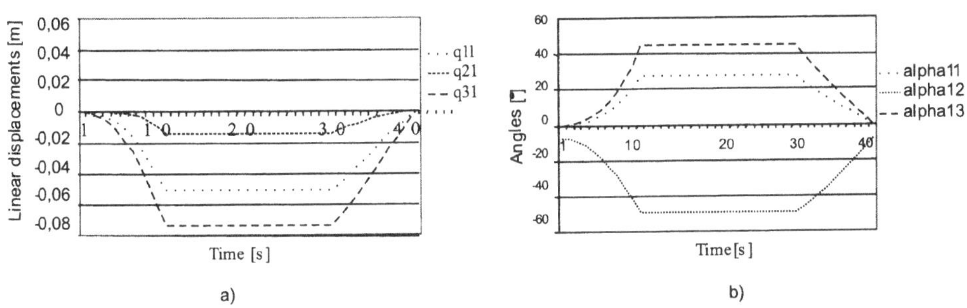

Figure 4. Generalised coordinates and corresponding passive angles

6 Conclusions

Main conclusions of the present work are:

a) based on assumed hypothesis, modular and reconfigurable parallel anthropomorphic robotic system can be developed; this was named **PARTNER** and the algorithm which can be used for development of all variants has been also presented.

b) both kinematics and dynamics of **PARTNER** Robotic System can be developed into modular manner; this mathematical modularity is based on the similarity of the loops and on their similar kinematic and dynamic characteristics.

c) the research work can be completed with the experimental component, based on the prototype which is started to be designed; this prototype will be designed also in modular manner and trying to use the same type of elements, actuators, joints etc.

Acknoledgment

This paper was financial supported by Alexander von Humboldt Foundation

Bibliography

Angeles J. The qualitative synthesis of parallel manipulators (2004), *ASME J. of Mechanical Design*, 126(4):617-624.

Brisan , C., Qian, X., Franitza, D., Hiller, M., (2003), Topology and Kinematics of Modular Reconfigurable Parallel Robots, *Proceedings of 11th World Congress in Mechanism and Machine Science*, China.

Brisan, C., Hiller, M., (2004), Virtual model and simulation for a new type of machine – tools, *IEEE-TTTC International Conference on Automation, Quality and Testing, Robotics*, Tome II, Cluj-Napoca, 55-60.

Cortés J. et Siméon T.(2003), Probabilistic motion planning for parallel mechanisms. *In IEEE Int. Conf. on Robotics and Automation*, pages 4354-4359, Taipei, 14-19.

Dash A.K. et al. (2004), Instantaneous kinematics and singularity analysis of three-legged parallel manipulators.*Robotica*, 22:189-203.

Franitza, D., Brisan , C., Qian, X., Hiller, M., (2003), Error Compensation for Parallel Reconfigurable Machine-Tools, *International Conference on Mechatronics, ICOM 2003*, Loughborough, 153-158.

Hesselbach J. et al. (2003), A micro-assembly-station based on a hybrid 4-dof robot. In *Int. Precision Assembly Seminar IPAS'2003*, pages 55-60, Bad Hofgastein.

Hiller, M., (1995), Multiloop Kinematic Chains and Dynamics of Multiloop Systems, in *Kinematics and Dynamics of Multi-Body Systems* (ed. By J.Angeles and A. Kecskemethy), CISM Courses and Lecture, nr. 360, Springer Verlag.

Kecskemethy, A.,(1999), Mobile, User's Guide, Gerhard-Mercator University, Duisburg.

Lenarcic J. et Stanisic M.M. (2003), A humanoid shoulder complex and the humeral pointing kinematics. *IEEE Trans. on Robotics and Automation*, 19(3):499-506.

Merlet J-P., (2003), Parallel robots. *Kluwer*, Dordrecht.

Pierrot F. (2003), Parallel mechanisms and redundancy. In *1st Int. Colloquium, Collaborative Research Centre 562*, pages 261-277, Braunschweig.

Zoppi M., Bruzzone L.E., et Molfino R.M.A (2004), Novel 5-dof interconnected-chains PKM for manufacturing revolute surfaces. In *4th Chemnitzer Parallelkinematik Seminar*, Chemnitz.

Workspace Comparison of Kinematically Redundant Planar Parallel Manipulators

Iman Ebrahimi[†], Juan A. Carretero[‡] and Roger Boudreau[§]
[†] University of New Brunswick, Department of Mechanical Engineering, Canada
E-mail: Iman.Ebrahimi@unb.ca
[‡] University of New Brunswick, Department of Mechanical Engineering, Canada
E-mail: Juan.Carretero@unb.ca
[§] Université de Moncton, Faculté d'ingénierie, Canada
E-mail: Boudrer@umoncton.ca

Abstract In the present work, the inverse displacement problem and the workspace of kinematically redundant planar parallel manipulators are discussed. Non-redundant and redundant planar parallel manipulators are compared. The advantages of redundant versus non-redundant parallel manipulators are discussed. Thereafter, two kinematically redundant planar parallel manipulators, one with 4 degrees of freedom (DOF), the other a new 6DOF design, are compared to similar non-redundant ones. It is shown here that both reachable and dexterous workspaces of the redundant manipulators are significantly larger compared to non-redundant manipulators of the same size. As a result, kinematic redundancy can be an appropriate solution for improving parallel manipulators' characteristics.

1 Introduction

In the last few decades, parallel manipulators have been studied intensely by many researchers. The reason is their advantages compared to serial ones; higher payload-to-weight ratio, higher accuracy, higher structural rigidity, location of motors at or close to the base and simpler solution of the inverse kinematics equations (Ben-Horin et al. (1998)). However, like every other type of mechanisms, these manipulators have some drawbacks. Smaller workspace, reduced dexterity and more complex kinematic and dynamic models are problems that parallel manipulators have as compared to their serial counterparts. Such added complexity comes from these mechanisms being more kinematically constrained compared to serial manipulators. Most of the research that has been done for parallel manipulators so far has focused on non-redundant manipulators (*e.g.*, Merlet (2006)).

Redundant parallel manipulators have been designed to overcome some of the aforementioned disadvantages of parallel manipulators (Jing and Gosselin (2004)). Basically, redundancy in parallel manipulators can be divided into two different types: a) Actuation redundancy and b) Kinematic redundancy.

Although redundancy has been investigated in parallel manipulators before, the majority of these studies have concentrated on actuator redundancy (Nakamura and Ghodoussi 1989, Merlet (1996), Kock and Schumacher (1998), Nokleby et al. (2005)). Actuator redundancy is defined as using more actuators than are strictly necessary to control the manipulator for a given task. Consequently, actuator redundancy does not change the mobility of manipulators. Generally, actuator redundancy can be obtained by changing some or all passive joints into active ones in an existing parallel manipulator. Adding actuators increases the chance of generating internal forces between links, therefore, controlling manipulators with actuation redundancies generally becomes more challenging. Kinematic redundancy, on the other hand, changes the mobility of the manipulator and adds degrees of freedom (DOF) to it. This idea has extensively been introduced before for serial manipulators (see, for instance, Nakamura (1991)). Conversely, for parallel manipulators there are few references that deal with the issue (*e.g.* Zanganeh and Angeles (1995), Jing and Gosselin (2004), Dasgupta and Mruthyunjaya (1998)).

Generally, kinematically redundant manipulators have more parameters or actuators for controlling the motion than the minimum required for a set of specified tasks. If the number of independent actuated joints is greater than the DOF required to perform a task, the manipulator is said to be kinematically redundant. Adding kinematic redundancy in parallel manipulators often entails several advantages such as the elimination of most kinematic singularities, an increased workspace as well as an improved dexterity.

In this paper, the inverse kinematics of redundant planar parallel manipulators are first discussed. The inverse displacement problem of a 4DOF kinematically redundant planar parallel manipulator first proposed by Jing and Gosselin (2004)(2-RPR+1-RRPR) is then analysed. Thereafter, a 3-PRRR, 6DOF, kinematically redundant parallel manipulator is introduced and a solution is proposed for its inverse displacement problem.

In the next section, the reachable and dexterous workspaces of redundant and non-redundant planar parallel manipulators are compared. First the workspace of the 3-RPR manipulator is compared to that of the manipulator introduced by Jing and Gosselin (2-RPR+1-RRPR). Then, the workspace of the 3-RRR manipulator, first analysed by Gosselin and Angeles (1988), is compared to that of the 3-PRRR manipulator. The workspace of the 3-RPR was studied by Sefrioui and Gosselin (1992), Merlet et al. (1998).

2 Inverse Displacement of Redundant Parallel Manipulators

2.1 Locus of Solutions

Manipulators with kinematic redundancy have an infinite number of solutions to the inverse displacement problem. Thus, instead of a limited number of solutions, if there are any, there may be one or more loci of solutions as opposed to a finite set of solutions. Figure 1 depicts an example for the locus of solutions for the given position and orientation of a 4DOF kinematically redundant planar parallel manipulator designed by adding an extra link L_1 and an active joint (θ) to one branch of the 3-RPR parallel manipulator (Jing and Gosselin (2004)). This manipulator will be deemed as the 4DOF kinematically redundant planar parallel manipulator. As shown in Figure 1 all possible solutions of the inverse displacement problem for point A_1 are on the arc $\widehat{MA_1N}$. Note that the shaded area in Figure 1 is the area that point A_1 as part of link B_1A_1 (pinned at point B_1) can cover based on the minimum and maximum actuator lengths of ρ_1. This

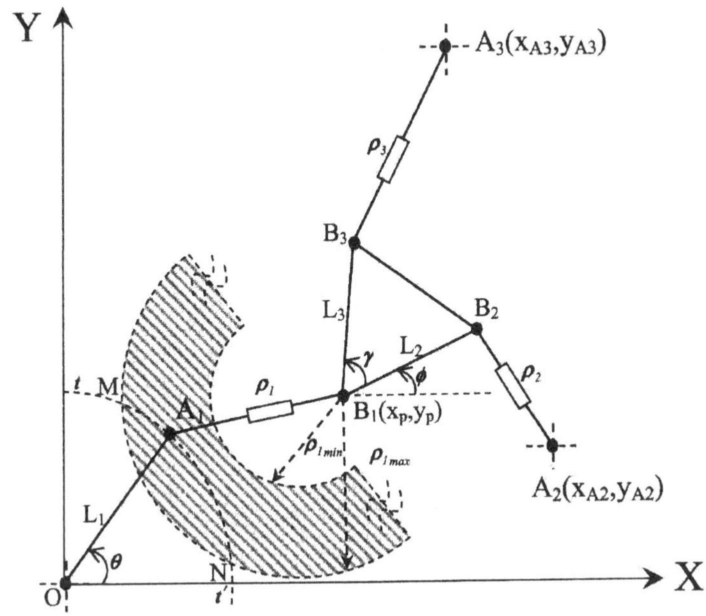

Figure 1. Planar 4DOF Kinematically redundant planar parallel manipulator(2-RP̲R+1-R̲RP̲R) proposed by Jing and Gosselin (2004).

shaded area should be a full annular region, but is shown as such to avoid overloading the figure.

Figure 2 illustrates a 6DOF redundant planar parallel manipulator, a 3-P̲R̲RR, that is proposed by the authors. A 3-P̲R̲RR is originally a 3-R̲RR that has three redundant actuated joints resulting in a 6DOF planar parallel manipulator. For a given position and orientation, each limb of the manipulator has a locus of possible solutions for the inverse displacement problem of the manipulator. The locus of all possible solutions for each limb consists of all points A_i on C_iO $(i = 1, 2, 3)$ that satisfy the following equation:

$$|\overline{B_iA_i}| < L_i + d_i \tag{2.1}$$

where lengths L_i and d_i, and points B_i and A_i are defined in Figure 2. Thus, for every given position and orientation of the end-effector, together referred to as the pose, there are three loci, one for each of the three limbs of the manipulator.

2.2 Optimised Motion

Since there is an infinite number of solutions, the problem is to determine which leg configurations should be chosen as solutions for the inverse displacement problem of the manipulator. To this end, a cost function may be defined based on goals such as

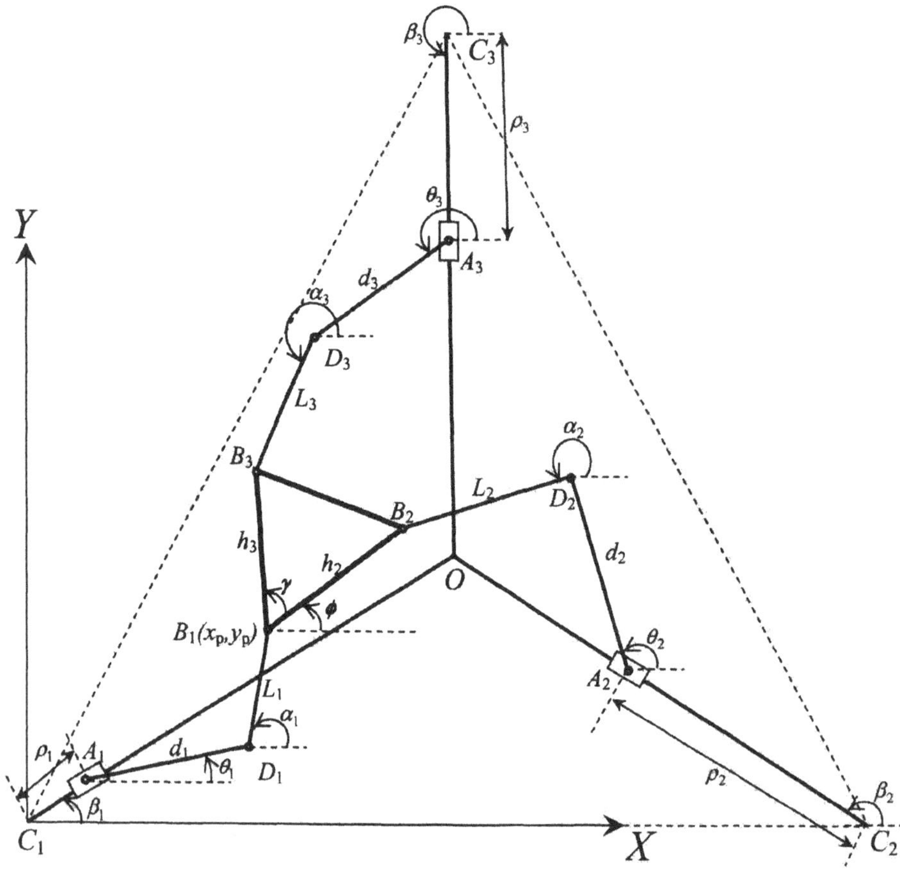

Figure 2. The 3-PRRR planar 6DOF kinematically redundant parallel manipulator

minimum energy, minimum displacement, minimum travel time and/or distance to the closest singular pose.

Regardless of the method that is chosen, the solution of the inverse displacement problem depends on the initial pose of the end effector and the configuration of each limb. Figure 3 depicts an example for the redundant limb of the 2-RPR+1-RRPR (shown in Figure 1). If a minimum difference principle is used as the cost function, the final configuration of the redundant limb depends on the initial pose, as shown in Figure 3. The figure shows how different initial poses (I and II) of the end effector and of the redundant limb give different solutions for the same final configuration of the end effector.

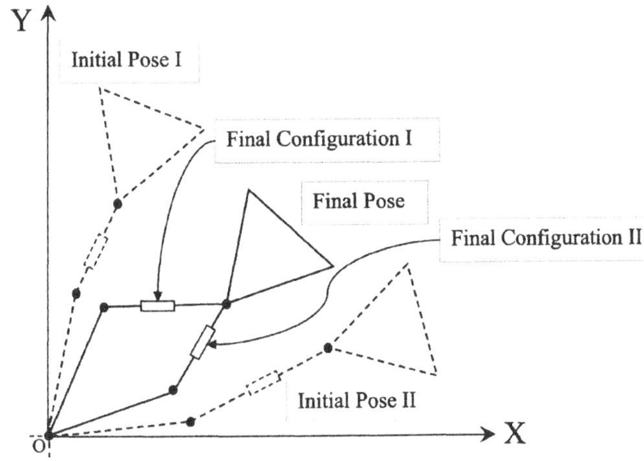

Figure 3. Effect of the initial pose on the final configuration of a redundant branch.

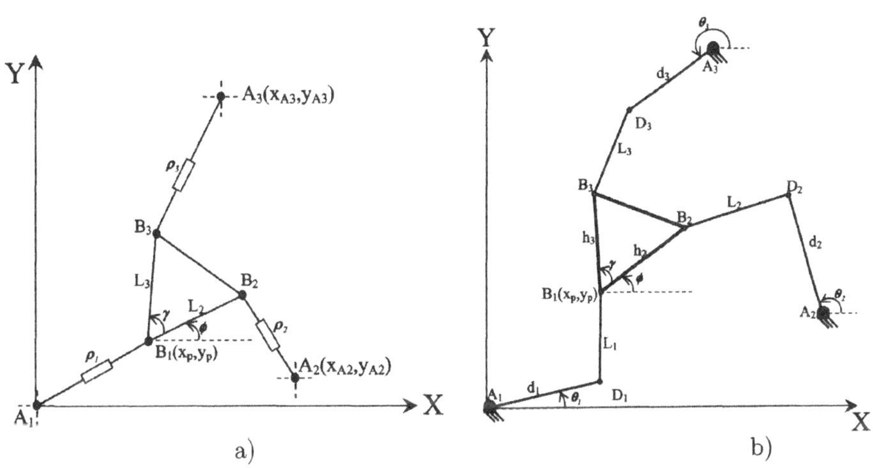

Figure 4. Non-redundant planar parallel manipulators a) the 3-R<u>P</u>R and b) 3-<u>R</u>RR

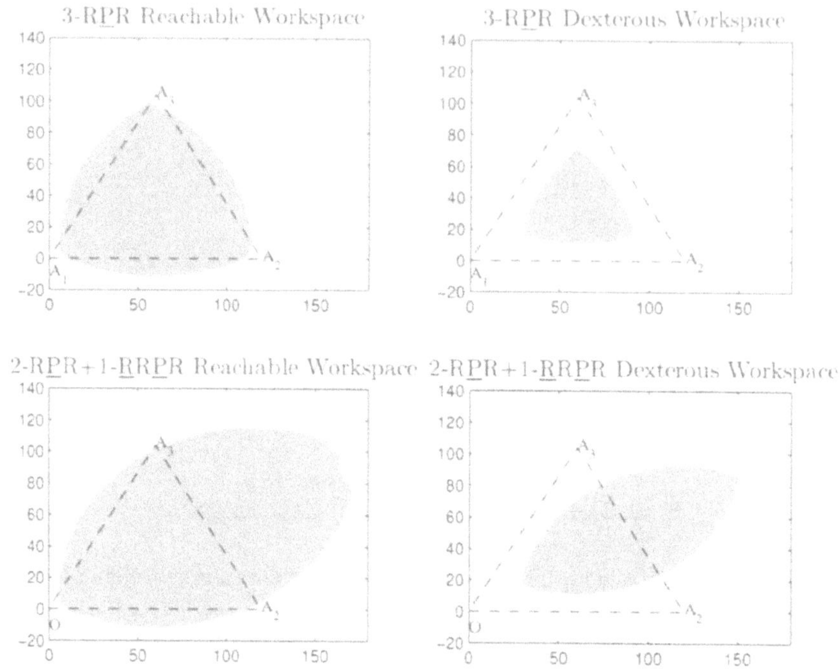

Figure 5. Workspace comparison between a 3-RPR planar parallel manipulator (non-redundant) and 2-RPR+1-RRPR(4 DOF redundant)

3 Workspace Comparison

Here, the reachable workspace is defined as the region within which every point can be reached by the end-effector in at least one orientation. A dexterous workspace is a region of the reachable workspace within which every point can be reached by the end-effector in all possible orientations. Here, the reachable and dexterous workspaces of the manipulators are obtained by discretizing the region and locating all points within the region that are inside the bounds of the corresponding workspace.

3.1 Comparison of the 3-RPR and 2-RPR+1-RRPR manipulators

The 3-RPR is shown in Figure 4a. By adding 1DOF of redundancy to the manipulator, it changes into the 4DOF manipulator shown in Figure 1. Figure 5 depicts the workspace of both the 3-RPR and the 2-RPR+1-RRPR. The geometric parameters of the 3-RPR are: $A_1A_2 = A_2A_3 = A_1A_3 = 120$ cm, $\rho_i|_{\max} = 103.92$ cm (the height of the equilateral triangle $A_1A_2A_3$), $\rho_i|_{\min} = 0$, $\gamma = \pi/3$, $i = 1,2,3$, $L_2 = L_3 = 20$ cm. For the 2-RPR+1-RRPR the geometric parameters are $OA_2 = A_2A_3 = OA_3 = 120$ cm, $L_1 = \rho_i|_{\max} = 103.92$ cm, $\rho_i|_{\min} = 0$, $\gamma = \pi/3$, $i = 1,2,3$ and $L_2 = L_3 = 20$ cm. As shown in Figure 5, both the reachable and dexterous workspaces are considerably larger by

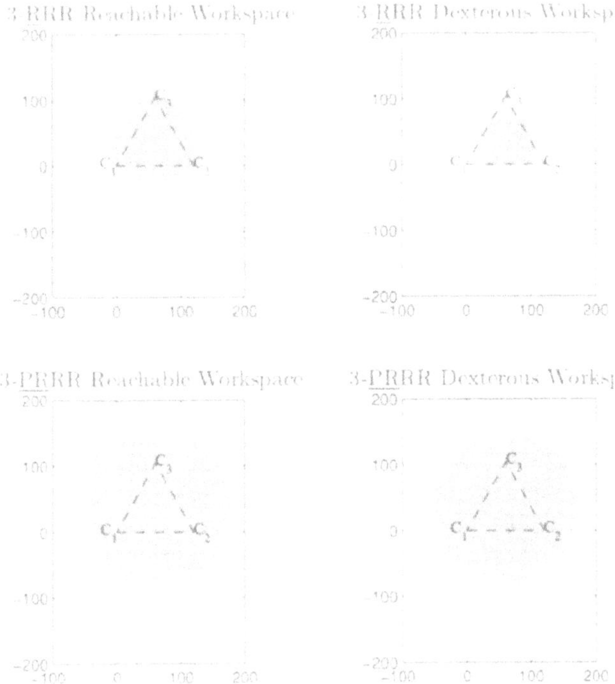

Figure 6. Workspace comparison between the 3-RRR (non-redundant) and the 3-PRRR (redundant)planar parallel manipulator

adding redundancy to the manipulator. It should be noted that both manipulators have the same base attachment points which are indicated by A_i, $i = 1, 2, 3$ for 3-RPR and OA_2A_3 for the 2-RPR+1-RRPR in the Figure 5.

3.2 Comparison of the 3-RRR and 3-PRRR manipulators

The 3-RRR and the 3-PRRR are shown in Figures 4b and 2, respectively. Figure 6 shows both reachable and dexterous workspaces of the two manipulators. The geometric parameters of the 3-RRR and 3-PRRR are: $C_1C_2 = C_2C_3 = C_1C_3 = 120$ cm, $\rho_i|_{\max} = C_iO = 69.28$ cm, $\rho_i|_{\min} = 0$, $\gamma = \pi/3$, $d_i = L_i = 60$ cm, $i = 1, 2, 3$ and $h_2 = h_3 = 20$ cm. Again, both manipulators have the same base attachment points, indicated by C_i, $i = 1, 2, 3$.

Both the reachable and dexterous workspaces of the 3-PRRR are significantly larger than the 3-RRR's. These significant improvements in workspace and maneuverability of the manipulator and the reduction of singularities when kinematic redundancy is added (Jing and Gosselin (2004)) may justify the fact that a 6DOF manipulator is used for a planar task.

4 Conclusion

In this paper, two kinematically redundant planar parallel manipulators have been studied and compared to their non-redundant original designs. When considering the inverse kinematic problem, instead of a limited number of solutions (if there are any), there is a locus of solutions. An appropriate solution can be selected by defining a cost function. Finally, the comparison of the workspaces of redundant and non-redundant planar parallel manipulators have been made in the Cartesian space. The comparison results have shown that kinematically redundant manipulators are very efficient in improving both reachable and dexterous workspaces. A new design, considering a 3-PRRR manipulator has been introduced. Compared to a similar sized non-redundant manipulator, the introduced 3-PRRR has a considerably larger workspace.

Bibliography

R. Ben-Horin, M. Shoham, and S. Djerassi. Kinematics, dynamics and construction of a planarly actuated parallel robot. *Journal of Robotics and Computer-Integrated Manufacturing*, 14(2):163–172, 1998.

B. Dasgupta and S. Mruthyunjaya. Redundancy in parallel manipulators: Theoretical and practical issues. *Mechanism and Machine Theory*, 33(6):727–742, 1998.

C. M. Gosselin and J. Angeles. Optimum kinematic design of a planar three-degree-of-freedom parallel manipulator. *Journal of Mechanisms, Transmissions, and Automation in Design*, 110(1):35–41, 1988.

W. Jing and C. M. Gosselin. Kinematic analysis and design of kinematically redundant parallel mechanisms. *Journal of Mechanical Design*, 126(1):109–118, 2004.

S. Kock and W. Schumacher. Parallel x-y manipulator with actuation redundancy for high-speed and active-stiffness applications. In *IEEE International Conference on Robotics and Automation*, pages 2295–2300, Leuven, Belgium, 1998.

J.-P. Merlet. Redundant parallel manipulators. *Laboratory Robotics and Automation*, 8 (1):17–24, 1996.

J.-P. Merlet. *Parallel Robots*. Springer, 2nd edition, 2006.

J.-P. Merlet, C. M. Gosselin, and N. Mouly. Workspaces of planar parallel manipulators. *Mechanism and Machine Theory*, 33(1):7–20, 1998.

Y. Nakamura. *Advanced Robotics: Redundancy and Optimization*. Addison-Wesley, 1991.

Y. Nakamura and M. Ghodoussi. Dynamics computation of closed-link robot mechanisms with non-redundant and redundant actuators. *IEEE Transactions of Robotics and Automation*, 5(3):294–302, 1989.

S. B. Nokleby, R. Fisher, and R. P. Podhorodeski. Force capabilities of redundantly-actuated parallel manipulators. *Mechanism and Machine Theory*, 40(5):578–599, May 2005.

J. Sefrioui and C. M. Gosselin. Singularity analysis and representation of planar parallel manipulators. *Journal of Robotics and Autonomous Systems*, 10(4):209–224, 1992.

K. Zanganeh and J. Angeles. Instantaneous kinematics of general hybrid parallel manipulators. *Journal of Mechanical Design*, 117(4):581–588, 1995.

Dexterity Analysis of Planar Parallel Manipulators

Geoffrey Pond[†] and Juan A. Carretero[‡]
[†] University of New Brunswick, Department of Mechanical Engineering, Canada
E-mail:Geoff.Pond@unb.ca
[‡] University of New Brunswick, Department of Mechanical Engineering, Canada
E-mail:Juan.Carretero@unb.ca

Abstract

Currently, two methods exist for obtaining constrained, dimensionally homogeneous Jacobian matrices for planar manipulators. These two methods differ in that a set of Cartesian velocities is expressed in either the moving or global frame which is then used to represent the end effector velocity. Therefore, the resulting Jacobian matrices are different depending on the method used. However, both methods formulate the Jacobian matrices based on an arbitrary choice of six potential Cartesian variables to define the dimensions of the task space.

This paper will obtain all six possible Jacobian matrices based on the six choices of available task space variables in the global frame. As an example, the Jacobian matrices of the 3-RRR planar parallel mechanism are formulated. These are then tested by comparing the resulting actuator velocities for a defined end effector path and velocity profile.

Each of the six Jacobian matrices is numerically unique and therefore have different singular values and condition numbers. Dexterity measures focusing on only one of the six matrices are potentially bias. Therefore, four proposed strategies for measuring the manipulator dexterity are then discussed. These are based on the singular values and condition number of the six constrained, dimensionally homogenous Jacobian matrices instead of a single matrix.

1 Introduction

The dexterity analysis of complex degree of freedom (DOF) mechanisms has thus far been problematic. A well accepted method of measuring the dexterity of spherical and translational manipulators has been the Jacobian matrix condition number as in Gosselin and Angeles (1989) and Badescu and Mavroidis (2004). Unfortunately, the inconsistent units between elements within the Jacobian of a complex-DOF parallel manipulator do not allow such a measure to be made as discussed by Tsai (1999) and Angeles (2003).

Gosselin (1992) introduced a method of obtaining a dimensionally homogeneous Jacobian matrix, based on either the Cartesian velocities of two points on the end effector for planar mechanisms, or three points on the end effector for spatial mechanisms. Gosselin (1992) also provides a formulation for the Jacobian matrix of a planar parallel manipulator based on a choice of three of the total four components of the Cartesian velocities of two points on the end effector platform. In Gosselin (1992), these Cartesian velocities are chosen with respect to the moving frame. However, a set of independent velocities chosen in the moving frame is not always possible as a result of potential dependencies between the various Cartesian velocities in the global frame.

Kim and Ryu (2003) map the Cartesian components of three points on the end effector, with respect to the global frame to actuator velocities, regardless of the degrees of freedom of the manipulator. However, the disadvantage of doing this is that the resulting Jacobian matrix is unconstrained, *i.e.*, for the 3-RRR planar manipulator, only three of the nine Cartesian component velocities are independent. The physical significance of the Jacobian matrix condition number and singular values is therefore lost. Recognising this, Pond and Carretero (2005) have developed a method of constraining the generalised Jacobian matrix published by Kim and Ryu (2003).

For the spatial 3-PRS parallel manipulator, Pond and Carretero (2005) map the velocities of three points along the same axis of the global frame to the actuator velocities. The resulting Jacobian matrix is constrained and dimensionally homogeneous. The condition number of this matrix is then proposed as a dexterity measure. In the planar case, choosing the velocity of three points along the same axis is no longer possible. This paper will further the work introduced by Pond and Carretero (2005) through examining the influence of the choice of independent variables, on the resulting Jacobian matrix and its condition number. This will be illustrated through the 3-RRR planar manipulator.

2 The 3-RRR Architecture and its Kinematics

The 3-RRR manipulator has been the subject of many studies. For example, inverse kinematics including velocity and acceleration, singularity analysis, and synthesis are provided by Gosselin (1988) and Arsenault and Boudreau (2004) respectively.

The symmetrical 3-RRR architecture is depicted in Figure 1. The degrees of freedom of the end effector are translations in x and y directions and a rotation ϕ, around an axis normal to the xy-plane, *i.e.*, parallel to all revolute axes.

As the inverse displacement kinematics of this manipulator are well published, no fur-

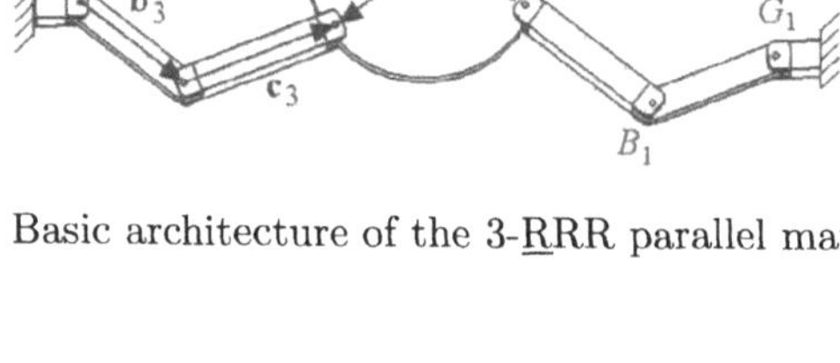

Figure 1. Basic architecture of the 3-RRR parallel manipulator.

ther discussion on the subject will be provided here. The Jacobian formulation provided for this manipulator by Gosselin (1988) and Arsenault and Boudreau (2004) is developed by differentiating the various inverse displacement equations, with respect to time. Tsai (1999) has also obtained the Jacobian matrix of the 3-RRR manipulator through the method of cross-products. Expressed in 2-D screw coordinates, their solutions are equivalent to the following direct ($\mathbf{J}_x$) and inverse ($\mathbf{J}_q$) Jacobians:

$$\mathbf{J}_x = \begin{bmatrix} (a_{1_x}c_{1_y} - a_{1_y}c_{1_x}) & c_{i_x} & c_{i_y} \\ (a_{2_x}c_{2_y} - a_{2_y}c_{2_x}) & c_{i_x} & c_{i_y} \\ (a_{3_x}c_{3_y} - a_{3_y}c_{3_x}) & c_{i_x} & c_{i_y} \end{bmatrix}_{3\times 3} \tag{2.1}$$

$$\mathbf{J}_q = \begin{bmatrix} b_{1_x}c_{1_y} - b_{1_y}c_{1_x} & 0 & 0 \\ 0 & b_{2_x}c_{2_y} - b_{2_y}c_{2_x} & 0 \\ 0 & 0 & b_{3_x}c_{3_y} - b_{3_y}c_{3_x} \end{bmatrix}_{3\times 3} \tag{2.2}$$

which can be combined as $\mathbf{J} = \mathbf{J}_q^{-1}\mathbf{J}_x$ to obtain a relation between end effector and actuator velocities $\dot{\mathbf{q}} = \mathbf{J}\dot{\mathbf{x}}$ where $\dot{\mathbf{q}} = [\ \dot{\theta}_{1,1} \quad \dot{\theta}_{1,2} \quad \dot{\theta}_{1,3}\ ]^T$, and $\dot{\mathbf{x}} = [\ \omega_z \quad \dot{x} \quad \dot{y}\ ]^T$.

In the following section, the Jacobian matrix $\mathbf{J}$ will be used to validate the Jacobian matrices formulated by more novel methods.

3 Jacobian Matrix Formulation for Dexterity Analysis

As mentioned, the Jacobian matrix $\mathbf{J}$ developed in the previous section is dimensionally inconsistent. Tsai (1999) and Angeles (2003) have outlined the importance in having a dimensionally homogeneous Jacobian matrix in dexterity analyses.

3.1 Dimensionally Homogeneous Jacobian Matrix

Kim and Ryu (2003) develop the following velocity relation:

$$\dot{\mathbf{q}} = \mathbf{J}_q^{-1}\mathbf{J}_x\dot{\mathbf{x}} \tag{3.1}$$

where letting $\tilde{\mathbf{k}} = [\ 0 \quad 0 \quad 1\ ]^T$, $\dot{\mathbf{x}}' = [\ \dot{A}_{1_x} \quad \dot{A}_{1_y} \quad \dot{A}_{2_x} \quad \dot{A}_{2_y} \quad \dot{A}_{3_x} \quad \dot{A}_{3_y}\ ]^T$ and $\dot{\mathbf{q}} = [\ \dot{\theta}_{1,1} \quad \dot{\theta}_{1,2} \quad \dot{\theta}_{1,2}\ ]^T$:

$$\mathbf{J}_q' = \begin{bmatrix} \mathbf{c}_1^T\left(\tilde{\mathbf{k}} \times \mathbf{b}_1\right) & 0 & 0 \\ 0 & \mathbf{c}_2^T\left(\tilde{\mathbf{k}} \times \mathbf{b}_2\right) & 0 \\ 0 & 0 & \mathbf{c}_3^T\left(\tilde{\mathbf{k}} \times \mathbf{b}_3\right) \end{bmatrix}_{3\times 3} \tag{3.2}$$

$$\mathbf{J}_x' = \begin{bmatrix} k_{1,1}\mathbf{c}_i^T & k_{1,2}\mathbf{c}_i^T & k_{1,3}\mathbf{c}_i^T \\ k_{2,1}\mathbf{c}_i^T & k_{2,2}\mathbf{c}_i^T & k_{2,3}\mathbf{c}_i^T \\ k_{3,1}\mathbf{c}_i^T & k_{3,2}\mathbf{c}_i^T & k_{3,3}\mathbf{c}_i^T \end{bmatrix}_{3\times 6} \tag{3.3}$$

The multiplication of $(\mathbf{J}'_q)^{-1}\mathbf{J}'_x$ produces the non-square dimensionally homogeneous Jacobian matrix $\mathbf{J}'$ which is equivalent to:

$$\mathbf{J}' = \begin{bmatrix} \frac{\partial\theta_1}{\partial A_{1_x}} & \frac{\partial\theta_1}{\partial A_{1_y}} & \frac{\partial\theta_1}{\partial A_{2_x}} & \frac{\partial\theta_1}{\partial A_{2_y}} & \frac{\partial\theta_1}{\partial A_{3_x}} & \frac{\partial\theta_1}{\partial A_{3_y}} \\ \frac{\partial\theta_2}{\partial A_{1_x}} & \frac{\partial\theta_2}{\partial A_{1_y}} & \frac{\partial\theta_2}{\partial A_{2_x}} & \frac{\partial\theta_2}{\partial A_{2_y}} & \frac{\partial\theta_2}{\partial A_{3_x}} & \frac{\partial\theta_2}{\partial A_{3_y}} \\ \frac{\partial\theta_3}{\partial A_{1_x}} & \frac{\partial\theta_3}{\partial A_{1_y}} & \frac{\partial\theta_3}{\partial A_{2_x}} & \frac{\partial\theta_3}{\partial A_{2_y}} & \frac{\partial\theta_3}{\partial A_{3_x}} & \frac{\partial\theta_3}{\partial A_{3_y}} \end{bmatrix} \quad (3.4)$$

3.2 Constrained Dimensionally Homogeneous Jacobian Matrix

Pond and Carretero (2005) have discussed the importance of mapping a set of only independent end effector velocities to the actuator velocities. The mapping being done in equation (3.1) maps six end effector velocities of which only three are independent (for the 3-DOF mechanism) to the three actuator velocities. Furthermore, Pond and Carretero (2005) propose to use a constraining matrix mapping the independent end effector velocities to the full set of both independent and dependent end effector velocities. For example, if all components of the Cartesian velocities in the xy plane of the three points could be obtained as functions of the Cartesian velocities $\dot{A}_{1_x}$, $\dot{A}_{2_x}$, $\dot{A}_{3_y}$, the following constraining matrix mapping $[\dot{A}_{1_x}\ \dot{A}_{2_x}\ \dot{A}_{3_y}]^T$ to $\dot{\mathbf{x}}'$ could be obtained:

$$\mathbf{P} = \begin{bmatrix} 1 & \frac{\partial A_{1_y}}{\partial A_{1_x}} & 0 & \frac{\partial A_{2_y}}{\partial A_{1_x}} & \frac{\partial A_{3_x}}{\partial A_{1_x}} & 0 \\ 0 & \frac{\partial A_{1_y}}{\partial A_{2_x}} & 1 & \frac{\partial A_{2_y}}{\partial A_{2_x}} & \frac{\partial A_{3_x}}{\partial A_{2_x}} & 0 \\ 0 & \frac{\partial A_{1_y}}{\partial A_{3_y}} & 0 & \frac{\partial A_{2_y}}{\partial A_{3_y}} & \frac{\partial A_{3_x}}{\partial A_{3_y}} & 1 \end{bmatrix}^T \quad (3.5)$$

The resulting multiplication of $\mathbf{J}'$ in equation (3.4) with $\mathbf{P}$ yields:

$$\mathbf{J}'\mathbf{P} = \begin{bmatrix} \frac{\partial\theta_1}{\partial A_{1_x}} & \frac{\partial\theta_1}{\partial A_{2_x}} & \frac{\partial\theta_1}{\partial A_{3_y}} \\ \frac{\partial\theta_2}{\partial A_{1_x}} & \frac{\partial\theta_2}{\partial A_{2_x}} & \frac{\partial\theta_2}{\partial A_{3_y}} \\ \frac{\partial\theta_3}{\partial A_{1_x}} & \frac{\partial\theta_3}{\partial A_{2_x}} & \frac{\partial\theta_3}{\partial A_{3_y}} \end{bmatrix} \quad (3.6)$$

In fact, it can be shown that $\dot{A}_{1_x}$, $\dot{A}_{2_x}$, and $\dot{A}_{3_y}$ may be considered as independent end effector velocities, from which the dependent velocities of $\dot{A}_{1_y}$, $\dot{A}_{2_y}$, and $\dot{A}_{3_x}$ may be obtained. Therefore, the constraining matrix $\mathbf{P}$ and 3×3 Jacobian matrix $\mathbf{J}'\mathbf{P}$ may also be obtained. Furthermore, instead of considering the set $\dot{A}_{1_x}$, $\dot{A}_{2_x}$, and $\dot{A}_{3_y}$ as independent velocities, five other potential sets could have been considered. Any subset consisting of three elements from the six elements of $\dot{\mathbf{x}}'$ which includes at least one x variable and at least one y variable may be used. These subsets are:

Case I:	$[\dot{A}_{1_x}\dot{A}_{2_x}\dot{A}_{3_y}]^T$	Case IV:	$[\dot{A}_{1_y}\dot{A}_{2_y}\dot{A}_{3_x}]^T$
Case II:	$[\dot{A}_{1_x}\dot{A}_{2_y}\dot{A}_{3_x}]^T$	Case V:	$[\dot{A}_{1_y}\dot{A}_{2_x}\dot{A}_{3_y}]^T$
Case III:	$[\dot{A}_{1_y}\dot{A}_{2_x}\dot{A}_{3_x}]^T$	Case VI:	$[\dot{A}_{1_x}\dot{A}_{2_y}\dot{A}_{3_y}]^T$

The same basic method explained earlier in this section for obtaining the square, constrained dimensionally homogeneous Jacobian matrix $\mathbf{J}'\mathbf{P}$, is used for each of the six cases corresponding to differently chosen independent variables.

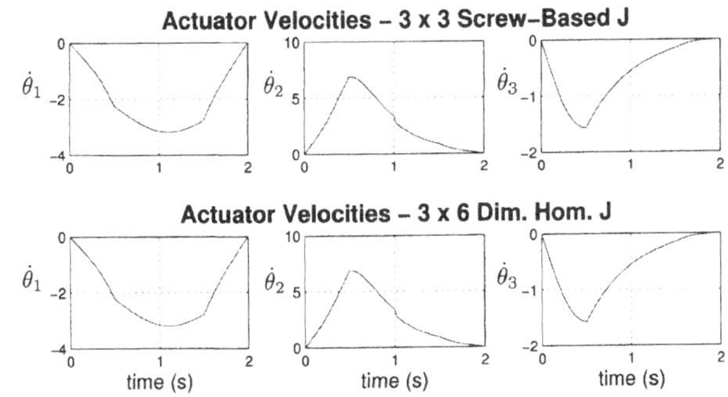

Figure 2. Required actuator velocities as determined by a trajectory planning algorithm.

3.3 Verification of Results

A trajectory planning program may be used to verify the results of each of the Jacobians formulated for Cases I-VI. Given a random starting pose and end pose for the end effector, the trajectory planning program obtains a straight line path between the two poses and, based on a desired end effector velocity curve, the required actuator velocities are obtained. The velocity curve used here completes the trajectory through an initial 0.5 second constant acceleration period, followed by a 1 second constant velocity period and 0.5 second constant deceleration period to rest, at the end pose. The actuator velocities are first obtained by the previously published Jacobians, *i.e.*, the screw based Jacobian $\mathbf{J}$, developed by Tsai (1999) and the dimensionally homogeneous Jacobian $\mathbf{J}'$ developed by Kim and Ryu (2003). For the random starting pose of [0.03 0.08 0.17] and end pose of [0.25 -0.15 0.2] ([x, y, ϕ]), these velocities are depicted in Figure 2. The architectural variables used are arbitrarily chosen as a base platform radius of 1, moving platform radius of 0.6 and unit link lengths *i.e.*, $|\mathbf{b}| = |\mathbf{c}| = 1$.

It is evident that the well established Jacobians provide similar velocity results in the actuator space for a given trajectory and velocity profile in the task space. The same program may be used to determine the required actuator velocities using each of the six constrained Jacobian matrices $\mathbf{J}'\mathbf{P}$. These results are depicted in Figure 3.

The maximum absolute difference between results obtained with the screw based Jacobian compared to results obtained by any of the other 7 Jacobian matrices presented in this paper was found to be on the order of 10^{-15}, well within what can be expected to result from computational error.

4 Dexterity Measurement

One of the objectives of performing dexterity analyses on parallel manipulators is to obtain an understanding of how sensitive the end effector pose is relative to the actuator displacement. As discussed, for some cases, this has historically been achieved through observation of the Jacobian matrix condition number.

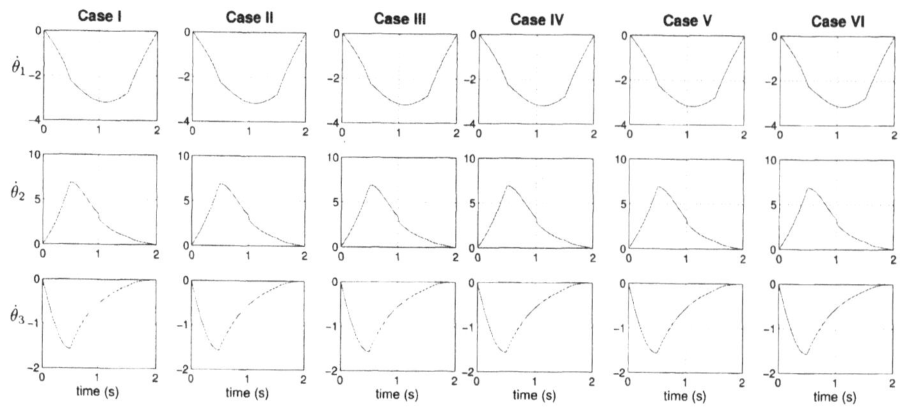

Figure 3. Required actuator velocities as determined by a trajectory planning algorithm using each of the six constrained dimensionally homogeneous Jacobian matrices.

The condition number of the screw based Jacobian matrix $\mathbf{J}$ and dimensionally homogeneous Jacobian matrix $\mathbf{J}'$ throughout this path are depicted in Figure 4. Clearly, the planned trajectory either passes through or very near a singular configuration as evidenced by the rapidly increasing condition number of the screw based Jacobian matrix at approximately $t = 0.9$ sec. In fact, it can be shown that the defined path passes through a direct singular configuration where the three vectors $\mathbf{c}_i$ depicted in Figure 1 intersect at a single point.

However, $\mathbf{J}'$, the 3×6 dimensionally homogeneous Jacobian matrix developed by Kim and Ryu (2003), does not suggest the same. Instead, its condition number gives the impression that the manipulator is relatively near isotropic condition throughout the defined path. Obviously then, the 3×6 dimensionally homogenous Jacobian matrix is not suitable as a dexterity measure. Because three of the six columns of $\mathbf{J}'$ are dependent on the other three columns, the eigenvalues of $\mathbf{J}'$ could correspond to velocity directions in the task space which are not obtainable. Therefore the eigenvalues and singular values are essentially meaningless.

Figure 4 also depicts the results obtained by observing the condition number of each of the six constrained dimensionally homogeneous Jacobian matrices. Each of the constrained Jacobian matrices clearly agree that the arbitrarily chosen trajectory has the manipulator passing near a singular configuration. The six matrices $\mathbf{J}'\mathbf{P}$ are constrained and therefore accurately predict singular configurations, as shown. Furthermore, their terms are dimensionally homogeneous. Therefore, their condition numbers allow a suitable means of measuring dexterity.

That being said, each of the six cases also gives a different value for the condition number of the Jacobian at any given pose. Therefore, a single measure of dexterity remains obscure for the manipulator. The following are four potential strategies proposed as dexterity measures:

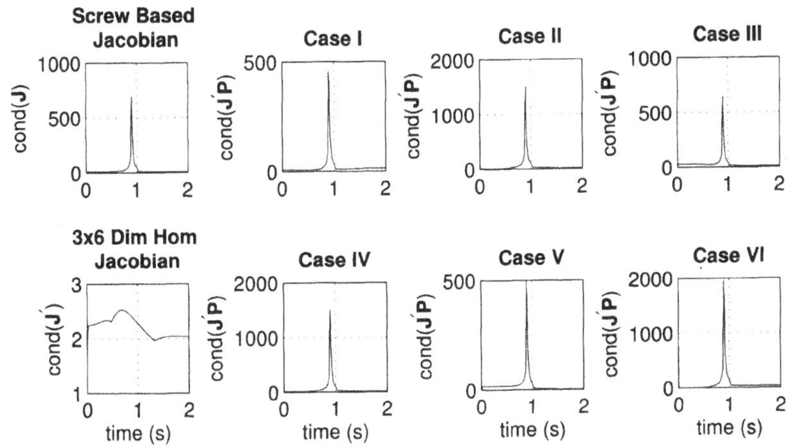

Figure 4. The condition number of each of the formulated Jacobian matrices throughout the planned trajectory.

a) The maximum condition number of the six Jacobian matrices. This measure is essentially the maximum ratio between the largest actuator effort required to move in a direction in one of the six defined task space variable sets, with the effort required to move in the easiest direction using the same task space variables.

b) The ratio of maximum singular value of all six Jacobian matrices to minimum singular value of all six Jacobian matrices. This compares the largest actuator effort required to move in any of the six defined task space variable sets, with the effort required to move in the easiest task space direction of any of the six task space sets.

c) The mean maximum singular value of each of the six Jacobian matrices. This is suggested in applications where high end effector velocities are desired. Alternatively, the absolute maximum singular value may be used.

d) Conversely, in applications where a high degree of accuracy is of greater importance than velocity capabilities, the mean or absolute minimum singular value may be used.

It is also suggested that measures using the singular values (such as in c) and d)), be used in conjunction with the condition number as in either a) or b). By doing so, both the velocity or accuracy characteristics of the manipulator are obtained, in addition to an indication of how 'near-isotropic' the architecture is at the studied pose. Future work on the subject will study the four proposed measures listed above in greater depth.

5 Conclusions

Through either method of obtaining a constrained dimensionally homogeneous Jacobian matrix (proposed by Gosselin (1992) and Pond and Carretero (2005), respectively) for planar mechanisms, a choice exists on which of the potential six Cartesian velocity components on the end effector be used to define the task space velocity variables. The choice has an influence on the resulting Jacobian matrix and therefore its condition number and singular values. Without constraining the Jacobian matrix, the condition number was demonstrated to be essentially meaningless, as in Kim and Ryu (2003).

Regardless of which of the potential six sets of variables are used to define the task space, a constrained dimensionally homogeneous Jacobian matrix may be obtained. Similar to that of the screw based Jacobian, the condition number of each of the six constrained dimensionally homogeneous Jacobian matrices climbs dramatically at direct singular configurations, whereas the condition number of the unconstrained 3×6 does not.

In terms of measuring dexterity, the constrained dimensionally homogeneous Jacobian matrices ($\mathbf{J}'\mathbf{P}$) are superior to the screw based Jacobian matrix ($\mathbf{J}$) in that they are dimensionally consistent. Furthermore, the six matrices ($\mathbf{J}'\mathbf{P}$) are superior to the 3×6 dimensionally homogeneous matrix ($\mathbf{J}'$) in that they are constrained, and therefore provide true dexterous information.

The condition number and singular values of each of the six matrices ($\mathbf{J}'\mathbf{P}$) are different for any given pose. Therefore, dexterity measures involving only one of the six ($\mathbf{J}'\mathbf{P}$) matrices are potentially bias. Four potential strategies for dexterity measurement have been proposed based on the condition number and/or singular values of the Jacobian matrices obtained in all six cases. Each measure has a distinct physical meaning, as discussed. Further work on the subject will explore the results of these measures in obtaining dexterous workspace size.

Bibliography

J. Angeles. *Fundamentals of Robotic Mechanical Systems*. Springer - Verlag, 2003.

M. Arsenault and R. Boudreau. The synthesis of three-degree-of-freedom planar parallel manipulators with revolute joints (3-RRR) for an optimal singularity-free workspace. *Journal of Robotic Systems*, 21(5):259–274, 2004.

M. Badescu and C. Mavroidis. Workspace optimization of 3-legged UPU and UPS parallel platforms with joint constraints. *Journal of Mechanical Design*, 126(2):291–300, 2004.

C. M. Gosselin. *Kinematic Analysis, Optimization and Programming of Parallel Robotic Manipualtors*. Ph.D. Thesis, McGill University, Montreal, Canada, 1988.

C. M. Gosselin. The optimum design of robotic manipulators using dexterity indices. *Journal of Robotics and Autonomous Systems*, 9(4):213–226, 1992.

C. M. Gosselin and J. Angeles. The optimum kinematic design of a spherical three-degree-of-freedom parallel manipulator. *Journal of Mechanisms, Transmissions, and Automation in Design*, 111:202–207, 1989.

S. G. Kim and J. Ryu. New dimensionally homogeneous Jacobian matrix formulation by three end-effector points for optimal design of parallel manipulators. *IEEE Transactions on Robotics and Automation*, 19(4):731–737, 2003.

G. Pond and J. A. Carretero. Formulating Jacobian matrices for the dexterity analysis of parallel manipulators. *Submitted to Mechanism and Machine Theory*, 2005.

L.-W. Tsai. *Robot Analysis: The Mechanics of Serial and Parallel Manipulators*. John Wiley and Sons, Inc., New York, 1999.

Stiffness experimental monitoring for WL-16RII Biped Locomotor during walking[1]

Carbone G.[1] Ceccarelli M.[1] Sugahara Y.[2] Lim H.O.[3,4] Takanishi A.[2,4]

[1] LARM: Laboratory of Robotics and Mechatronics, University of Cassino, Cassino, Italy
[2] Graduate School of Science and Engineering, Waseda University, Tokyo, Japan
[3] Kanagawa University, Department of Mechanical Engineering, Yokohama, Japan
[4] Humanoid Robotics Institute, Waseda University, Tokyo, Japan

Abstract. In this paper, the importance of experimental evaluation of stiffness performance is pointed out for evaluating positioning errors and guaranteeing the effectiveness of a robotic walking system. Thus, a simple reliable procedure is proposed as based on a simplified version of Milli-CaTraSys (Milli Cassino Tracking System). Experimental tests are carried out on a biped locomotor that is named as WL-16RII (Waseda Leg No.16 Refined II) in dynamic conditions. The experimental tests provide useful information for both design and control purposes.

1 Introduction

Stiffness properties of a robot can significantly affect its accuracy and payload performance, as pointed out for example in (Duffy, 1996; Merlet, 1997; Tsai, 1999; Rivin, 1999; Ceccarelli, 2004). In fact, if stiffness properties are inadequate, external wrenches may cause undesired large compliant displacements at the robot end-effector. Thus, numerical and experimental evaluation of stiffness performance should be part of a general procedure for evaluating positioning errors and guaranteeing the effectiveness of a robotic system for given tasks, as proposed for example in (Vaishnav and Magrab, 1987; Ceccarelli and Carbone, 2005).

Several stiffness analyses and numerical evaluations of stiffness performance have been carried out for different robotic architectures such as in (Gosselin, 1990; Alici and Shirinzadeh, 2003; Carbone, 2003). But few preliminary comparisons of numerical results with experimental experience have been proposed like for example in (Kim and Streit, 1995; Ceccarelli and Carbone, 2002; Carbone et al. 2003; Carbone, 2003). The limited number of experimental results is mainly due to the lack of a simple reliable standard procedure for carrying out the necessary experimental tests. In fact, basic experimental approach consists in measuring a compliant displacement in one direction for a step external force that is applied along the same direction in static conditions. This approach can be tedious and affected by trial-to-trial repeatability errors.

---

[1] A part of this study has been carried out as part of the humanoid project at the Humanoid Robotics Institute, Waseda University, supported by the "The innovative research on symbiosis technologies for human and robots in the elderly dominated society", 21st Century Center of Excellence Program, Japan Society for the Promotion of Science.

A general procedure for experimental evaluation of stiffness performance has been proposed in (Carbone and Ceccarelli, 2004) as based on Milli-CaTraSys (Milli Cassino Tracking System). In this paper, a simplified version of Milli-CaTrasys has been installed on the WL-16RII Biped Locomotor in order to measure the linear compliant displacements occurring to a point on its foot during dynamic walking experiments.

2 Multi-purpose Bipedal Locomotor WL-16RII

WL-16RII (Waseda Leg No.16 Refined II), Fig.1, is one of the WL (Waseda Leg) series of bipedal robots that has been developed at Takanishi Lab., Waseda University, (Takanishi Lab, 2005; Sugahara et al., 2003; Sugahara et al., 2004; Sugahara et al., 2005). It has two 6-DOF parallel mechanism legs with only lower-limbs and a waist that can walk independently, as shown in Fig.1b). This biped locomotor is intended to be used in the future, for example, as a walking wheelchair that is able to climb and descend stairs carrying or assisting a human. In fact, WL-16 has been the first biped robot in the world able to perform a dynamic walking by carrying an adult woman or a 60 kg adult man (Sugahara et al., 2003), earlier than the release of Toyota Motor Corporation's Toyota Personal Robot (Toyota, 2004). More recently, WL-16R (WL-16 Refined) has been able to carry a human up to 94 kg weight (Sugahara et al., 2004), and WL-16RII has been able to ascend and descend stairs carrying a human of 60 kg weight (Sugahara et al., 2005).

Considering the proposed application task, one can easily understand the importance of the stiffness performance for WL-16RII, since any undesired motion could injure the human that is going to be carried or assisted. Therefore, numerical and experimental tests have been performed on a previous version (WL-15) in order to validate its stiffness performance in static conditions as reported for example in (Carbone et al., 2003). Nevertheless, dynamic conditions can strongly affect the stiffness performance. Thus, a new experimental set-up has been proposed in order to measure the stiffness performance also in dynamic conditions.

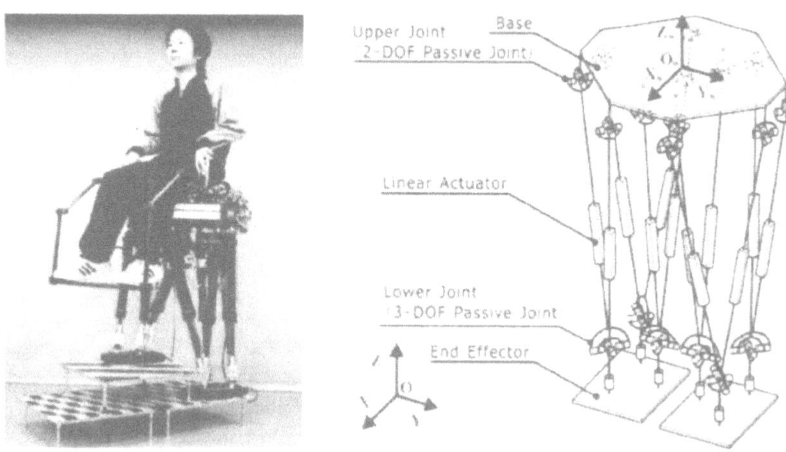

Figure 1. WL-16RII biped locomotor: a) a built prototype ascending stairs by carrying a human; b) a simplified kinematic scheme.

3 Experimental set-up

Figure 2 shows a scheme of the proposed experimental set-up. Its design and operation has been derived from Milli-CaTraSys design (Carbone and Ceccarelli, 2003; Carbone, 2003). It is composed of three Mutoh D-5400 linear encoder wire sensors that are attached on the pelvis of WL-16RII, as shown in Fig.3a). These sensors have an accuracy of about 0.05% of full scale. Moreover, a Nitta IFS67M25T50 force/torque sensor is embedded in each foot body as shown in Fig.3b). This sensor has an accuracy of about 0.01% of full scale.

The linear encoder wire sensors can measure the length of three wires, whose free end is connected with a fixed point H. This point H is located on the robot foot, as shown in Fig.3b). The three sensors are connected with a Contec 24Bit PCI Counter Board. This counter board is managed by means of a proper virtual instrument that has been developed in LabView environment.

The output of this virtual instrument is a data file with the length of the wires as function of time. If one considers the trilateration technique, the position of H on the robot end-effector can be defined as the position of the point at which three arcs intersect.

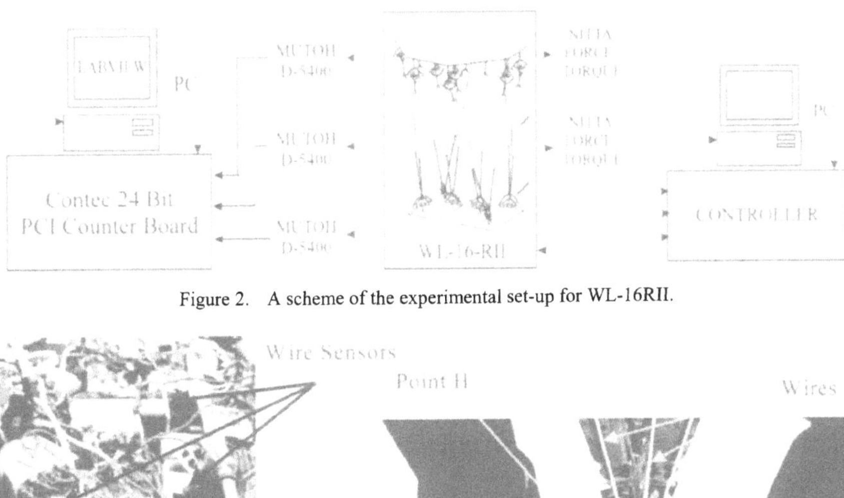

Figure 2. A scheme of the experimental set-up for WL-16RII.

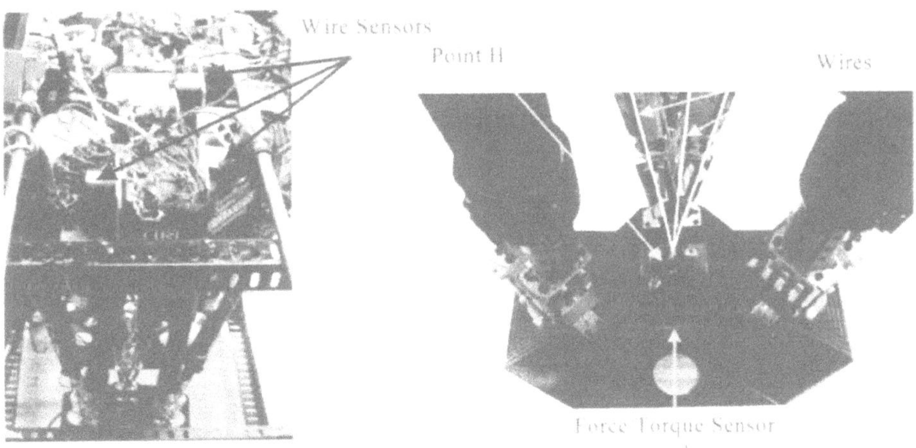

Figure 3. Location of the sensors on WL-16RII: a) the three wire sensors installed on the pelvis; b) the force/torque sensor on the foot.

Thus, referring to Fig.4, the position of point H with coordinates x_H, y_H and z_H can be determined from equations

$$u_i^2 = (x_H - x_{oi})^2 + (y_H - y_{oi})^2 + (z_H - z_{oi})^2 \tag{1}$$

where i=1,2,3 and x_H, y_H and z_H are the Cartesian coordinates of point H and x_{oi}, y_{oi}, z_{oi} are the coordinates of the points O_i, (i=1,2,3), which are the source of the u_i wire lengths. It is worth noting that these coordinates are defined with respect to a fixed frame OXYZ that is attached to the robot pelvis.

Suitable algebraic manipulations of Eq.(1) provide the position of the point H with respect to the fixed frame OXYZ as reported for example in (Carbone et al., 2003). Then, the data file with the length of the wires as function of time is used for computing the vector $\mathbf{S}=(x_H; y_H; z_H)^T$ that gives the position of point H as function of time in Matlab environment.

The force/torque sensors provide the values for the six components of the wrench F_x, F_y, F_z, T_x, T_y, and T_z that are acting on the foot as function of time. Thus, one can compute also the vector $\mathbf{W}=(F_x; F_y; F_z)^T$ as function of time.

Any increment in the wrench $\Delta\mathbf{W}$ produces a vector of compliant displacements $\Delta\mathbf{S}$. In the case that the compliant displacements are small enough, one can write the relationship between the vectors $\Delta\mathbf{W}$ and $\Delta\mathbf{S}$ as

$$\mathbf{W} = \mathbf{K}\ \Delta\mathbf{S} \tag{2}$$

where **K** is a 3x3 Stiffness Matrix. The 9 components of this matrix can be computed provided that the results of at least three experiments with different applied wrenches are available for a given robot pose as reported for example in (Carbone et al., 2003; Carbone and Ceccarelli, 2004). Thus, the 3x3 Stiffness Matrix of WL-16RII can be computed for any pose by carrying out at least three experimental tests for the same given walking pattern with different applied wrenches.

Once the stiffness matrix has been derived, it is necessary to be able to compare different stiffness matrices and estimate the stiffness performance of the overall system. A suitable comparison method could be obtained by using for example the determinant, trace, or eigenvalues of stiffness matrices. Also graphic representation of stiffness performance can be used through the definition of compliance ellipses and ellipsoids or through iso-stiffness curves, [Carbone, 2003].

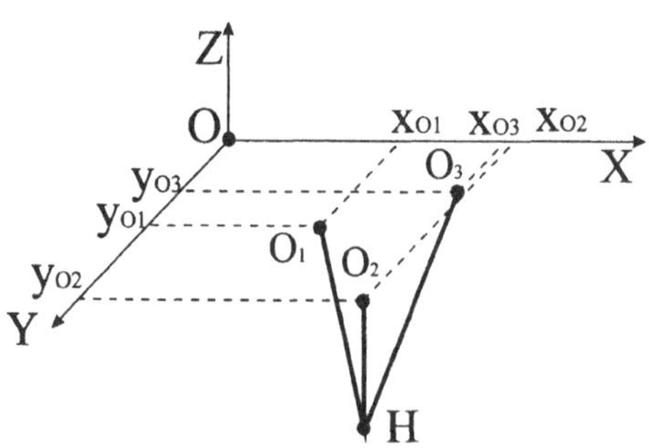

Figure 4. A scheme of the reference frame location.

Among the possible method we select to use in this paper the determinant of **K** for analyzing the stiffness behavior. In fact, the compliant displacements can be computed as

$$\mathbf{\Delta S} = \mathbf{K}^{-1} \quad \mathbf{W} \tag{3}$$

and each term k_{ij}^{-1} of the inverse matrix of **K** can be computed as

$$k_{ij}^{-1} = \frac{(\mathbf{K})_{ji}}{\det \mathbf{K}} \tag{4}$$

where $(\mathbf{K})_{ji}$ is the algebraic complement of the element k_{ij} of the matrix **K** with i, j=1,2,...,6. Thus, if the determinant det **K** is zero, the Eq.(4) gives singular values and Eq.(3) cannot be computed. Therefore, the determinant of **K** can be used as a performance index to investigate synthetically the effect of the design parameters on the stiffness behavior, since it is easy computable and particularly significant for determining stiffness singularity properties.

4 Experimental results

The experimental set-up that has been described in the previous section has been used for several experimental tests in different conditions and for different walking patterns. In particular, forward, backward, sideward walking, marching in place and turning experiments have been carried out. Each experiment has been carried out several times in order to provide statistically meaningful results. Moreover, experimental tests have been carried out for the same walking patterns with the waist of WL16-RII locked. In this way, it has been possible to eliminate the effect of the robot weight.

Figures 5 and 6 show an example of the experimental results that have been obtained. In particular, Fig.5 shows the components of the wrench that is applied to the right leg of WL16-RII during a three steps forward walking with a speed of 0.96 sec/step. The maximum measured force is about 700 N as shown in the plot of Fig.5a). This value is about 10% higher the WL16-RII due to dynamic effects.

Figure 6 shows the plots of x, y and z coordinates for point H versus time. In particular, the continuous line shows the desired position of point H as it is set in the pattern planning algorithm in the WL16-RII controller. The continuous dotted line is the measured position of point H for the same walking pattern during a walking experiment. The dotted line is the measured position for the same walking pattern but with the waist of WL16-RII locked. The comparison between the desired and measured walking pattern shows that there are significant overshooting phenomena due to the dynamic behavior of WL16 RII. Moreover, the comparison of the position of H with and without the waist locked shows that the effect of the weight produce a maximum increase in the compliant displacements that is of millimeter order. Although this value can be considered acceptable for the operation of WL16-RII a compliance compensation algorithm and design enhancements should be implemented in a near future.

The measured wrenches and compliant displacements as in Fig.5 and 6 have been used for computing the 3x3 stiffness matrix of WL-16RII at any pose. In particular, the stiffness matrix of WL-16RII can be computed as

$$\mathbf{K} = 10^5 \begin{bmatrix} -0.0223 & -0.0579 & -2.1145 \\ -0.0039 & -0.0172 & -0.0585 \\ 0.0713 & 0.2925 & 3.9867 \end{bmatrix} \tag{5}$$

when WL-16RII is close to its vertical configuration. Moreover, by using the experimental results in Figs.5 and 6 it has been possible to compute the determinant of the stiffness matrix **K** of WL-16RII as function of time during a 4 steps forward walking motion. This determinant is computed for each pose assume by WL-16RII during 4 steps forward walking motion as reported in Fig.7. The results of Fig.7 show that the stiffness matrix **K** of WL-16RII is singular when the leg is exactly in its vertical configuration. One can also note that the closer is the leg to a singular configuration the higher is the value of computed determinant. Moreover, minimum value of determinant is obtained when the leg is far from its singular configurations. In fact, singular configurations are usually used in human-like walking since they are the stiffest configurations.
Experimental tests have been carried out also in static conditions by applying different external wrenches to one leg of WL-16RII. Main results of these static experiments are reported in Tab.1. These experiments show that maximum compliant displacements about y direction can reach 3 mm when a maximum force of 100 N is applied about y direction. This value is not negligible and should be carefully considered for future control and design enhancements of WL-.16RII.

Table 1. Average values of compliant displacements that have been measured in static conditions.

F_x [N]	F_y [N]	F_z [N]	T_x [Nm]	T_y [Nm]	T_z [Nm]	Δx [mm]	Δy [mm]	Δz [mm]
91.7	-7.9	-14.1	-0.18	0.97	-1.10	1.81	-0.16	0.03
94.9	-12.0	-9.42	-0.18	0.33	-1.60	2.05	-0.11	0.03
82.7	-10.1	-12.0	-0.181	0.654	-1.323	1.05	-0.09	0.03
6.6	-86.2	-7.0	-2.07	-0.09	-1.35	0.30	-3.00	-0.06
5.5	-77.0	-7.0	-1.90	-0.10	-1.32	0.39	-2.50	-0.05
-2.2	-50.3	-4.8	-1.060	0.444	-5.898	0.15	-1.38	-0.03

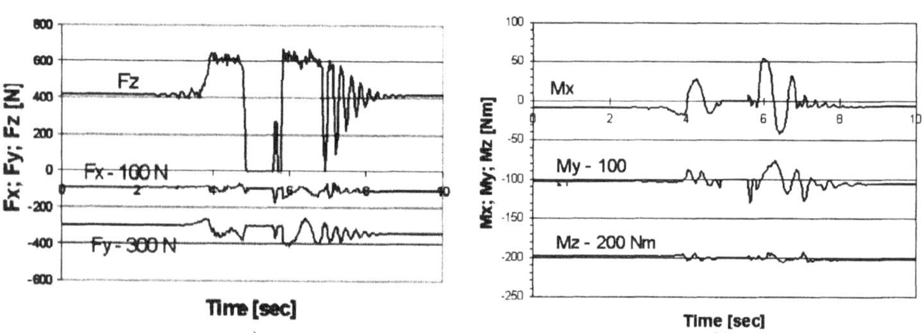

a) b)

Figure 5. Forces and torques measured on the right foot of WL-16RII during a three steps forward walking experiment (constant values of 100 and 200 have been subtracted to Fx, My, and Fy Mz for spreading out the various plot lines): a) Fx; Fy; Fz; b) Mx; My; Mz.

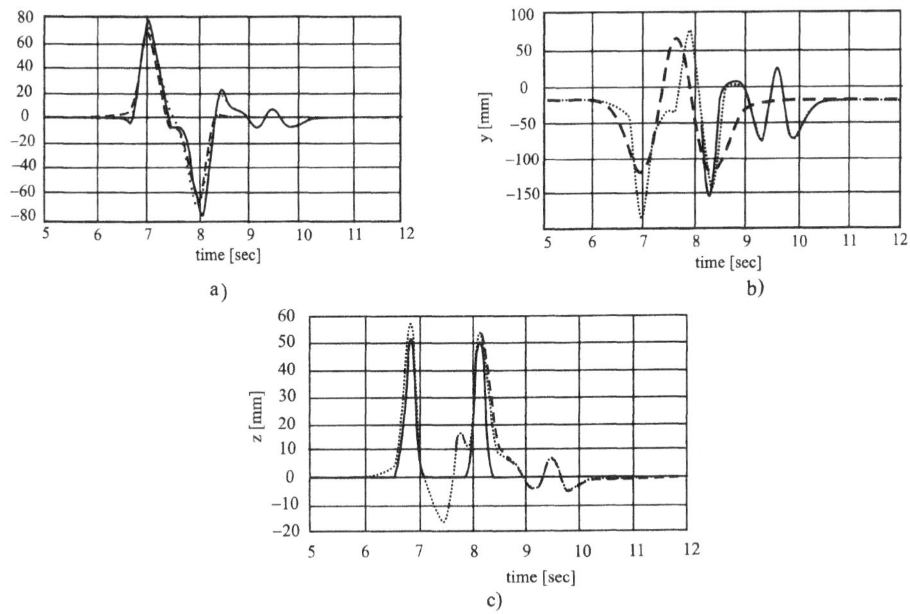

Figure 6. Displacements of point H on right foot of WL-16RII (continuous line theoretical; continuous dotted line no load experiment; dotted line same walking experiment of Fig.5): a) X-component; b) Y-component; c) Z-component.

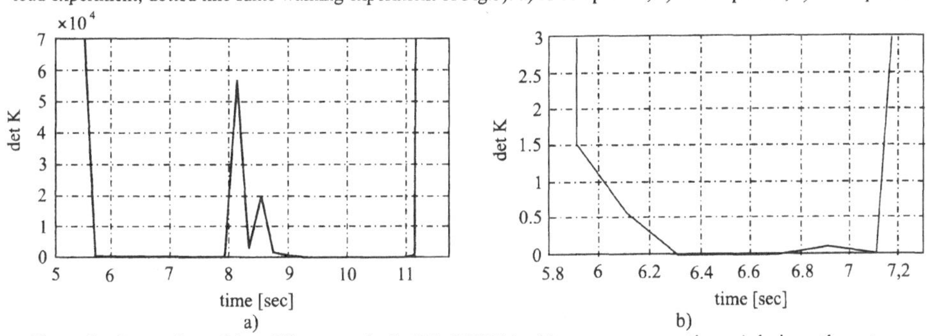

Figure 7. Determinat of the stiffness matrix for WL-16RII biped locomotor versus time: a) during a three steps forward walking motion; b) a zoomed view.

Conclusions

In this paper, an experimental set-up and procedure is proposed for monitoring the stiffness performance for the biped locomotor WL-16RII during dynamic walking experiments. Experimental tests have been carried out in several walking conditions. A 3x3 stiffness matrix has been computed for any pose assumed by WL-16RII during the walking and its determinant has been computed as stiffness performance index. Although the measured stiffness performance can be considered acceptable for the operation of WL16-RII a compliance compensation algorithm and design enhancements should be implemented in a near future.

Bibliography

J. Duffy, "Statics and kinematics with applications to robotics", Cambridge University Press, Cambridge, pp.153-169, 1996.

M. Ceccarelli, "Fundamentals of Mechanics of Robotic Manipulation", Kluwer Academic Publishers, Dordrecht, 2004.

J.-P. Merlet, "Les robots paralleles", Hermes, Paris, 1997.

L.W. Tsai, "Robot analysis: the mechanics of serial and parallel manipulators", John Wiley & Sons, New York, pp.260-297, 1999.

E.I. Rivin, "Stiffness and damping in mechanical design", Marcel Dekker Inc., New York, 1999.

R.N. Vaishnav, E.B. Magrab, "A general procedure to evaluate robot positioning errors", International Journal of Robotics Research, vol.6, n.1, pp.59-74, 1987.

C. Gosselin, "Stiffness mapping for parallel manipulators", IEEE Transactions on Robotics and Automation, vol.6, n.3, pp.377-382, 1990.

G. Alici, B. Shirinzadeh, "Exact stiffness analysis and mapping for a 3-SPS+S parallel manipulator", 7th International Conference on Automation Technology AUTOMATION 2003, Taiwan, paper F120, 2003.

G. Carbone, "Stiffness evaluation of multibody robotic systems", PhD dissertation, LARM, University of Cassino, Cassino, 2003.

M. Ceccarelli, G. Carbone, "A stiffness analysis for CaPaMan (Cassino Parallel Manipulator)", Mechanism and Machine Theory, vol.37, n.5, pp.427-439, 2002.

Carbone G., Sugahara Y., Lim H.O., Takanishi A., Ceccarelli M., "Stiffness Performances Estimation for Biped Locomotor WL-15", IEEE/ASME International Conference on Advanced Intelligent Mechatronics AIM 03, Kobe, pp.956-961, 2003.

G. Carbone, M. Ceccarelli, "A Procedure for Experimental Evaluation of Cartesian Stiffness Matrix", 15th CISM-IFToMM Symposium on Robot Design, Dynamics and Control ROMANSY 2004, Montreal, CD Proceedings, paper Rom04-24, 2004.

Ceccarelli M., Carbone G, "Numerical and experimental analysis of the stiffness performance of parallel manipulators", 2nd International Colloquium Collaborative Research Centre 562, Braunschweig, pp.21-35, 2005.

H.Y. Kim, D.A. Streit, "Configuration dependent stiffness of the Puma 560 manipulator: analytical and experimental results", Mechanism and Machine Theory, vol.30, n.8, pp.1269-1277, 1995.

Y. Sugahara, et al., "Control and Experiments of a Multi-purpose Bipedal Locomotor with Parallel Mechanism," Proc. of the IEEE ICRA 2003, pp. 4342-4347, Taipei, 2003.

Y. Sugahara, et al., "Realization of Dynamic Walking by a Biped Locomotor Carrying a Human," Proc. of the 9th RSJ/JSME/SICE Robotics Symposia, pp. 96-101, Okinawa, 2004.

Y. Sugahara, et al., "Realization of Dynamic Human-Carrying Walking by a Biped Locomotor," Proc. of the IEEE ICRA 2004, pp. 3055-3060, New Orleans, 2004.

Y. Sugahara, et al., "Walking Up and Down Stairs Carrying a Human by a Biped Locomotor with Parallel Mechanism," Proc. of the IEEE/RSJ IROS 2005, Edmonton, 2005.

Toyota Motor Corporation webpage, http://www.toyota.co.jp/en/special/robot/index.html, 2004.

Takanishi Lab. Webpage, http://www.takanishi.mech.waseda.ac.jp/, 2005.

LARM Lab. Webpage, http://webuser.unicas.it/weblarm/larmindex.htm, 2005.

Estimation of Leg Stiffness Parameters of a 6DOF Parallel Mechanism

Józef Knapczyk [1] and Michał Maniowski [1]

[1] Cracow University of Technology, Institute of Automobiles and Internal Combustion Engines, Poland, e-mail: J_kn@mech.pk.edu.pl, mmaniowski@poczta.onet.pl

Abstract. The paper describes a novel algorithm for estimation of the stiffness parameters of a 6DOF parallel mechanism with a known topology and geometry. The mechanism is composed of a platform guided by six or more legs with Spherical-Prismatic-Universal joints. A substitute compliance of each leg is modelled by linear springs in the P-joints. Such structure has a broad application in robotics, machining tools, and car suspension systems. The stiffness parameters of the legs are considered as unknown or hard to measure directly. These parameters are estimated using as an input data the spatial displacements of the platform from an initial pose, measured under specified quasi-static loads. The presented method is exemplified on a five-rod wheel guiding mechanism of an actual passenger car. Results of the mechanism measurements carried out on a test stand are used. Five static rates of elastomeric bushings installed in the mechanism rods are determined taking into account the influence of the measurement noise.

1 Introduction

Establishing the relationship between the mechanism parameters and the mechanism characteristics is the fundamental issue in solving many technical problems. When designing a mechanism to perform a given task the object parameters (geometric, stiffness, damping, mass, etc.) are usually adjusted using the synthesis procedure. For example, in the dimensional synthesis a desired load-free trajectory can be achieved by optimising the geometric parameters (Knapczyk and Maniowski, 2003). A desired load-displacement characteristic of the mechanism can be obtained applying the stiffness synthesis (Knapczyk and Maniowski, 2006-a, Thomas et al., 1985). When accuracy is an important performance specification of a mechanism, the kinematic calibration is a practical and economical way of enhancing this characteristic. In the calibration process the end-effector position error is predicted and compensated using the mechanism model described by parameters with direct or substitute character. These parameters (geometric and/or stiffness) are identified on the basis of results of the end-effector pose measurements under selected conditions (Daney and Emiris, 2004, Maggiolaro et al., 2005). In the case when some parameters are unknown or hard to measure directly, procedure similar to that used for calibration can be applied to estimate them.

In this paper a novel algorithm for estimation of the stiffness parameters of a 6DOF parallel mechanism with a given topology and geometry is presented. The considered mechanism belongs to the group of Stewart/Gough platforms (see Davidson and Hunt, 2004), which have found a broad application in robotics, motion simulators, machining tools, measuring machines, and car suspension systems among the others. Kinematic scheme of this spatial mechanism is composed of a rigid platform guided by six legs with Spherical-Prismatic-

Universal joints. The P-joint can be driven by an actuator or supported by a spring element. The S and U-joints are usually passive.

Depending on the mechanism application, its structural compliance is influenced by different design parameters. The mechanism leg design usually plays the most significant role in this regard. Using the algorithm described herein, it is possible to estimate a substitute stiffness of any leg for different positions of the mechanism and load conditions.

The presented procedure may be separated into three steps: (i) modelling, (ii) measurements and data preparation, and (iii) identification. A numerical example is solved for a wheel guiding mechanism with five rods made of steel sheet and ended by elastomeric bushings. The bushings stiffness parameters are estimated using as an input data the spatial displacements of the wheel carrier from an initial pose under specified quasi-static loads. Sensitivity of the algorithm on data/measurements errors is shown.

2 Problem Statement and Solution

Some compliant structures or mechanisms can be modelled as a parallel system, presented in Figure 1, with the moving platform (reference system $\{0^b x^b y^b z^b\}$) constrained with respect to the fixed platform (with reference system $\{0xyz\}$) by n-legs ($n \geq 6$ for a spatial case). Each leg is considered as a serial kinematic chain with S-P-U joints. The P-joint with a translational spring substitutes a longitudinal compliance of the whole leg.

It is assumed that all the joints and links are ideal and fully dimensioned; the springs are not preloaded and can be described by linear force-deflection characteristics. The moving platform is initially at rest in a given pose, described by a position vector ($\boldsymbol{o}^b$) and orientation matrix ($\boldsymbol{R}$), relatively distant from a singular configuration. When the load state exerted on the platform is changed with a selected magnitude and in quasi-static fashion, the moving platform undergoes a displacement dependent on the structural compliance.

For simplicity, it is assumed that the base reference frame initially coincides with the frame of the moving platform (Figure 1). The following spatial vectors (six-by-one): twist and wrench, are used to describe an infinitesimal displacement and quasi-static increment of the external load, respectively (Davidson and Hunt, 2004, Knapczyk and Maniowski, 2006-b):

$$\hat{\boldsymbol{t}} = [\boldsymbol{q}^T \, \boldsymbol{\varepsilon}^T]^T \, ; \; \hat{\boldsymbol{w}} = [\boldsymbol{f}^T \, \boldsymbol{m}^T]^T \tag{1}$$

where the Cartesian vectors (three-by-one) describe respectively: $\boldsymbol{q}$ – translation, $\boldsymbol{\varepsilon}$ – infinitesimal rotation, $\boldsymbol{f}$ – force increment, $\boldsymbol{m}$ – torque increment. The twist is represented in *axis-coordinates* , and the wrench in *ray-coordinates*. Both twist and wrench in equation (1) can be represented by the *screw coordinates* (Davidson and Hunt, 2004).

Static spring rates of the mechanism legs can be collected in a column vector $\boldsymbol{k}$, with the following two parts separated as follows:

$$\boldsymbol{k} = [k_1 \; k_2 \ldots k_n]^T = [\, \boldsymbol{k}_\alpha^{\;T} \; \boldsymbol{k}_\beta^{\;T}]^T ; \quad n \geq 6; \quad n = \alpha + \beta \tag{2}$$

where $\boldsymbol{k}_\alpha$ includes coefficients that are to be determined, and $\boldsymbol{k}_\beta$ – that are known.

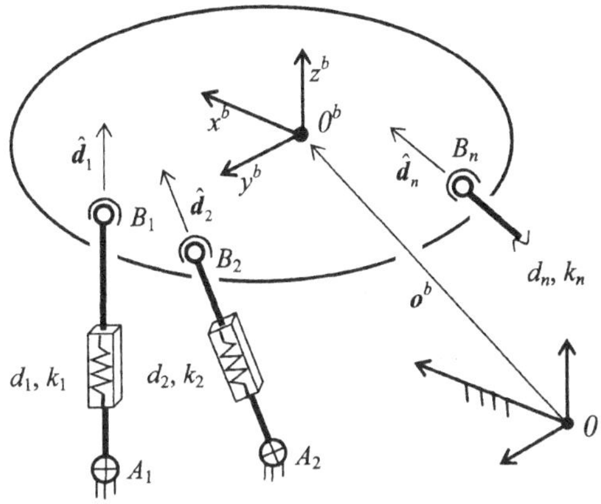

Figure 1. Rigid platform constrained by S-P-U legs.

Each leg (Figure 1) can be described by *line coordinates* (Davidson and Hunt, 2004), considered as the screw with zero pitch, which are given in *ray-coordinates* as follows:

$$\widehat{\boldsymbol{l}}_i = [\hat{\boldsymbol{d}}_i^T \;\; (\boldsymbol{b}_i \times \hat{\boldsymbol{d}}_i)^T]^T; \quad i = 1, 2 \ldots n \tag{3}$$

The inverse problem of the first-order kinematics for an in-parallel actuated mechanism (Davidson and Hunt, 2004, Knapczyk and Dzierzek, 1995) can be described by:

$$\delta\boldsymbol{d} = \boldsymbol{J}\,\widehat{\boldsymbol{t}} \tag{4}$$

where $\delta\boldsymbol{d}$ is a vector (n-by-one) of incremental changes of each leg-length. Successive i-th row of the Jacobian matrix $\boldsymbol{J}$ (n-by-six), in Eq. (4), contains the transposed line coordinates (Eq. 3) of the i-th leg. Therefore, corresponding change of the i-th leg-length (δd_i) is a scalar product of the i-th leg-line (Eq. 3) with the platform twist (Eq. 1).

Using the principle of virtual work, the stiffness problem of the platform support (Figure 1) (Carbone and Ceccarelli, 2004, Knapczyk and Maniowski, 2006-b) can be expressed as follows:

$$\widehat{\boldsymbol{w}} - \boldsymbol{K}\,\widehat{\boldsymbol{t}} \tag{5}$$

where $\boldsymbol{K}$ is the position dependent stiffness matrix, describing a linear mapping of the body twist into the wrench. The stiffness matrix is defined as follows:

$$\boldsymbol{K} = \boldsymbol{J}^T \mathrm{diag}(\boldsymbol{k})\,\boldsymbol{J} = \sum_i \widehat{\boldsymbol{l}}_i \,\widehat{\boldsymbol{l}}_i^T k_i\,; \quad i = 1, 2 \ldots n \tag{6}$$

Equation (5), when properly rearranged, can be rewritten as:

$$\hat{w} = \boldsymbol{H}\,\boldsymbol{k} = \boldsymbol{H}_{\alpha}\boldsymbol{k}_{\alpha} + \boldsymbol{H}_{\beta}\boldsymbol{k}_{\beta} \tag{7}$$

where $\boldsymbol{H}$ is a design matrix (six-by-n). This position and twist dependent matrix is composed of n columns defined as:

$$\hat{l}_i \hat{l}_i^T \hat{t}\,; \quad i = 1, 2 \ldots n \tag{8}$$

The matrices $\boldsymbol{H}_{\alpha}$ and $\boldsymbol{H}_{\beta}$ in Eq. (7) correspond to the two partitions of the vector $\boldsymbol{k}$ in Eq. (2).

Equation (7) describes an analytical relation between unknown and given parameters, therefore it can be used to solve the formulated problem. In order to determine α spring rates ($\boldsymbol{k}_{\alpha}$), suitable number (p) of independent components of the measured platform twist ($\hat{t}$) under specified load ($\hat{w}$) must be given. Solution of Eq. (7) can be directly determined by a linear solver (e.g., Matlab, 2000). However, in order to avoid impractical solutions the problem is solved by using the constrained linear least-squares algorithm. Depending on the quantity of the input information the system can be underdetermined ($p < \alpha$), determined ($p = \alpha$) or overdetermined ($p > \alpha$). The last case, due to its averaging character, is especially recommended for dealing with uncertain data, e.g., from measurements.

3 Numerical Example

In order to illustrate the formulated algorithm a numerical example is solved for a wheel guiding mechanism with the suspension coil spring, pneumatic tire, and five rods made of steel sheet and ended by elastomeric bushings.

3.1 The Mechanism Description

The elastokinematical model assumed for the studied five-rod mechanism for guiding a car wheel (Knapczyk and Dzierzek, 1995, Knapczyk and Maniowski, 2006-b), is shown in Figure 2. It can be considered as a special case of the model from Figure 1. The reference system $\{O^b x^b y^b z^b\}$, with its origin at the wheel centre point, is attached to the wheel carrier, where y^b is the wheel spin axis.

Main (gross) displacement of the wheel carrier (moving platform) relative to the car body (fixed platform) depends on geometric constraints related to the five rigid binary rods with S-U joints with centres at points B_i and A_i respectively, i = 1, 2...5. Magnitude of this displacement is controlled by two elastic forces in the suspension spring (steel coil spring) and pneumatic tire. The suspension spring is considered as the compliant leg no 6 (Figure 2) with the spring rate k_6. The road wheel is instantaneously modelled as the leg no 7 (Figure 2). Radial elasticity of the tire is modelled by a linear spring element with the static rate k_7 (wheel ride rate). Both spring rates (k_6 and k_7) are assumed to be known in the analysis.

The elastomeric bushings of the actual rods are responsible for an additional small displacement of the wheel carrier under action of an external load. Radial stiffnesses of the rod-bushings are reduced to the rod longitudinal axis (Figure 2) with substitute spring rates k (i = 1, 2…5), which are to be determined by the estimation procedure.

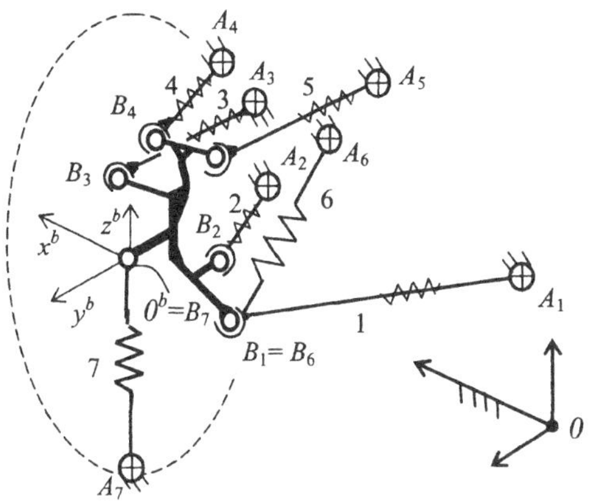

Design position, $\boldsymbol{R} = \boldsymbol{I}$, $\boldsymbol{o}^b = \boldsymbol{b}_7$, $\boldsymbol{b}_6 = \boldsymbol{b}_1$			
$\boldsymbol{a}_1$	−103.2	−490.6	−65.6
$\boldsymbol{a}_2$	309.3	−204.6	−49.6
$\boldsymbol{a}_3$	200.8	−283.6	−11.6
$\boldsymbol{a}_4$	205.8	−269.6	95.4
$\boldsymbol{a}_5$	−1.7	−349.6	118.4
$\boldsymbol{a}_6$	−87.7	−55.7	154.1
$\boldsymbol{a}_7$	0	5.3	−284.9
$\boldsymbol{b}_1$	−39.6	−52.9	−90.3
$\boldsymbol{b}_2$	46.8	−82.6	−126.7
$\boldsymbol{b}_3$	141.8	−50.4	−26.2
$\boldsymbol{b}_4$	67.2	−57.1	95.3
$\boldsymbol{b}_5$	−6.3	−47.7	121.7
$\boldsymbol{b}_7$	0	0	0
$k_6 = 0.015$; $k_7 = 0.171$ [kN/mm]			

Figure 2. Elastokinematical model of the five-rod suspension (left side, rear wheel). Dimensions in [mm].

Further, the following assumptions are made: (i) an external load is reduced on the wheel centre (O^b), and (ii) the car body is fixed with respect to the road/base. This kind of the axle support complies with typical conditions of measurements at Kinematic and Compliance (K&C) test stand. Under these conditions, the wheel carrier with 6 DOF is supported with respect to the base on seven legs with substitute stiffness (like in Figure 1 when $n = 7$), corresponding to: the five rods (k_1 to k_5); the main spring (k_6); and the tire (k_7). Characteristic feature of the considered suspension mechanism is a significant diversification of the stiffness parameters ($k_7/k_6 \approx 10^1$; $k_i/k_7 \approx 10^1$; $i = 1, 2\ldots5$).

Dimensions and coordinates of the mechanism links and joints were estimated (see Knapczyk and Dzierzek, 1995, Knapczyk and Maniowski, 2006-b) by measurements of the individual components. This data is given in Figure 2. In order to make possible the evaluation of the presented algorithm accuracy, the substitute spring rates (k_i, $i = 1, 2\ldots5$) of the bushings were also estimated on the basis of direct measurements of each rod dismounted from the mechanism. The determined force-radial deflection characteristics can be described in normal operating range by constant static spring rates. These values are used further for reference only.

3.2 Input Data Preparation

The input data to the estimation algorithm is composed of spatial displacements of the wheel carrier caused by quasi-static loads exerted on the mechanism at K&C test rig, which is described in detail in Knapczyk and Maniowski, 2006-b. At the test rig, the road wheel was supported on a rig platform that allows inducing horizontal forces at the tire contact patch. The wrench ($\hat{w}$) applied to the wheel carrier was estimated on the basis of the measured components of the load. The wheel carrier displacement ($\hat{t}$) with respect to the rig base was estimated on the basis of measurements with use of a 6-DOF sensor with a serial kinematic chain. All rig experiments were performed with the wheel and carrier fixed together by a brake. The range of the load was selected properly in order to maintain the analysis assumptions.

Table 1. Mean values and tolerances (3σ confidence interval) for the wheel knuckle twist caused by the applied wrench (SI units).

No	Wrench	Twist
1.	$\widehat{w}_{m,1} = [-1000\ \ 0\ \ 0\ \ 0\ \ 280\ \ 0]^T$	$\widehat{t}_{m,1} = [0.0004\ 0.0003\ 0.0021\ {-0.0024}\ 0.0139\ {-0.0010}]^T$
2.	$\widehat{w}_{m,2} = [\ 0\ \ 1000\ \ 0\ \ {-280}\ \ 0]^T$	$\widehat{t}_{m,2} = [0.0005\ 0.0001\ {-0.0009}\ {-0.0041}\ 0.0032\ {-0.0002}]^T$
	$e_f = [I]_{3\times1}\,5$ [N]; $e_m = [I]_{3\times1}\,1$ [Nm];	$e_q = [I]_{3\times1}\,1e-4$ [m]; $e_\varepsilon = [I]_{3\times1}\,8.7e-4$ [rad];

Two samples of the obtained wrench-twist (mean values) are given in Table 1. The first one corresponds to a car braking manoeuvre, the second – cornering. In order to study an influence of input data errors on the estimation results an additive noise is taken into account, as expressed in Eq. (9):

$$\widehat{t}_i = \widehat{t}_{m,i} + \left[e_q^{\ T}\ e_\varepsilon^{\ T}\right]^T \zeta\ ;\quad \widehat{w}_i = \widehat{w}_{m,i} + \left[e_f^{\ T}\ e_m^{\ T}\right]^T \zeta\ ;\ i = 1, 2, \ldots g \tag{9}$$

It is assumed that the noise on the measurements can be modelled as a normal random process with zero mean and constant standard deviation (σ). The assumed tolerances for each component of the wrench and twist are given in Table 1 as 3σ confidence intervals. Additionally, a noise amplification coefficient (ζ), which ranges from 0 to 1, is included in Eq. (9) for changing a noise level.

3.3 Estimation Results

The described algorithm was implemented in Matlab. The problem parameters are the following: $n = 7$; $\alpha = 5$; $\beta = 2$; $g = 2$; $p = 12$. For each mean value from the measurements (Tab. 1) and the selected noise level (ζ) a set of 50 twists and wrenches was generated according to the formula (9). Each time, Eq. (7) is solved to obtain the vector $\boldsymbol{k}_\alpha$ (k_i, $i = 1, 2 \ldots 5$), which is further processed. The procedure is analogous to the Monte Carlo analysis (Scheaffer and McClave, 1986). For a Pentium PC 700 MHz, total computing time was about a few seconds.

The estimation results are presented in Figure 3 using the so-called box and whiskers plot (Scheaffer and McClave, 1986). Each set of the observations is described by a vertically arranged box, which is bounded from top by upper quartile (Q_3), and from bottom by lower quartile (Q_1) of the data. The height of the box is therefore equal to the interquartile range (IQR). The line inside the box corresponds to the data median (Q_2). The whiskers, i.e., the vertical lines ended by short horizontal segments, cover a range of data from minimum value to maximum, but with the condition that the whiskers cannot be longer than Q_3 + 1.5 IQR for upper whisker, and Q_1 – 1.5 IQR for lower whisker. When some observations stand out of the whiskers, they are plotted separately (marked by crosses) and are treated as outliers.

Probability distribution of the obtained results (Figure 3) slightly differs from the normal one due to nonlinearities in the estimation algorithm. In the case of *exact* input data (for $\zeta = 0$) the computed spring rates approach the mean values determined on the basis of direct measurements (horizontal lines in Figure 3). As the noise level increases ($\zeta > 0$) scatter of the results grows proportionally.

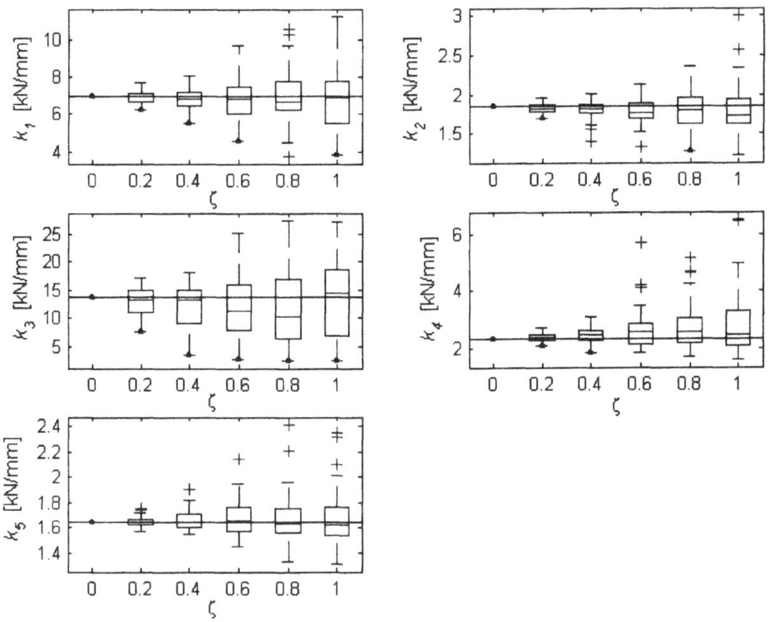

Figure 3. Box and whiskers plots of spring rates (k_i, $i = 1, 2 \ldots 5$) of the mechanism legs determined by the estimation procedure for increasing level of measurements noise (ζ). Horizontal lines correspond to the mean values from direct measurements.

Uncertainty of some estimates (k_1 and k_3) is higher than the rest. It is related to the problem conditioning. The condition number of the design matrix $\boldsymbol{H}$ in Eq. (7) depends on the mechanism position, a direction of the exerted wrench, and a number of different load scenarios.

4 Conclusions

A novel algorithm for estimation of the stiffness parameters of 6DOF parallel mechanism with six or more legs with S-P-U joints is presented. A substitute compliance of any leg is modelled by a linear spring in the P-joint. The formulated algorithm was successfully used for estimation of static rates of the elastomeric bushings installed in rods of the multilink suspension mechanism with given structure and dimensioning. Using the described approach it is possible to determine the bushings rates for different positions of the mechanism and load conditions. The measured spatial displacements of the wheel carrier caused by quasi-static loads exerted on the mechanism at K&C are needed only.

The algorithm is formulated using the principle of virtual work and screw calculus. Main assumptions concern infinitesimal displacements and linear force-deflection characteristics of the mechanism elastic components. The implemented algorithm is simple and computationally effective. Therefore it is well suited for an on-line estimation. In the case of input data with errors (measurement noise) the algorithm demonstrates proportional to the noise level increase of the estimates uncertainty.

The described approach can be useful for various applications of Stewart/Gough platform, e.g.: machining tools; force sensors with spring-loaded legs; or a passive compliance device with six or more legs. The leg stiffness parameters, estimated using the presented procedure, can be used to predict elastic deflections of the platform, or decrease the positioning error.

Currently, an algorithm for simultaneous estimation of stiffness and geometric parameters for similar mechanisms is being developed.

Notations

T	(upper right-hand) transposition operator;
diag($\boldsymbol{c}$)	diagonal matrix with entities composed of some vector $\boldsymbol{c}$;
$\hat{c}$	unit vector;
$\widehat{c}$	spatial vector;
$\times$	cross product.

Bibliography

Carbone, G., and Ceccarelli, M. (2004). A procedure for Experimental Evaluation of Cartesian Stiffness Matrix of Multibody Robotic Systems. In *Proc. ROMANSY*, Montreal, Canada, 1-9.

Daney, D., and Emiris, I.Z. (2004). Identification of Parallel Robot Kinematic Parameters under Uncertainties by Using Algebraic Methods. Proc. of *the 11th World Congress in Mechanism and Machine Science*, Tianjin, China.

Davidson, K., and Hunt, K. (2004). *Robots and Screw Theory*. Oxford University Press.

Knapczyk, J., and Dzierżek, S. (1995). Displacement and force analysis of five-rod suspension with flexible joints. *Trans. ASME, Journal of Mech. Design*, Vol. 117, 532-538.

Knapczyk, J., and Maniowski, M. (2006-a). Synthesis of a five-rod suspension for given load-displacement characteristics. *Proc. Instn. Mech. Engrs, Part D: J. Automobile Eng*, paper submitted for publication.

Knapczyk, J., and Maniowski, M. (2003). Dimensional synthesis of a five-rod guiding mechanism for car front wheels. *The Archive of Mechanical Engineering*, Vol. 50, No 1, 89-116.

Knapczyk, J., and Maniowski, M. (2006-b). Elastokinematic modeling and study of five-rod suspension with subframe. *Mechanism and Machine Theory*, paper accepted for publication in vol. 41.

Maggiolaro, M.A., Dubowski S., and Mavroidis C. (2005). Geometric and elastic error calibration of a high accuracy patient positioning system. *Mechanism and Machine Theory*, 29, 415-427.

Thomas, M., Yuan-Chou, H.C., and Tesar, D. (1985). Optimal Actuator stiffness distribution for robotic manipulators based on local dynamic criteria. *Proc. of IEEE Int. Conf. on Robotics and Automation*, 275-281.

Using MATLAB, Version 6 (2000). *The Math Works Inc.*

Scheaffer, R., and McClave J. (1986). *Probability and Statistics for Engineers*. PWS-KENT, Boston.

Invariant Error Dynamics Controller for a 6-dof Electrohydraulic Stewart Platform

Ioannis Davliakos and Evangelos Papadopoulos [1]

National Technical University of Athens, Department of Mechanical Engineering
15780 Athens Greece
e-mail: {gdavliak | egpapado@central.ntua.gr}

Abstract. The study of an invariant error dynamics controller for a six-degrees-of-freedom (dof) electrohydraulic Stewart platform is presented. Rigid body and electrohydraulic models, including servovalve dynamics are employed. Friction is also included in the mechanical model. The developed controller employs the dynamic and hydraulic model of the system and yields the six servovalve input current vector, in analytical form. Using mechanism inverse kinematics, the desired Cartesian trajectories yield desired actuator length trajectories. Simulations with typical desired trajectories are presented and a good performance of the controller is obtained.

1 Introduction

Stewart platform type (Stewart, 1965-66) parallel manipulators have been studied in their kinematics (see, e.g., Shim et al., 1997, Liu et al., 2000, Gao et al., 2005) and dynamics (Lebret et al., 1993, Tsai, 1999, 2000). Although electrohydraulic Stewart platforms have been used in the past, even commercially–mainly in aircraft simulations, not much published work on their dynamics and control (see, e.g., Li and Salcudean, 1997, Sirouspour and Salcudean, 2000, Kim et al., 2000) exists.

Hydraulics science combined with controls, has given new thrust to hydraulics applications. The main reasons why hydraulics are preferred to electromechanical drives in some industrial and mobile applications, include their ability to produce large forces at high speeds, their high durability and stiffness, and their rapid response (Merritt, 1967). Hydraulic systems differ from electromechanical ones, in that the force or torque output is not proportional to actuator current and therefore, hydraulic actuators cannot be modelled as force /torque sources. As a result, controllers that have been designed for robot control, assuming the capability of setting actuator force/torque, cannot be used here.

Control techniques are used to compensate for the nonlinearities of electrohydraulic servosystems. Nonlinear adaptive control techniques for hydraulic servosystems have been proposed by researchers assuming linearization (Garagic and Srinivasan, 2004) and backstepping (Sirouspour and Salcudean, 2001), approaches. The modelling of an experimental hydraulic robot arm and the implementation of a model-based motion controller that compensates for dynamic

[1] Support by the EPAN Cooperation Program 4.3.6.1 (Greece-Poland) of the Hellenic General Secretariat for Research and Technology and the NTUA Senate Committee of Basic Research, Programme "Protagoras", R.C. No. 10, is acknowledged.

forces have been presented by Honegger and Corke (2001). A tracking controller for electrohydraulic servosystems has been developed (Davliakos and Papadopoulos, 2005) including a fast model-based force tracking loop.

Nguyen et al. (1992) have developed a joint-space adaptive control scheme applied to an electromechanically driven Stewart platform-based manipulator, using the Lyapunov direct method under the assumption that Stewart platform motion was slow compared to the controller adaptation rate. Also, Kim and Lee (1998), studied and applied a high speed tracking control of a 6-6 electric Stewart platform, using an enhanced sliding mode control approach.

Further, the modelling and control of an inverted, ceiling-mounted electrohydraulically driven Stewart platform has been researched (Li and Salcudean, 1997), using the virtual work principle and a pressure feedback control approach. Extended work of the same mechanism has been studied and applied (Sirouspour and Salcudean, 2000), in which the Lyapunov analysis approach has been used for a nonlinear controller. A robust tracking control design for a 6-dof hydraulically driven Stewart type mechanism has been developed (Kim et al., 2000) and has been achieved, using two types of controllers, which were based on Lyapunov analysis.

In this paper, an invariant error dynamics controller for a 6-dof electrohydraulic Stewart platform is developed. Dynamic models that describe the rigid body equations of motion and the hydraulic dynamics of the main elements are given. A detailed friction model is included. The developed control scheme employs the dynamic and hydraulic model of the system and yields the six servovalve input current vector, in analytical form. Using mechanism inverse kinematics, the desired Cartesian trajectories yield desired actuator length trajectories. Simulations with typical desired trajectories are presented and a good performance of the controller is obtained. The approach can be applied to hydraulic manipulators and aircraft simulator control systems.

2 Dynamic Modelling of the Six - Dof Electrohydraulic Stewart Platform

In this section, the dynamic model of a 6-dof electrohydraulic Stewart platform servomechanism (Stewart, 1965-66) is developed. This is a six dof closed kinematic chain mechanism consisting of a fixed base and a movable platform with six linear actuators supporting it. The mechanism is illustrated schematically in Figure 1.

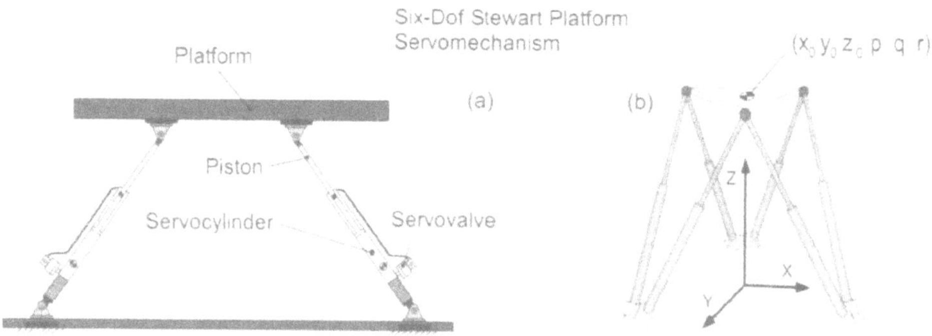

Figure 1. (a) Drawing of two out of the six servoactuators, (b) Schematic of a six-dof Stewart Platform.

A full servosystem model includes the moving mass equation of motion. This system provides a relation between the actuator torques/forces and the resulting motion. The equation of motion for the Stewart platform system is derived applying a Lagrangian formulation and is written as

$$\mathbf{M}(\boldsymbol{l})\ddot{\boldsymbol{l}} + \boldsymbol{V}(\boldsymbol{l},\dot{\boldsymbol{l}}) + \boldsymbol{G}(\boldsymbol{l}) + \boldsymbol{F}_{fr}(\dot{\boldsymbol{l}}) = \boldsymbol{F}_p \tag{1}$$

where $\boldsymbol{l}$ is the 6×1 displacement vector of the mechanism actuators, $\mathbf{M}(\boldsymbol{l})$ is the 6×6 positive definite mass matrix of the system, the 6×1 vector $\boldsymbol{V}(\boldsymbol{l},\dot{\boldsymbol{l}})$ represents forces/torques arising from centrifugal and Coriolis forces, the 6×1 vector $\boldsymbol{G}(\boldsymbol{l})$ represents torques due to gravity, $\boldsymbol{F}_{fr}(\dot{\boldsymbol{l}})$ is the 6×1 vector of the forces/torques due to friction and $\boldsymbol{F}_p$ is the 6×1 vector of the actuator forces $F_{p,j}$, $j = 1,2,...,6$ (see section 3).

A number of methods exists, that model the friction vector $\boldsymbol{F}_{fr}(\dot{\boldsymbol{l}})$, (Helouvry et al., 1994). A widely used method to compute its elements is used here, (Papadopoulos and Chasparis),

$$F_{fr}(\dot{l}_j) = \begin{cases} f_{C,j}\,\mathrm{sgn}(\dot{l}_j) + b_j\dot{l}_j, & \dot{l}_j \neq 0 \\ f_{ext,j}, & |f_{ext,j}| < f_{s,j},\ \dot{l}_j = 0,\ \ddot{l}_j = 0 \\ f_{s,j}\,\mathrm{sgn}(f_{ext,j}), & |f_{ext,j}| > f_{s,j},\ \dot{l}_j = 0,\ \ddot{l}_j \neq 0 \end{cases} \tag{2}$$

where l_j is the j actuator displacement, b_j, $f_{C,j}$, $f_{ext,j}$, $f_{s,j}$ are the j parameters of viscous friction coefficient, Coulomb friction, external force and breakaway force, which is the limit between static and kinetic friction, respectively, for $j = 1,2,...,6$, and sgn is the sign function.

If the motion of the platform is specified in the 6-dof Cartesian space (x_0, y_0, z_0, p, q, r), where x_0, y_0, z_0 are the Cartesian generalized coordinates of the platform centre of mass (CM) and p, q, r are the Euler angles, then inverse kinematics must be used to determine the required leg lengths that correspond to the desired motion.

3 Electrohydraulic Servosystem Modelling

In this section, the dynamic modelling of high performance electrohydraulic servocylinders is presented briefly. An electrohydraulic servosystem consists of a servomechanism, including servovalves, servoactuators, controllers, mechanical loads and a hydraulic power supply. Next, simple models of major components are described.

An ideal single rod hydraulic cylinder is described by

$$Q_r = A_r\dot{l}, \quad r = 1,2 \tag{3a}$$

$$p_1A_1 - p_2A_2 = F_p \tag{3b}$$

where Q_r are the flows through its two chamber ports, p_1, p_2 are the chamber pressures, A_1 is the piston side area, A_2 is the rod side area, l is the piston displacement and F_p is the piston output force. A real cylinder model also includes chamber oil compressibility, chamber leakages and other effects. However, these can be neglected at an initial stage.

A typical hydraulic servovalve consists of four symmetric and matched servovalve orifices making up a four-legged flow path of four nonlinear resistors, modulated by the input voltage.

Thereby, the servovalve is modelled as the hydraulic equivalent of a Wheatstone bridge, see Figure 2. When the servovalve input current is positive, $i > 0$, flow passes through the orifices 1 and 3 (path $P \to A \to B \to T$), while flow leakages through valve orifices 2 and 4. Respectively, when the servovalve input current is negative, $i < 0$, flow passes through the path $P \to B \to A \to T$, and flow leakages through valve orifices 1 and 3. This model is described by

$$Q_{v1} = f_1(i)\sqrt{p_s - p_1},\ Q_{v2} = f_2(i)\sqrt{p_s - p_2},\ Q_{v3} = g_1(i)\sqrt{p_2 - p_T},\ Q_{v4} = g_2(i)\sqrt{p_1 - p_T} \quad (4)$$

where p_s and p_T are the supply and return pressure of the servosystem, respectively, i is the servovalve motor current (control command), $f_1(i)$, $f_2(i)$, $g_1(i)$ and $g_2(i)$ are servovalve nonlinear orifice conductances, functions of the servovalve motor current. Because of servovalve symmetry, the current functions are given by

$$f_1(i) = g_1(i) = f_2(-i), \qquad f_2(i) = g_2(i) = f_1(-i) \quad (5)$$

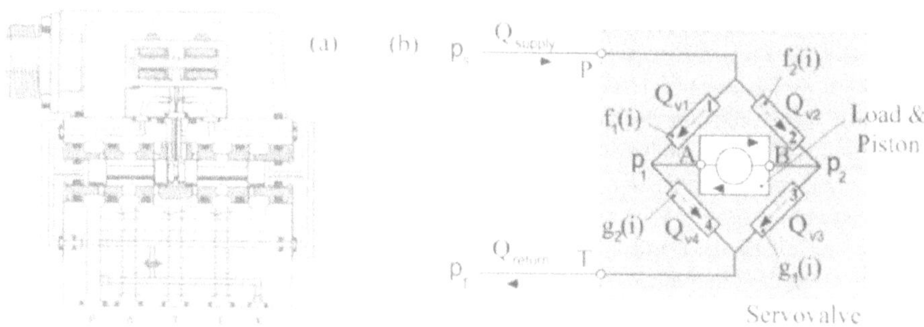

Figure 2. (a) Drawing of a real servovalve, (b) Schematic model of a servovalve.

A good approximation is that the servovalve orifice conductances can be assumed as linear functions of the input current. Therefore, current functions are written as

$$f_1(i) = K_1 i + K_{0,1}, \qquad f_2(i) = K_2 i + K_{0,2} \quad (6)$$

where K_1, $K_{0,1}$, K_2 and $K_{0,2}$ are positive constants, which correspond to the main and leakage valve flow path and because of valve symmetry, see Eq. (5), the following must hold,

$$K_1 = -K_2, \qquad K_{0,1} = K_{0,2} \quad (7)$$

Further, neglecting leakage flows, the flows through the servovalve orifices in Eqs. (4), are equal to the flows through the cylinder chamber ports, see Eq. (3a), and are written as

$$Q_{v1} = A_1 \dot{x}, \qquad Q_{v3} = A_2 \dot{x} \quad (8)$$

4 Invariant Error Dynamics Controller

To control the platform in operational (Cartesian) space, first, manipulator inverse kinematics is solved to transform motion requirements from operational-space into the joint-space. Then, an invariant error dynamics control scheme is designed that allows tracking of the reference inputs.

In electromechanical systems, the force acting on moving masses is proportional to actuator current. This simplifies their control laws and allows one to achieve second order error dynamics converging exponentially to zero. However, a simple relationship between force and current does not exist in electrohydraulic systems. Despite this, we are interested in studying whether such a system can be described by error dynamics such as

$$\ddot{e} + \mathbf{K}_{\mathbf{v}}\dot{e} + \mathbf{K}_{\mathbf{p}}e = 0 \tag{9}$$

where $e = l_{des} - l$ is the 6×1 vector position error of the actuator displacements, l_{des} is the 6×1 desired vector of the actuator displacements, and $\mathbf{K}_{\mathbf{p}}$ and $\mathbf{K}_{\mathbf{v}}$ are 6×6 diagonal matrices, which represent the control gains of the system and are given by,

$$\mathbf{K}_{\mathbf{p}} = \mathrm{diag}\{\omega_1^2, \omega_2^2, ..., \omega_6^2\}, \quad \mathbf{K}_{\mathbf{v}} = \mathrm{diag}\{2\zeta_1\omega_1, 2\zeta_2\omega_2, ..., 2\zeta_6\omega_6\} \tag{10}$$

where $\omega_j, \zeta_j, j=1,2,...,6$ are the closed-loop natural frequencies and the damping for the six linear actuators, respectively. Usually, all ω_j are taken to be equal, and $\zeta_j = 1$.

Using Eqs. (4), (6) and (8), the servocylinder chamber pressures are computed,

$$p_1\big|_j = [p_s - \frac{A_1^2}{(K_1 i + K_{0,1})^2}\cdot i^2]\Big|_j, \quad p_2\big|_j = [p_T + \frac{A_2^2}{(K_1 i + K_{0,1})^2}\cdot i^2]\Big|_j, \quad j=1,2,...,6 \tag{11}$$

Substituting Eqs. (11) into Eq. (3b), the hydraulic forces of the servoactuators are computed as,

$$[p_1 A_1 - p_2 A_2]\big|_j = [A_1 p_s - A_2 p_T - \frac{A_1^3 + A_2^3}{(K_1 i + K_{0,1})^2}\cdot i^2]\Big|_j, \quad j=1,2,...,6 \tag{12}$$

In Eq. (12), i_j is the control input for the jth valve/ actuator and $[p_1 A_1 - p_2 A_2]\big|_j$ is the resulting actuator force. However, Eq. (12) is also function of the velocity of the actuators, $\dot{l}_j$. Substituting Eq. (12) in the system equation of motion, see Eqs. (1) - (2), the following equations of motion are derived,

$$\mathbf{M}(l)\ddot{l} + V(l,\dot{l}) + G(l) + F_{fr}(\dot{l}) = \begin{bmatrix} [A_1 p_s - A_2 p_T - \frac{A_1^3 + A_2^3}{(K_1 i + K_{0,1})^2}\cdot i^2]\Big|_1 \\ \cdots \\ [A_1 p_s - A_2 p_T - \frac{A_1^3 + A_2^3}{(K_1 i + K_{0,1})^2}\cdot i^2]\Big|_6 \end{bmatrix} \tag{13}$$

Solving Eq. (13) for the input commands, i_j, $j=1,2,...,6$, the components of the servovalve current vector $\boldsymbol{i}=\begin{pmatrix} i_1 & i_2 & \dots & i_6 \end{pmatrix}^{\mathrm{T}}$ are computed as,

$$i_j = \left[\frac{\dot{l}}{K_1 \sqrt{\dfrac{1}{A_1^3 + A_2^3} [A_1 p_s - A_2 p_T - \langle \mathrm{F} \rangle]}} - \frac{K_{0,1}}{K_1} \right] \Bigg|_j , \qquad j = 1,2,...,6 \tag{14}$$

where $\langle \mathrm{F} \rangle |_j$ represents the jth element of the vector $\mathbf{M}(l)\ddot{l} + V(l,\dot{l}) + G(l) + F_{fr}(\dot{l})$.

Further, assuming that Eq. (9) has been satisfied, this equation is solved for $\ddot{l}$ which is substituted in the elements $\langle \mathrm{F} \rangle |_j$. Then, these elements are obtained as,

$$\langle \mathrm{F} \rangle |_j = [\mathbf{M} \cdot (\ddot{l}_{des} + \mathbf{K}_v \dot{e} + \mathbf{K}_p e) + V + G + F_{fr}] |_j \tag{15}$$

Finally, the servovalve current vector, the elements of which are given in Eq. (14) describes the invariant error dynamics control law scheme for the 6-dof electrohydraulic Stewart platform. Note that the control law requires feedback of both the position and velocity length errors, as well as it includes the mechanism dynamics and a servovalve model.

Substituting Eq. (14) in Eq. (1), an equation of the form of Eq. (9) results, which also shows that system is asymptotically stable. The control law is illustrated schematically in Figure 3.

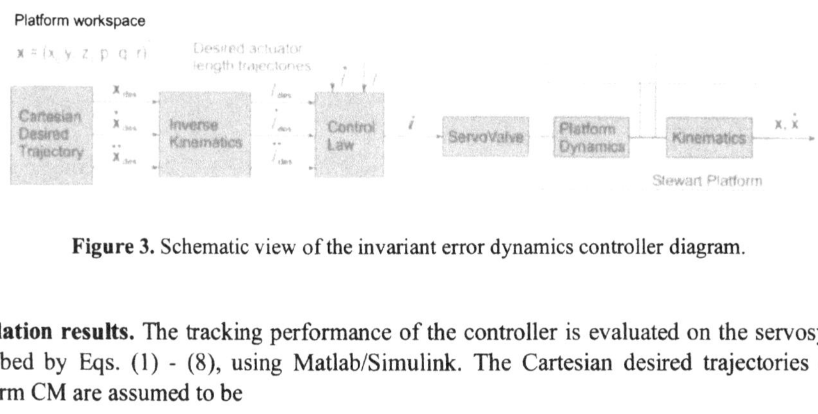

Figure 3. Schematic view of the invariant error dynamics controller diagram.

Simulation results. The tracking performance of the controller is evaluated on the servosystem, described by Eqs. (1) - (8), using Matlab/Simulink. The Cartesian desired trajectories of the platform CM are assumed to be

$$x_0(t) = x_c \sin(2\pi f t), \quad y_0(t) = y_c \cos(2\pi f t), \quad z_0(t) = z_{c1} + z_c \sin(2\pi f t) \tag{16a}$$

$$p(t) = p_c \cos(2\pi f t), \quad q(t) = q_c \sin(2\pi f t), \quad r(t) = r_c \sin(2\pi f t) \tag{16b}$$

where x_c, y_c, z_c, z_{c1}, p_c, q_c and r_c are trajectory constants.

Simulation runs were obtained using a number of desirable trajectories. As an example, Figure 4 and Figure 5 show typical results, in which the desired trajectories are given by Eqs. (16) with platform mass $m = 300\,kg$, moments of inertia about the centre of platform mass of x, y, z axes $I_{xx} = I_{yy} = 25\,kgm^2$, $I_{zz} = 50\,kgm^2$, respectively, $f = 0.5\,Hz$, $x_c = 0.1\,m$, $y_c = 0.2\,m$,

$z_c = 0.1\,m$, $z_{c1} = 1.25\,m$, $p_c = 30^\circ$, $q_c = 20^\circ$, $r_c = 15^\circ$, $\omega_j = \pi\ rad/s$, $\zeta_j = 1$ and friction parameters $b_j = 200\,\mathrm{Ns/m}$, $f_{C,j} = 20\,\mathrm{N}$, and $f_{s,j} = 50\,\mathrm{N}$, $j = 1,2,...,6$. The platform displacements in the three Cartesian axes and orientation, and the position and orientation errors of the platform are shown in Figure 4. Also, one of the six leg lengths of the mechanism, the force acting on the platform, and the input signal for the same actuator, are depicted in Figure 5. The position errors converge to zero, as were expected, in settling time $t_{s,j} = 6/\omega_j\,[s]$, $j = 1,2,...,6$. The remaining position errors of the steady state are due to the inertial forces caused by the harmonic trajectory.

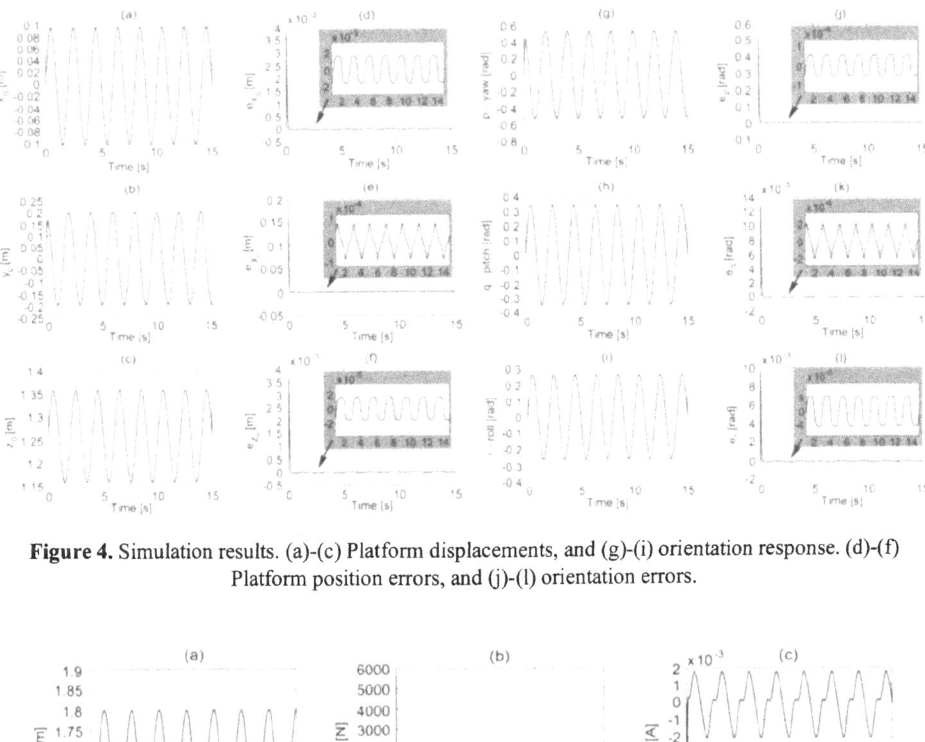

Figure 4. Simulation results. (a)-(c) Platform displacements, and (g)-(i) orientation response. (d)-(f) Platform position errors, and (j)-(l) orientation errors.

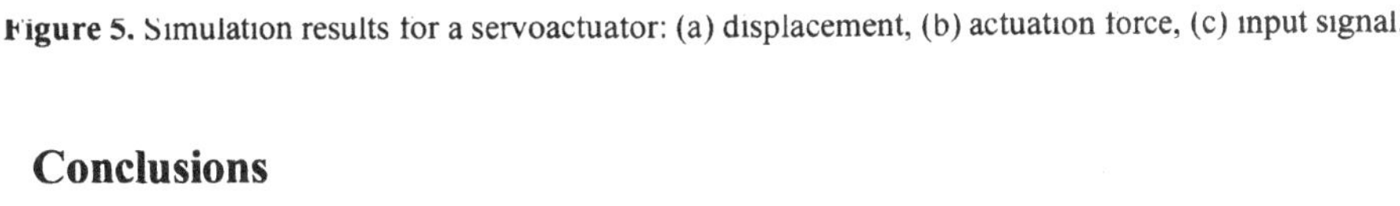

Figure 5. Simulation results for a servoactuator: (a) displacement, (b) actuation force, (c) input signal.

5 Conclusions

This paper focused on the development of the invariant error dynamics controller for a 6-dof electrohydraulic Stewart platform. Rigid body and electrohydraulic models, which included servovalve dynamics and friction, were employed. The developed controller employs the dynamic and hydraulic model of the system and yields the six servovalve current inputs, in analytical form. Using mechanism inverse kinematics, the desired Cartesian trajectories yield desired actuator length trajectories. Simulations with typical desired trajectories were presented and a good performance of the controller was obtained. The approach can be further extended to hydraulic manipulator and simulator control.

Bibliography

Davliakos, I., and Papadopoulos, E. (2005). Development of a Model-Based Nested Controller for Electrohydraulic Servos. In *Proceedings 13th Mediterranean Conference on Control and Automation*, June 27-29, Limassol, Cyprus, 107–112.

Gao, X., S., Lei, D., Liao, Q., and Zhang, G., F. (2005). Generalized Stewart-Gough Platforms and Their Direct Kinematics. *IEEE Transactions on Robotics*, vol. 21, No. 2, 141–151.

Garagic, D., and Srinivasan, K. (2004). Application of Nonlinear Adaptive Control Techniques to an Electrohydraulic Velocity Servo-mechanism. *IEEE Trans. on Control Systems Tech.*, vol. 12, 303–314.

Helouvry, B., A., Dupont, P., and De Wit, C., C. (1994). A Survey of Models, Analysis Tools and Compensation Methods for the Control of Machines with Friction. *Automatica*, 30: 7, 1083–1138.

Honegger, M., and Corke, P. (2001). Model-Based Control of Hydraulic Actuated Manipulators. In *Proceedings 2001 IEEE Int. Conf. on Robotics & Automation*, Seoul, Korea, 2553–2559.

Kim, D., H., Kang, J., Y., and Lee, K., I. (2000). Robust Tracking Control Design for a 6 DOF Parallel Manipulator *Journal of Robotic Systems*, Vol. 17, No. 10, 527–547.

Kim, N., I., and Lee, C., W. (1998). High Speed Tracking Control of Stewart Platform Manipulator via Enhanced Sliding Mode Control. *IEEE Conf. On Robotics and Automation*, Belgium, 2716–2721.

Lebret, G., Liu, K., and Lewis, F., L. (1993). Dynamic Analysis and Control of a Stewart Platform Manipulator. *Journal of Robotic Systems*, 10(5), 629–655.

Li, D., and Salcudean, S., E. (1997). Modeling, Simulation, and Control of a Hydraulic Stewart Platform. In *Proceedings of the IEEE Int. Conf. on Robotics and Automation*, 3360–3366.

Liu, M., J., Li, C., X., and Li, C., N. (2000). Generalized Stewart-Gough Platforms and Their Direct Kinematics. *IEEE Transactions on Robotics and Automation*, vol. 16, No. 1, 94–98.

Merritt, H. E. (1967). Hydraulic Control Systems. J. Wiley.

Nguyen, C., C., Antrazi, S., S., Zhou, Z., L., and Campbell, C., E. (1992). Adaptive Control of a Stewart Platform-Based Manipulator. *Journal of Robotic Systems*, 10(5), 657–687.

Papadopoulos, E. and Chasparis, G., (2004). Analysis and Model-based Control of Servomechanisms with Friction. *ASME Journal Dynamic Systems, Measurement and Control*, Vol. 126, No. 4, 911-915.

Shim, J., H., Park, J., Y., Kwon, D., S., and Kim, S. (1997). Kinematic Design of a Six Degree-of-Freedom In-Parallel Manipulator for Probing Task. In *Proceedings of the IEEE Int. Conf. On Robotics and Automation*, New Mexico, 2967–2973.

Sirouspour, M. R., and Salcudean, S. E. (2000). A New Approach to the Control of a Hydraulic Stewart Platform. *ISER 2000*, 447–460.

Sirouspour, M. R., and Salcudean, S. E. (2001). Nonlinear Control of Hydraulic Robots. *IEEE Transactions on Robotics and Automation*, vol. 17, No. 2, 173–182.

Stewart, D. (1965-66). A platform with six degrees of freedom. In *Proceedings of the IMechE*, Vol. 180, Pt. 1, No 15, 371–385.

Tsai, L., W. (1999). Robot Analysis, The Mechanics of Serial and Parallel Manipulators. J. Wiley & Sons.

Tsai, L., W. (2000). Solving the Inverse Dynamics of a Stewart-Gough Manipulator by the Principle of Virtual Work. *Journal of Mechanical Design, Transactions of the ASME*, vol. 122, 3–9.

Chapter III

Motion Planning and Synthesis

Hyperbolic-linear, Extended Jacobian Inverse Kinematics Algorithm for Doubly Nonholonomic Mobile Manipulators

Krzysztof Tchoń[‡] and Janusz Jakubiak[‡*]
[‡] Wrocław University of Technology, Institute of Computer Engineering, Control and Robotics, Wrocław, Poland, email: {tchon|jjakubia}@ict.pwr.wroc.pl

Abstract. A doubly nonholonomic mobile manipulator is defined as a mobile manipulator consisting of a nonholonomic stationary manipulator mounted on a nonholonomic mobile platform. Exploiting two augmenting kinematics maps, called hyperbolic and linear, we propose an extended Jacobian inverse kinematics algorithm for doubly nonholonomic mobile manipulators. By definition, our algorithm is repeatable. Performance of the algorithm is examined by computer simulations.

1 Introduction

Mobile manipulators are robotic systems of growing significance and applicability. A topical literature referring to mobile manipulators has been reviewed in Tchoń and Jakubiak (2003, 2005b); for the sake of space limitations we cannot re-review it here. In this paper we are concerned with doubly nonholonomic mobile manipulators that are composed of a nonholonomic mobile platform and a nonholonomic onboard manipulator. The design of the nonholonomic manipulator has been described in Sordalen et al. (1994); its control is studied in Nakamura et al. (2001); Chung and Nakamura (2002). Due to the use of sophisticated nonholonomic gears the motion of the nonholonomic manipulator is controlled by two actuators. In Tchoń et al. (2004) we have derived a mathematical model of kinematics, proposed a Jacobian inverse kinematics algorithm, and defined dexterity measures for doubly nonholonomic mobile manipulators. An extended Jacobian inverse kinematics algorithm for doubly nonholonomic mobile manipulators has been designed in Tchoń and Jakubiak (2005a). By design, the extended Jacobian inverse kinematics algorithm is repeatable Tchoń (2002).

The present paper develops a new, hyperbolic-linear, extended Jacobian inverse kinematics algorithm for doubly nonholonomic mobile manipulators, based on a pair of augmenting kinematics maps: a map with hyperbolic and a map with linear invariant subspaces. The degree of hyperbolicity or linearity of the algorithm is a design parameter. In limit cases the algorithm becomes either totally hyperbolic or totally linear. Performance of a specific hyperbolic-linear algorithm will be illustrated with computer simulations.

*This research has been supported by the Foundation for Polish Science.

This paper is organized in the following way. Section 2 collects basic facts concerned with the kinematic modeling of doubly nonholonomic mobile manipulators. Section 3 presents a general derivation of the extended Jacobian inverse kinematics algorithm. A hyperbolic-linear algorithm is set forth in section 4. Section 5 includes results of computer simulations. The paper is concluded with section 6.

2 Basic Concepts

In acordance with references Tchoń et al. (2004); Tchoń and Jakubiak (2005a) the kinematics of the doubly nonholonomic mobile manipulator can be represented as a smooth driftless control system with outputs

$$\begin{cases} \dot{q} = G(q)u^1 = \sum_{i=1}^{m_1} g_i(q)u_i^1, \quad \dot{p} = F(p)u^2 = \sum_{i=1}^{m_2} f_i(p)u_i^2, \\ y = k(q,p) = (k_1(q,p),\ldots,k_r(q,p)), \end{cases} \tag{2.1}$$

where $q \in R^{n_1}$, $p \in R^{n_2}$ denote platform coordinates and joint positions of the onboard manipulator, $u^1 \in R^{m_1}$, $u^2 \in R^{m_2}$ are platform and manipulator control variables defined over a time interval $[0,T]$, $y \in R^r$ denotes end effector position and orientation in the taskspace. The endogenous configuration space $\mathcal{X} = L^2_{m_1}[0,T] \times L^2_{m_2}[0,T]$ is an inifitely dimensional Hilbert space. An initial state (q_0,p_0) of the mobile manipulator, and an endogenous configuration $(u^1(\cdot),u^2(\cdot)) \in \mathcal{X}$ determine the state $(q(t),p(t)) = (\varphi_{q_0,t}(u^1(\cdot)),\psi_{p_0,t}(u^2(\cdot)))$, $t \in [0,T]$, and the end effector $y(t) = k(q(t),p(t))$ trajectories of the mobile manipulator. The instantaneous kinematics of the doubly nonholonomic mobile manipulator

$$K_{q_0,p_0,T}(u^1(\cdot),u^2(\cdot)) = y(T) = k(\varphi_{q_0,T}(u^1(\cdot)),\psi_{p_0,T}(u^2(\cdot)), \tag{2.2}$$

are defined as the value of the system output at T. The derivative of (2.2)

$$\begin{aligned} J_{q_0,p_0,T}(u^1(\cdot),u^2(\cdot))(v^1(\cdot),v^2(\cdot)) &= \left.\frac{d}{d\alpha}\right|_{\alpha=0} K_{q_0,p_0,T}(u^1(\cdot)+\alpha v^1(\cdot),u^2(\cdot)+\alpha v^2(\cdot)) \\ &= C_1(T)\int_0^T \Phi(T,s)B_1(s)v^1(s)ds + C_2(T)\int_0^T \Psi(T,s)B_2(s)v^2(s)ds \end{aligned} \tag{2.3}$$

will be called the analytic Jacobian of the doubly nonholonomic mobile manipulator. The matrices appearing in (2.3) originate from the linear approximation of the system (2.1) along $(u^1(t),u^2(t),q(t),p(t))$, and can be computed as follows:

$$\begin{aligned} A_1(t) &= \frac{\partial\left(G(q(t))u^1(t)\right)}{\partial q}, \quad B_1(t) = G(q(t)), \quad C_1(t) = \frac{\partial k(q(t),p(t))}{\partial q} \\ A_2(t) &= \frac{\partial\left(F(p(t))u^2(t)\right)}{\partial p}, \quad B_2(t) = F(p(t)), \quad C_2(t) = \frac{\partial k(q(t),p(t))}{\partial p}. \end{aligned}$$

Matrices $\Phi(t,s), \Psi(t,s)$ are transition matrices of the linear approximation to (2.1)

$$\begin{cases} \dot{\xi} = A_1(t)\xi + B_1(t)v^1(t), \quad \dot{\zeta} = A_2(t)\zeta + B_2(t)v^2(t), \\ \eta = C_1(T)\xi + C_2(T)\zeta. \end{cases} \tag{2.4}$$

The analytic Jacobian distinguishes regular and singular endogenous configurations of the doubly nonholonomic mobile manipulator. An endogenous configuration $(u^1(\cdot), u^2(\cdot))$ is regular provided that the Jacobian $J_{q_0,p_0,T}(u^1(\cdot), u^2(\cdot))$ is a surjective map of the endogenous configuration space onto the taskspace; otherwise the configuration is singular.

3 Extended Jacobian Inverse Kinematics Algorithm

The extended Jacobian inverse kinematics algorithm is based on a specific right inverse of the analytic Jacobian (2.3). The main step in designing this inverse lies in defining an augmenting kinematics map in such a way that the original kinematics and this map form a local diffeomorphism of the endogenous configuration space. This means that the augmenting kinematics map should transform the endogenous configuration space into a quotient of this space by the taskspace, $H_{q_0,T} : \mathcal{X} \longrightarrow \mathcal{X}/R^r$. We can conveniently define the quotient space $\mathcal{X}/R^r$ by introducing orthonormal bases $\{\varphi_i(t)e_k | i \geq 0, k = 1, \ldots, m_1\}$ and $\{\psi_j(t)e_l | j \geq 0, l = 1, \ldots, m_2\}$, e_k, e_l standing for unit vectors in R^{m_1} and R^{m_2}, in platform control functions $L^2_{m_1}[0,T]$ and in manipulator control functions $L^2_{m_2}[0,T]$. Then we select integers $k_1, k_2, \ldots, k_{m_1} \geq -1$ and $l_1, l_2, \ldots, l_{m_2} \geq -1$, and make an identification

$$(u^1(\cdot), u^2(\cdot)) \cong (x^1, x^2, \tilde{u}^1(\cdot), \tilde{u}^2(\cdot)), \tag{3.1}$$

where the vectors of coefficients $x^1 \in R^{s_1}$, $x^2 \in R^{s_2}$, for $s_1 = m_1 + \sum_{i=1}^{m_1} k_i$, $s_2 = m_2 + \sum_{i=1}^{m_2} l_i$, and truncated control functions $\tilde{u}^1(\cdot) \in L^2_{k_1,k_2,\ldots,k_{m_1}}[0,T]$, $\tilde{u}^2(\cdot) \in L^2_{l_1,l_2,\ldots,l_{m_2}}[0,T]$. These last space are isomorphic with the quotients $L^2_{m_1}[0,T]/R^{s_1}$, $L^2_{m_2}[0,T]/R^{s_2}$, and consist of control functions whose *ith* component representation at the suitable basis is lacking, respectively, $k_i + 1$ and $l_i + 1$ first terms. If a certain k_i or l_i is equal to -1, all terms of the corresponding component of the control function are present. By using matrices of basic functions $P_{s_1}(t) = blockdiag\{P_{k_1}(t), \ldots, P_{k_{m_1}}(t)\}$, $P_{s_2}(t) = blockdiag\{P_{l_1}(t), \ldots, P_{l_{m_2}}(t)\}$, with respective block entries $P_k = [\varphi_0(t), \ldots, \varphi_k(t)]$ and $P_l = [\psi_0(t), \ldots, \psi_l(t)]$, we define the representations

$$u^1(t) = P_{s_1}(t)x^1 + \tilde{u}^1(t) \quad \text{and} \quad u^2(t) = P_{s_2}(t)x^2 + \tilde{u}^2(t). \tag{3.2}$$

Components of x^1 and x^2 will be called coordination variables, while $\tilde{u}^1(t)$, $\tilde{u}^2(t)$ are referred to as control variables. Under assumption $s_1 + s_2 = r$, we obtain the quotient space

$$\mathcal{X}/R^r \cong L^2_{k_1,k_2,\ldots,k_{m_1}}[0,T] \oplus L^2_{l_1,l_2,\ldots,l_{m_2}}[0,T].$$

Now, let us choose an augmenting kinematics map

$$H_{q_0,p_0,T} = (H_{1\,q_0,p_0,T}, H_{2\,q_0,p_0,T}) : \mathcal{X} \to L^2_{k_1,k_2,\ldots,k_{m_1}}[0,T] \times L^2_{l_1,l_2,\ldots,l_{m_2}}[0,T] \tag{3.3}$$

composed of a platform map $H_{1\,q_0,p_0,T}$ and an onboard manipulator map $H_{2\,q_0,p_0,T}$. Outside algorithmic singularities the extended Jacobian

$$\begin{aligned}\bar{J}_{q_0,p_0,T}(u^1(\cdot), u^2(\cdot))(v^1(\cdot), v^2(\cdot)) = \big(&J_{q_0,p_0,T}(u^1(\cdot), u^2(\cdot)),\\ &DH_{1\,q_0,p_0,T}(u^1(\cdot), u^2(\cdot)), DH_{2\,q_0,p_0,T}(u^1(\cdot), u^2(\cdot))\big)(v^1(\cdot), v^2(\cdot)),\end{aligned} \tag{3.4}$$

is a linear isomorphism of $\mathcal{X}$ whose inverse yields for every $\eta \in R^r$ an extended Jacobian inverse Tchoń and Jakubiak (2005a,b)

$$J^{E\#}_{q_0,p_0,T}(u^1(\cdot),u^2(\cdot))\eta = \bar{J}^{-1}_{q_0,p_0,T}(u^1(\cdot),u^2(\cdot))(\eta,0(\cdot)). \tag{3.5}$$

The corresponding extended Jacobian inverse kinematics algorithm for doubly nonholonomic mobile manipulators is defined as a dynamic system in $\mathcal{X}$

$$\frac{d}{d\theta}\begin{pmatrix} u^1_\theta(t) \\ u^2_\theta(t) \end{pmatrix} = -\gamma\left(J^{E\#}_{q_0,p_0,T}(u^1_\theta(\cdot),u^2_\theta(\cdot))(K_{q_0,p_0,T}(u^1_\theta(\cdot),u^2_\theta(\cdot)) - y_d)\right)(t). \tag{3.6}$$

The solution of the inverse kinematics problem is obtained as the limit of the trajectory of (3.6) at $\theta \to +\infty$. Using results of Tchoń (2002), it is easily checked that the algorithm (3.6) is repeatable. Furthermore, during the algorithm's operation the map (3.3) remains constant.

4 Hyperbolic-linear Algorithm

Specifically, we shall consider the following pair of platform and onboard manipulator augmenting maps being a fusion of the hyperbolic map and of the linear map considered in Tchoń and Jakubiak (2005a)

$$\begin{aligned} H_{1\,q_0,p_0,T}(u^1(\cdot),u^2(\cdot))(t) = \Big(& x^1_{i_1}\tilde{u}^1_1(t),\ldots,x^1_{i_{\alpha_1}}\tilde{u}^1_{\alpha_1}(t), x^1_{j_1}\tilde{u}^2_1(t),\ldots,x^1_{j_{\alpha_2}}\tilde{u}^2_{\alpha_2}(t)\,, \\ & \frac{\tilde{u}^1_{\alpha_1+1}(t)}{x^1_{k_1}},\ldots,\frac{\tilde{u}^1_{\alpha_{13}}(t)}{x^1_{k_{\alpha_3}}},\frac{\tilde{u}^2_{\alpha_2+1}(t)}{x^1_{l_1}},\ldots,\frac{\tilde{u}^2_{\alpha_{24}}(t)}{x^1_{l_{\alpha_4}}}, x^2_{i_1}\tilde{u}^1_{\alpha_{13}+1}(t),\ldots,x^2_{i_{\alpha_5}}\tilde{u}^1_{\alpha_{135}}(t), \\ & x^2_{j_1}\tilde{u}^2_{\alpha_{24}+1}(t),\ldots,x^2_{j_{\alpha_6}}\tilde{u}^2_{\alpha_{246}}(t),\frac{\tilde{u}^1_{\alpha_{135}+1}(t)}{x^2_{k_1}},\ldots,\frac{\tilde{u}^1_{\alpha_{1357}}(t)}{x^2_{k_{\alpha_7}}},\frac{\tilde{u}^2_{\alpha_{246}+1}(t)}{x^2_{l_1}},\ldots,\frac{\tilde{u}^2_{\alpha_{2468}}(t)}{x^2_{l_{\alpha_8}}}\Big) \end{aligned} \tag{4.1}$$

and

$$\begin{aligned} & H_{2\,q_0,p_0,T}(u^1(\cdot),u^2(\cdot))(t) = \\ & \Big(x^1_{i_1}\tilde{u}^1_{\alpha_{1357}+1}(t),\ldots,x^1_{i_{\beta_1}}\tilde{u}^1_{\alpha_{1357}+\beta_1}(t), x^1_{j_1}\tilde{u}^2_{\alpha_{2468}+1}(t),\ldots,x^1_{j_{\beta_2}}\tilde{u}^2_{\alpha_{2468}+\beta_2}(t)\,, \\ & \frac{\tilde{u}^1_{\alpha_{1357}+\beta_1+1}(t)}{x^1_{k_1}},\ldots,\frac{\tilde{u}^1_{\alpha_{1357}+\beta_{13}}(t)}{x^1_{k_{\beta_3}}},\frac{\tilde{u}^2_{\alpha_{2468}+\beta_2+1}(t)}{x^1_{l_1}},\ldots,\frac{\tilde{u}^2_{\alpha_{2468}+\beta_{24}}(t)}{x^1_{l_{\beta_4}}}, \\ & x^2_{i_1}\tilde{u}^1_{\alpha_{1357}+\beta_{13}+1}(t),\ldots,x^2_{i_{\beta_5}}\tilde{u}^1_{\alpha_{1357}+\beta_{135}}(t), x^2_{j_1}\tilde{u}^2_{\alpha_{2468}+\beta_{24}+1}(t),\ldots,x^2_{j_{\beta_6}}\tilde{u}^2_{\alpha_{2468}+\beta_{246}}(t), \\ & \frac{\tilde{u}^1_{\alpha_{1357}+\beta_{135}+1}(t)}{x^2_{k_1}},\ldots,\frac{\tilde{u}^1_{\alpha_{1357}+\beta_{1357}}(t)}{x^2_{k_{\beta_7}}},\frac{\tilde{u}^2_{\alpha_{2468}+\beta_{246}+1}(t)}{x^2_{l_1}},\ldots,\frac{\tilde{u}^2_{\alpha_{2468}+\beta_{2468}}(t)}{x^2_{l_{\beta_8}}}\Big) \end{aligned} \tag{4.2}$$

where $u^1(t) = P_{s_1}(t)x^1 + \tilde{u}^1(t)$, $u^2(t) = P_{s_2}(t)x^2 + \tilde{u}^2(t)$. Integers $\alpha_i, \beta_j \geq 0$ sum up as $\sum_{i=1}^8 \alpha_i = m_1$ and $\sum_{j=1}^8 \beta_j = m_2$. To simplify notations, in (4.1), (4.2) we use abbreviations of the form $\alpha_{ijkl} = \alpha_i + \alpha_j + \alpha_k + \alpha_l$, similarly for β_{ijkl}. For definiteness we need that $\alpha_{1357} + \beta_{1357} = m_1$ and $\alpha_{2468} + \beta_{2468} = m_2$.

Taking into account this augmenting map, after necessary mathematical developments resulting from (3.5), we arrive at the following extended Jacobian inverse

$$(J^{E\#}_{q_0,p_0,T}(u^1(\cdot),u^2(\cdot))\eta)(t) = \begin{pmatrix} v^1(t) \\ v^2(t) \end{pmatrix} = \begin{bmatrix} [P_{s_1}(t),0] + M_1(t) \\ [P_{s_2}(t),0] + M_2(t) \end{bmatrix} F^{-1}\eta, \tag{4.3}$$

where $F = C_1(T)\int_0^T \Phi(T,s)B_1(s)([P_{s_1}(s),0]+M_1(s))ds + C_2(T)\int_0^T \Psi(T,s)B_2(s)([0,P_{s_2}(s)] + M_2(s))ds$ and

$$M_1(t) = \begin{bmatrix} -diag\{\frac{\tilde{u}^1_k(t)}{x^1_{i_k}}\}_{\alpha_1} S_{\alpha_1,r} \\ diag\{\frac{\tilde{u}^1_{\alpha_1+k}(t)}{x^1_{i_k}}\}_{\alpha_3} S_{\alpha_3,r} \\ -diag\{\frac{\tilde{u}^1_{\alpha_{13}+k}(t)}{x^2_{i_k}}\}_{\alpha_5} S_{\alpha_5,r} \\ diag\{\frac{\tilde{u}^1_{\alpha_{135}+k}(t)}{x^2_{i_k}}\}_{\alpha_7} S_{\alpha_7,r} \\ -diag\{\frac{\tilde{u}^1_{\alpha_{1357}+k}(t)}{x^1_{i_k}}\}_{\beta_1} S_{\beta_1,r} \\ diag\{\frac{\tilde{u}^1_{\alpha_{1357}+\beta_1+k}(t)}{x^1_{i_k}}\}_{\beta_3} S_{\beta_3,r} \\ -diag\{\frac{\tilde{u}^1_{\alpha_{1337}+\beta_{13}+k}(t)}{x^2_{i_k}}\}_{\beta_5} S_{\beta_5,r} \\ diag\{\frac{\tilde{u}^1_{\alpha_{1357}+\beta_{135}+k}(t)}{x^2_{i_k}}\}_{\beta_7} S_{\beta_7,r} \end{bmatrix}, \quad M_2(t) = \begin{bmatrix} -diag\{\frac{\tilde{u}^2_k(t)}{x^1_{i_k}}\}_{\alpha_2} S_{\alpha_2,r} \\ diag\{\frac{\tilde{u}^2_{\alpha_2+k}(t)}{x^1_{i_k}}\}_{\alpha_4} S_{\alpha_4,r} \\ -diag\{\frac{\tilde{u}^2_{\alpha_{24}+k}(t)}{x^2_{i_k}}\}_{\alpha_6} S_{\alpha_6,r} \\ diag\{\frac{\tilde{u}^2_{\alpha_{246}+k}(t)}{x^2_{i_k}}\}_{\alpha_8} S_{\alpha_8,r} \\ -diag\{\frac{\tilde{u}^2_{\alpha_{2468}+k}(t)}{x^1_{i_k}}\}_{\beta_2} S_{\beta_2,r} \\ diag\{\frac{\tilde{u}^2_{\alpha_{2468}+\beta_2+k}(t)}{x^1_{i_k}}\}_{\beta_4} S_{\beta_4,r} \\ -diag\{\frac{\tilde{u}^2_{\alpha_{2468}+\beta_{24}+k}(t)}{x^2_{i_k}}\}_{\beta_6} S_{\beta_6,r} \\ diag\{\frac{\tilde{u}^2_{\alpha_{2468}+\beta_{246}+k}(t)}{x^2_{i_k}}\}_{\beta_8} S_{\beta_8,r} \end{bmatrix}. \tag{4.4}$$

Coordination variables that appear in formulas (4.1), (4.2), (4.4) are defined by means of selector matrices $S_{\alpha_i,r}$ or $S_{\beta_j,r}$, $i,j = 1,\ldots,8$, picking α_i or β_j elements of an r dimensional vector, while the diagonal matrix $diag\{a_k\}_n$ is defined as $diag\{a_1, a_2, \ldots, a_n\}$.

The inverse (4.3) defines a repeatable extended Jacobian inverse kinematics algorithm in acordance with the scheme (3.6). The algorithm is well defined provided that $rank\, F = r$. For both augmenting maps (4.1) and (4.2) are constant on trajectories of system (3.6), there is a specific hyperbolic-linear dependence between control and coordination variables. For this reason the algorithm will be called hyperbolic-linear. The contribution of the hyperbolic and the linear component is regulated by the values of α_i and β_j. In the limit when α_{3478} and $\beta_{3478} = 0$, the algorithm becomes totally hyperbolic, whereas for α_{1256} and $\beta_{1256} = 0$ the algorithm reduces to totally linear.

5 Simulations

In order to check the performance of the hyperbolic-linear extended Jacobian inverse kinematics algorithm we have solved a collection of inverse kinematic problems for the doubly nonholonomic mobile manipulator shown in Figure 1. The kinematics of this

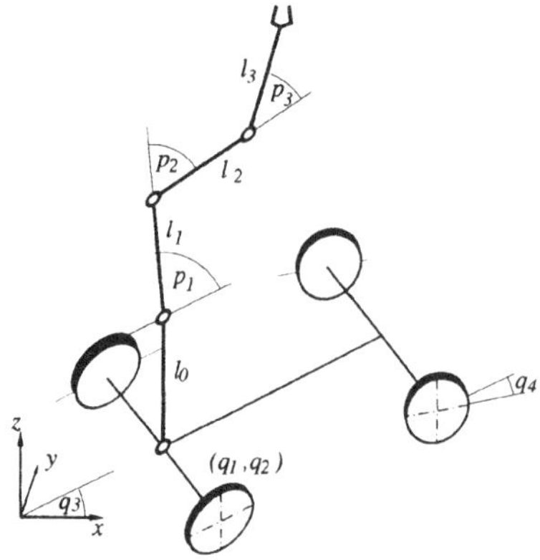

Figure 1. 3 d.o.f nonholonomic manipulator mounted on kinematic car.

manipulator have the following control system representation

$$\begin{cases} \dot{q}_1 = u_1^1 \cos q_3 \cos q_4, \quad \dot{q}_2 = u_1^1 \sin q_3 \cos q_4, \quad \dot{q}_3 = u_1^1 \sin q_4, \quad \dot{q}_4 = u_2^1, \\ \dot{p}_1 = u_1^2, \quad \dot{p}_2 = a_2 u_2^2 \sin p_1, \quad \dot{p}_3 = a_3 u_2^2 \sin p_2 \cos p_1, \end{cases}$$
$$y = \begin{pmatrix} y_1 \\ y_2 \\ y_3 \end{pmatrix} = \begin{pmatrix} q_1 + (l_1 \cos p_1 + l_2 \cos(p_1 + p_2) + l_3 \cos(p_1 + p_2 + p_3)) \cos q_3 \\ q_2 + (l_1 \cos p_1 + l_2 \cos(p_1 + p_2) + l_3 \cos(p_1 + p_2 + p_3)) \sin q_3 \\ l_0 + l_1 \sin p_1 + l_2 \sin(p_1 + p_2) + l_3 \sin(p_1 + p_2 + p_3) \end{pmatrix}. \tag{5.1}$$

Hereabove $q \in R^4$ refers to the platform coordinates, $p \in R^3$ describes joint positions of the onboard manipulator, $y \in R^3$ denotes Cartesian position of the end effector. Band-limited control functions in the form of truncated Fourier series $u^1(t) = P(t)\lambda$, $u^2(t) = P(t)\mu$, are used in computations, where $\omega = 2\pi/T$, the matrix $P(t) = \begin{bmatrix} 1 & \sin\omega t & \cos\omega t & 0 & 0 & 0 \\ 0 & 0 & 0 & 1 & \sin\omega t & \cos\omega t \end{bmatrix}$, control coefficients $\lambda = (\lambda_0^1, \lambda_1^1, \lambda_2^1, \lambda_0^2, \lambda_1^2, \lambda_2^2)$, $\mu = (\mu_0^1, \mu_1^1, \mu_2^1, \mu_0^2, \mu_1^2, \mu_2^2) \in R^6$. As a result of truncation, the band-limited endogenous configuration space becomes finite dimensional $\tilde{\mathcal{X}} \cong R^{12}$. For further considerations we choose $x^1 = (\lambda_0^1, \lambda_0^2)$ and $x^2 = \mu_0^1$, so $s_1 = 2$, $s_2 = 1$, and $s_1 + s_2 = r = 3$. Also, we set $\alpha_5 = \alpha_7 = \beta_2 = \beta_4 = 1$, and let all remaining α_i and β_j be zero. For selector matrices $S_{\alpha_5,r} = S_{\alpha_7,r} = \begin{bmatrix} 0 & 0 & 1 \end{bmatrix}$ and $S_{\beta_2,r} = \begin{bmatrix} 1 & 0 & 0 \end{bmatrix}$ and $S_{\beta_4,r} = \begin{bmatrix} 0 & 1 & 0 \end{bmatrix}$ the band-limited augmenting maps (4.1) and (4.2) are defined as

$$\tilde{H}_{1\, q_0,p_0,T}(\lambda,\mu) = \left(\mu_0^1\lambda_1^1, \mu_0^1\lambda_2^1, \frac{\lambda_1^2}{\mu_0^1}, \frac{\lambda_2^2}{\mu_0^1}\right), \quad \tilde{H}_{2\, q_0,p_0,T}(\lambda,\mu) = \left(\lambda_0^1\mu_1^1, \lambda_0^1\mu_2^1, \frac{\mu_0^2}{\lambda_0^2}, \frac{\mu_1^2}{\lambda_0^2}, \frac{\mu_2^2}{\lambda_0^2}\right).$$

Relying on results of section 4 we derive a band-limited extended Jacobian inverse

$$\tilde{J}^{E,\#}_{q_0,p_0,T}(\lambda,\mu) = \begin{bmatrix} E_1(\lambda,\mu) \\ E_2(\lambda,\mu) \end{bmatrix} F^{-1}(\lambda,\mu),$$

for $F(\lambda,\mu) = C_1(T)\int_0^T \Phi(T,s)B_1(s)P(s)dsE_1(\lambda,\mu) + C_2(T)\int_0^T \Psi(T,s)B_2(s)P(s)dsE_2(\lambda,\mu)$ and

$$E_1(\lambda,\mu) = \begin{bmatrix} 1 & 0 & 0 \\ 0 & 0 & -\frac{\lambda_1^1}{\mu_0^1} \\ 0 & 0 & -\frac{\lambda_2^1}{\mu_0^1} \\ 0 & 1 & 0 \\ 0 & 0 & \frac{\lambda_1^2}{\mu_0^1} \\ 0 & 0 & \frac{\lambda_2^2}{\mu_0^1} \end{bmatrix}, \quad E_2(\lambda,\mu) = \begin{bmatrix} 0 & 0 & 1 \\ -\frac{\mu_1^1}{\lambda_0^1} & 0 & 0 \\ -\frac{\mu_2^1}{\lambda_0^1} & 0 & 0 \\ 0 & \frac{\mu_0^2}{\lambda_0^2} & 0 \\ 0 & \frac{\mu_1^2}{\lambda_0^2} & 0 \\ 0 & \frac{\mu_2^2}{\lambda_0^2} & 0 \end{bmatrix}$$

A discrete, band-limited, hyperbolic-linear extended Jacobian inverse kinematics algorithm for the doubly nonholonomic mobile manipulator takes the following form

$$\begin{pmatrix}\lambda_{\theta+1}\\ \mu_{\theta+1}\end{pmatrix} = \begin{pmatrix}\lambda_{\theta}\\ \mu_{\theta}\end{pmatrix} - \gamma \tilde{J}^{E\#}_{q_0,p_0,T}(\lambda_\theta,\mu_\theta)(\tilde{K}_{q_0,p_0,T}(\lambda_\theta,\mu_\theta) - y_d). \tag{5.2}$$

In computer simulations we have assumed the platform length as the measure unit, set $l_0 = 1$, $l_1 = l_2 = l_3 = 0.2$, and chosen joint gear ratios $a_2 = a_3 = 0.1$. A sample of performance of this algorithm is presented in Figures 2-3.

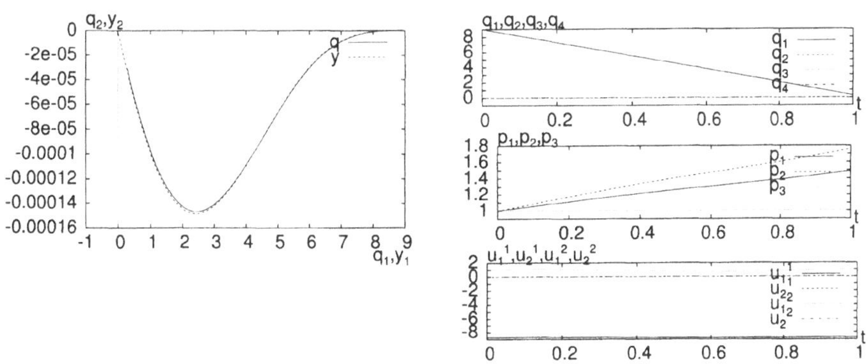

Figure 2. Initial data: $q_0 = (9,0,0,0)$, $p_0 = (1,1,1)$, $\lambda_0 = \mu_0 = (1,-0.01,0.01,0.01,-0.0001,-0.0001)$. Convergence: no. of iterations 17, final taskspace error $5.20183e-07$.

6 Conclusion

Using a combination of augmenting maps with hyperbolic and linear invariant subspaces we have derived a hyperbolic-linear, extended Jacobian, repeatable inverse kinematics algorithm for doubly nonholonomic mobile manipulators. The algorithm has regulated degree of hyperbolicity and linearity, as well as adjusted dependence between coordination

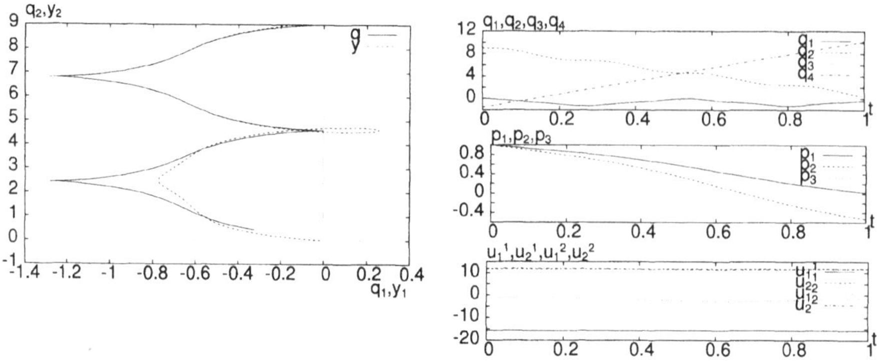

Figure 3. Initial data: $q_0 = (0, 9, 0, -\pi/2)$, $p_0 = (1, 1, 1)$, $\lambda_0 = \mu_0 = (1, 0.01, 0.01, -0.01, -0.0001, -0.0001)$. Convergence: no. of iterations 92, final taskspace error $9.87297e-07$.

and control variables. Computer simulations involving a band-limited version of this algorithm have demonstrated its satisfactory performance.

Bibliography

W. Chung and Y. Nakamura. Design and control of a chained form manipulator. *Int. Journal Robotics Res.*, 21:389–408, 2002.

Y. Nakamura, W. Chung, and O. Sordalen. Design and control of the nonholonomic manipulator. *IEEE Trans. Robot. Automat.*, 17:48–59, 2001.

O. J. Sordalen, Y. Nakamura, and W. Chung. Design of a nonholonomic manipulator. In *Proc. 1994 IEEE Int. Conf. Robot. Automat*, pages 8–13, 1994.

K. Tchoń. Repeatability of inverse kinematics algorithms for mobile manipulators. *IEEE Trans. Autom. Control*, 47:1376–1380, 2002.

K. Tchoń and J. Jakubiak. Endogenous configuration space approach to mobile manipulators: a derivation and performance assessment of Jacobian inverse kinematics algorithms. *Int. J. Control*, 76:1387–1419, 2003.

K. Tchoń and J. Jakubiak. An extended Jacobian inverse kinematics algorithm for doubly nonholonomic mobile manipulators. In *Proc. 2005 IEEE Int. Conf. Robot. Automat*, pages 1560–1565, 2005a.

K. Tchoń and J. Jakubiak. A repeatable inverse kinematics algorithm with linear invariant subspaces for mobile manipulators. *IEEE Trans. Syst., Man, and Cybernet., - Part B: Cybernetics*, 35:1051–1057, 2005b.

K. Tchoń, J. Jakubiak, and K. Zadarnowska. Doubly nonholonomic mobile manipulators. In *Proc. 2004 IEEE Int. Conf. Robot. Automat*, pages 4590–4595, 2004.

Local Variation Method to Determine Cheap Path for Nonholonomic Systems

Ignacy Duleba[†] and Pawel Ludwików[†]

[†] Inst. of Computer Engng., Control and Robotics, Wrocław University of Technology, Poland, e-mail: ignacy.duleba@pwr.wroc.pl

Abstract. In this paper a local variation method is proposed to determine a cheap path for nonholonomic motion planning. It can be used as a preliminary step in motion planning to find the path to be followed with the minimal control effort. The algorithm based on the method is detailed and parameters that influence its work discussed. Simulations performed on the unicycle mobile robot confirm efficiency of the method.

1 Introduction

One of the most interesting subjects of research in contemporary robotics is motion planning for nonholonomic systems. The systems are frequently encountered while analyzing mobile robots, underactuated manipulators, free floating robots, underwater vessels. Many methods to plan a motion of the systems have been proposed. They differ in the range of planning (local vs. global) the class of admissible models (specialized vs. general), an approach used (numeric (e.g. graph searching methods) vs. analytical), accepting other constraints (control, state space restrictions) or not. In (Duleba, 1999) an overview of attributive methods of motion planning is provided. It is well recognized that nonholonomic motion planning is not governed by the Euclidean geometry (Montgomery, 1994) what means that the minimal effort motion between two given configurations likely will not follow a straight line joining the states (car parking maneuver is an illustrative example). Some methods (Lafferriere and Sussmann, 1991) of the global motion planning require an initial path to work, some others (local Lie algebraic approaches (Duleba, 1997b)) need a local reliable direction to follow. Therefore, it is necessary to find a cheap path (not necessary directly realized with controls) that can serve as a reference path. It is a reasonable requirement that determination of the path must be computationally cheap as it is preliminary step in motion planning and must be based on a firm physical background preferring energy inexpensive paths and punishing demanding (possibly Euclidean short) ones. Duleba and Khefifi (2006) have shown that there is an analytical expression (a measure) to evaluate locally, around a current state, directions of motion with respect to their efficiency in translating the state of the system using constant amount of energy. The measure can be globalized easily, by summing up costs of local pieces of a given path, aimed at evaluating the whole path instead of its local tangents (directions). Equipped with a tool to evaluate a given path it is an interesting task to find such a path that minimizes (at least locally) the global measure. In this paper the local variation method (LVM) is proposed to solve this task. It requires any initial path that is improved in an iterative process. The path is discretized and in any iteration a local variation of the path is applied, i.e. a small piece of the path is slightly modified. If the modified part is better (smaller cost) than it was before it replaces the old part. The clear advantage of the method, inherited from all iterative methods, is that at any time there is a path to be supplied to motion planning algorithms.

The paper is organized as follows. Section 2 recalls Lie algebraic concepts required to introduce the Lie algebraic measure to evaluate paths. In Section 3 the local variation algorithm is presented and some factors influencing its performance are discussed. In Section 4 simulations of the algorithm are provided for the unicycle robot. Section 5 concludes the paper.

2 Lie algebraic preliminaries

Driftless nonholonomic systems of robotics are described by the formula

$$\dot{q} = \sum_{i=1}^{m} g_i(q)u_i \quad \dim q = n > m = \dim u \tag{2.1}$$

where g_i, $i = 1, \ldots, m$ are real analytic vector fields (called generators). At any point of the state space $Q \ni q$, admissible directions of motion are given by linear combinations of vector fields generated directly with generators or obtained, with the use of a Lie bracket, from their off-springs. The Lie bracket produces a vector field $[X, Y]$ from two other vector fields X and Y. In coordinates, the Lie bracket is defined by the formula:

$$[X, Y] = \frac{\partial Y}{\partial q}X - \frac{\partial X}{\partial q}Y.$$

The degree of a vector field counts the number of generators the field is composed of

$$\text{degree}(g_i) = 1, \ \ i = 1, \ldots, m, \qquad \text{degree}([V, Z]) = \text{degree}(V) + \text{degree}(Z).$$

For example: $\text{degree}(g_2) = 1$, $\text{degree}([[g_2, g_1], g_1]) = 3$. All vector fields with the same degree belong to a layer. The Lie algebra of system (2.1) is composed of linear combinations over R of vector fields (generators and all their Lie brackets). According to Chow (1939) theorem, system (2.1) is small time locally controllable (STLC) if and only if its Lie algebra spans R^n at each point of the state space, i.e. at each state a motion in any direction is possible. Instead of working with all Lie brackets, some (dependent on each other, due to the antisymmetry property and the Jacobi identity, Duleba (1997a)) can be excluded. Consequently, their number is shrunk and the basis of the algebra is considered. Usually it is the Ph. Hall basis although other bases are also permitted (e.g. the Lyndon basis). Further on the Ph. Hall basis will be exploited.

Having given vector fields produced (locally directions of motion), it is necessary to generate the fields with admissible controls. For this purpose the generalized Campbell-Baker-Hausdorff-Dynkin formula (Strichartz, 1987) can be applied. The formula allows to express a state translation at any point of the state space as a linear combination of vector fields

$$\Delta q = \sum_{i=1}^{k} \varphi(Z_i, \tau) Z_i \tag{2.2}$$

where Z_i are vector fields (usually obtained from the Ph. Hall basis (Duleba, 1997a)). For practical (numeric) reasons the value of the upper limit k in Eq. (2.2) is restricted and as small as possible (however not too small to avoid violating the Chow's controllability condition). Duleba and Khefifi (2006) have shown that each coefficient $\varphi(Z, \tau)$ of motion along the vector field Z is related to the degree of the vector field as follows

$$\varphi(Z, \tau) \sim \tau^{\text{degree}(Z)/2}, \tag{2.3}$$

where τ is a small time of motion. Coefficients $\varphi(Z_i, \tau)$ depend also on amplitudes of controls $u(\cdot)$ applied to the system. Consequently, motion planning is to purposefully impact coefficients $\varphi(Z_i, \tau)$ in (2.2) having given desired state translation Δq.

Let the path $q(t)$, $t \in [0, 1]$ be given. Since system (2.1) is assumed to be small time controllable then, at each point, a set of vector fields can be determined

$$G(q) = \{g_1(q), \ldots, g_m(q), g_{m+1}(q), \ldots, g_s(q)\},$$

where $s \geq n$ and $g_{m+1}(q), \ldots, g_s(q)$ are higher-order Lie brackets derived from generators $g_1(q), \ldots, g_m(q)$, that satisfy the Lie algebra rank condition

$$\operatorname{rank} G(q) = n. \tag{2.4}$$

If any vector field has been selected to $G(q)$, all other vector fields with the same degree that belong to the Ph. Hall basis of Lie algebra spanned by the generators g_i, $i = 1, \ldots, n$ must be included to the set as their impact on the resulting motion is of the same order, cf. Eq. (2.3). Therefore frequently the number of vector fields in G exceeds dimensionality of the state space. Another message from Eq. (2.3) is that, if it is only possible, small degree vector fields should enter the matrix G because they are more effective in state translation than vector fields with higher degrees. Vector fields that contribute to the matrix $G(q)$ may vary from one point of the state space to another. Respecting (2.4), around a current state q, velocity $\dot{q}$ can be expressed in terms of extended (virtual) controls v

$$\dot{q} = G(q)v.$$

Extended controls are derived using the weighted pseudo-inverse (Nakamura, 1991)

$$v = Q_0 (G^\star(q))^\# \dot{q}, \tag{2.5}$$

where Q_0 is a positive defined weighting matrix

$$G^\star(q) = G(q) Q_0^{-1}, \quad (G^\star)^\#(q) = (G^\star(q))^T (G^\star(q)(G^\star(q))^T)^{-1}.$$

If $n = s$ then $Q_0(G^\star(q))^\# = G^{-1}(q)$. More computationally demanding weighted pseudo-inverse of G is used, instead of the standard Moore-Penrose inverse, because matrix G

collects vector fields which differ in efficiency of motion (Eq. (2.3)). Therefore motion along higher degree vector fields should be dumped and lower degree ones preferred. To increase the impact of low degree vector fields on the resulting motion, the weighting matrix Q_0 was introduced.

A local cost of the trajectory $q(t)$ evaluated over short interval of time $[t, t+\Delta T]$ is described by the formula

$$k(t, q(t), \Delta T) = \sum_{i=1}^{s} \varphi\left(g_i(q(t)), \Delta T\right) |v_i(t)|. \tag{2.6}$$

The global cost of $q(\cdot)$ is determined by summing up local costs (2.6) along the discretized path

$$K(q(\cdot), \Delta T) = \sum_{r=0}^{1/\Delta T} k(r\Delta T, q(r\Delta T), \Delta T).$$

Similar measure was considered by Sekhavat and Chyba (1999).

3 Algorithm

The idea of the local variation method, given in the introduction section, will be presented in the form of an algorithm. The symbol $\leftarrow$ is reserved for substitutions, $c(q_i)$ denotes the cost of motion, based on formula (2.6), from the point q_{i-1} to q_{i+1} via q_i

$$c(q_i, \Delta T) = k(t_{i-1}, q(t_{i-1}), \Delta T) + k(t_i, q(t_i), \Delta T). \tag{3.1}$$

The algorithm runs as follows:

Step 1 Read in the initial discretized path q_i, $i \in \{0, \dots, l\}$, discretized uniformly in time with the step ΔT, $1 = l \cdot \Delta T$. Set the value of the vector s, dim q_i = dim $s = n$ used to vary coordinates of the modified point.

Step 2 Select one intermediate point $q_i = q(i \cdot \Delta T)$, $i \in \{1, l-1\}$.

Step 3 Vary the selected point

$$q_i^\star \leftarrow q_i + (r_1 s_1, \dots, r_n s_n)^T, \tag{3.2}$$

where the vector $s = (s_1, \dots, s_n)^T$ collects maximal amplitudes of coordinates' changes while the vector of coefficients $r = (r_1, \dots, r_n)^T \in [-1, 1]^n$ = Cube, impacts the amplitude of the state modification.

Step 4 Compare costs of the current path and just modified (the modification is local so only small parts of the paths are compared):

$$\text{if} \quad c(q_i^\star, \Delta T) < c(q_i, \Delta T) \quad \text{then} \quad q_i \leftarrow q_i^\star.$$

Step 5 Check an appropriate stop condition. If the condition is satisfied stop the algorithm and output the current path otherwise process with Step 2.

The algorithm requires fixing some details. Selecting an intermediate point (Step 2) can be done in many ways. Straightforward choices are presented below (stochastic and deterministic):

- the intermediate point is chosen at random,
- in consecutive iterations points along the path are selected in the forward direction, i.e. the index of selected point varies from 1 to $l-1$,
- as in the previous selection method but in the backward direction (the index goes from $l-1$ down to 1),
- the point $q_{i^\star}$ is taken that is currently the most expensive with respect to the cost (3.1) along the path: $c(q_{i^\star}, \Delta T) = \max_{j\in\{1,l-1\}} c(q_j, \Delta T)$. This choice must be equipped with a mechanism to avoid repeating the same unsuccessful trial.

Vector s of the maximal admissible coordinate modifications should depend on the nature of coordinates (angular, translational) and the range of the motion. The vector of coefficients r can select points from

1. vertexes of the cube; r_j $(j \in \{1, n\})$ picked at random from the set $\{-1, 1\}$
2. inside the cube; all r_j $(j \in \{1, n\})$ picked at random from the range $[-1, 1]$
3. the sphere, r picked at random, but satisfying $r^T r = 1$.

Above the simplest (and numerically effective) methods of varying the selected point were presented. More complex approach is also possible when the point is selected from the R^{n-1} space anchored at q_i and perpendicular to the path on its interval $q_{i-1}q_iq_{i+1}$.

There is no natural condition to stop computations as optimality, even local, can not be shown. Therefore the stop condition should be chosen arbitrary, likely to set reasonable trade-off between the time of computations and the quality of the current best solution. Two stop conditions can be proposed:

- computations are stopped after N iterations (the fixed parameter). This condition does not take into account dynamics of the decrease of the path cost,
- the algorithm is terminated when N_p (the fixed parameter) iterations passed without satisfying condition (3.2).

4 Simulation

The proposed algorithm of the path optimization was tested on the unicycle robot described by the following equation

$$\dot{q} = \begin{bmatrix} \dot{x} \\ \dot{y} \\ \dot{\theta} \end{bmatrix} = \begin{bmatrix} \cos\theta \\ \sin\theta \\ 0 \end{bmatrix} u_1 + \begin{bmatrix} 0 \\ 0 \\ 1 \end{bmatrix} u_2 = g_1(q)u_1 + g_2(q)u_2, \tag{4.1}$$

where x, y denote the Cartesian location of the mobile robot while θ describes its orientation. Step vector was set to $s = (0.05, 0.05, 0.05)$ The paths were composed of $l = 21$ points, $\Delta T = 1/l$. In computing the extended controls v, Eq. (2.5), the matrix $Q_0 = diag((\Delta T)^{-1/2}, (\Delta T)^{-1/2}, (\Delta T)^{-1})$ was employed and $\dot{q}$ was approximated as $\dot{q}_i = ((q_{i+1} - q_i)/\Delta T + (q_i - q_{i-1})/\Delta T)/2 = (q_{i+1} + q_{i-1})/(2\Delta T)$. Four initial paths were considered:

Path 1 straight-line motion with the fixed orientation from the point $(0, 0, 45°)$ to $(1, 0, 45°)$

Path 2 motion along the straight line joining the positions $(0, 0)$ and $(1, 0)$ with linear orientation changes from $90°$ to $270°$

Table 1. Simulation results for random and forward strategies for different values of N

Path	N	random				forward			
		cost		improvements		cost		improvements	
		best	average	best	average	best	average	best	average
1	500	2508	3815	130	119	2492	3798	124	121
	1000	1221	1673	158	166	1139	1652	169	168
	5000	900	1037	231	198	884	1029	200	200
	10000	850	984	213	207	841	981	218	207
2	500	1714	2778	100	97	1574	2773	107	97
	1000	853	1330	126	132	837	1323	137	133
	5000	598	739	100	165	592	726	153	167
	10000	562	691	184	173	557	679	178	175
3	500	1526	2830	98	89	1935	2847	86	90
	1000	822	1428	132	128	908	1419	140	130
	5000	572	728	178	165	561	718	174	166
	10000	500	677	197	174	509	671	195	175
4	500	1787	3300	118	115	1654	3185	130	115
	1000	986	1480	158	151	942	1436	156	150
	5000	725	850	186	183	707	838	169	181
	10000	684	805	198	191	691	792	176	189

Path 3 the straight line motion from $(0, 0, 90°)$ to $(1, 1, 0°)$,

Path 4 Path 3, with the extra point $(2, 0, -90°)$ added as the goal point.

To average fluctuations (possible for each single task) and to draw statistically sound conclusions, 1000 simulations were performed for any tested task. In all simulations the second method of varying selected point was exploited. Due to the lack of space only two strategies were presented: the deterministic forward direction and the random selection method. In Table 1 data obtained from simulation were collected. Beside costs also the number of improvements was shown (split into two columns presenting the average number of improvements and the number for the optimal path obtained). As one can expect, the increase of the number of iterations improves the quality of resulting path (its cost decreased). The decrease is linear at the beginning (2000 iterations) and then constantly slows down until plateau of the characteristics is reached. It appears that the number of iterations to stop the algorithm should not be larger than a few thousands. To get insight into the computational complexity of the algorithm, the run time was collected from 4000 tasks with fixed initial conditions (the initial path, the strategy of selecting points). The average time for $N = 500, 1000, 5000, 10000$ iterations was $0.012, 0.022, 0.101, 0.200[sec]$ respectively. The algorithm implemented in C programming language was run on PC clocked with 1.4[GHz]. It is very fast as computations involved in each path modification are cheap. It can be run even in on-line (real time) mode. In Fig. 1 the best obtained paths were depicted on the xy plane. Most resulting paths presented are quite intuitive and the experienced driver can generate them. Maybe the first one corresponding to the parking maneuver needs some explanations. One can

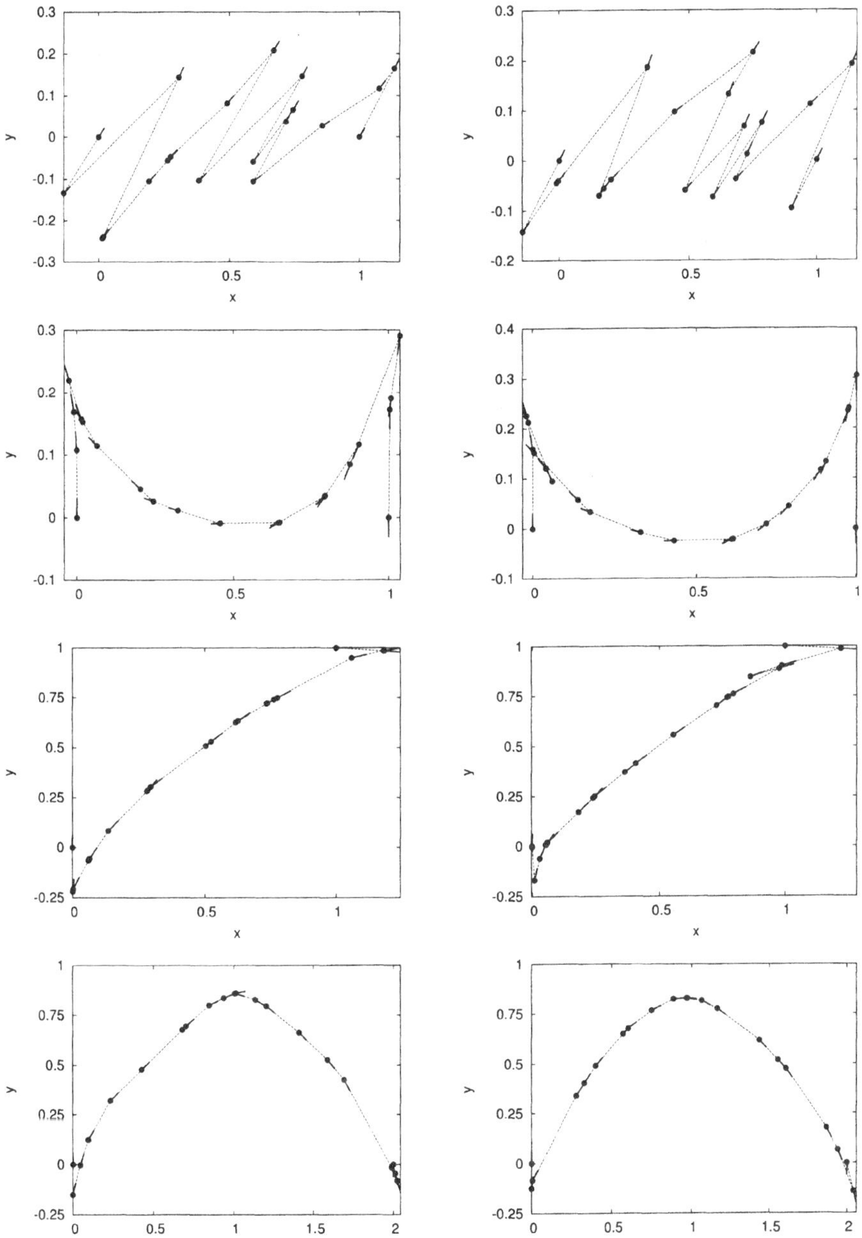

Figure 1. The best paths obtained by LVM initialized with Paths 1-4. Two strategies of selecting the modified point were applied: random (the left panel) and the forward direction (the right panel). Black dots depict some intermediate points on the path while short line segments anchored at the dots show the unicycle orientation.

expect the sub-optimal motion scenario should pass through three phases: reorienting the robot towards the goal point, motion along a straight line and finally reorienting to get the desired angular position. This scenario requires global information because at the very beginning of the motion its cost must increase slightly. Because, due to its definition, the local variation method does not accept any increase of the cost along the path, this scenario of motion can not be generated. However, the robot tries to minimize the cost by doing a sequence of parking maneuvers. The maximal amplitude of deviation from the straight line between the initial and the final state set some kind of equilibrium between the Euclidean length of the path increased and gains from the local cost decrease.

5 Conclusions

In this paper the local variation method for computation of cheap paths for nonholonomic motion planning was introduced. It uses the Lie-algebraic measure to iteratively evaluate local costs between neighboring points of the path. It is computationally feasible and can be run even in real-time mode. The method admits setting some parameters to accommodate it to a particular vector of state coordinates (units and domains). Iterative methods provide a path at any stage of computations to be issued to the motion planer module. The method of local variations can be used either to set consecutive targets for local methods of motion planning (for example Lie-algebraic methods) or to supply the target path to be generated with controls for some global methods.

Bibliography

W.L. Chow. Über Systeme von linearen partiellen Differentialgleichungen erster Ordnung. *Math. Ann.*, 117(1):98–105, 1939.

I. Duleba. Checking controllability of nonholonomic systems via optimal Ph. Hall basis generation. In *Proc. IFAC SyRoCo Conf.*, pages 485–490, 1997a.

I. Duleba. Locally optimal motion planning of nonholonomic systems. *Journ. of Rob. Syst.*, 14(11):767–788, 1997b.

I. Duleba. *Algorithms of motion planning for nonholonomic robots.* Wroclaw Univ. of Technology Publ., 1999.

I. Duleba and W. Khefifi. Pre-control form of the generalized Campbell-Baker-Hausdorff-Dynkin formula for affine nonholonomic systems. *Syst. Contr. Lett.*, 55:146–157, 2006.

G. Lafferriere and H. Sussmann. Motion planning for controllable systems without drift. In *Proc. IEEE Conf. on Rob. and Autom.*, pages 1148–1153, 1991.

R. Montgomery. Abnormal minimizers. *SIAM Jorn. Contr. and Optimization*, 32(6): 1605–1620, 1994.

Y. Nakamura. *Advanced Robotics: Redundancy and Optimization.* Addison Wesley, 1991.

S. Sekhavat and M. Chyba. Nonholonomic deformation of a potential field for motion planning. In *Proc. IEEE Conf. on Rob. and Autom.*, pages 817–822, 1999.

R.S. Strichartz. The Campbell-Baker-Hausdorff-Dynkin formula and solutions of differential equations. *Journ. of Funct. Anal.*, 72:320–345, 1987.

Nonlinearity detection and reduction based on unnormalized quasi-velocities

Przemyslaw Herman ‡ and Krzysztof Kozlowski ‡

‡ Chair of Control and Systems Engineering, Poznan University of Technology, Poznan, Poland

Abstract. In this paper an analysis concerning dynamical couplings between manipulator links is proposed. In order to detect nonlinearities we utilize equations of motion in terms of unnormalized quasi-velocities (UQV) introduced originally by Jain and Rodriguez (1995). Based on the equations some observations which are helpful for nonlinearity investigation and reduction were made. The presented strategy was tested by simulation on a 3 *d.o.f.* spatial manipulator.

1 Introduction

Classical equations of motion are second-order and highly nonlinear. It is well known fact that satisfactory control objective can be achieved if the nonlinearities arising from a manipulator design are not very high. If their level is high then the obtained performance are deteriorated. In spite of that there are many algorithms (e.g. PD control, inverse dynamic control) used for robotic manipulators (Sciavicco and Siciliano, 1996, Slotine and Li, 1991). But the problem of nonlinear decoupling of the system dynamics under control law is very important and still under investigation (Cotsaftis et al., 1998). The best situation is when complete decoupling of dynamic equations i.e. eliminating Coriolis, centripetal, gravitational and cross inertia terms is possible. It can be even achieved for a planar 2 *d.o.f.* manipulator (Coelho et al., 2004). In other cases reduction of nonlinearities caused in the system is indispensable (e.g. based on the method described in (Pons et al., 1997)). Couplings are used also for control purpose of underactuated manipulators (Bergerman et al., 1995).

On the other hand we can decompose the inertia matrix in order to obtain first-order equations of motion. One of known methods (Jain and Rodriguez, 1995) allows one to diagonalize dynamic equations of a manipulator based on introducing the unnormalized quasi-velocities (UQV).

The objective of this paper is to present a scheme for detecting of dynamical couplings in order to reduce them. This problem is important mainly in direct drive robots. We utilize the approach given by Jain and Rodriguez (1995). If the manipulator is described in terms of the UQV then some observations which allow us to evaluate the couplings can be made. Based on dynamic properties it is possible to eliminate cross inertia terms and decrease values of Coriolis and gravitational terms. All quantities concern each individual link (i.e. an imaginary link which contains dynamical couplings between the k-th link and others) and depend on the manipulator parameters. It is shown that the UQV are useful for evaluation of the system dynamics. The proposed strategy can serve as a tool for decreasing of nonlinearities in the manipulator.

The paper is organized as follows. The second Section shows equations of motion in terms of UQV. In the third Section a strategy for nonlinearity investigation and reduction is proposed. Equations of motion in terms of UQV based on optimal solution are given. Simulation results concerning a 3 *d.o.f.* system are presented in the fourth Section. The last section offers conclusions.

2 Equations of motion in terms of UQV

The dynamic model of a manipulator is given by Sciavicco and Siciliano (1996):

$$M(\theta)\ddot{\theta} + C(\theta,\dot{\theta}) + G(\theta) = \tau, \tag{2.1}$$

where:

$\theta, \dot{\theta}, \ddot{\theta} \in R^N$ – vectors of generalized positions, velocities and accelerations, respectively,

N – number of degrees of freedom,

$M(\theta) \in R^{N\times N}$ – system mass matrix,

$C(\theta,\dot{\theta}) = (\dot{M}\dot{\theta} - \frac{1}{2}\dot{\theta}^T M_\theta \dot{\theta}) \in R^N$ – vector of Coriolis and centrifugal forces in standard equations of motion, where the expression $\dot{\theta}^T M_\theta \dot{\theta}$ is the column vector $col(\dot{\theta}^T M_{\theta_k} \dot{\theta})$ ($M_{\theta_k} = \frac{\partial M}{\partial \theta_k}$ denotes the partial derivative of the mass matrix $M(\theta)$ with respect to the joint coordinate θ_k) and $\dot{M}$ is the time derivative of the matrix $M(\theta)$ (derivation and explanation of the term $C(\theta,\dot{\theta})$ was given e.g. by Koditschek (1984, 1985)),

$G(\theta) \in R^N$ – vector of gravitational forces in standard (classical) equations of motion,

$\tau \in R^N$ – vector of applied moments.

In accordance with the notation given by Jain and Rodriguez (1995) links are numbered from the tip to the base of a manipulator and the symbols mean:

$\xi \in R^N$ – vector of unnormalized quasi-velocities (UQV), i.e. $\xi = [I + H\phi K]^T \dot{\theta}$,

$H \in R^{N\times 6N}$ – projection operator for all joint axes,

$\phi \in R^{6N\times 6N}$ – rigid manipulator force transformation matrix, item $K \in R^{6N\times N}$ – shifted Kalman gain matrix,

$P \in R^{6N\times 6N}$ – articulated inertia matrix,

$D = HPH^T \in R^{N\times N}$ – articulated inertia about joint axes matrix $D = D(\theta)$,

$\dot{D} \in R^{N\times N}$ – time derivative of the matrix $D = D(\theta)$,

$C_\xi(\theta,\xi) \in R^N$ – vector of Coriolis and centrifugal forces in diagonalized normalized equations of motion,

$G_\xi(\theta) \in R^N$ – vector of gravitational forces in diagonalized unnormalized equations of motion,

$L(\theta) \in R^{N\times N}$ – spatial operator arising from a mass matrix $M(\theta)$ decomposition, namely $M(\theta) = L(\theta)D(\theta)L^T(\theta)$,

$\dot{L}(\theta) \in R^{N\times N}$ – time derivative of the matrix $L(\theta)$,

$\kappa \in R^N$ – vector of unnormalized quasi-moments,

$(.)^T$ – transpose operation.

Equations of motion in terms of UQV can be written as follows:

$$D\dot{\xi} + C_\xi(\theta, \xi) + G_\xi(\theta) = \kappa \tag{2.2}$$

$$\xi = L^T\dot{\theta} \tag{2.3}$$

where $\dot{\xi} = \dot{L}^T\dot{\theta} + L^T\ddot{\theta}$, $C_\xi(\theta,\xi) = L^{-1}C(\theta,\dot{\theta}) - D\dot{L}^T\dot{\theta}$, $G_\xi(\theta) = L^{-1}G(\theta)$ and $\kappa = L^{-1}\tau$. Note that Eq.(2.2) describes dynamics of a manipulator whereas Eq.(2.3) gives transformation between the joint velocities vector $\dot{\theta}$ and the vector of quasi-velocities ξ. The physical units of both vector elements are the same. Elements of the matrix L^T are unitless.

3 Nonlinearity investigation

Note that the terms in Eqs.(2.2) and (2.3) can be considered separately for $k = 1, 2, \ldots, N$ joints:

$$M_{\xi k} = D_k\dot{\xi}_k, \;\; C_{\xi k} = C_{\xi k}(\theta, \xi), \;\; G_{\xi k} = G_{\xi k}(\theta), \;\; \kappa_k, \;\; \xi_k = \sum_{i=k}^{N} L^T_{ki}\dot{\theta}_i. \tag{3.1}$$

Several observations concerning these terms can be made.
(A 1). Quantities $M_{\xi k}$, $C_{\xi k}$, $G_{\xi k}$, κ_k and ξ_k are calculated individually for each k-th joint. All of them are related to the motion of the k-th link and contain dynamical couplings between links. We can imagine that each link (containing the couplings) rotates about the k-joint axis. Besides, a part of couplings (cross inertia terms) of $M_{\xi k}$ is shifted to the term $C_{\xi k}$ and next transform into the UQV space. In this way we ensure eliminating of the cross inertia terms. The system dynamics is hidden also in ξ_k. Recalling Eq.(2.2) one can observe that vectors C_ξ and G_ξ depend on the transformation matrix L^{-1} (C_ξ additionally on $\dot{L}^T$). If the matrix $M(\theta)$ will be diagonal then the matrix L is the identity one. As a consequence nonlinearities resulting from decomposition of the matrix $M(\theta)$ do not exist. However, each element behind of the diagonal causes that the matrix L introduces a nonlinearity which can be detected using elements of vectors C_ξ and G_ξ. In such case dynamical couplings between links appear. Hence, if we want to decrease nonlinearities then values $M_{\xi k}$, $C_{\xi k}$, $G_{\xi k}$ should be as small as possible because the control effort measured by sum of all κ_k (and also by τ) will be reduced and at the same time accelerations will be decoupled. If the quantities $C_{\xi k}$, $G_{\xi k}$, κ_k, ξ_k are close to C_k, G_k, τ_k, $\dot{\theta}_k$, respectively, then we can say that the manipulator links are structurally decoupled. We can give a more formal measure of maximal nonlinearity e.g.

$$C_{\max\ k} = \max|C_k - C_{\xi k}|, \;\; G_{\max\ k} = \max|G_k - G_{\xi k}|, \;\; Q_{\max\ k} = \max|\tau_k - \kappa_k|. \tag{3.2}$$

The selection of the criteria is motivated as follows. The proposed quantities allows one to evaluate differences between forces expressed in classical equations of motion and the analogous quantities presented in equations in terms of UQV. The differences arise as a result of mass matrix diagonalization and represent influence of dynamical couplings for the manipulator behavior.

(A 2). Elements of the diagonal matrix D allow one to compute the inertia which rotates about each manipulator joint axis during the motion. For the k-th joint each value D_k takes into account parameters of all links closer to the manipulator tip. The quantity D_k gives information how much 'total' rotational inertia corresponds to each joint axis.
(A 3). From reference (Jain and Rodriguez, 1995) arises that the kinetic energy of the system is expressed as follows:

$$K_E(\theta, \xi) = \frac{1}{2}\xi^T D\xi = \frac{1}{2}\sum_{k=1}^{N} D_k \xi_k^2. \tag{3.3}$$

From Eq.(3.3) results that D_k and ξ_k can be treated as independent variables in terms of UQV. Based on expression $\frac{1}{2}D_k\xi_k^2$ we can detect the kinetic energy which is related to each ξ_k (i.e. to individual link) while by calculating the sum (3.3) we recognize how much of the kinetic energy is dissipated through the manipulator if it moves. Note, that using classical state equations (2.1) such information is not given.
(A 4). Elements of the UQV vector ξ represent some velocities which take into account both all joint velocities closer to the manipulator base and mechanical parameters of the system (inertial, geometrical etc.). Each quasi-velocity ξ_k depends on the system configuration. If the manipulator moves then dynamical couplings between links deform the desired (or actual) velocity profile $\dot{\theta}_k$. Therefore we can propose a velocity deformation for the k-th joint as follows:

$$V_{dk} = \max|\xi_k - \dot{\theta}_k|. \tag{3.4}$$

This quantity can serve as a tool for detection of mechanical couplings in the manipulator. If the couplings are high then values of the index (3.4) is big and the quasi-velocity profile will be different than the velocity in the k-th joint.
(A 5). Based on results obtained from equations of motion in terms of UQV it is possible to judge which link parameters causes nonlinearities in the manipulator. In general we propose the following algorithm.
(1) Assume the parameters set of the original manipulator.
(2) Consider possible parameter modifications based on matrices containing masses, inertias, static moments and geometrical parameters.
(3) Determine quantities and indexes from Eqs.(3.1) - (3.4) and compare time responses.
(4) Change one subset of parameters and repeat the step (3).
(5) Assume the parameters set which guarantees the greatest reduction of nonlinearities.

Because, in the presented method both dynamics of each link and of the whole system are taken into account it seems useful if changes of many parameter are allowed.

4 Simulation results

The first aim of simulation is to point out some advantages presented in Section 3. As an example a spatial 3 *d.o.f.* Yasukawa-like manipulator was chosen. Its kinematic scheme is depicted in Figure 1. The used dynamical equations were derived by Kozlowski (1992). It was assumed that for *Case 1* the set of manipulator parameters is as in Table 1.

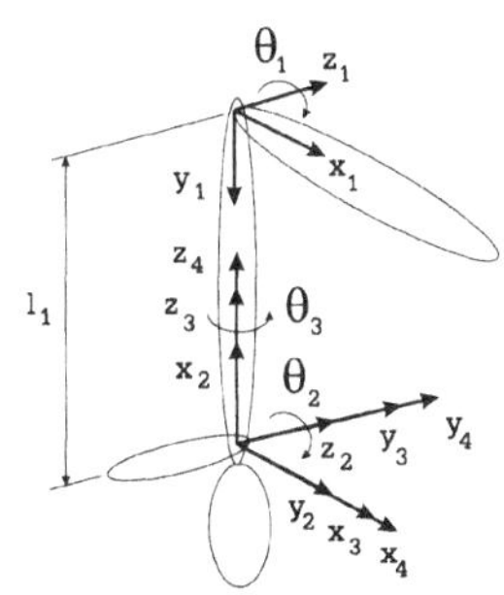

Figure 1. Kinematic scheme of the manipulator.

Table 1. Parameters of the spatial manipulator.

Link number k		1	2	3
m_k	kg	10	30	60
l_{k-1}	m	0	0.8	0
p_{xk}	m	0.1	0.1	0
p_{yk}	m	0.01	0.01	0
p_{zk}	m	0.01	0	0
J_{xx}	kgm^2	0.4	0.2	0
J_{xy}	kgm^2	0	0.01	0
J_{xz}	kgm^2	0	−0.01	0
J_{yy}	kgm^2	0.04	0.7	0
J_{yz}	kgm^2	−0.01	−0.01	0
J_{zz}	kgm^2	0.5	1.0	1.5
α_k	deg	0	−90	0

For the simulation fifth-order polynomials according the following general formula ($\Delta\theta = \theta_f - \theta_i$) i.e. $\theta(t) = \theta_i + (10\Delta\theta/t_f^3)t^3 - (15\Delta\theta/t_f^4)t^4 + (6\Delta\theta/t_f^5)t^5$ were chosen. The polynomial trajectories were described with initial points $\theta_{i1} = (1/3)\pi\ rad$, $\theta_{i2} = \pi\ rad$, $\theta_{i3} = -(1/2)\pi\ rad$, final points $\theta_{f1} = -(2/3)\pi\ rad$, $\theta_{f2} = 0\ rad$, $\theta_{f3} = (1/2)\pi\ rad$, and time duration $t_f = 1\ s$. All simulations realized in MATLAB were performed with the fixed step size $0.005\ s$.

Simulation results for *Case 1* are presented in Figure 2. In Figure 2(a) joint velocities $\dot{\theta}_k$ ($d\theta_k$) and quasi-velocities ξ_k are compared. As a consequence of great nonlinearities in the system the first UQV is the most deformed. We conclude that the first link is sensitive to mechanical couplings. Indexes of deformation are $V_{dk} = 8.83, 0.30$ and 0. Differences between Coriolis components $C_k - C_{\xi k}$ are shown in Figure 2(b). As a result of couplings between links inertial forces are accumulated mainly in the third and in the first joint. The Coriolis term indexes are $C_{\max\ k} = 43.9, 19.1, 73.8$. Figure 2(c) presents comparison between gravitational terms $G_k - G_{\xi k}$. We see that nonlinearities are noticeable above all for the first joint. The gravitational term indexes are $G_{\max\ k} = 0, 17.0$ and 86.8. Next, in Figure 2(d) inertial terms $M_{\xi k}$ for each joint are given. They represent the

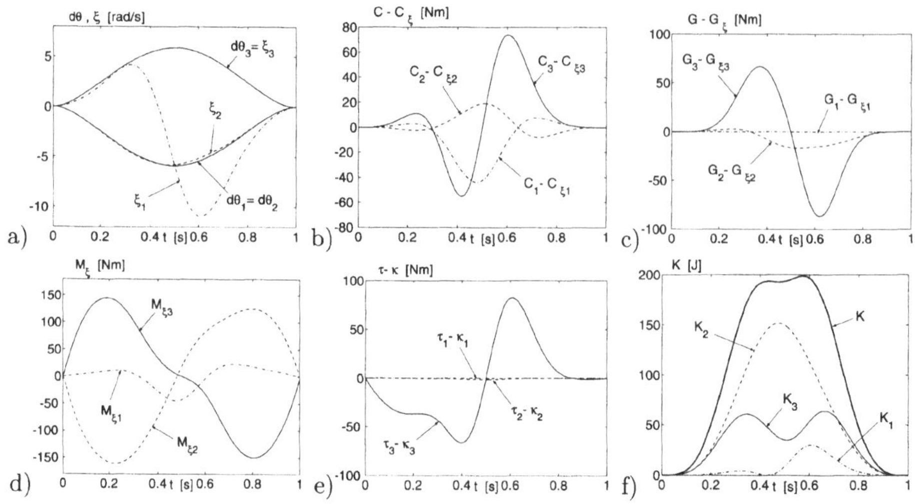

Figure 2. Case 1: a) quasi-velocities ξ_k and joint velocities $d\theta_k$; b) differences between elements of k-th Coriolis term $C_k - C_{\xi k}$; c) differences between elements of k-th gravitational term $G_k - G_{\xi k}$; d) inertial terms $M_{\xi k}$; e) differences between k-th applied and quasi-moment $\tau_k - \kappa_{\xi k}$; f) kinetic energy K_E for all links and for the whole manipulator.

inertia which rotates about each joint axis. The quantities have comparable values for the second and for the third joint. Note, that big values in the second joint arise from couplings (compare m_2, J_{zz2} and m_3, J_{zz3}). Figure 2(e) compares differences between applied moments and quasi-moments $\tau_k - \kappa_{\xi k}$. As can be expected (from Figures 2(b)-2(d)) the greatest values concern the third joint. We obtained $Q_{\max\ k} = 0, 1.4$ and 82.3. It is a result of dynamical couplings between the third link and others. Figure 2(f) shows the kinetic energy dissipated by the manipulator and by each link. As arises from this figure the biggest kinetic energy is reduced by the second link whereas by the third and the first less. Therefore, the existing couplings increase the energy which has to be reduced by the second link.

Based on these results we conclude that the first joint velocity is the most sensitive for nonlinearities. The effect of internal couplings is observable for the third joint. Moreover, coupling inertia, which has to be dissipated, is contained mainly in $\frac{1}{2}D_2\xi_2^2$.
Therefore we propose to redesign some of manipulator parameters according to the algorithm given in *(A 5)*.
Case 2. The following manipulator parameters were changed: $m_1 = 6kg$, $m_2 = 40kg$, $J_{zz2} = 1.2kgm^2$, $p_{x1} = -0.05m$, $p_{x2} = 0.05m$ and $l_1 = 0.9m$. All of them are related to the first and the second link.
In Figure 3(a) joint velocities $\dot{\theta}_k$ ($d\theta_k$) and quasi-velocities ξ_k are given. Note, that as a result of nonlinearities reduction ξ_1 is closer to $\dot{\theta}_1$ than previously (in spite of that m_2 and J_{zz2} increased). As results from Figure 3(b) the Coriolis term differences were decreased and couplings arising from decomposition of the mass matrix are not so big

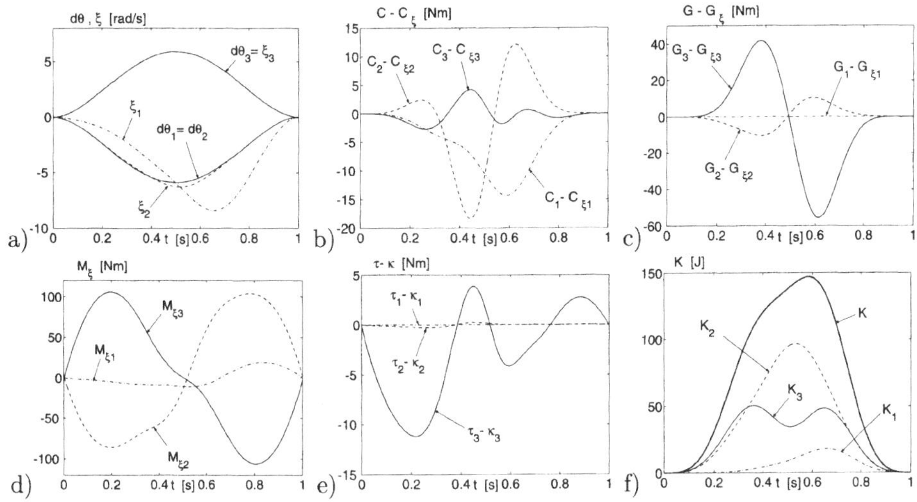

Figure 3. Case 2: a) quasi-velocities ξ_k and joint velocities $d\theta_k$; b) differences between elements of k-th Coriolis term $C_k - C_{\xi k}$; c) differences between elements of k-th gravitational term $G_k - G_{\xi k}$; d) inertial terms $M_{\xi k}$; e) differences between k-th applied and quasi-moment $\tau_k - \kappa_{\xi k}$; f) kinetic energy K_E for all links and for the whole manipulator.

in the third joint as previously. Similarly, differences between gravitational terms were reduced (Figure 3(c)). However, for the third joint the value $G_3 - G_{\xi 3}$ is still big. The same effect of nonlinearities reduction is observable in Figure 3(d). At present less inertia rotates about each joint axis than in *Case 1*. Differences $\tau_k - \kappa_{\xi k}$ are shown in Figure 3(e). One can see that the difference for the third joint is quite small because of decreasing of mutual couplings between links. Furthermore, both the kinetic energy dissipated by the whole manipulator and especially the energy $\frac{1}{2}D_2\xi_2^2$ was reduced. The above observations mean that if a control algorithm will be used one obtains more individual regulation of each link than in *Case 1*.

For *Case 2* we obtain reduced values of indexes concerning deformation of quasi-velocity trajectory, Coriolis term, gravitational term and torques term, respectively. They are as follows: $V_{dk} = 3.95, 0.42, 0$, $C_{\max\ k} = 14.3, 18.3, 4.2$, $G_{\max\ k} = 0, 10.6, 55.6$ and $Q_{\max\ k} = 0, 0.29, 11.2$. As a result of the manipulator design improvement we received also considerable decoupling between its links.

5 Conclusions

Some properties which can be detected based on analysis of equations of motion in terms of the UQV have been presented in this paper. It was shown that dynamics of a manipulator represented by two first-order equations of motion gives information which allows one to evaluate mechanical couplings between links. Advantages of the UQV representation were summarized in several observations. Based on the introduced indexes and using the UQV vector it is possible to calculate the inertia rotating by each manipulator joint axis (and the corresponding inertial forces), deformation of the joint velocity trajectory, the kinetic energy dissipated by each link and the separate quasi-moment which corresponds to the appropriate quasi-velocity. Moreover, simulation results for a 3 *d.o.f.* spatial manipulator with rotational joints lead to conclusion that changing the system parameters set we can reduce the existing nonlinearities. The proposed strategy can be utilized for evaluation and reduction of nonlinearities in the manipulator. As a consequence of couplings evaluation we are able to improve the performance of the controller even though the algorithm corresponds to classical second-order equations of motion. Therefore, the presented approach can be used in order to gain further insight into the manipulator dynamics. It was shown that based on the introduced quantities, indexes and simulation results it was possible to evaluate mechanical couplings between manipulator links and to reduce them changing the set of parameters. As a result of the approach expensive experimental investigations can be avoided.

Bibliography

M. Bergerman, M. Lee, and Y.S. Xu. A Dynamic Coupling Index for Underactuated Manipulators. *J. Robotic Systems*, 10:693-707, 1995.

T.A.H. Coelho, L.Yong, and V.F.A. Alves. Decoupling of dynamic equations by means of adaptive balancing of 2-dof open-loop mechanisms. *Mechanism and Machine Theory*, 39:871-881, 2004.

M. Cotsaftis, J. Robert, M. Rouff, and C. Vibet. Applications and prospect of the nonlinear decoupling method. *Comput. Methods Appl. Engrg.*, 154:163-178, 1998.

A. Jain and G. Rodriguez. Diagonalized Lagrangian Robot Dynamics. *IEEE Transactions on Robotics and Automation*, 11:571-584, 1995.

D. Koditschek. Natural Motion for Robot Arms. In *Proceedings of the 23rd IEEE Conference on Decision and Control*, pages 733-735, 1984.

D. Koditschek. Robot Kinematics and Coordinate Transformations. In *Proceedings of the 24th IEEE Conference on Decision and Control*, pages 1-4, 1985.

K. Kozlowski. *Mathematical Dynamic Robot Models and Identification of Their Parameters.* Technical University of Poznan Press, Poznan, 1992 (in Polish).

J.L. Pons, R. Ceres, A.R. Jimenez, L. Calderon, and J.M. Martin. Nonlinear Performance Index (npi): A tool for Manipulator Dynamics Improvement. *Journal of Intelligent and Robotic Systems*, 18:277-287, 1997.

L. Sciavicco and B. Siciliano. *Modeling and Control of Robot Manipulators.* The McGraw-Hill Companies, Inc., New York, 1996.

J.J. Slotine and W. Li. *Applied Nonlinear Control.* Prentice Hall, New Jersey, 1991.

Motion Planning for Collision Avoidance via Cylindrical Models of Rigid Bodies

Pierre M. Larochelle *‡ *and John S. Ketchel †‡
* Associate Professor, pierrel@fit.edu
† Doctoral Candidate, jketchel@fit.edu
‡ Robotics and Spatial Systems Lab, Florida Institute of Technology, Melbourne, FL, USA

Abstract. This paper presents a novel methodology for motion planning to avoid collisions. Right circular cylinders are used to model the bodies of interest in the workcell. This algorithm uses line geometry and dual number algebra to exploit the geometry of right circular cylindrical objects to facilitate the detection of collisions. First, the rigid bodies are modelled with infinite length cylinders and an efficient necessary condition for collision is evaluated. If the necessary condition is not satisfied then the two bodies do not collide. If the necessary condition is satisfied then a collision between the bodies may occur and we proceed to the next stage of the algorithm. In the second stage the bodies are modelled with finite cylinders and a definitive necessary and sufficient collision detection algorithm is employed. The result is a straight-forward and efficient means of detecting collisions of cylindrically shaped bodies moving in three dimensions. This methodology is then employed in a traditional motion planning algorithm. A case study is included.

1 Introduction

Collision detection is vital for real world implementation of robotic mechanical systems. Collision detection assists in motion planning, real-time control, digital prototyping and motion simulation of these systems. Zsombor-Murray (1992) presents the visualization of the shortest distance between two lines in space. His constructive geometry and algebraic solutions to the problem motivated the work presented here. Xavier (2000) correctly states that failure to detect a collision is less acceptable than false positives, which can be further checked and that for the sake of speed exact or accurate collision detection is often sacrificed. Prior works on the collision detection problem have yielded many software packages such as: VEGAS(Johnson and Vance, 2001), V-Clip(S. Caselli and Mazzoli, 2002), RAPID(S. Caselli and Mazzoli, 2002), SOLID(S. Caselli and Mazzoli, 2002), I-Collide(J. Cohan and Ponamgi, 1995), V-Collide(T. Hudson and Manocha, 1997) and PQP(S. Caselli and Mazzoli, 2002). They vary in their modelling and in mathematical

*This material is based upon work supported by the National Science Foundation under grant No. #0422705. Any opinions, findings, and conclusions or recommendations expressed in this material are those of the authors and do not necessarily reflect the views of the National Science Foundation.

method for determining if a collision has occurred but none seek to exploit line geometry to facilitate collision detection as is done here.

In this paper we present a methodology for motion planning to avoid collisions that uses a novel algorithm for determining quantitatively if two bodies moving in three dimensional space collide. We present a two stage algorithm for determining if two bodies moving in collide. In the first stage, infinite length cylinders are used to model the objects, then line geometry is used to determine if the cylinders intersect. If these infinite cylinders do not intersect then the two bodies do not collide and no further testing is required. If the two infinite cylinders do intersect then further testing is necessary. We proceed to the second stage where cylinders of finite length are used to model the objects and they are tested to determine quantitatively if they collide.

The methodology presented here is general and can be used to detect collisions between any rigid bodies moving in three dimensions provided that the bodies are predominantly cylindrical in shape. We focus upon such bodies because i is commonly found in industrial robots and parallel kinematic machines.

2 Collision Detection

2.1 Infinite Cylinder Testing

Initially, each rigid body in the workcell is modelled by a cylinder of infinite length and finite radius. If the distance between the two cylinders is less than the sum of their two radii then the infinite cylinders have collided. Hence, if the actual finite cylindrical objects have collided it is *necessary* that the minimum distance between their associated infinite cylinders be less than the sum of their radii.

We use Plücker coordinates and dual vectors to represent the major axis of an infinite cylinder (McCarthy, 2000). Line S_1 can be defined by points $\vec{c}$ and $\vec{f}$ or by point $\vec{c}$ and direction vector $\vec{s}$ (see Fig. 1 and Eqs. 2.1 & 2.2).

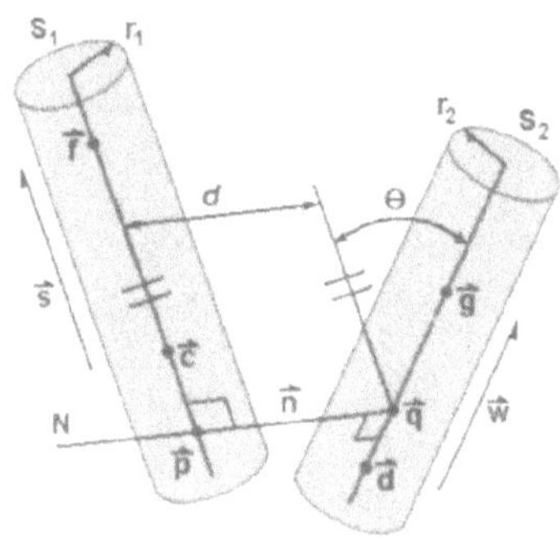

Figure 1. Infinite Cylinders

$$S_1 = \left(\frac{\vec{f}-\vec{c}}{\|\vec{f}-\vec{c}\|}, \vec{c} \times \frac{\vec{f}-\vec{c}}{\|\vec{f}-\vec{c}\|}\right) = (\vec{s}, \vec{c} \times \vec{s}) \tag{2.1}$$

$$S_2 = \left(\frac{\vec{g}-\vec{d}}{\|\vec{g}-\vec{d}\|}, \vec{d} \times \frac{\vec{g}-\vec{d}}{\|\vec{g}-\vec{d}\|}\right) = (\vec{w}, \vec{d} \times \vec{w}) \tag{2.2}$$

We also use the dual vector representation of the lines and dual vector algebra (see Fischer, 1995 and McCarthy, 2000) where $\epsilon^2=0$.

$$\hat{S}_1 = (\vec{s}, \vec{c} \times \vec{s}) = a + \epsilon a^0 \text{ and } \hat{S}_2 = (\vec{w}, \vec{d} \times \vec{w}) = b + \epsilon b^0 \tag{2.3}$$

Line dot product:

$$\hat{S}_1 \cdot \hat{S}_2 = (a, a^0) \cdot (b, b^0) = \cos\theta - \epsilon d \sin\theta = \cos\hat{\theta} \tag{2.4}$$

Line cross product:

$$\hat{S}_1 \times \hat{S}_2 = (a, a^0) \times (b, b^0) = (\sin\theta + \epsilon d \cos\theta)\hat{N} = \sin\hat{\theta}\hat{N} \tag{2.5}$$

where $\hat{N}$ is the common normal line to $\hat{S}_1$ and $\hat{S}_2$. The above operations are useful for calculating the distance d and the angle θ between two lines. If $d \sin\theta \neq 0$ then the lines do not intersect ($d \neq 0$) and are not parallel ($\sin\theta \neq 0$). If $d \sin\theta = 0$ and $\cos\theta \neq 1$ then the lines intersect ($d = 0$) and are not parallel. If the $\cos\theta$ term of the dot product is equal to 1 then the lines are parallel and the resultant dual vector of the cross product will have a 0 real component. The cross product's dual component ($d \cos\theta$) will be 0 when the lines are identical. If $d \cos\theta \neq 0$ then the distance d can be calculated. Fig. 2 shows a detailed flow chart of the necessary condition for a collision. If the necessary condition is satisfied then additional testing is required.

2.2 Finite Cylinder Testing

If a possible collision has been detected by the infinite cylinder test then further testing is required to determine if an actual collision has occurred. The model is modified from cylinders of infinite length to cylinders of finite length. This changes the approach from testing lines to testing line segments. The finite cylinder testing procedure is presented in detail in Ketchel and Larochelle (2005).

3 Case Study

To demonstrate the application of the collision detection methodology presented above to robot motion planning we designed a robotic workcell (Fig. 3) for a pick and place application. This is a typical application found in the automotive industry. In this task the robot is to pick-up a dashboard, maneuver it into an auto body, place it, and then retraces its motion out of the auto body. The model of the workcell includes a robot, a tool, a workpiece (dashboard) and the drop location (automobile frame). The robot selected for the case study is the ABB6000 (Fig. 4) and it is modelled by 9 cylinders shown in shades of orange and grey of varying radii. The tool is modelled by 9 cylinders shown in shades of light grey of varying radii. The dashboard is modelled by 15 green cylinders of varying radii. The drop-off location is modelled by 25 red and blue cylinders.

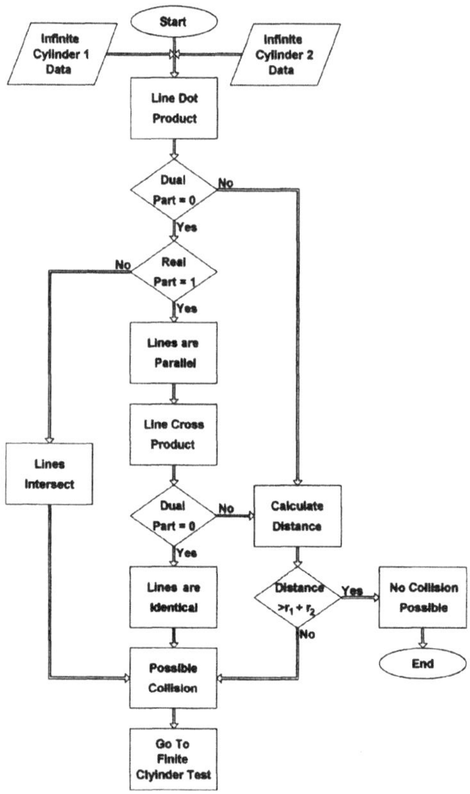

Figure 2. Infinite Cylinder Testing

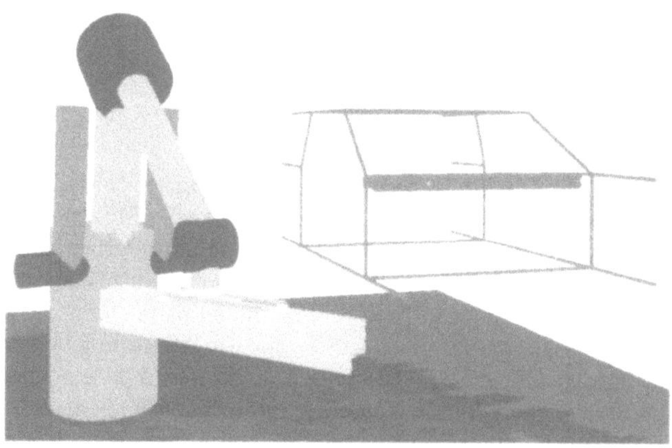

Figure 3. Workcell Model

Figure 4. ABB 6000: Photo (From Directory (2005)) & Model

3.1 Robot Motion Planning

A simple motion planner was created to move the robot incrementally through the desired motion. Euler parameters for orientation (yaw, pitch and roll) and position (X, Y and Z) are used to describe the desired via points of the robot's motion. For a desired orientation there are two yaw, pitch and roll parameter sets that yield the same orientation. The motion planner enables the user to select which to use to represent the orientation. For each via point the finite difference between the current and next set of Euler parameters are found. This difference is divided by the number of steps for the desired move. The planner uses linear interpolation to step the motion incrementally from one via point to the next. Since this is not a real-time application the number of incremental steps between via points is prescribed by the user. The orientation parameters require that the shortest path is taken from one via point to the next. If the previous commanded yaw is 170 (deg) and the next commanded yaw is -170 (deg) the motion planner takes the shortest route; incrementing 20 (deg) instead of 340 (deg). The motion planner does this by checking if the difference between two sequential orientation parameters is > 180 (deg). If it isn't then the target via point is modified by adding or subtracting 360 (deg) until the difference is < 180 (deg). The position parameters are easily interpolated for smooth motion.

At each increment of motion planning every cylinder must be tested for a possible collision with every other cylinder. Since speed is important for the calculations, we look to reduce the number of tests that must be run for each incremental step. First, cylinders can not collide with themselves since they are rigid links. Second, if cylinder 5 has already been tested for a collision with cylinder 1 then testing cylinder 1 for a possible collision with cylinder 5 is redundant. The number of tests required here for each step is 1032 (in this model the tool and the dashboard can not collide). At each increment all 1032 tests are preformed using infinite cylinders to quickly check for possible collisions. Any infinite cylinders that are identified as possibly colliding are sent to finite cylinder testing to definitively determine if an actual collision has occurred.

3.2 Collision Case

A robot motion was planned for the task. A set of via points for the tool center point to follow and the number of intermediate steps were determined (Tbl. 1). At every incremental step all of the 1032 possible collisions were tested. Several collisions occurred. The first collision (shown in Tbl. 2) occurs between the 2nd and 3rd via points. Fig. 5 shows where the dashboard side (segment 56) collides with the car frame right windshield support (segment 16). Hence this motion is unsuitable and additional motion planning was performed.

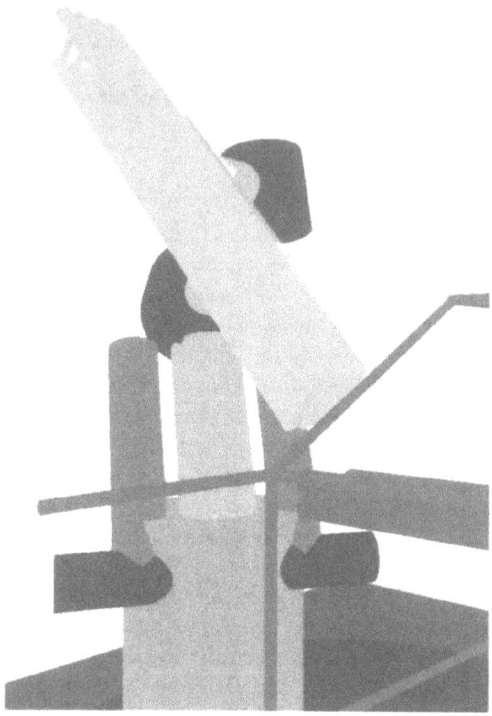

Figure 5. Collision of the Dashboard and Car Frame

3.3 Non-Collision Case

Here a collision free motion has been planned by altering the 2nd and 3rd via points as follows: X was decreased to 800, Z decreased to 950, and the pitch was increased to -120 to have the tool center point closer to the robot base and the dash rotated slightly more vertically. These alterations provide the dashboard clearance from the car frame. The final position of the dashboard places it within the car frame with $\frac{1}{2}$ (inch) clearance (Fig. 6).

Table 1. Collision Case Study - Via Points

X mm	Y mm	Z mm	Yaw deg	Pitch deg	Roll deg	# of Increments
500	−1400	0	0	−180	−45	10
900	−1200	1200	90	−135	0	20
900	−375	1200	90	−135	0	20
1300	−375	521.5	90	−135	0	20
2150	−375	521.5	90	−180	0	40
2150	−725	521.5	90	−180	0	40

Table 2. Collision Case Study - Output

Collision Type		Seg. No.		Seg. No.	
Two Hit		16		56	
X	Y	Z	Yaw	Pitch	Roll
900	−750	1200	90	−135	0
mm	mm	mm	deg	deg	deg
Axis 1	Axis 2	Axis 3	Axis 4	Axis 5	Axis 6
−28.08	−30.81	20.46	−84.53	62.41	−56.69
deg	deg	deg	deg	deg	deg

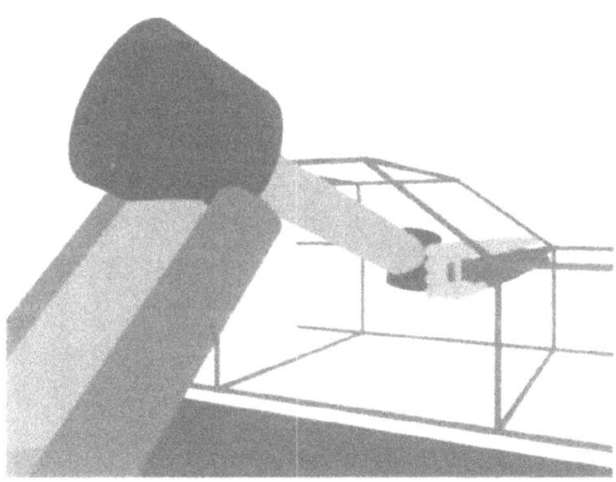

Figure 6. Successful Dashboard Placement

4 Conclusion

We have presented a novel methodology for motion planning to avoid collisions. Right circular cylinders were used to model all of the bodies of interest in the robot workcell. The proposed algorithm uses line geometry and dual number algebra to exploit the geometry of right circular cylindrical objects to facilitate the detection of collisions. The result is a straight-forward and efficient means of detecting collisions of cylindrically shaped bodies that is applicable to robot motion planning. An illustrative case study was presented.

Bibliography

Surplus Record Machinery & Equipment Directory. www.surplusrecord.com, 2005.

I. Fischer. *Dual-Number Methods in Kinematics, Statics, and Dynamics.* CRC Press LLC, 1995.

D. Manocha J. Cohan, M. Lin and M. Ponamgi. I-collide: An interactive and exact collision detection system for large-scale environments. In *Symposium on Interactive 3D Graphics*, 1995.

T. Johnson and J. Vance. The use of voxmap pointshell method of collision detection in virtual assembly methods planning. In *Proceedings of the ASME 2001 Design Engineering Technical Conferences and Computers and Information Conference*, number DET2001/DAC-21137. ACM Press, 2001.

J. Ketchel and P. Larochelle. Collision detection of cylindrical rigid bodies using line geometry. In *Proceedings of the ASME 2005 Design Engineering Technical Conferences and Computers and Information Conference*, number DETC05/MECH-84699, 2005.

J. M. McCarthy. *Geometric Design of Linkages.* Springer-Verlag, 2000.

M. Reggiani S. Caselli and M. Mazzoli. An experimental evaluation of collision detection packages for robot motion planning. In *Proceedings of the IEEE International Conference on Intelligent Robots and Systems*, 2002.

J. Cohen S. Gottschalk T. Hudson, M. Lin and D. Manocha. V-collide: Accelerated collision detection for vrml. In *Proceedings of the 2nd Symposium on VRML.* ACM Press, 1997.

P. Xavier. Implict convex-hull distance of finite-screw-swept volumes. In *Proceedings of the 2002 IEEE International Conference on Robotics & Automation*, 2000.

P.J. Zsombor-Murray. Spatial visualization and the shortest distance between two lines problem. In *Proceedings of the 5th ASEE International Conference ECGDG*, 1992.

On the Dynamic Stability of Off-Road Vehicles

Moshe P. Mann * and Zvi Shiller † *

* Graduate Student, Technion, Israel Institute of Technology, Mechanical Engineering, Haifa, Israel

† College of Judea and Samaria, Mechanical Engineering-Mechatronics, Ariel, Israel

Abstract. Dynamic stability reflects the vehicle's ability to traverse uneven terrain at high speeds. It is determined from the set of admissible speeds and tangential accelerations of the center of mass along the path, subject to the ground force constraints and the geometric path constraints. This paper presents an analytical method for computing the stability margins of a planar all-wheel drive vehicle. It consists of mapping the ground force constraints to constraints on the vehicle's speeds and accelerations along the path. The boundaries of the set of admissible speeds and accelerations determines the static and dynamic stability margins, used to gauge the traversability of the vehicle along the path. The first is the maximum feasible acceleration at zero speed, whereas the second is the maximum feasible speed. Both stability margins are demonstrated for a planar vehicle moving on a sinusoidal path.

1 Introduction

Stability during motion on rough terrain is essential for autonomous vehicles in space as well as in terrestrial military applications. It is commonly measured by stability margins that take into account various geometric and physical attributes of the vehicle and terrain.

Several vehicle stability margins were previously proposed (Garcia, 2002; Lin and Song, 1993; McGhee and Frank, 1968; Messuri, 1985; Papadopulos and Rey, 2000; Yoneda and Hirose, 1997). These stability margins, however, either do not consider vehicle dynamics or do not consider friction constraints, and hence are not suitable to evaluate vehicle stability during high speed motion on rough terrain.

A different approach (Shiller and Gwo, 1991) that accounts for vehicle dynamics offers a *dynamic* stability margin that consists of the maximum allowable speed under which the vehicle remains stable along its path. It was first derived for a suspended point mass model, and then extended to a planar rigid body (Shiller and Mann, 2004) and to a Rocker Bogie vehicle (Mann and Shiller, 2005). Because of the indeterminate nature of the vehicle system with respect to the ground forces (3 equations and 4 unknowns), these

*This work was performed at the Paslin Robotics Research Laboratory at the College of Judea and Samaria. It was supported partly by grant 01-99-08430 of the Israeli Space Agency through the Ministry of Science Culture and Sports of Israel.

stability margins were computed numerically, which is time consuming. The pseudoinverse method proposed in (Klein and Chung, 1987), (Shiller and Mann, 2004), (Mann and Shiller, 2005), is deterministic and therefore more efficient, but was shown to provide a too conservative stability margin (Shiller and Mann, 2004) (Mann and Shiller, 2005).

In this paper, we propose an efficient method to compute the dynamic stability margins of a planar rigid body vehicle. To this end, we wish to determine the feasible range of speed and acceleration along the path that does not violate the vehicle's contact constraints. This is similar to the classical problem in grasping, where given a rigid body with multiple contact points, one determines the range of net forces that may act upon the object and its accelerations without violating its contact and friction constraints. The two are similar since the net force acting on the object is related to the tangential speed and acceleration of the object's center of mass. The method is developed for a longitudinal planar vehicle with two wheels. The results are then compared to that of the point mass used in (Shiller and Gwo, 1991).

2 Vehicle Model

The dynamic model used for this method is a planar two-wheel all-wheel drive vehicle as shown in Figure 1. The frictional contacts between the massless wheels of radius r and the ground obey Coulomb friction, such that the ground forces $\mathbf{f}_i$ are bounded by their respective friction cones

$$\mathbf{f}_{ti} \leq \mu \mathbf{f}_{ni} \tag{2.1}$$

$$\mathbf{f}_{ti} \geq -\mu \mathbf{f}_{ni}, i = 1, 2 \tag{2.2}$$

as shown in Figure 2, where the forces f_{ni} and f_{ti} represent the two components of $\mathbf{f}_i$ in directions normal, $\mathbf{n}_i$, and tangent, $\mathbf{t}_i$, to the ith contact:

$$\mathbf{f}_i = f_{ni}\mathbf{n}_i + f_{ti}\mathbf{t}_i, i = 1, 2 \tag{2.3}$$

The position vectors from the c.g. to the back and front contact points, $\mathbf{r}_1$ and $\mathbf{r}_2$, respectively, and the gravity force acting on the vehicle, are expressed in the vehicle's $x - y$ frame, as shown in Figure 1. The vehicle's orientation θ is measured relative to the inertial frame $X - Y$.

Kinematics

The orientation of the vehicle at any point is computed by modelling the vehicle and its contact points as a closed kinematic chain. The terrain profile is assumed known, so for a given back wheel contact point $\mathbf{x}_1$, we can compute the contact point $\mathbf{x}_2$ of the front wheel. Denoting d as the distance between the two wheel axles and r as the wheel radius, $\mathbf{x}_2$ is found numerically by solving:

$$\| - r\mathbf{n}_1 + \mathbf{x}_2 - \mathbf{x}_1 + r\mathbf{n}_2 \| = d \tag{2.4}$$

where $\| - \|$ denotes the Euclidean norm. Once $\mathbf{x}_2$ is known, the vehicle's center of mass $\mathbf{x}$, and orientation θ are easily found. The path constraint leaves the vehicle one degree

Figure 1. A planar vehicle model

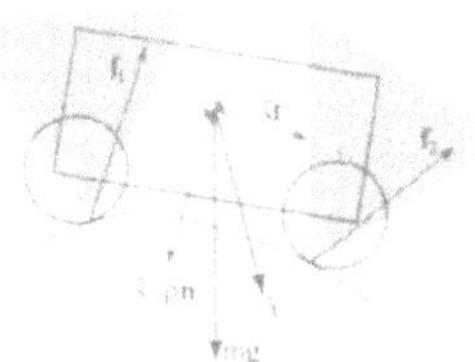

Figure 2. Forces acting on the vehicle

of freedom, which can be represented by the arc length, s, of the trajectory followed by the center of mass. The vehicle's linear and angular acceleration are thus expressed in terms of the speed $\dot{s}$ and acceleration $\ddot{s}$ along the path.

The vehicle acceleration $\ddot{\mathbf{x}}$ can be decomposed into its tangent and normal components to the path, $\ddot{s}\mathbf{t}$ and $\dot{s}^2/\rho\mathbf{n}$, as shown in Figure 2, where $\mathbf{n}$ points toward the instantaneous center of curvature, and ρ is the radius of curvature. Substituting κ for $1/\rho\mathbf{n}$, this transformation to path coordinates is expressed in matrix form:

$$\ddot{\mathbf{x}} = \begin{bmatrix} t_x & \kappa_x \\ t_y & \kappa_y \end{bmatrix} \begin{bmatrix} \ddot{s} \\ \dot{s}^2 \end{bmatrix} = K(s) \begin{bmatrix} \ddot{s} \\ \dot{s}^2 \end{bmatrix} \tag{2.5}$$

The angular acceleration may be similarly expressed as:

$$\ddot{\theta} = \theta_s \ddot{s} + \theta_{ss} \dot{s}^2 \tag{2.6}$$

where θ_s and θ_{ss} are the 1st and 2nd derivatives of θ with respect to s.

Vehicle Dynamics

The equations of motion for the vehicle are:

$$\mathbf{f}_1 + \mathbf{f}_2 + m\mathbf{g} = m\ddot{\mathbf{x}} \tag{2.7}$$

$$\mathbf{r_1} \times \mathbf{f_1} + \mathbf{r_2} \times \mathbf{f_2} = I\ddot{\theta} \tag{2.8}$$

Substituting (2.5) into (2.7) and (2.6) into (2.8) yields the moment around the c.g. Since the vehicle model is planar, the moment equation is considered as a scalar:

$$\mathbf{f}_1 + \mathbf{f}_2 + m\mathbf{g} = mK(s) \begin{bmatrix} \ddot{s} \\ \dot{s}^2 \end{bmatrix} \tag{2.9}$$

$$\mathbf{r_1} \times \mathbf{f_1} + \mathbf{r_2} \times \mathbf{f_2} = I(\theta_s \ddot{s} + \theta_{ss} \dot{s}^2) \tag{2.10}$$

Equations (2.9) and (2.10) relate any applied forces $\mathbf{f_1}$ and $\mathbf{f_2}$ to the corresponding $\dot{s}$ and $\ddot{s}$. The constraints on $\mathbf{f_1}$ and $\mathbf{f_2}$ thus map to constraints on $\dot{s}$ and $\ddot{s}$, which determine the static and dynamic stability margins derived next.

Note that Eqs. (2.7) and (2.8) are coupled since the vehicle is traversing a given terrain profile and thus the angular acceleration is determined from its linear acceleration. The

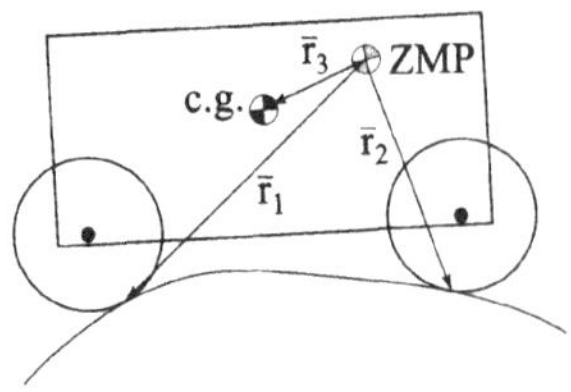

Figure 3. The Zero Moment Point.

moment equation (2.8) hence imposes a constraint on the ground forces. This constraint can be derived as an equality constraint by substituting (2.5) into (2.6) to express $\ddot{\theta}$ in terms of $\ddot{\mathbf{x}}$, then replacing $\ddot{\mathbf{x}}$ with the left hand side of (2.7) to yield:

$$\begin{aligned}\mathbf{r_1} \times \mathbf{f_1} + \mathbf{r_2} \times \mathbf{f_2} &= I \left[\begin{array}{cc} \theta_s & \theta_{ss} \end{array} \right] K^{-1} \left(\frac{\mathbf{f}_w}{m} + \mathbf{g} \right) \\ &= \left[v_1 \quad v_2 \right] \cdot \left(\frac{\mathbf{f}_w}{m} + \mathbf{g} \right)\end{aligned} \tag{2.11}$$

where v_1 and v_2 are the two scalar components of the vector $\{I \left[\begin{array}{cc} \theta_s & \theta_{ss} \end{array} \right] K^{-1}\}$, and the sum of the ground forces is

$$\mathbf{f}_w = \mathbf{f}_1 + \mathbf{f}_2 \tag{2.12}$$

Rearranging (2.11) yields:

$$\mathbf{f}_1 + \mathbf{f}_2 + m\mathbf{g} = m\ddot{\mathbf{x}} \tag{2.13}$$

$$\bar{\mathbf{r}}_1 \times \mathbf{f}_1 + \bar{\mathbf{r}}_2 \times \mathbf{f}_2 + \bar{\mathbf{r}}_3 \times m\mathbf{g} = 0 \tag{2.14}$$

where

$$\bar{\mathbf{r}}_1 = \mathbf{r}_1 + \bar{\mathbf{r}}_3 \tag{2.15}$$

$$\bar{\mathbf{r}}_2 = \mathbf{r}_2 + \bar{\mathbf{r}}_3 \tag{2.16}$$

$$\bar{\mathbf{r}}_3 = \left[-v_2 \quad v_1 \right] \tag{2.17}$$

Equation (2.14), in comparison to (2.8), represents the moment equation of an equivalent system that has a zero moment of inertia, and a virtual point that we call ZMP (zero moment point) where the moments of all forces applied on the system are zero. The ZMP is located at $-\bar{\mathbf{r}}_3$ from the center of mass, as is shown in Figure 3. Note that $\bar{\mathbf{r}}_3$ reflects the body's inertial parameters as well as the path curvature. The forces acting upon the equivalent system, however, have not changed, and so Equation 2.7 still holds. Our goal now is to find the set of admissible accelerations that are produced by the ground forces that satisfy the friction constraints and the equality constraint (2.14). The set of admissible accelerations would yield the desired dynamic stability margins.

3 Computing the Set of Admissible Accelerations

The set of admissible accelerations is a convex polygon in the plane of $\mathbf{f}_1$ and $\mathbf{f}_2$ since both are constrained by a set of linear inequality and equality constraints (de Berg,

2000). This polygon is the intersection of half planes, each satisfying two of the four inequality constraints (2.1) and (2.2) and the equations of motion (2.13) and (2.14). The boundary of each half plane is determined by considering the equality part of the respective inequality constraints.

Consider for example the two ground forces at the front edge of their friction cones:

$$f_{t1} = \mu f_{n1} \tag{3.1}$$

$$f_{t2} = \mu f_{n2} \tag{3.2}$$

These forces can be written as

$$\mathbf{f}_1 = f_1 \mathbf{e}_1 \tag{3.3}$$

$$\mathbf{f}_2 = f_2 \mathbf{e}_2 \tag{3.4}$$

where $\mathbf{e}_1$ and $\mathbf{e}_2$ are unit vectors in the directions of the respective boundaries of the two friction cones. Substituting (3.3) and (3.4) into (2.11) yields:

$$f_1 \bar{\mathbf{r}}_1 \times \mathbf{e}_1 + f_2 \bar{\mathbf{r}}_2 \times \mathbf{e}_2 + \bar{\mathbf{r}}_3 \times m\mathbf{g} = 0 \tag{3.5}$$

which represents a line in the $f_1 - f_2$ plane. To transform this line into a line in the $\ddot{x} - \ddot{y}$ plane, we substitute (3.3) and (3.4) into (2.7) to obtain:

$$\begin{bmatrix} f_1 \\ f_2 \end{bmatrix} = \begin{bmatrix} \mathbf{e}_1 & \mathbf{e}_2 \end{bmatrix}^{-1} (m\ddot{\mathbf{x}} - m\mathbf{g}) \tag{3.6}$$

which in turn is substituted into (3.5) to obtain:

$$\begin{bmatrix} \bar{\mathbf{r}}_1 \times \mathbf{e}_1 & \bar{\mathbf{r}}_2 \times \mathbf{e}_2 \end{bmatrix} \begin{bmatrix} \mathbf{e}_1 & \mathbf{e}_2 \end{bmatrix}^{-1} (\ddot{\mathbf{x}} - \mathbf{g}) + \bar{\mathbf{r}}_3 \times \mathbf{g} = 0 \tag{3.7}$$

Equation (3.7) may be written in the scalar form:

$$a\ddot{x} + b\ddot{y} + c = 0 \tag{3.8}$$

Equation (3.8) represents a line in the $\ddot{x} - \ddot{y}$ plane that bounds a half plane, where:

$$\begin{bmatrix} a & b \end{bmatrix} = \begin{bmatrix} \bar{\mathbf{r}}_1 \times \mathbf{e}_1 & \bar{\mathbf{r}}_2 \times \mathbf{e}_2 \end{bmatrix} \begin{bmatrix} \mathbf{e}_1 & \mathbf{e}_2 \end{bmatrix}^{-1} \tag{3.9}$$

$$c = \bar{\mathbf{r}}_3 \times \mathbf{g} - \begin{bmatrix} \bar{\mathbf{r}}_1 \times \mathbf{e}_1 & \bar{\mathbf{r}}_2 \times \mathbf{e}_2 \end{bmatrix} \begin{bmatrix} \mathbf{e}_1 & \mathbf{e}_2 \end{bmatrix}^{-1} \mathbf{g} \tag{3.10}$$

and $\ddot{x}$ and $\ddot{y}$ are the Cartesian components of $\ddot{\mathbf{x}}$.

This formulation enables us to map the entire region of attainable ground forces to the region of feasible forces acting on the ZMP. The intersections of all the half planes produced by every pair of inequality constraints (2.1) provides the set of admissible accelerations, ACC, as illustrated in Figure 4. Also shown in Figure 4 are the tangent and normal vectors to the path followed by the center of mass. The boundaries of ACC determine the range of tangent and normal accelerations, and in turn the stability margins derived next.

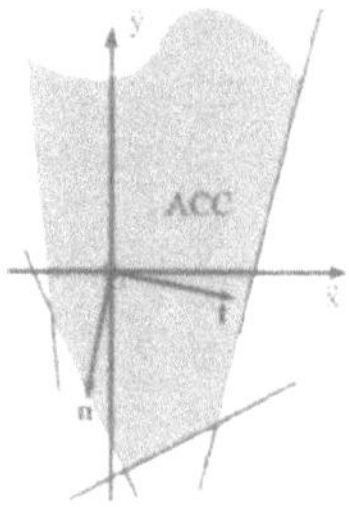

Figure 4. The range of attainable accelerations ACC

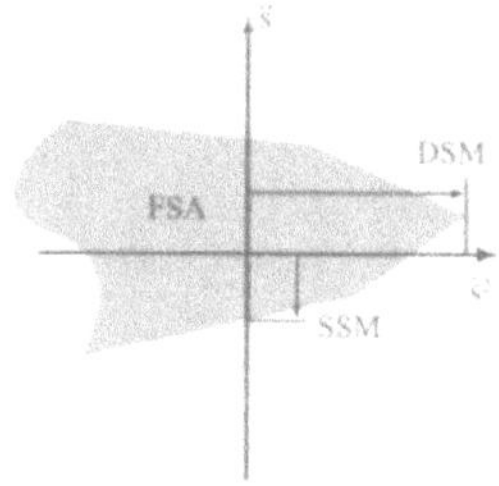

Figure 5. The Feasible range of Speed and Acceleration (FSA).

4 Stability Margins

Because the vehicle traverses along a specified path, its acceleration can be expressed in terms of the time derivatives of the path arc length, s, $\ddot{s}$ and $\dot{s}^2$ (2.5). The set of all feasible speeds and accelerations, FSA, is thus simply obtained by a linear transformation of ACC using (2.5):

$$FSA = K(s)^{-1}ACC \tag{4.1}$$

Figure 5 shows schematically the FSA obtained for the set of admissible acceleration shown in Figure 4. The FSA includes all path accelerations and speeds attainable by the feasible ground forces. Any given pair of speed (squared) and acceleration that lies outside of the FSA region is obviously unattainable. We define the dynamic stability margin, the DSM, as the maximum feasible speed, and the static stability margin, the SSM, as the minimum feasible acceleration range at zero speed along every point of the path, s, as shown in Figure 5, where the stability margins are indicated by the boundaries of the FSA:

$$DSM(s) = \begin{cases} max(\dot{s}), \dot{s}^2 \in FSA & \text{if } max(\dot{s}) \geq 0 \\ 0 & \text{otherwise} \end{cases} \tag{4.2}$$

$$SSM(s) = \begin{cases} min(|\ddot{s}_{min}|, |\ddot{s}_{max}|)_{\dot{s}=0} & \text{if } \ddot{s}_{max} > 0,\ \ddot{s}_{min} < 0 \\ = 0 & \text{otherwise} \end{cases} \tag{4.3}$$

The SSM and DSM are important indicators of the stability of a vehicle traversing a preset trajectory. For example, a vehicle on a steep incline will have a low SSM, indicating that slippage is likely to occur if the vehicle accelerates too rapidly. A vehicle on a convex surface will have a low DSM, which indicates that a sufficiently high velocity will result in the vehicle losing contact with the surface.

5 Examples and Results

The method for computing the FSA presented in this paper is demonstrated for a planar vehicle traversing a sinusoidal track. Figure 6 shows the vehicle at one point along that track, and the resulting FSA's at that point for three vehicle models: a rigid body, a suspended point mass (rigid body with zero moment of inertia), and a point mass rolling on the surface. The static and dynamic stability margins for the entire path for the three models are shown in Figure 7.

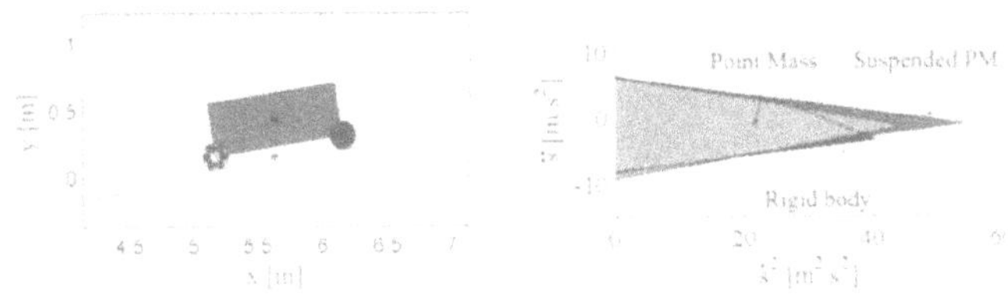

Figure 6. The FSA for a vehicle at one point along a sinusoidal path.

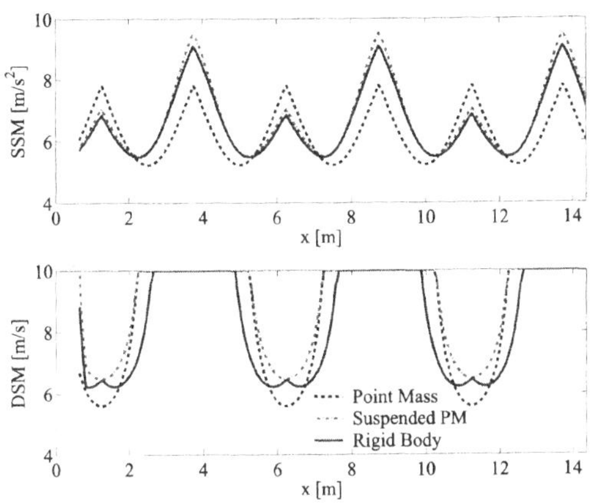

Figure 7. The SSM and DSM for a vehicle traversing a sinusoidal path.

Generally, the DSM is higher at the bottom of the hill, due to the concavity of the surface, where higher speeds elicit greater traction. On the other hand, the DSM is lower at the top of the hill, where the ground forces are reduced due to the radial acceleration. The SSM is low along the inclines of the hill, where the acceleration is reduced to the gravity force. It is higher at the top and bottom of the hills where the vehicle is horizontal and its acceleration not affected by the gravity force.

Comparing the vehicles with and without moment of inertia, the DSM is always higher for the zero inertia because a greater moment of inertia requires a greater external wrench to maintain rotational equilibrium. For the point mass case, the DSM is lower at the top of the hill than that for the zero inertia case because the curvature of the path traced by the point mass on the sinusoidal surface is greater than the curvature of the path traced by the suspended point mass.

6 Conclusion

This paper presented a simple procedure to determine the dynamic stability of a two dimensional vehicle travelling on general terrain. The ground friction constraints are mapped into the range of admissible accelerations using the vehicle's equations of motion. This range is then mapped into the set of feasible speeds and tangential accelerations, from which we determine the static and dynamic·stability margins. The first is the maximum acceleration at zero speed, whereas the latter is the maximum feasible speed without violating ground force constraints. The dynamic stability margins have been shown to be an effective metric for computing optimal trajectories over general terrain (Shiller and Gwo, 1991). The proposed rapid computation enables autonomous vehicles to quickly and safely traverse rough terrain, an essential feature for future robotic missions.

Bibliography

M. de Berg, M. van Kreveld. M. Overmars, O. Schwarzkopf, "Computational Geometry", Springer, 2000.

E. Garcia, J. Estremera, P.G. de Santos, "A Comparative Study of Stability Margins For Walking Machine",*Robotica* Vol. 30, pp. 595-606, 2002.

C.A. Klein, T.S. Chung, "Force Interaction and Allocation for the Legs of a Walking Vehicle", *IEEE Transactions of Robotics and Automation*, Vol RA-3, No. 6, Dec. 1987.

B. Lin, S. Song, "Dynamic Modelling, Stability, and Energy Efficiency of a Quadrupedal Walking Machine", Proc. IEEE Conf. on Robotics and Automation, Atlanta, Georgia, 1993, pp. 367-373.

M. Mann, Z. Shiller: "Dynamic Stability of a Rocker Bogie Vehicle: Longitudinal Motion", Proc. IEEE Conf. on Robotics and Automation, Barcelona, Spain, April 2005.

R.B. McGhee, A.A. Frank: "On the Stability Properties of Quadruped Creeping Gait", *Mathematical Biosciences*, Vol. 3, No. 2, pp. 331-351, 1968.

D. Messuri, "Optimization of the Locomotion of a Legged Vehicle with respect to Maneuverability", PhD Thesis, Ohio State University, 1985.

Murray, Li, Sastry, "Introduction to Robotic Grasping," CRC Press, 1994.

J.S Pang, J. Trinkle: "Stability Characterizations of Rigid Body Contact Problems with Coulomb Friction", *Zeithschrift fur Angewandte Mathematik und Mechanik*, Vol. 80, No. 10, pp. 643-663, 2000.

E. Papadopulos, D. Rey: "The Force-Angle Measure of Tipover Stability Margin for Mobile Manipulators", *Vehicle System Dynamics*, Vol. 33, No. 1, pp. 29-48, 2000.

Z. Shiller, Y.R. Gwo: "Dynamic Motion Planning of Autonomous Vehicles", *IEEE Transactions on Robotics and Automation*, Vol. 7, No. 2, April 1991.

Z. Shiller, M. Mann: "Dynamic Stability of Off Road Vehicles", Proc. IROS 2004, Sendai, Japan, October 2004, pp. 1849-1853.

K. Yoneda, S. Hirose, 1997:Three Dimensional Stability Criterion of Integrated Locomotion and Manipulation, *Journal of Robotics and Mechatronics*, Vol. 9, No. 4, pp. 267-274.

Sub-Optimal Motion Planner of Mobile Manipulators in Generalized Point-to-Point Task With Stability Constraint

M. Haddad, T. Chettibi, T. Saidouni, S. Hanchi and H. E. Lehtihet

Laboratory of Structure Mechanics, E.M.P. BP 17, 16111, Algiers, Algeria.
HaddadMoussa2003@yahoo.fr, he.lehtihet@yahoo.fr

Abstract. A trajectory planning method is presented for wheeled mobile manipulators in a structured workspace. The problem is known to be complex, particularly when dynamics is considered. Our approach is based on the simultaneous search for the robot path and for the time history on this path. The whole problem is formulated in such a way that kinodynamic constraints and stability conditions are transformed to bounds on admissible travel time.

1 Introduction

A Wheeled Mobile Manipulator (WMM) typically consists of a robotic manipulator mounted on a mobile platform. Such a system combines the dexterous capability of fixed-base manipulators with the high mobility of wheeled platform. Thus, a WMM covers a considerably larger workspace than does a fixed-base manipulator and has many potential applications in manufacturing, construction, planetary exploration and nuclear reactor maintenance. This makes the motion planning phase very important and requires enormous attention. This importance is naturally reflected in the number and the variety of research activities.

Due to the constant need to increase the productivity in various industries, several approaches have been proposed to solve the Minimum-Time Trajectory Planning Problem (MTTPP) (Perrier et al., 1998) and the minimum acceleration sum trajectory planning problem (Huang and Sugano, 1996) for a WMM while considering cinematic constraints. It is clear however that dynamics is also important, particularly in the case of high velocities or when there is a need to compute correctly inputs of the system control. Few are works that treat the MTTPP while handling dynamics and stability constraint.

Techniques that treat the trajectory problem for a WMM while handling dynamics and/or stability constraint can be grouped in two families: indirect and direct methods. In general, the former are applications of the optimal control theory. They use particularly a variational method (Mohri et al., 2001, Furuno et al., 2003) or Pontryagin's Minimum Principle (Desai and Kumar, 1997). In contrast, direct methods use different schemes to discretize the system state and control variables to convert the difficult infinite-dimensional optimization problem to a more tractable finite-parameter optimization problem (Chen and Zalzala, 1997).

Here, we focus on one of the most important problems in mobile manipulators: planning trajectories while considering constraints on geometry, kinematics, dynamics and stability. The method we describe is an adaptation to the case of a WMM of the Random Profile Approach (RPA) applied in (Haddad et al., 2005) to wheeled mobile robots. The RPA was initially proposed in (Chettibi and Lehtihet, 2002) to treat trajectory problems for fixed-base manipulators. It uses the concept of time-scaling (Hollerbach, 1983) and the concept of trajectory profiles. It converts kinodynamic constraints to bounds on the travel time then reduces the dynamic problem to a constrained-profile optimization subsequently solved via a stochastic technique. Although it generally yields sub-optimal results, the RPA is remarkably flexible. It is applicable in various types of problems and for various types of cost functions and constraints.

2 Dynamic Model

The WMM considered hereafter consists of a series-chain multi-link manipulator mounted on a Wheeled Mobile Platform subjected to move on a flat floor. The platform type is not specified here. It may belong to any of the nonholonomic WMP discussed in (Campion et al., 1996).

Let $\boldsymbol{q}_p$ be the n_p-vector of the platform generalized coordinates, $\boldsymbol{q}_a$ the n_a-vector of the arms generalized coordinates and $\boldsymbol{q} = [\boldsymbol{q}_p^t,\ \boldsymbol{q}_a^t]^t$ the n-vector of the whole system generalized coordinates, so that $n = n_p + n_a$. Note that vector $\boldsymbol{q}_p$ already encompasses the configuration parameters (x, y, θ) of the platform in the referential frame in addition to the generalized coordinates related to the wheels. The platform is subjected to m nonholonomic constraints that can be written as follows (Bloch et al., 1992):

$$A(q_p)\cdot\dot{q} = 0 \tag{1}$$

where $\boldsymbol{A}(\boldsymbol{q}_p)$ is a full rank $(m\times n)$ matrix. Then, using the Lagrangian approach and considering (1), we can model the WMM dynamics by (see Bloch et al., 1992):

$$\boldsymbol{M}(\boldsymbol{q})\cdot\ddot{q} + c(q,\dot{q}) + \boldsymbol{g}(\boldsymbol{q}) = \boldsymbol{B}(\boldsymbol{q})\cdot\tau + \boldsymbol{A}^t(\boldsymbol{q})\cdot\eta \tag{2}$$

where $\boldsymbol{M}(\boldsymbol{q})$ is the $(n\times n)$ inertia matrix, $c(\boldsymbol{q},\dot{\boldsymbol{q}})$ is the n- vector of centrifugal and Coriolis forces, $\boldsymbol{g}(\boldsymbol{q})$ is the n-vector of gravity forces, η is the m-vector of Lagrange's multipliers associated to the m nonholonomic constraints, τ is the $(n - m)$-vector of active joint torques and $\boldsymbol{B}(\boldsymbol{q})$ is the $(n\times(n - m))$ input transformation matrix.

Following Bloch et al. (1992), we write $\boldsymbol{A}(\boldsymbol{q}_p) = [A_1\ A_2]$, where A_1 is a $(m\times(n-m))$ matrix and A_2 is a $(m\times m)$ non singular matrix. Next, if we define a full rank matrix $\boldsymbol{R}(\boldsymbol{q}_p)$ in terms of $\boldsymbol{I}$ the $((n-m)\times(n-m))$ identity matrix and A_1, A_2 :

$$\boldsymbol{R}(q_p) = \begin{bmatrix} \boldsymbol{I} \\ -A_2^{-1}\cdot A_1 \end{bmatrix}$$

then it follows that $\boldsymbol{A}(\boldsymbol{q}_p)\cdot\boldsymbol{R}(\boldsymbol{q}_p) = 0$. Consequently, as the $((n-m)\times(n-m))$ square matrix $\boldsymbol{R}^t(\boldsymbol{q}_p)\boldsymbol{B}(\boldsymbol{q})$ has full rank for any feasible $\boldsymbol{q}$ (Bloch et al., 1992), we may pre-multiplying (2) by $[\boldsymbol{R}^t(\boldsymbol{q}_p)\boldsymbol{B}(\boldsymbol{q})]^{-1}$ to eliminate η and to deduce the vector of active joint torques as follows:

$$\tau = \left[\boldsymbol{R}^t(q_p)\boldsymbol{B}(q)\right]^{-1}\boldsymbol{R}^t(q_p)\left[\boldsymbol{M}(q)\ddot{q} + c(q,\dot{q}) + g(q)\right] \tag{3}$$

3 Problem Statement

The WMM is required to move freely from initial to final states q^{init} and q^{fin}. In addition to the vector $\tau(t)$ of actuator efforts and the travel time T, we must find the trajectory $q(t)$ such that the initial and final states are matched, constraints are respected and a given cost function is minimized.

The following boundary conditions inherent to the achievement of the desired task must be taken into account:

– Position $\quad \boldsymbol{q}(0) = \boldsymbol{q}^{init}$ and $\boldsymbol{q}(T) = \boldsymbol{q}^{fin}$ (4*a*)

– Velocity $\quad \dot{\boldsymbol{q}}(0) = \vec{0}$ and $\dot{\boldsymbol{q}}(T) = \vec{0}$ (4*b*)

When obstacles are present in the workspace, the following constraints will hold during the motion:

$$Col(\boldsymbol{q}(t)) = false \qquad (0 < t < T) \tag{4c}$$

The function *Col* indicates whether the robot at configuration $\boldsymbol{q}$ is in collision with an obstacle.

Additional constraints that may need to be included represent physical limitations on the WMM performances. For example, we may incorporate any of the following constraints on:

– joint positions: $\quad |q_i(t)| \le q_i^{\max}$ (5*a*)

– active-joint velocities: $\quad |\dot{q}_i(t)| \le \dot{q}_i^{\max}$ (5*b*)

– active-joint accelerations: $\quad |\ddot{q}_i(t)| \le \ddot{q}_i^{\max}$ (5*c*)

– active-joint torques: $\quad |\tau_i(t)| \le \tau_i^{\max}$ (5*d*)

Here, the index *i* refers to the ith component of the considered vector. Of course, non-symmetrical bounds on the above physical quantities can be handled also.

In order to ensure the stability of the WMM, the following inequality must be verified:

$$sd(\boldsymbol{q}(t), \dot{\boldsymbol{q}}(t), \ddot{\boldsymbol{q}}(t)) \ge 0 \tag{6}$$

where *sd* is the stability degree defined using the Zero Moment Point (ZMP) criterion (Kim et al., 2002).

The goal function *J* to be minimized represents the travel cost between initial and final states. It is generally an expression containing significant physical parameters related to the WMM behavior and also to the productivity of the system. For simplicity, we will limit the discussion to the case of a balance between the travel time *T* and the quadratic average of actuator efforts:

$$J = (1-\alpha)\,T + \alpha \int_0^T \sum_{i=1}^{n-m} \left(\frac{\tau_i(t)}{\tau_i^{\max}} \right)^2 dt \tag{7}$$

α is a weight coefficient chosen in [0,1] according to the relative importance we would like to give to the minimization of the travel time *T*. The case $\alpha = 0$ corresponds to the MTTPP.

4 Proposed Method

Let $\lambda \in [0, 1]$ be an independent real parameter, *T* the unknown travel time and ξ a normalized time scale: $\xi = t / T$. Hereafter, the prime symbol will indicate a derivative with respect to ξ.

With λ varying from 0 (initial state) to 1 (final state), the geometry of the robot path in the generalized coordinate space is completely represented by the following continuous time-independent sequence $\boldsymbol{P}(\lambda) = [\mathbf{X}^t(\lambda), \boldsymbol{q}_a^t(\lambda)]^t$, where $\mathbf{X}(\lambda) = [x(\lambda), y(\lambda)]^t$. Indeed, once $\mathbf{X}(\lambda)$ is given, the other generalized coordinates of $\boldsymbol{q}_p$ can be deduced directly from (1). As for the shape of the time-history of such a sequence, it can be represented by a monotonically increasing motion function $\lambda(\xi)$. The key point is that the travel time T becomes a dependent parameter. Namely, if we choose arbitrarily a geometrically feasible profile $\boldsymbol{P}(\lambda(\xi))$, then the cost function J can be written explicitly in terms of the single variable T. Therefore, the optimal travel time $T_{\boldsymbol{P}}$ corresponding to the chosen profile $\boldsymbol{P}$ can be deduced via a straightforward (albeit constrained) one–dimensional minimization of J.

Here, the problem is transformed into a parametric optimization where the only unknown are three sets $\boldsymbol{S}_p$, $\boldsymbol{S}_a$ and $\boldsymbol{S}_m$ of free control nodes. $\boldsymbol{S}_p$ is composed of N_p control points in the robot workspace (Figure 1*a*). $\boldsymbol{S}_a$ is composed of N_a control points in the manipulator *C*-space (Figure 1*b*). $\boldsymbol{S}_m$ is composed of N_m collocation points in the (ξ, λ) plane (Figure 1*c*).

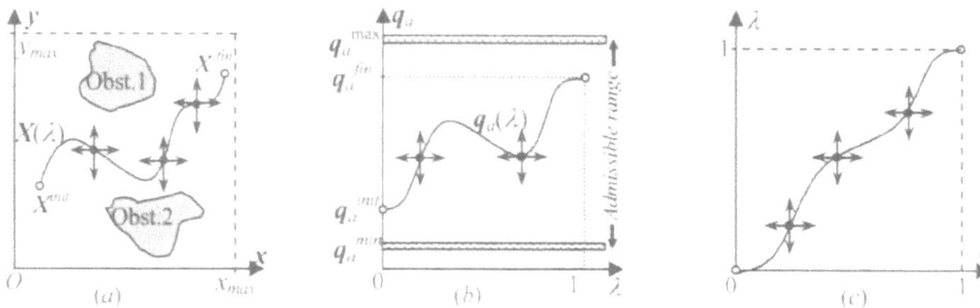

Figure 1. (*a*) A path $\mathbf{X}(\lambda)$ in the workspace, (*b*) a path $\boldsymbol{q}_a(\lambda)$ in the manipulator C-space, (*c*) a motion function.

With $\boldsymbol{S}_p$ and $\boldsymbol{S}_a$, we can build a trial geometric path $\boldsymbol{P}(\lambda)$ using B-spline functions that take into account constraints (4*a*); whereas with $\boldsymbol{S}_m$ we can specify a trial monotonically increasing motion function $\lambda(\xi)$ using a clamped cubic spline model that takes into account constraints (4*b*).

The method adopted here uses a stochastic process that scans simultaneously the available solution space of the three sets $\boldsymbol{S}_p$, $\boldsymbol{S}_a$ and $\boldsymbol{S}_m$ to propose candidate trajectory profiles $\boldsymbol{P}(\lambda(\xi))$ for a global minimization of J. As explained next, care must be taken to express the cost function in terms of T and to account for constraints and for obstacles that might be present in the workspace.

4.1 Evaluation of the cost function

First, we rewrite (3) in terms of T and derivatives of $\boldsymbol{q}$ with respect to ξ as follows:

$$\boldsymbol{\tau} = \boldsymbol{D}_\tau(\boldsymbol{q},\boldsymbol{q}',\boldsymbol{q}'')\,T^{-2} + \boldsymbol{Q}_\tau(\boldsymbol{q}) \tag{8}$$

where $\boldsymbol{D}_\tau(\boldsymbol{q},\boldsymbol{q}',\boldsymbol{q}'') = \left[\boldsymbol{R}^t(\boldsymbol{q})\boldsymbol{B}(\boldsymbol{q})\right]^{-1}\boldsymbol{R}^t(\boldsymbol{q})\left[\boldsymbol{M}(\boldsymbol{q})\boldsymbol{q}'' + \boldsymbol{c}(\boldsymbol{q},\boldsymbol{q}')\right]$, $\quad \boldsymbol{Q}_\tau(\boldsymbol{q}) = \left[\boldsymbol{R}^t(\boldsymbol{q})\boldsymbol{B}(\boldsymbol{q})\right]^{-1}\boldsymbol{R}^t(\boldsymbol{q})\boldsymbol{g}(\boldsymbol{q})$.

Then, given a candidate profile $\boldsymbol{P}(\lambda(\xi))$, the cost function (7) becomes:

$$J = T\cdot\left(C_0 + C_2 T^{-2} + C_4 T^{-4}\right) \tag{9}$$

where $C_0 = (1-\alpha) + \alpha\int_0^1 \sum_{i=1}^{n-m} \overline{Q_{\tau i}}^2 d\xi$, $C_2 = 2\alpha \int_0^1 \sum_{i=1}^{n-m} \overline{D_{\tau i}}\,\overline{Q_{\tau i}}\, d\xi$, $C_4 = \alpha\int_0^1 \sum_{i=1}^{n-m} \overline{D_{\tau i}}^2\, d\xi$.

$$\overline{D_{\tau i}} = D_{\tau i}(\boldsymbol{q},\boldsymbol{q}',\boldsymbol{q}'')\big/\tau_i^{\max}, \qquad \overline{Q_{\tau i}} = Q_{\tau i}(\boldsymbol{q})\big/\tau_i^{\max}$$

C_0, C_2 and C_4 are real coefficients that do depend implicitly on the proposed candidate $\boldsymbol{P}$ but that do not depend on T. Note also that C_0 and C_4 are always positive. Expression (9) represents a family of curves whose general shape, for a feasible candidate is shown in Figure 2. The minimum of each of these curves is reached when T takes on the value T_m given by:

$$T_m = \left(\frac{C_2 + \sqrt{C_2^2 + 12C_0C_4}}{2C_0} \right)^{1/2} = \left(\frac{6C_4}{\sqrt{C_2^2 + 12C_0C_4} - C_2} \right)^{1/2} \qquad (10)$$

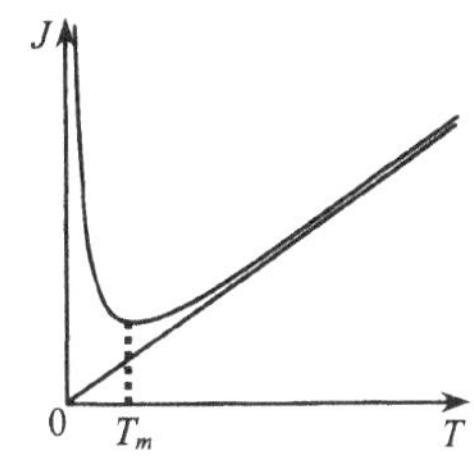

Figure 2. General shape of the cost function J vs. T for a given candidate P.

Although these two forms are mathematically equivalent, we should use the first (second) if C_2 is positive (negative) to reduce round-off errors. In any case, as kinodynamic constraints have not been included yet, T_m might not coincide with the optimal travel time T_p of candidate $\boldsymbol{P}$.

4.2 Treatment of constraints

Given a candidate $\boldsymbol{P}$, most constraints simply translate to bounds on admissible values of the travel time of that candidate. Here, constraints are grouped in three categories:

The first category concerns geometric constraints. They will not yield any restriction on T. For example, any candidate that infringes (4*c*) or (5*a*) will be rejected early in the process of selection since it will not lead to a feasible solution.

The second category concerns kinematic constraints (5*b*) and (5*c*). They translate to an explicit lower bound on T:

$$T \geq \max\left(T_{\mathrm{v}},\ T_{\mathrm{A}}\right) \qquad (11)$$

where $T_{\mathrm{v}} = \max\limits_i \left[\dfrac{\max |q_i'(\xi)|}{\dot{q}_i^{\max}} \right] \quad \xi \in [0 \;\; 1], \quad T_{\mathrm{A}} = \max\limits_i \left[\dfrac{\max |q_i''(\xi)|}{\ddot{q}_i^{\max}} \right]^{1/2} \quad \xi \in [0 \;\; 1]$

As for dynamic constraints (5*d*), since they may be rewritten via (8) under the form:

$$-\tau_i^{\max} \leq D_{\tau\, i} T^{-2} + Q_{\tau\, i} \leq \tau_i^{\max} \qquad i = 1 \ldots, n - m,$$

they usually translate to bracketing bounds on T: $\quad T \in [T_{\mathrm{L}}, T_{\mathrm{R}}] \qquad (12)$

The stability constraint (6) may be treated similarly and transformed into a set of admissible subintervals for T. For simplicity, the stable region is defined by the largest circle inscribed in the WMM sustentation polygon (Figure 3). The center of this circle is called the Most Stable Point (MSP) and the circle radius is noted r_{sa}. The ZMP is defined as a point on the floor under the moving platform where the sum of the moments of all external, gravity and inertial forces applied on the system vanishes (Kim et al., 2002). The ZMP coordinates can be computed as follows:

$$x_{zmp} = -Mp_y / fp_z \,; \quad y_{zmp} = Mp_x / fp_z$$

where $\boldsymbol{M_p} = [Mp_x, Mp_y, Mp_z]^t$ and $\boldsymbol{f_p} = [fp_x, fp_y, fp_z]^t$ are the overall moment and force applied to the platform expressed in the platform frame attached to MSP. If d_{zmp} is the distance from MSP to ZMP then inequality (6) becomes as follows:

$$r_{sa} - d_{zmp} \le 0 \tag{13}$$

Now, with the time scaling ($\xi = t / T$), components Mp_x, Mp_y and fp_z can be written as:

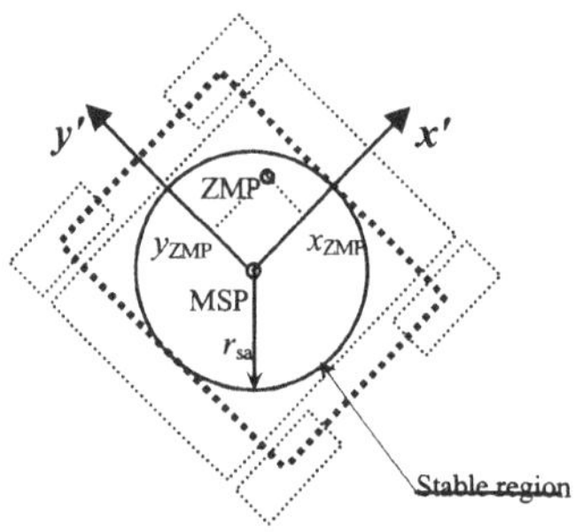

Figure 3. Stable region and ZMP.

$$\begin{cases} Mp_x = D_{Mx}(\boldsymbol{q},\boldsymbol{q}',\boldsymbol{q}'')T^{-2} + Q_{Mx}(\boldsymbol{q}) \\ Mp_y = D_{My}(\boldsymbol{q},\boldsymbol{q}',\boldsymbol{q}'')T^{-2} + Q_{My}(\boldsymbol{q}) \\ fp_z = D_{fz}(\boldsymbol{q},\boldsymbol{q}',\boldsymbol{q}'')T^{-2} + Q_{fz}(\boldsymbol{q}) \end{cases} \tag{14}$$

In the above expressions, the first terms represent the effect of inertial, Coriolis and centrifugal forces while the second regroup the effect of gravitational and external forces. Using (14), the stability constraint (13) may be written under the following form:

$$A_0 + A_2 T^{-2} + A_4 T^{-4} \le 0 \tag{15}$$

where $A_0 = r_{sa}^2 Q_{fz}^2 - Q_{Mx}^2 - Q_{My}^2$, $A_2 = 2\left(r_{sa}^2 D_{fz} Q_{fz} - D_{Mx} Q_{Mx} - D_{My} Q_{My}\right)$, $A_4 = r_{sa}^2 D_{fz}^2 - D_{Mx}^2 - D_{My}^2$.

Hence, inequality (15) may yield at most to two distinct intervals of admissible values of T:

$$T \in [T_{L1}, T_{R1}] \cup [T_{L2}, T_{R2}] \tag{16}$$

However, intersecting (11), (12) and (16) along the trajectory for ξ in [0, 1] may yield to a set of k distinct admissible intervals: $\Pi = [T_1, T_2] \cup [T_3, T_4] \cup \ldots [T_{k-1}, T_k]$. Consequently, for any chosen candidate $\boldsymbol{P}$, the corresponding optimal travel T_p must minimize J within Π. As for the value of T_m given by (10) for the unconstrained case, it will coincide with T_p only if it belongs to Π. Otherwise, T_p will be one of the T_i ($i = 1 \ldots k$).

5 Numerical results

We consider the MTTPP with and without stability constraint for a 2-link planar WMM (Furuno et al., 2003). Parameters are grouped in Table 1. Constraints on joints position and driving torques are given as follows:

$$-135° \le q_{a1}, q_{a2} \le 135°, \; -8.0 \le \tau_{p1}, \tau_{p2} \le 8.0\ (N.m), \; -4.0 \le \tau_{a1} \le 4.0\ (N.m), \; -2.0 \le \tau_{a2} \le 2.0\ (N.m)$$

Table 1. Parameters of the Robot

$m_p = 50.0\ kg$	$I_{zp} = 2.08\ \ kg.m^2$	$L_p = 0.5\ m$
$m_w = 2.0\ kg$	$I_{zw} = 0.006\ kg.m^2$	$L_w = 0.5\ m$
$m_1 = 20.5\ kg$	$I_{yw} = 0.01\ \ kg.m^2$	$L_1 = 0.4\ m$
$m_2 = 20.0\ kg$	$I_{z1} = 0.273\ kg.m^2$	$L_2 = 0.4\ m$
$r = 0.5\ m$	$I_{z2} = 0.273\ kg.m^2$	$B = 0.0\ m$
$z = 0.5\ m$	$z_1 = z_2 = 1.0\ m$	$r_{sa} = 0.16\ m$

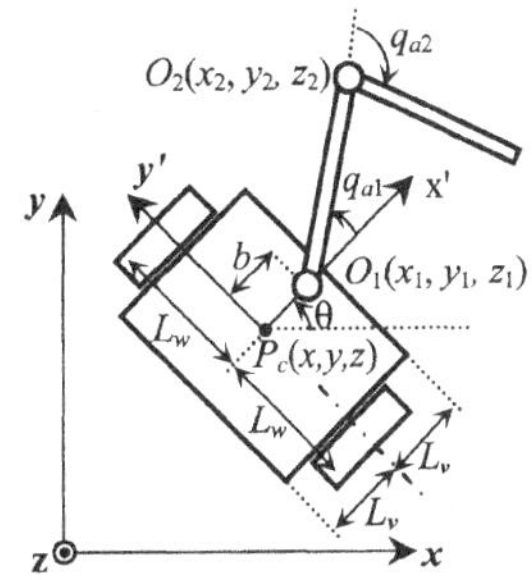

Figure 4. Planar mobile manipulator

Figure 5. Results obtained without stability constraint: (*a*) path, (*b*) platform joint torques, (*c*) manipulator joint torques, (*d*) stability degree.

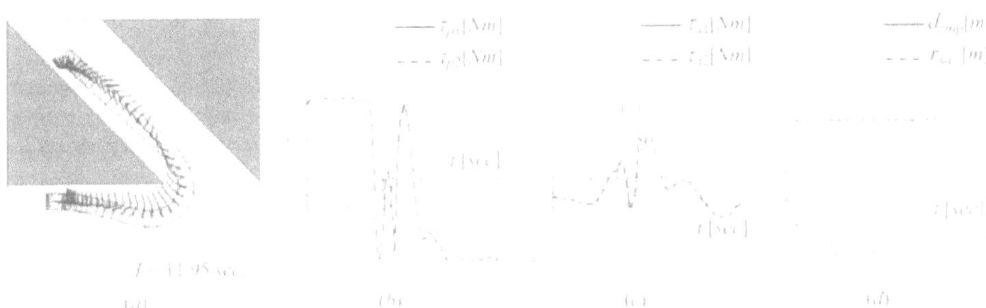

Figure 6. Results obtained without stability constraint: (*a*) path, (*b*) platform joint torques, (*c*) manipulator joint torques, (*d*) stability degree.

The workspace is a 8m×8m flat floor (Figure 5a). The motion starts at (1.5, 2.5, 0°, 0°, 90°) and ends at (2.0, 6.8, 135°, 90°, -90°). For this problem we have adopted fourth-order B–spline models with $N_p = N_a = 6$ control points and a clamped cubic spline model with $N_m = 3$

interpolation points. Computations have been done on a 2.4 GHz P4 using a simulated annealing optimization code written in Fortran 5.0.

Results obtained without stability constraint are shown in Figure 5. The calculated score is 11.92 seconds for a runtime of 148 seconds. We note that Figure 5d indicates that $d_{ZMP} > r_{sa}$. Results obtained for the same problem with the stability constraint are shown in Figure 6. The calculated score is now 11.95 seconds for a runtime of 192 seconds.

6 Conclusion

We have proposed a versatile motion planner for wheeled mobile manipulators. The problem treated here consists of defining the optimal time history of a continuous sequence of configurations between two limit configurations while considering kinodynamic constraints and stability issues. The proposed method is an iterative process that tries to improve solutions by searching simultaneously for the robot path and for the time evolution on this path. The use of parametric functions to model the path and the motion transforms the optimal motion planning problem, which is a complex optimal control problem, to a constrained parametric optimization problem. Furthermore, kinodynamic constraints have been conveniently translated to bounds on admissible values of the traveling time and have been treated sequentially. Additional constraints can be handled similarly without inducing any significant modification of the method.

Bibliography

Bloch, A. M., Reyhanoglu, R., and McClamroch, N. H. (1992). Control and Stabilization of Nonholonomic Dynamic Systems. In *IEEE Transactions on Automatic Control*, Vol. 37, 1746–1757.

Campion, G., Bastin, G., and d'Andrea-Novel, B. (1996). Structural Properties and Classification of kinematics and Dynamic Models of Wheeled Mobile Robots. In *IEEE Trans. on Robotics and Automation*, Vol. 12, No.1, 47-62.

Chen, M. W., and Zalzala, A. M. S. (1997). Dynamic modeling and Genetic-Based Trajectory Generation for Non-Holonomic Mobile Manipulators", *Control Eng. Practice*, Vol.5, No.1, 39-48.

Chettibi, T., and Lehtihet, H. E. (2002). A new approach for point to point optimal motion planning problems of robotic manipulators. In *Proc. Of 6th Biennial Conf. on Engineering System Design and Analysis*, APM10.

Desai, J. P., and Kumar, V. (1997). Nonholonomic Motion Planning for Multiple Mobile Manipulators. In *Proc. IEEE Int. Conf. on Robotics and Automation*, 3409-3414.

Furuno, S., Yamamoto, M., and Mohri, A. (2003). Trajectory Planning of Mobile Manipulator with Stability Considerations. In *Proc. IEEE Int. Conf. on Robotics and Automation*, 3403-3408.

Haddad, M., Chettibi, T., Lehtihet, H. E., and Hanchi, S. (2005). A New Approach for Minimum-Time Motion Planning Problem of Wheeled Mobile Robots. In 16th *IFAC World Congress*.

Hollerbach, J. M. (1983). Dynamic Scaling of Manipulator Trajectories. *M.I.T.A.I. Lab Memo* 700.

Huang, Q., and Sugano, S. (1996). Motion Planning of Stabilization and Cooperation of a Mobile Manipulator. In *Proc. IEEE Int. Conf. IROS*, 568-575.

Kim, J., Chung, W. K., Youm, Y., and Lee, B. H. (2002). Real time ZMP Compensation Method using Null Motion for Mobil Manipulator. In *Proc. IEEE Int. Conf. on Robotics and Automation*, 1967-1972.

Mohri, A., Furuno, S., Iwamura, M., and Yamamoto, M. (2001). Sub-optimal Trajectory Planning of Mobile Manipulator. In *Proc. IEEE Int. Conf. on Robotics and Automation*, 1271-1276.

Perrier, C., Dauchez, P., and Pierrot, F. (1998). A Global Approach for Motion Generation of Non-holonomic Mobile Manipulators. In *Proc. IEEE Int. Conf. on Robotics and Automation*, 2971-2976.

A Decoupled Approach to Optimal Time Energy Trajectory Planning of Parallel Kinematic Machines

Amar Khoukhi , Luc Baron and Marek Balazinski

Mechanical Engineering Department,
École Polytechnique de Montréal,
C. P. 6079, Succ. CV
Montreal PQ, Canada H3C 3A7 ,
(amar.khoukhi, luc.baron, marek.balazinski)@polymtl.ca

Abstract. This paper deals with a new optimal dynamic trajectory planning approach for Parallel Kinematic Machines (PKM) under task, workspace, and manipulator constraints. From kinematic, and dynamic models based on Euler-Lagrange formalism, we develop a multi-objective trajectory planning, optimizing electrical energy and robot traveling time and satisfying several technological constraints such as task requirements, actuator, link length, and workspace limitations. A new decoupled formulation is considered in order to cope with the difficulty of dynamic parameters computation. A systematic implementation procedure is then given along with simulation results.

1 Introduction

The path planning problem of parallel manipulators is receiving an increasing interest by the robotic community. Bhattacharya *et al.* (1998) had developed a scheme for avoiding singularities of PKM by restructuring a pre-planned path in the vicinity of a singularity. In (Sen *et al.*, 2003), a variational approach is reported for planning a singularity-free minimum-energy path between two end-points for Gough-Stewart platforms. These trajectory planners are however essentially kinematic based schemes, while others consider only static effects. For weakly coupled mechanical structures where static forces dominate over inertial ones, and tasks requiring low velocities, such kinematic approaches like the inverse rate control algorithm (Craig, 2005) can assure satisfactory results. But for strongly non-linear, coupled mechanical structures, and for high velocity and precision tasks, a proved fact for serial manipulators is that these approaches fail to give acceptable results (Vukobratovic *et al.*, 1985), (Khoukhi, 2002). When the linear speed goes over 1.5m/s at the end-effector, good performance can not be achieved without taking into account of the Coriolis and centrifugal forces. So, how do these dynamic parameters affect the performance of parallel robots?

In this paper, an integrated optimal control system is designed for PKM dynamic trajectory planning. From a state space representation by a system of differential equations, the trajectory planning is solved with an optimization procedure by applying optimal control techniques (Khouhki *et al.*, 2005). The robot traveling time separating two sampling periods is optimized as well as the actuator torques. This multi-objective aspect allows achieving smooth trajectories satisfying several technological constraints such as actuator, link length, task, and workspace limitations. The resulting constrained multi-objective optimal control problem is solved through the augmented Lagrangean optimization technique (Polak, 1997). A new decoupling approach is introduced here in order to handle some inextricable computation issues arising in the original non decoupled formulation. Section 2 introduces the kinematic and dynamic models of the PKM. Section 3, gives the discrete time dynamic model. In section 4, the optimal control problem is formulated in a non-linear programming framework. Section 5 develops the new decoupled formulation to the control problem, and in section 6 the Augmented Lagrangean technique is used to solve the resulting linear and decoupled problem. In section 7, preliminary simulation results are reported. Finally, section 8 concludes this article.

2 Kinematic and Dynamic Models

The kinematic and dynamic models are briefly introduced here. The reader is referred to (Baron *et al.*, 2000), and (Harib *et al.*, 1997) for the details. From Fig.1, the closure equations are:

$$a_i = x + {}^{W}R_P\,{}^{P}a_i \tag{1}$$

while the inverse rate kinematic model is given by:

$$\dot{l} = J^{-1}\dot{q} \tag{2}$$

where ${}^{W}R_P$ is the rotation matrix describing the orientation of the frame $\boldsymbol{P}$ with respect to the base frame $\boldsymbol{W}$, J is the robot Jacobian, l represents the joint coordinates and $q(t)=(x(t), y(t), z(t), \varphi(t), \theta(t), \psi(t))$ constitutes the Cartesian coordinates of $\boldsymbol{P}$, and the Euler angles describing the position and orientation of the platform with respect to the frame $\boldsymbol{W}$.

The dynamic model of the PKM is derived using Euler-Lagrange formalism (Angeles 2003), (Harib *et al.*, 1997). In the operation space this dynamic equation is given in a canonical form as:

$$\tau = M(q)\ddot{q} + N(q,\dot{q}) + G(q) \tag{3}$$

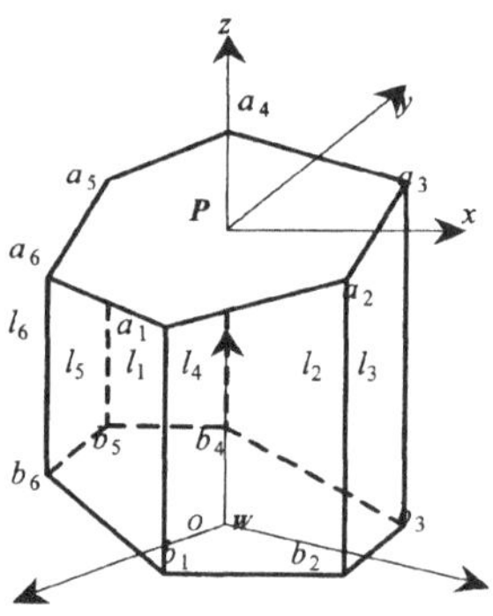

Figure 1. Geometry of a PKM

3 Discrete Time Operation Space Dynamic Model

To develop the discrete-time operation space dynamic model, we introduce the variables:

$$X(t) = (X_1(t), X_2(t))^T \text{, with } X_1(t) = q(t)\text{, } X_2(t) = \dot{q}(t) \quad (4)$$

Hence, the discrete-time state-space dynamic model can be written in the following form (Khoukhi, 2002):

$$[X_{k+1}] = \begin{bmatrix} I_{6\times6} & h_k I_{6\times6} \\ O_{6\times6} & I_{6\times6} \end{bmatrix}[X_k] - \begin{bmatrix} \frac{h_k^2}{2} I_{6\times6} \\ h_k I_{6\times6} \end{bmatrix}\left[M^{-1}(X_{1k})\left[N(X_{1k}, X_{2k}) + G(X_{1k})\right]\right] + \begin{bmatrix} \frac{h_k^2}{2} I_{6\times6} \\ h_k I_{6\times6} \end{bmatrix}\left[M^{-1}(X_{1k})\right][\tau]_k \quad (5)$$

4 Optimal Time-Energy Trajectory Planning Problem

The performance index relates time and energy consumption. The approach used to solve for time minimization consists to fix the number of discretisations N and vary the sampling periods h_k. The discrete optimal control problem consists in finding the optimal sequences $((\tau_1^1, \tau_1^2,...,\tau_1^6), (\tau_2^1, \tau_2^2,...,\tau_2^6), (\tau_3^1, \tau_3^2,...,\tau_3^6),...,(\tau_N^1, \tau_N^2,...,\tau_N^6))$ and $(h_1, h_2,...,h_N)$ allowing the robot to move from an initial state $X_0 = X_S$ to a given final state $X_N = X_T$, while minimizing the cost:

$$C_d = \underset{\substack{\tau\in\Re^{6\times N} \\ h\in\Re^{N+}}}{\text{Min}}\left\{\sum_{k=1}^{N}\left[(\tau_k R \tau_k^T + \iota + X_{2k} Q X_2^T) h_k\right]\right\} \quad (6)$$

where $R, Q,$ and ι are respectively electric, and kinetic energy and time level-headnesses.

This minimization is made subject to dynamic eq. (5) and next equality and inequality constraints:

- Initial and target state attainability:

$$X_0 = X_S,\ X_N = X_T \quad (7)$$

- Link intermediate length limits:

$$l_{Min}^i < l_k^i < l_{Max}^i \quad i = 1,...,6, \text{ and } k = 1,2,...,N\text{, with } l_{Max}^i = \sqrt{L_{iMax} \cdot L_{iMax}} = \Theta_{iMax}(X) \quad (8)$$

- Torque limits:

The robot torques is assumed to belong to a compact and bounded set $D_{\tau ad} \subset R^{2N}$:

$$D_{\tau ad} = \left\{\tau_{k,i}, \text{such that: } \tau_{\min,i} < \tau_{k,i} < \tau_{\max,i}\text{, } i \in \{1,2,...,6\}, k=0,...,N-1\right\} \quad (9)$$

- Sampling period limits:

If the total traveling time T is too small, there may be no admissible solution to the optimal control problem. On the other hand, h_k must be smaller than the system smallest time constant in order to prevent the system from being uncontrollable between two control times. A tradeoff is made through bounding the sampling periods within an admissible domain such that:

$$D_{had} = \left\{h_k\text{, such that : } h_{\min} < h_k < h_{\max}\right\} \quad (10)$$

5 Constrained Linear-Decoupled Formulation

A major computational difficulty in the right hand side of equation (5) is that it contains the inverse of the total inertia matrix of the moving platform and the struts as well as their Coriolis and centrifugal wrenches. These matrices might take several pages long to display. By performing with the original non linear formulation, when one computes the co-states λ_k associated to the dynamic states X_k, he or she will have to perform derivatives of the inverse of this total inertia matrix with respect to state components. The difficulty of these calculations has motivated a linear-decoupled formulation similar to that used in decoupled non linear control systems (Isidori 1995). This formulation reduces drastically the computations, by alleviating us performing the mentioned hard calculations.

Theorem:
Under the invertibility condition of the inertia matrix, the control law defined as:

$$\tau = M(q)v + N(q, q) + G(q) \tag{11}$$

allows the robot to have a linear and decoupled behavior as: $\ddot{q} = v$ (12)
where v is an auxiliary input.
Proof: See for instance (Isidori 1995).

This brings the robot to have the decoupled and linear behavior described by the following simple dynamic equation written in discrete form as:

$$X_{k+1} = A(h_k)X_k + B(h_k)v_k = f^D_{d_k}(X_k, v_k, h_k) \tag{13}$$

with

$$f^D_{d_k}(X_k, v_k, h_k) = \begin{bmatrix} I_{6\times6} & h_k I_{6\times6} \\ O_{6\times6} & I_{6\times6} \end{bmatrix} \begin{bmatrix} X_k \end{bmatrix} + \begin{bmatrix} \frac{h_k^2}{2} I_{6\times6} \\ h_k I_{6\times6} \end{bmatrix} \begin{bmatrix} v_k \end{bmatrix}_{6\times1}$$

Note that the control inputs for the decoupled problem become the acceleration and sampling period variables.

6 Augmented Lagrangian for the Decoupled Formulation

The augmented Lagrangian associated to the decoupled formulation is written as:

$$L^D_\mu(X, v, h, \lambda, \alpha, \beta, \gamma, \delta, \rho, \zeta, \varsigma) = \sum_{k=0}^{N-1}\Big\{[[(M(X_{1k})v_k + N(X_{1k}, X_{2k}) + G(X_{1k}))^T R$$

$$(M(X_{1k})v_k + N(X_{1k}, X_{2k}) + G(X_{1k}))] + X_{2k}^T Q X_{2k} + \iota]h_k\Big\} + \sum_{k=0}^{N-1}\Big\{\lambda_{k+1}^T(X_{k+1} - f^D_{d_k}(X_k, v_k, h_k)\Big\} +$$

$$\sum_{k=0}^{N-1} h_k\Big\{\Phi_\mu(\alpha_k, X_k - X_{Max}) + \Phi_\mu(\beta_k, X_{Min} - X_k)\Big\} +$$

$$\sum_{k=0}^{N-1} h_k\Big\{\Phi_\mu(\delta_k, \tau_{Min} - (M(X_{1k})v_k + N(X_{1k}, X_{2k}) + G(X_{1k}))\Big\} +$$

$$\sum_{k=0}^{N-1} h_k\Big\{\Phi_\mu(\gamma_k, (M(X_{1k})v_k + N(X_{1k}, X_{2k}) + G(X_{1k}) - \tau^j_{Max}))\Big\} +$$

$$\sum_{k=0}^{N-1} h_k\{\Phi_\mu(\rho_k, \Theta(X_k) - l_{Max}) + \Phi_\mu(\zeta_k, l_{Min} - \Theta(X_k))\} + h_N\{\Psi_\mu(\varsigma_N, X_{1N} - X_{1T})\} \tag{14}$$

where the function $f^D_{d_k}(X_k, \tau_k, h_k)$ is defined by the state equation (13), N is the total number of sampling points, $\alpha, \beta \in \Re^{12N+}$, $\gamma, \delta, \rho, \zeta \in \Re^{6N+}$, $\varsigma \in \Re^{N+}$ are the Lagrange multipliers. The vector $\lambda \in R^{12N}$ designates the co-states. Ψ_μ and Φ_μ are penalty functions associated to equality and inequality constraints. For the control sequences, $v_0, \ldots, v_{N-1}$, $h_0, \ldots, h_{N-1}$ and the corresponding trajectory states $X_0, \ldots, X_N$ to be optimal solution to the problem, there must exist some positive Lagrange multipliers

$$P_k = (\alpha_k, \beta_k, \gamma_k, \delta_k, \rho_k, \zeta_k, \varsigma_k),\ k = 0, \ldots, N, \tag{15}$$

and a finite positive penalty coefficient μ satisfying the first order Karush-Kuhn-Tucker (KKT) optimality necessary conditions. The development of these conditions and various related expressions and gradients are quite long and not given here. For more details, the reader is referred to (Khoukhi *et al.*, 2006).

A systematic procedure for the augmented Lagrangian implementation appears in the Appendix, where notations $\Xi(X)$ and $I(X)$ had been generically used to define equality and inequality constraints. Also, the Lagrange multipliers are divided into two sets *P1* and *P2* corresponding respectively to equality and inequality constraints. In this procedure, after selecting robot parameters, task definition, (such as starting and target poses), workspace limitations and simulation parameters, the kinematic unit defines a feasible solution satisfying initial and target poses. Then the inner optimization including energy and time minimization units give a dynamic solution, from which the states are computed, and tested. If feasible, a further convergence test is made. If convergence holds, then the program stops and displays the optimal solution. Otherwise, if the feasibility test fails, restart inner optimization unit. Else, if the feasibility test succeeds, go further to the dual part of the augmented Lagrangian and test for constraints satisfaction and update penalty and tolerance parameters. Then re-iterate until the number of permitted iterations is reached.

7 Simulation Results

A simulation program has been implemented in Matlab to test the proposed optimal trajectory planning approach on a case study parallel robot reported in (Liu *et al* 2004), and preliminary results are encouraging. This robot consists of two degrees of freedom planar parallel manipulator. A schematic of the manipulator is given in Figure 2. The following numeric values of the parameters are used: The number of sampling points is 10. The radius is $r = 75mm$, $R = 1203.0mm$ and $L = 1972.5mm$. The mass of the platform is $m_p(kg) = 200$, that of each leg is $m_l(kg) = 575$, and that of the slider is $m_s(kg) = 60$. The maximum velocity is $V_{max} = 200\,(mm/s)$. The workspace is defined as: ($x \in [-800mm, 800mm]$ and $y \in [-1718.14mm, -718.14mm]$). As for the augmented Lagrangian, the following parameter values had been taken $\omega_s = 0.5$, $\eta_s = 0.5$, $\alpha_w = \alpha_\eta = 0.4$, $\gamma_1 = 0.25$, $\gamma_2 = 1.2$, $w^* = 10^{-4}$, $\upsilon = 0.01$, $\eta^* = \eta_1^* = 10^{-3}$, $\beta_w = \beta_\eta = 0.4$, $\upsilon = 0.01$, $\bar{\upsilon} = 0.3$.

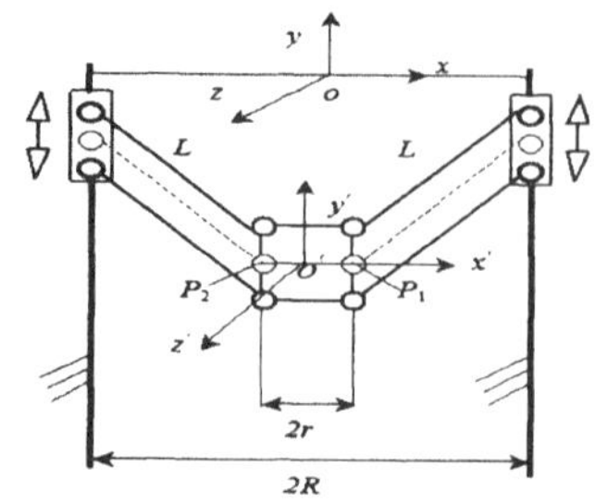

Figure 2. A geometric representation of the planar parallel manipulator

The time and energy level-headedness are $R = 1$ and $\iota = 1$. The initial Lagrange multipliers $P_0^T = (\alpha_0, \beta_0, \gamma_0, \delta_0, \rho_0, \varsigma_0, \upsilon_0)$ are set to zero. The trajectory consists of a straight line from an initial state position $x_0 = -0.48$, $y_0 = -1.48$ to a target position $x_F = 0.15$, $y_F = -0.40$ (all in meters). Note that at this level of simulation, the constraints (8) mentioned in section 4 is not considered. Figures 3 and 4 show the simulation results for both initial kinematic solution and augmented Lagrangian solution. The first plots from the top shows the 2-dimension linear displacement of the platform. The second plots show the consumed time to achieve the trajectory. The third and fourth plots show the instantaneous variations of the joint torques. While the last ones show the consumed energy. One notes that although the initial solution is cinematically feasible, when the corresponding torques is computed considering the dynamic model and forces, one gets several times torque values outside the admissible domain resulting in high values for the cost function. However, with the augmented lagrangian, the variations of the energy consumption are increasing smoothly and monotonously. Finally, in terms of CPU processing time, for seven dual iterations and four inner iterations performed to find a minimum time-energy trajectory, with a precision of 0.5×10^{-3} in target attainability, the total CPU time is of order of 75sec.

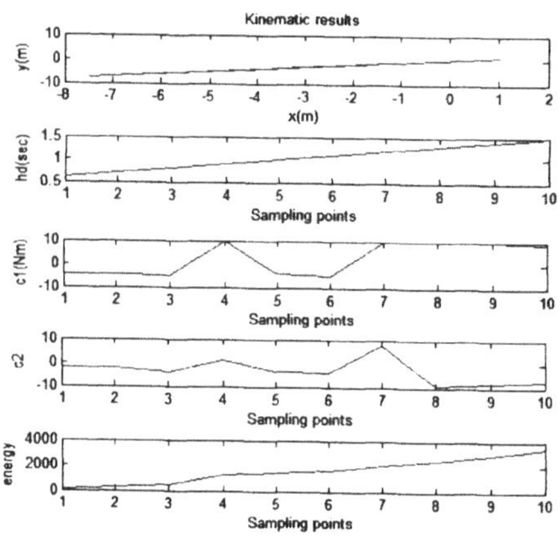

Figure 3. kinematic simulation results

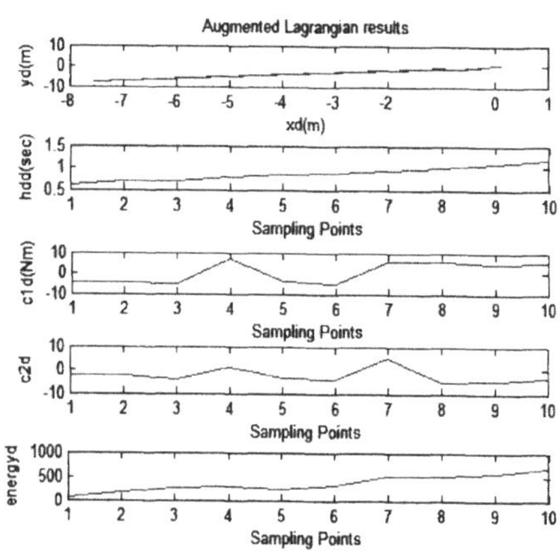

Figure 4. Simulation results using the augmented Lagrangian

8 Conclusions

In this paper, a new design to optimal multi-objective trajectory planning is presented. This design involves several criteria and constraints related to a global performance index integrating the interactions of the robot kinematics, dynamics, task, and environment requirements. A decoupling formulation had been also considered in order to tackle several inextricable computational issues. The augmented Lagrangian algorithm had been used to solve the resulting non linear and non convex optimal control problem. A major advantage of this approach is that one can introduce any type of constraints related to robot, task or environment, like actuators and friction models, obstacles and singularity avoidance by deriving the corresponding constraint expressions and adding them naturally in the Lagrangian in order to have them included in the trajectory planning system. Some of these topics are being developed in an ongoing work.

Acknowledgment: The authors would like to gratefully thank the Natural Science and Engineering Research Council of Canada for supporting the present work.

Bibliography

J. Angeles (2003). Fundamentals of Robotic Mechanical Systems, Theory, Methods and Algorithms. Second Ed., Springer Verlag.

L. Baron and J. Angeles (2000). The Kinematic Decoupling of Parallel Manipulators Using Joint-Sensor Data, *IEEE Transactions on Robotics and Automation* 16(6): 644-651.

S. Bhattacharya, H. Hatwal, A. Ghosh (1998). Comparison of an Exact and Approximate Method of Singularity Avoidance in Platform Type Parallel Manipulators. Mech. Mach. Theory 33 (7): 965-974.

J. J. Craig (2005). Introduction to Robotics: Mechanics and Control. 3rd Ed, Addison-Wesley.

A. Isidori. Non Linear Control Systems, Springer; 3 edition

K. Harib, K. Srinivasan. (1997). Kinematic and Dynamic Analysis of Stewart Platform-Based Machine Tool Structures. Robotica 21: 541-551.

A. Khoukhi (2002). An Optimal Time-energy Control Design for a Prototype Educational Robot. Robotica 20: 661-671.

A. Khoukhi, L. Baron, M. Balazinski (2005). Programmation Dynamique Time-Énergie Minimum des Robots Parallèles par Lagrangien Augmenté. CCToMM SM3, May 26-27, Canadian Space Agency, Saint Hubert QC, Canada, : 147-148.

A. Khoukhi, L. Baron, M. Balazinski (2006). Multi-Objective Off-line Programming of Parallel Kinematic Machines, Submitted to Robotics and Computer-Integrated Manufacturing.

X. J. Liu, Q. M. Wang, J. Wang (2004). Kinematics, Dynamics and Dimentional Synthesis of a Novel 2-DoF Translational Manipulator", Journal of Intelligent and Robotic Systems, 41, 205-224, 2004.

E. Polak. (1997). Optimization, Algorithms and Consistent Approximation. Springer-Verlag, New York.

S. Sen, B. Dasgupta, A. K. Mallik. (2003). A Variational Approach for Singularity-Free Path-Planning of Parallel Manipulators. Mec. and Machine Theory 38: 1165-1183.

M. Vukobratovic, D. Stotic, (1985). Is Dynamic Control Needed in Robotics System? and If So to What Extent? ", Int'l Jou. Of Robotics Research 2(2): 18-35.

Appendix: Augmented Lagrangian algorithm

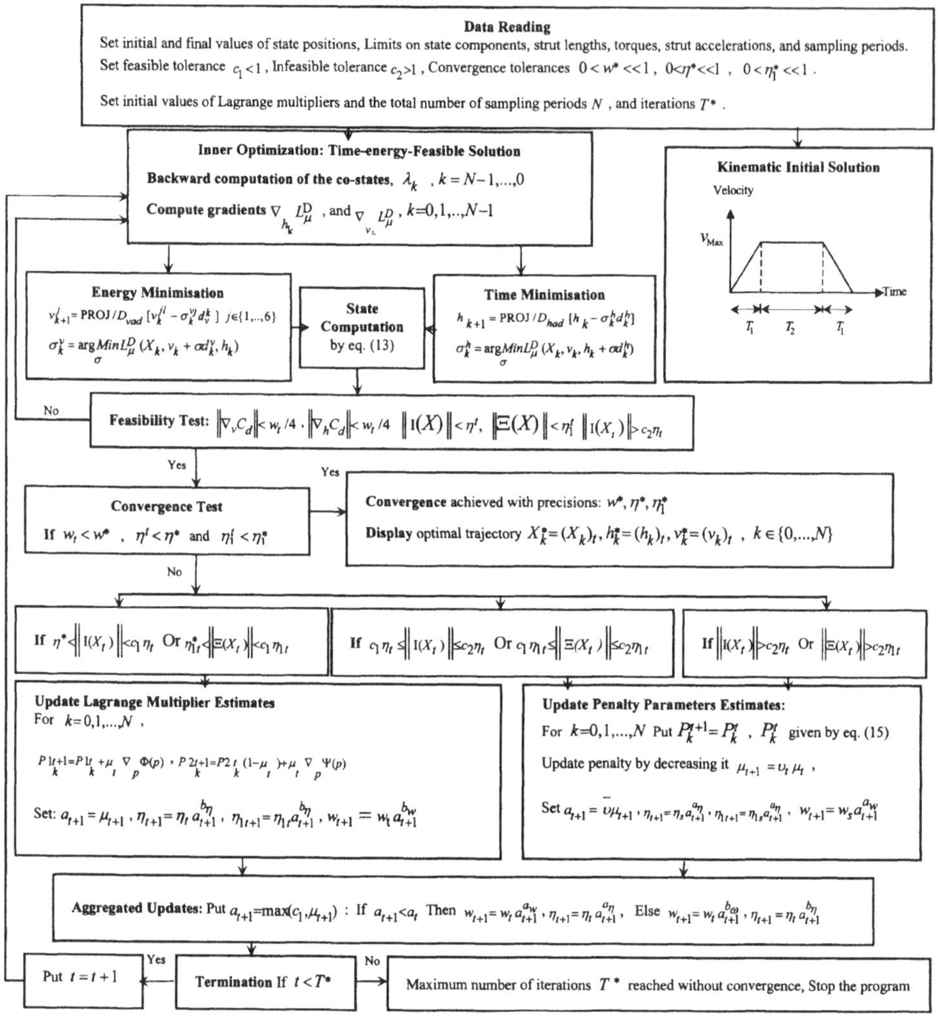

Flowchart for the Augmented Lagrangian algorithm and architecture

Reliable and Adjustable Biped Gait Generation for Slopes Using a GA Optimized Fourier Series Formulation

Lin Yang [1], Chee Meng Chew [1], Teresa Zielinska [2], Aun Neow Poo [1]

[1] The National University of Singapore, Faculty of Engineering,
Department of Mechanical Engineering, Control and Mechatronics Lab
e-mail: yanglin@nus.edu.sg mpeccm@nus.edu.sg mpepooan@nus.edu.sg

[2] Warsaw University of Technology
Institute for Aircraft Engineering and Applied mechanics
e-mail: teresaz@meil.pw.edu.pl

Abstract. This paper presents an online-adjustable method for generating reliable and optimal stable human-like gaits suitable for inclined surfaces for a biped robot based on a Tailored Fourier Series Formulation with coefficients obtained using Genetic Algorithm. The generated gaits are optimized with respect to posture stability, even walking speed and low foot strike velocities. The walking rhythm and walking pattern are designed to be online adjustable using feedback information from the environment. This Genetic Algorithm Optimized Fourier Series Formulation *(GAOFSF)* approach is a general gait generation method not limited to any specific kinematic structure of the robot and can be applied to motions of different requirements. In this paper, the results of the application of the *GAOFSF* approach, including dynamic simulation, for walking on terrains with slopes are presented.

1 Introduction

Recently, Humanoid Robots have attracted the interest of many researchers. Bipedal locomotion is an important area of research that needs to be undertaken when developing humanoids. To reduce the complexity of bipedal motion analysis, simplified dynamic models are often applied. An example is the inverted pendulum model which makes certain assumptions about the robot's motion and structure (Kajita et al., 2001). Some researchers made use of recorded human walking data (Nakanishi et al., 2004) as a reference for robot gait learning. There are also many gait generation methods developed for bipedal walking robots. Examples of these include the trajectory planning approach using inverse kinematics with the *ZMP* (Zero Moment Point) stability indicator (Kagami, 2002), a biologically inspired approach using the idea of *CPG* (Central Pattern Generator) (Kimura & Cohen, 2004; Zielinska, 2005), the artificial intelligence based approaches i.e. Reinforcement Learning (Chew & Pratt, 2002), and passive walking (Ohta et al., 2001).

Motion assumptions inevitably make the results from the ensuing analysis or simulation not so reflective of the actual robot dynamics when the derived algorithms are implemented on real robots, and limit the application to robots of other structures. Human walking data is a useful reference for gait analysis and synthesis, but there still exist significant differences between the dynamics of robots and their human counterparts (Chew & Pratt, 2002). Although the *CPG*-based approach is very attractive because it can result in closed-loop trajectory tracking, it is difficult to design interconnection and feedback pathways, and to manually tune all open parameters in order to achieve the desired behavior. For *CPG* models (e.g. Neural Oscillator or Van der Pol equation), the relationship between the properties of the model and robot dynamics is not so obvious and difficulties arise in searching for suitable parameters. Many *ZMP*-related methods are model-based with trajectory planning performed in Cartesian Space. As such, they require precise modelling and inverse kinematics computations and the generated motions result in a typical "bent-knee" posture which theoretically avoids Jacobian singularity. The bent-knee posture also requires more energy and the resulting motion is not as natural as the human gait (Nakanishi et al., 2004).

In this paper we proposed the Genetic Algorithm Optimized Fourier Series Formulation (*GAOFSF*) approach and shown its applicability and robustness for slope-terrain walking implementations. This *GAOFSF* approach has several advantages. Firstly, due to the characteristics of the Fourier Series used, the period of the gait can be specified directly. Through the use of constrained multiple-objective optimization, concurrently feasible, stable and natural human-like motion patterns can be achieved. Feasible and stable gaits are generated automatically based on specified requirements without the need for human gait data or manual planning. The walking rhythm, step-length and the walking pattern, can be easily adjusted in real-time, adapted for terrains of different inclinations either through the adjustment of simple functions or by the change to another record in the generated motion database (will be discussed in a subsequent paper). The *GAOFSF* approach can be applied to various motion conditions, including terrains of different slopes, different elastic properties, for robots of different sizes and inertia properties, and even for stairs or *3D* walking.

The *GAOFSF* approach is applied to the generation of feasible and stable biped gaits for both uphill and downhill walking in this paper. The feasibility and stability of the generated gaits are confirmed with dynamic simulations using Yobotics!. Section 2 introduces the *GAOFSF* model while the objective functions, constraints, and optimization approach using *GA* are discussed in Section 3. Section 4 presents the results of the gaits generated for two examples of walking on terrains with uphill and downhill slopes. In Section 5, the results of simulation performed using the Yobotics! dynamic simulation software are discussed which confirms the applicability of the approach.

2 GAOFSF Model

We define the following angles, all projected onto the sagittal plane, for the right leg as: right hip pitch angle, θ_{rh} = angle in the clockwise direction according to the body link; right knee pitch angle, θ_{rk} = angle in the clockwise direction according to the thigh link.

The corresponding angles for the left leg, θ_{lh} and θ_{lk}, are similarly defined. In the work presented here, the *GAOFSF* approach is applied on the 7-link planar robot whose properties are shown in Table 1.

Table 1. Robot modelling properties.

body	thigh	shank	foot	mass distribution
m=12kg; L=0.2m W=0.2m;H=0.45m	m=6kg; L=0.3m D=0.04m	m=5kg; L=0.3m D=0.04m	m=1kg; L=0.3m W=0.1m;H=0.02m	All are uniformed

Gait synthesis for this bipedal robot requires the determination of the trajectories of the hip-pitch and knee-pitch angles, θ_{rh}, θ_{rk}, θ_{lh} and θ_{lk}. The ankle-pitch angles are set to be such that both feet are always maintained parallel to the ground.

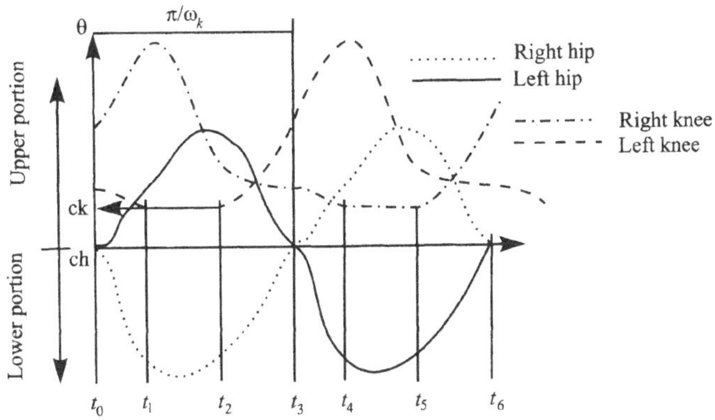

Figure 1. Typical shapes of hip and knee angle trajectories for a walking cycle.
Right support: $[t_2, t_5)$; Right swing: $[t_0,t_2)$ and $[t_5, t_6]$;
Left support: $[t_0,t_2)$ and $[t_5, t_6]$; Left swing: $[t_2, t_5)$;

c_h is a hip angle offset taken as that value at which the left and right hip joint trajectories intersect each other. c_k is the support knee angle during the lock phase and c_k=0 if the support knee is completely straightened during this phase. From Figure 1, it can be seen that the hip angle trajectories are not symmetrical about the axis θ=c_h but the trajectories for both legs are identical in shape but shifted in time, relative to each other, by half the walking period.

Let's first consider the hip angle trajectories. These can be conveniently analyzed by differentiating the upper portion, defined as $\theta_h^+ \geq c_h$, from the lower portion, defined as $\theta_h^- < c_h$. From Figure 1, it can be noted that all the upper and lower portions of the trajectories resemble part of a sinusoid. As such, a Fourier Series representation for these curves is chosen as a direct approach which will not involve too many higher order terms. The general *FSF* (Fourier Series Function) is expressed as

$$f(x) = \frac{1}{2}a_0 + \sum_{i=1}^{\infty} a_i \sin(ix) + \sum_{i=1}^{\infty} b_i \cos(ix) \tag{1}$$

To simplify the *FSF*, each of the curves is modelled as half of an odd function. Thus, the values of the coefficients a_0 and b_i in (1) will be all zero resulting in the simpler specific form

$$f(t) = \sum_{i=1}^{n} a_i \sin i\omega(t - t_f) + c_f \tag{2}$$

where a_i, n, t_f, and c_f are constants to be determined and ω is the fundamental frequency determined by the period of the walking cycle.

Using (2), and by inspection of the curves in Figure 1, the *FSF* for the hip angles are derived as

$$\theta_{rh},\theta_{lh} = \begin{cases} \theta_h^+ = \sum_{i=1}^{n} A_i \sin i\omega_h (t - t_h^+) + c_h, & t_h^+ \le t \le (t_h^+ + \frac{\pi}{\omega_h}) \\ \theta_h^- = \sum_{i=1}^{n} B_i \sin i\omega_h (t - t_h^-) + c_h, & t_h^- \le t \le (t_h^- + \frac{\pi}{\omega_h}) \end{cases} \tag{3}$$

with $\omega_h = \pi/(t_3 - t_0) = \pi/(t_6 - t_3)$

and where A_i and B_i are constant coefficients, θ_h^+ and θ_h^- the upper and lower portion respectively of the hip angle trajectory and t_h^+ and t_h^- are time-shift values. ω_h determines the period of the walking cycle. For walking on slopes and stairs, the appropriate value of c_h depends on the steepness of the slopes or the height of the stairs. In the proposed approach, *GA* is used to determine the optimal value of c_h to give a suitable gait and robot posture. This value will be a constant for the same walking situation.

For the knee joints, the trajectories are described similarly as

$$\theta_{rk},\theta_{lk} = \begin{cases} \theta_{k1} = \sum_{i=1}^{n} C_i \sin i\omega_k (t - t_k) + c_k, & t_k \le t \le t_k + \frac{\pi}{\omega_k} \\ \theta_{k2} = c_k \ge 0 \end{cases} \tag{4}$$

with $\omega_k = \pi/((t_6 - t_2) + (t_1 - t_0))$

and C_i is a set of constant coefficients, t_k a time-shift for the knee trajectory and θ_{k1} and θ_{k2} are the knee joint angle for the changeable phase and locked phase respectively.

Compared with other approximation approaches (Rastogi 2005), the advantages of using the tailored *FSF* to synthesize walking gaits for robots include

1. It is accurate and convenient for approximating the shapes of joint trajectories for human gait where the upper and the lower portions are not symmetrical but individually are similar to a sinusoid.
2. Each tailored *FSF* used here is a simple expression that makes no mathematic constraint when considered as half of an odd function. The gait period is included directly.
3. Key parameters such as c_h c_k and ω_h can be adjusted online. c_h and c_k vary according to the grounds profile while ω_h determines the walking rhythm.

3 Implementation Tools for GAOFSF

3.1 Dynamic stability criterion—ZMP

The *ZMP* is used in this work to determine the physical feasibility of the gaits generated using the Fourier Series Formulation. For an *n*-link robot, the coordinates, (P_x, P_y) in the *ZMP* frame coordinate, of the *ZMP* is given by (Sardain & Bessonnet, 2004)

$$P_x = (\sum_{i=1}^{n} m_i(\ddot{z}_i + g)x_i - \sum_{i=1}^{n} m_i \ddot{x}_i z_i - \sum_{i=1}^{n} I_{iy}\dot{\Omega}_{iy}) / \sum_{i=1}^{n} (\ddot{z}_i + g)m_i \tag{5.1}$$

$$P_y = (\sum_{i=1}^{n} m_i(\ddot{z}_i + g)y_i - \sum_{i=1}^{n} m_i \ddot{y}_i z_i + \sum_{i=1}^{n} I_{ix}\dot{\Omega}_{ix}) / \sum_{i=1}^{n} (\ddot{z}_i + g)m_i \tag{5.2}$$

where, for link *i*, (x_i, y_i, z_i) are the coordinates of the centroid, m_i the mass, I_{ix} and Ω_{ix} the centroidal moment of inertia and angular acceleration respectively about the *X*-axis, I_{iy} and Ω_{iy}

the corresponding parameters about the *Y*-axis, and *g* the acceleration due to the gravity. Here, only (5.1) is used since we consider only the motion in the sagittal plane.

For any given robot gait generated, if the *ZMP* as computed by (5.1) is always restricted to within the area covered by the "footprint" of the robot, the generated gait will be a physically realizable and stable. *ZMP* localization within the footprint is thus used as a primary objective for the Genetic Algorithm (*GA*) optimization strategy which follows.

3.2 Optimization strategy—GA

In order to generate the desired trajectories for stable walking, a suitable set of coefficients A_i, B_i, C_i and the parameters c_h, t_1 and t_2 need to be obtained. t_1 is the instant when the stance knee starts to be locked and t_2 is the instant when the walking phases of the two legs are switched as indicated in Figure 1. *GA* (Michalewicz 1994) was used to search for the desired values within suitable bounds of the parameters (A_i, B_i, C_i , c_h, c_k, t_1, t_2) for walking on slopes or stairs. To achieve a natural walking gait while maintaining adequate walking stability, a multi-objective function to be minimized is defined in the form

$$f_T = w_1 f_1 + w_2 f_2 + w_3 f_3 + w_4 f_4 \tag{6}$$

where f_1 = sum of the absolute value of the distances from the *ZMP* to the centre of the footprint over a walking cycle;

f_2 = standard deviation of the *ZMP* over a walking cycle;

f_3 = root mean square of the error of the speed of the trunk from the desired speed over a walking cycle;

f_4 = leg strike speed;

and w_1, w_2, w_3, and w_4 are the corresponding assigned weighting factors. Six motion constraints are also defined to ensure a valid and natural walking motion. These are:

s_1 constrains the *ZMP* to be within the supporting footprint,

s_2 constrains the swing height to be above a specified minimum value;

s_3 constrains the trunk's speed to be always positive;

s_4 constrains the swing foot's speed to be always positive except for the touching down motion;

s_5 constrains the deviation of the step length from that desired to within a small specified value;

s_6 constrains the deviation of the foot's touch-down instant, t_d, from the planned phase switching time, t_2, to within a small specified value.

Based on these constraints, a penalty function is defined as

$$P = \sum_{i=1}^{6} p_i s_i \tag{7}$$

where $p_i, i = 1,\ldots,6$ are assigned penalty weighting factors.

The conceptual difference between objectives and motion constraints is that the objectives tend to minimize the corresponding function values while constraints are limiting the function values within a desired range. That means there is no penalty for constraints function value even though all constraints are marginally achieved. However, for objectives, the smaller the objective function values are the better is the optimization performed.

Using (6) and (7), the fitness function for the *GA* is established as

$$F = \begin{cases} 0, & \text{if } \sum_{i=1}^{n}|A_i| = 0 \text{ or } \sum_{i=1}^{n}|B_i| = 0 \text{ or } \sum_{i=1}^{n}|C_i| = 0 \\ C_{\max} - f_T - P, & \text{otherwise} \end{cases} \tag{8}$$

The selection operator is the normalized geometric ranking method whose possibility of generation evolution is according to the fitness ranking. Therefore, the fitness value can be negative. However, if the fitness of some chromosome is very low, in case that after mutation, a bad chromosome can be a good one, the fitness can be manually increased a bit. Therefore, C_{max} is used. It should be suitably chosen to ensure most possible solutions are of positive fitness value. If C_{max} becomes negative, a small positive value is assigned to the chromosome to give small chance to survive to the next generation.

4 Gaits Generated by GAOFSF

Fifth-order FSF, with $n=5$, were used for the trajectories described in (3) and (4). The *GAOFSF* was applied to generate gaits for walking on the two slopes listed in Table 2.

Table 2. Specifications for slope walking.

Type of slope	10-degree up	10-degree down
Step length	0.26m	0.3m
Desired walking speed	0.27m/s	0.63m/s

For the up-slope and down-slope examples in given Table 2, the following chromosome solutions x_1 and x_2 respectively, in the format of $(A_i, B_i, C_i, c_h, c_k, t_1, t_2)$ with the times measured from t_0=0, were obtained:

x_1 = [0.22165 -0.040608 -0.010511 0.013618 -0.0047791 -0.49998 -0.035155 0.096458 0.039317 0.0037056 0.65021 -0.096304 -0.086896 0.0851 0.024097 -0.14372 0.26648 0.0001 0.54]

with C_{max}=3000 and fitness value=2860.

x_2 = [0.31239 -0.076315 0.01822 -0.0048476 -0.00448 -0.32653 -0.18954 -0.095695 -0.03138 -0.0089343 0.14377 0.11785 -0.0066839 -0.11365 -0.032868 -0.19356 0.79946 0.14727 0.2]

with C_{max}=3000 and fitness value=2935.

These generated gaits are discussed in the following two sections. GA parameters can be referred to our *GAOFSF* project report (Yang, et al., 2005).

4.1 Ten-degree Up-slope Walking

Figure 2 to Figure 4 shows the results for the gait generated, for one walking cycle, for the 10-degree up-slope, Figure 2 shows the trajectory of the *ZMP*. It is clear that the *ZMP* remains within the supporting region (-0.0985m<P_x<0.197m) of the footprint. Figure 3 shows the joint angle trajectories which can be seen to be similar to the general form shown in Figure 1.

For this solution, the swing foot strike velocity is found to be v_x= 0.042m/s, v_z=-0.28m/s which will not cause an undue ground impact as confirmed by the subsequent dynamic simulation (see Section 5). The step length obtained of 0.261m is also very close to the target step length. Figure 4 shows the stick diagram of the locomotion. From that, the desired even walking speed, 0.27m/s, has also been achieved. The sampling time is 0.05s.

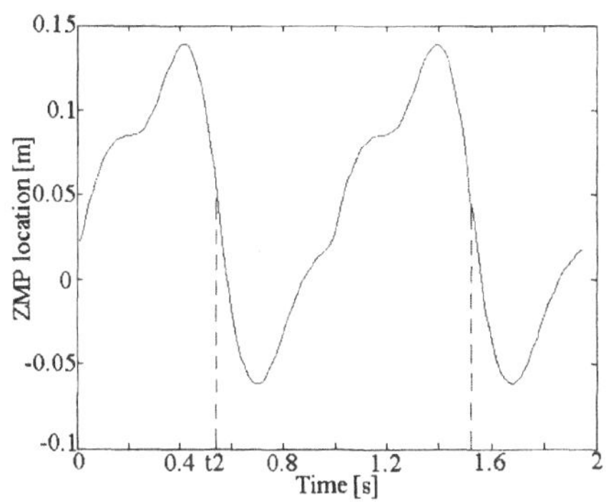

Figure 2. *ZMP* trajectory.

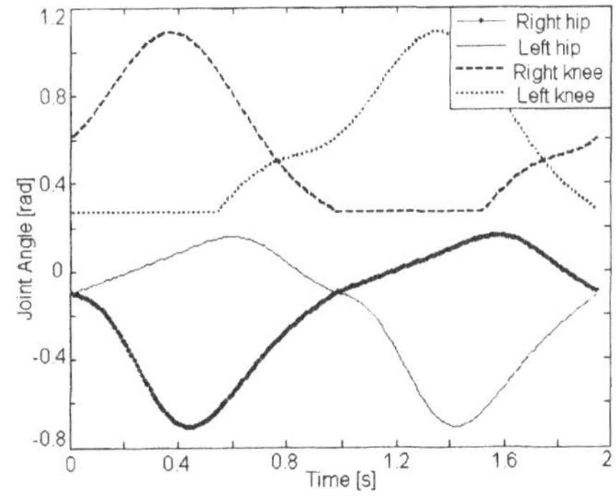

Figure 3. Hip and knee joint angle trajectory.

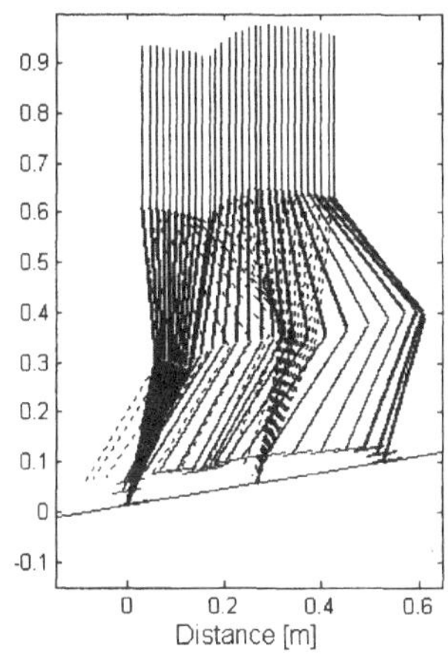

Figure 4. Stick diagram of the generated motion for 10-degree up-slope motion.

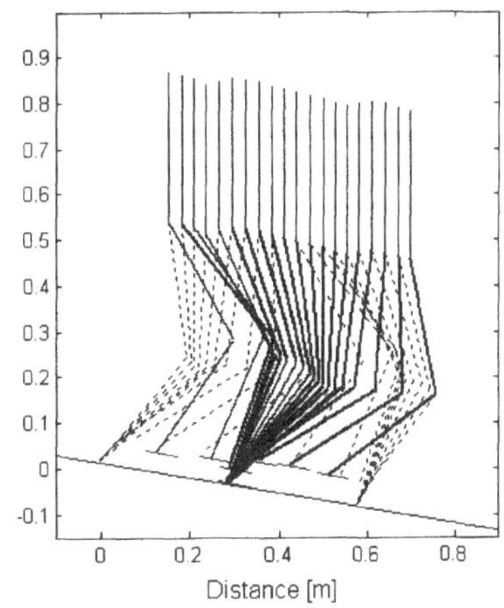

Figure 5. Stick diagram of the generated motion for 10-degree down-slope motion.

4.2 Ten-degree Down-slope Walking

The resulting motion obtained for the 10-degree down-slope motion, given by the solution x_2 shown in Figure 5. For this motion, the step length 0.3m, the swing foot strike velocity is v_x=-0.15m/s and v_z=-0.5m/s, the walking speed is around 0.625m/s, and the first leg strike instant is at t_d=0.21s, which is also close to t_2 in the *GA* solution.

5 Dynamic Simulations

Using the trajectories generated above, dynamic simulation of the robot was performed in Yobotics! (Yobotics! 2000-2003). Figure 6 and 7 show the stick diagrams of the up-slope and down-slope dynamic walking respectively. The sampling time used for the stick diagrams is 0.15s.

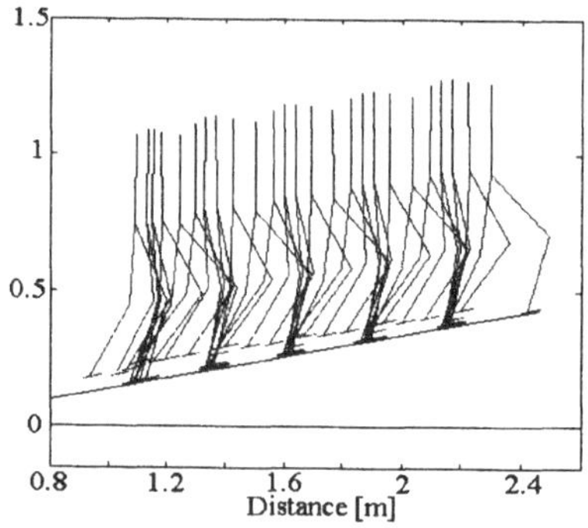

Figure 6. Stick diagram (up-slope).

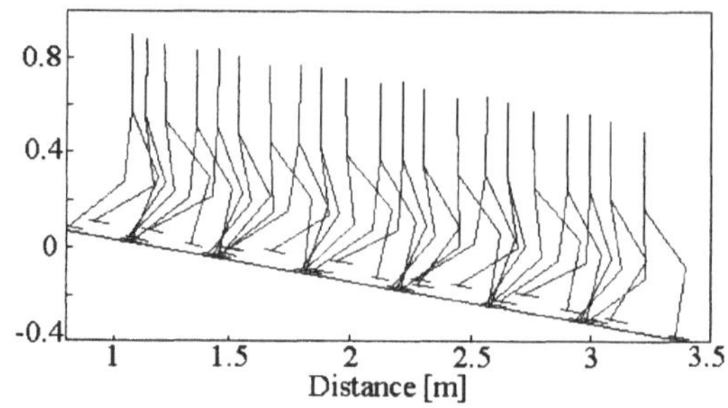

Figure 7. Stick diagram (down-slope).

6 Conclusion

The effectiveness of the *GAOFSF* approach for bipedal robot gait generation was proven through up-slope and down-slope motions and confirmed with dynamic simulations. The gaits generated are stable and reliable with human-like natural walking features. The method is not constrained by either the geometrical or inertial properties of the robot. It is also able to adjust gait in real-time for an irregular terrain by adjusting walking rhythm, step-length and updating the walking pattern. It has been done and will be talked in the following paper. Future work will focus also on the use of *GAOFSF* for the generation of *3D* motions.

Bibliography

Kajita, S., Matsumoto, O. and Saigo, M. (2001), "Real-time 3D walking pattern generation for a biped robot with telescopic legs", In *Proc. of the IEEE Int. Conf. on Robotics & Automation, Seoul, Korea.*

Nakanishi, J., Morimoto, J., Endo, G, Cheng, G, Schaal, S. & Kawato, M. (2004), "Learning from demonstration and adaptation of biped locomotion", *Robotics & Autonomous Systems* 47, 79–91.

Kagami, S. (2002), "A Fast Dynamically Equilibrated Walking Trajectory Generation Method of Humanoid Robot", *Autonomous Robots* 12, 71–82.

Kimura, H. and Cohen A. H. (2004), "Biologically Inspired Adaptive Dynamic Walking of a Quadruped Robot", In *Proc. of the 8th Int. Conf. on the Simulation of Adaptive Behaviour, Los Angeles,* 201–210.

Zielinska, T. (2005), "Biological Inspirations in Robotics: Motion Planning" In *Proc. of 4th Asian Conf. on Industrial Automation and Robotics, Bangkok, Thailand.*

Chew, C. M. and Pratt, G A. (2002), "Dynamic bipedal walking assisted by learning",*Robotica* 20, 477–491.

Ohta, H., Yamakita, M. and Furuta, K. (2001), "From passive to active dynamic walking", *Int. J. Robust & Nonlinear Control* 11, 287–303.

Rastogi, R. K. (2005), "Task oriented stable gait synthesis in biped locomotion", *Master thesis, Indian Institute of Information Technology, Allahabad.*

Sardain, P. and Bessonnet, G (2004), "Forces Acting on a Biped Robot. Center of Pressure—Zero Moment Point", *IEEE Trans. on Systems, Man, & Cybernetics—part A: systems and humans,* vol. 34, No. 5.

Michalewicz, Z. (1994), "Genetic Algorithms+Data Structures =Evolution Programs." *AI Series.Springer-Verlag.*

Yang, L., Chew, C. M. and Poo, A. N. (2005), "GAOFSF project report" *Mechanical engineering, NUS.*

Users Guide for Yobotics! Menu (2000-2003), *Yobotics, Inc.*

A Simplified Method for Generating 3D Gait Using Optimal Sagittal Gait

Tarik Saidouni[1] and Guy Bessonnet[2]

[1] Laboratory of Mechanics of Structures, EMP, Algeria, e-mail: tsaidouni@yahoo.fr
[2] Laboratoire de Mécanique des Solides, University of Poitiers, France, e-mail: bessonnet@lms.univ-poitiers.fr

Abstract. A simplified method for generating three-dimensional bipedal gait is presented. The approach consists in combining a lateral sway motion of the biped with an optimal sagittal movement. A method for computing the latter was presented in an earlier paper. The lateral movement is generated on the base of a simplified kinematic model and using a stochastic search method minimizing a dynamic performance criterion. Sagittal and frontal movements are successfully associated to produce well-balanced 3D-gait.

1 Introduction

The interest of researchers for legged locomotion has been increasing since the feasibility of walking machines was confirmed thanks to Japanese outstanding humanoid prototypes (Hirai et al., 1998, and Yamaguchi et al., 1998, and Kaneko et al., 2004). Gait synthesis using dynamic optimization is aimed both at improving the biped control and at reducing the energy expenditure which should result in increasing autonomy. Most papers dealing with dynamic optimization of gait are mainly devoted to the sagittal movements during the swing phase (Channon et al., 1992, Chevallereau and Aoustin, 2001, Bessonnet and Rostami, 2001).

The present paper is related to earlier researches on optimal gait generation for the bipedal robot BIP, described in reference (Sardain and Espiau, 2000). In a previous work (Saidouni and Bessonnet, 2004) a method for generating optimal cyclic sagittal gait was proposed for this biped modeled as a seven-link planar mechanical system. Dynamic models of the single support phase (SSP) and the double support phase (DSP) of gait were extensively developed. Transition configurations from one phase to the other were partly specified and a variety of constraints defining feasible walk was taken into account. Joint trajectories were approximated using cubic splines functions connected at uniformly distributed knots. A Sequential Quadratic Programming (SQP) method was used to solve the resulting algebraic optimization problem. Numerical simulations concerning slow and fast optimal gaits were carried out.

In this paper, we propose a method for combining sagittal gait with the sway motion of three-dimensional walk. The model obtained in this way constitutes a good approximation of real walk. Indeed, even if the movements in the frontal plane have reduced ranges, they should not be neglected when one wants to generate reference gait for an anthropomorphic biped robot, because the sway motions are required to provide lateral balance for the biped. The lateral stability is imperative. Thus, many works were focused on the fulfillment of this requirement for the swing phase using the concept of zero moment point (or center of pressure).

Moreover, although mechanical biped prototypes are inspired by the kinematic structure of human locomotion system, together with its body dimensions and mass distribution, human walk patterns can not be reproduced exactly by mechanical bipeds. We speculate that the determinants of human walk (McMahon, 1984) in the frontal plane such as pelvic rotation and pelvic tilt have not absolutely to be reproduced by mechanical walkers.

Both the kinematic simplicity and the dynamic balance of mechanical walk constitute the two essential points of our approach to optimal synthesis of three-dimensional gait. As a first step, we are interested in the swing phase during which the dynamic balance is harder to achieve.

2 Simplified 3D-gait model

A simplified approach for the gait-generation problem is to deal separately with the movements in frontal and sagittal planes. The decoupling between these two movements allows better knowledge about the effects of the lateral movement on its sagittal counterpart. This approach is also helpful to design the control algorithms having to ensure the dynamic stability of the biped.

2.1 Anthropomorphic biped model and constraints

The leg kinematics of the biped BIP is designed to mimic the human locomotion system. Thus, the legs exhibit three concurrent actuated axes at hip, two at ankle and one at knee (Fig. 1). Three additional active pelvis-trunk joint axes complete the biped kinematics. During the swing phase, the biped is modeled as a rooted kinematic chain with tree-topology, having fifty joints.

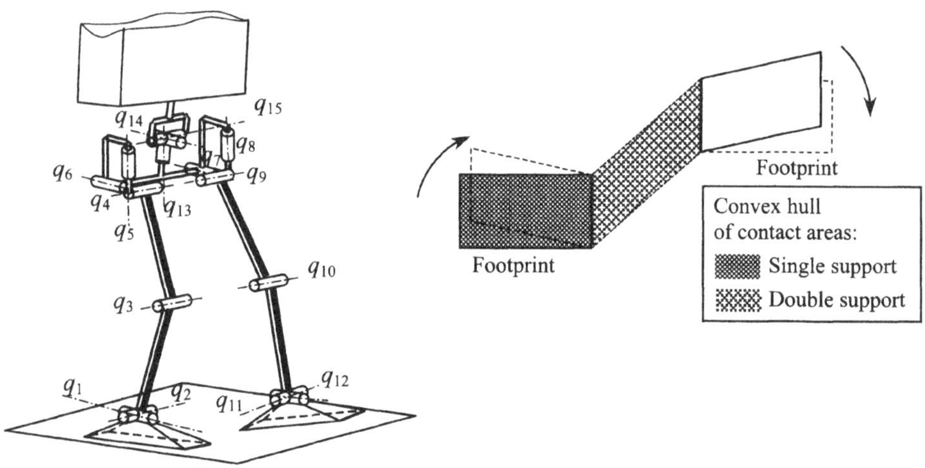

Figure 1. Robot kinematics and supporting areas

The dynamic balance is obtained by keeping the position of the center of pressure inside the convex hull of contact surfaces. This condition is guaranteed by the unilaterality requirement for the contact forces.

Let u_i the actuating torque applied by the joint of coordinate q_i, the performance criterion is defined as the integral of quadratic actuating torques :

$$J = \int_0^T \sum_{i=1}^{15} [u_i(t)]^2 \, dt \tag{1}$$

It is computed, together with dynamic constraints expressing unilaterality and non-sliding conditions of contact, using a recursive Newton–Euler algorithm.

Indeed, the constraints to be fulfilled by a feasible walk are of three types: the geometrical constraints of intermediate contact avoidance between the swing foot and the ground, then, kinematics constraints such as joint velocity and acceleration limitations; and finally, dynamic constraints prescribing actuator torque limitations, dynamic stability, and non-sliding conditions. Kinematic and dynamic constraints delimit the range of admissible range durations of movements.

2.2 Straight line walking

It is obvious that the lateral angular tilt of the support leg is necessary to keep the balance of the biped in the frontal plane. However, we can imagine a model that prescribes a movement of the swing foot parallel to the sagittal plane. This requirement results in the condition:

$$z_9 \,//\, z_0 \;\Leftrightarrow\; z_9 \times z_0 = \mathbf{0} \tag{2}$$

where $\mathbf{z}_i$ ($i = 0, 9$) stands for the unit vector of the ith axis of rotation.

From the constraint (2), we get two relationships of dependency on the joint variables (q_i, $i =$ 1 to 8). These equations can be solved in q_7 and q_8, respectively, describing the movements of adduction-abduction and the external-internal rotation of the hip of the swing leg. The following solutions are obtained, within the joint limits of the biped BIP, as follows:

$\forall . t \in [0, T]$,

$$q_7 = -q_6 + arctg(-tg(q_1)sin(Q), -cos(Q)cos(q_5)tg(q_1) - sin(q_5)) \tag{3}$$

$$q_8 = arctg\left(\sqrt{A + tg^2(q_1)}, \sqrt{1-A}\right) \tag{4}$$

where $Q = q_2 + q_3 + q_4$, and $A = sin^2(q_5)\left(1 - tg^2(q_1)\, cos^2(Q)\right) + tg(q_1)\, cos(Q)\, sin(2\, q_5)$.

As condition (2) is met both at the beginning and the end of the swing phase, the frontal joint of the swing ankle can be set to $q_{12} = 0$, in order to ensure a linear contact heel-strike. Moreover, as the angle Q is close to $\pi/2$, if we assume that:

$$q_6 = -q_1 \text{ and } q_5 = -\pi/2, \tag{5}$$

Then, the lateral tilt of the trunk is practically compensated.

2.3 Change in walking direction

It is possible to define a condition leading to the gradual change of walking direction over several steps by acting essentially on the joint coordinate q_8 of the hip. The change in walking direction may be completed by swiveling the trunk in direction of the front foot during the stance phase by acting on the external-internal rotation of the hip of the new stance foot (rotation in q_5).

During the swing phase, we can thus specify a rotation angle ψ of the supporting limb in the transverse plane:

$$z_9 \times z_0 = (0, sin(\psi), 0)^T \quad (6)$$

The angle $\psi(\tau)$ may vary gradually from zero until reaching a maximum value Ψ^{max} at the end of the swing phase. The change in direction can be achieved through a few steps. As in the above section, we solve in q_7 and q_8 with $q_6 = -q_1$ and $q_5 = -\pi/2$, the two equations resulting from condition (6).

3 Resolution strategy

Motions in sagittal plane corresponding to the joint coordinates q_2, q_3, q_4, q_9, q_{10}, q_{11} and q_{12} are early obtained by parametric optimization of the BIP sagittal model gait dynamics (see Saidouni and Bessonnet, 2004). Frontal motion is then obtained using a stochastic search method of which basic idea was introduced by Chettibi and Lehtihet (2002). This method dissociates the search for optimal time from that of motion parameters that are cubic spline knots positions (see Saidouni and Bessonnet 2002).

In our case, the optimal joint trajectory q_1 is obtained by disturbing randomly the best current trajectory in term of minimization of the performance criterion (1). The current trajectory q_1 is updated at each improvement of the performance criterion without constraint violations unless it is rejected. Moreover, the search for optimal trajectories is initiated with only three knots cubic spline approximation function to explore quickly the space of the admissible gaits. The insertion of new knots makes it possible to refine the search and is accompanied by a reduction of the search neighborhood by a factor varying between 2 and 10. Figure 2 summarizes the evolution of the search process.

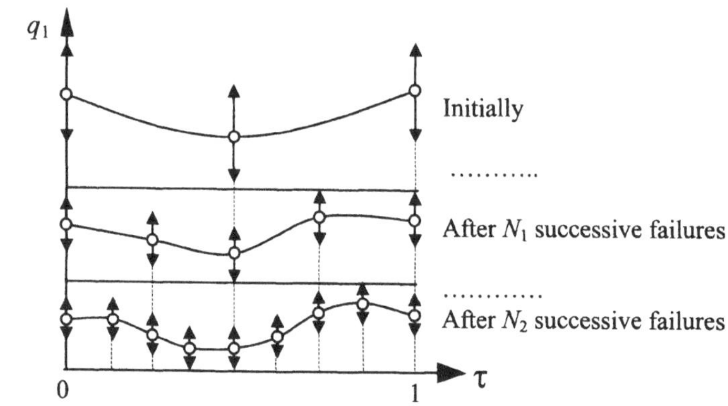

Figure 2. Evolution of the stochastic process used for trajectory q_1 search

The various constraints are checked sequentially at each step of the numerical integration of the performance criterion, in the order of ascending difficulty of realization. If at a given time, one of the constraints is violated or if the feasible range of the durations is lost, the solution or the gait is immediately rejected.

4 Numerical simulations

The sagittal movements are provided by gait optimization of a planar model of the biped. Only a simple correction of the position of the swing foot is required because the leg tilt $q_1 \neq 0$ modifies the length of the step and violates the foot-ground anti-collision constraints at initial and final times. When the tilt q_1 has a reduced range, the angular corrections are negligible and thus have not significant effects on gait dynamics. This correction is possible by readjusting the joint coordinates at knee and ankle of the swing leg.

4.1 Straight-line walking

The stochastic search process can be initialized by a laterally unstable gait. Initial and final velocities of the frontal joints are set to zero. The lateral balance is obtained using the frontal joint of the stance ankle (Figure 3). The trunk and the swing foot are set almost entirely again parallely to the sagittal plane by an opposite rotation of the hip frontal joint of the stance leg ($q_6 = -q_1, \forall \tau$). Thus, the variations of the joint coordinates q_7 and q_8 are lower than 0.5 degree. The other joints are fixed in their initial positions.

Figure 3. Frontal view of the 3D-model of the robot BIP during the swing phase

The center of pressure remains inside the support surface, close to the internal edge of the foot (figure 4). The projection of the trajectory of the center of gravity on the ground is not necessarily contained in the perimeter of the footprint. Thus, the gait is dynamic, not static. Moreover, numerical simulations showed that the lateral deviation range of the center of gravity is decreasing versus the walk speed. This is in agreement with human gait.

The center of pressure as well as the projection of the center of gravity on the ground remains close to the internal edge of the stance foot. The reasons of this fact is the antagonistic effects of the condition of lateral stability and anti-collision ground-foot constraints ones as well as the penalization of the performance criterion through the adduction-abduction hip torques due to the pelvic tilt increasing. Indeed, the maximum tilt angle of the ankle is less than 6.5 degrees. Also, the sway motion is reduced, and then the global stability of the biped improves on several steps.

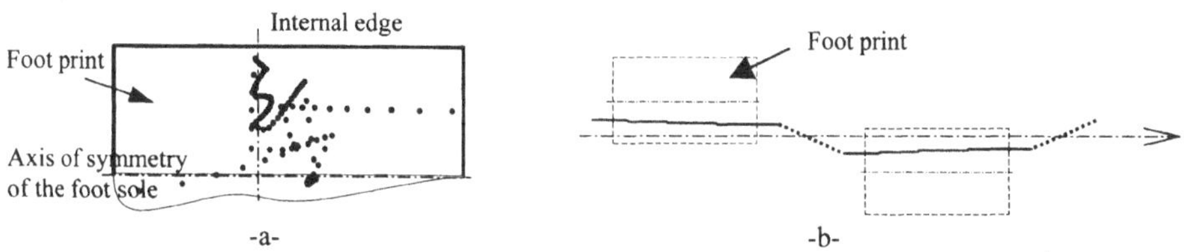

Figure 4. a- Trajectory of the center of pressure during the swing phase at moderate speed.
b- Projection of the center of gravity on the foot prints

The table 1 shows a comparison between the planar model of the biped and its simplified 3D one. Sagittal movements correspond to mean walking-speeds of 0.4, 0.8 and 1.3 m/s for the complete sagittal cycle. Slow, moderated, and fast speed walk refer to the obtained 3D-walk movements.

Table 1. Main characteristics of three gaits during the swing phase: the planar model versus the simplified 3D-model

Walking speed	Swing phase duration (s)	Minimized value of the criterion	Mechanical work (J)	Mean power (W)
Slow	0.967 / 0.877	0.19 / 1.076	73 / 96	75 / 109
Moderate	0.488 / 0.512	0.229 / 0.747	150 / 162	308 / 316
Fast	0.379 / 0.414	0.7 / 1.088	298 / 269	786 / 649

We noted also that actuating torques at non-sagittal joints influence significantly the values of the quadratic torque criterion unlike that based on mechanical energy expenditure. These former remain very sensitive to the duration of the movement. The swing phase duration raising decreases the energy expenditures.

Figure 5.a shows that the variations of the actuator torques of the sagittal plane joints are on the whole saved. In the case of the gait at slow speed, the three sub-phases of the swing phase described in Saidouni and Bessonnet (2004), are always clearly visible. Nevertheless, we note that torques are not cancelled completely at mid-swing. The gait takes fewer advantages from the pendulum effect, the main reason of the energy expenditure reduction for the slow gait.

We observe in figure 5.b that the joint torque u_6 actuating the movement of adduction-abduction of the stance hip maintains the trunk in position by supporting the moment exerted by the supported bodies. There is a great similarity between the variations of this torque and the vertical contact force exerted on the foot.

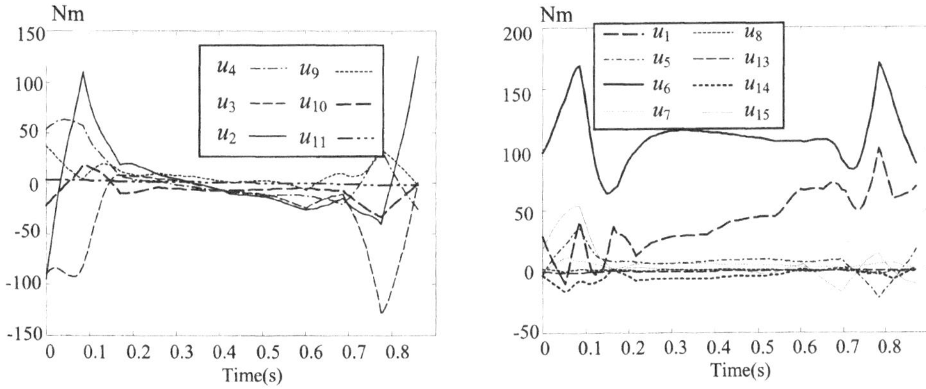

Figure. 5. Evolution of the joint torques at slow speed: a- in sagittal plane, b- in other planes

4.2 Change in walking direction

A progressive internal-external rotation $\psi(\tau)$ of the swing leg can be achieved by prescribing a movement with zero velocity at beginning and end, defined by the polynomial law :

$$\forall\tau \in [0,1], \quad \psi(\tau) = 3\Psi\,\tau^2 - 2\Psi\,\tau^3 \tag{7}$$

Figure 6 illustrates how the rotation of the foot is carried out during its swing movement. This action neither involves notable additional energy expenditure, nor influences greatly the dynamics of the balance.

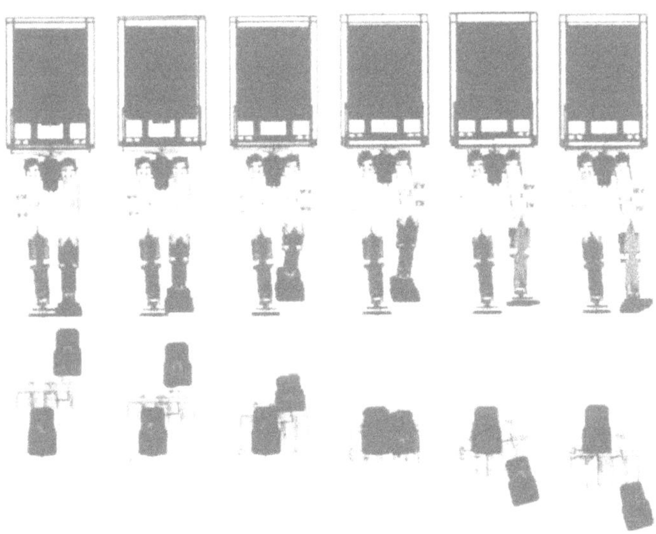

Figure 6. Cinematography of the vertical rotation movement of the swing foot (front and top views). Walking direction change is completed by swiveling the trunk during the stance phase

5 Conclusion

Optimal three dimensional gait synthesis can be achieved by splitting up the underlying optimization problem in two parts, and using a simplified walking model. First, a planar model is used to optimize the sagittal movement of gait, while taking into account transition constraints between phases. Then, the frontal movement during the swing phase is generated by a stochastic search method and is combined with its sagittal counterpart.

In this simplified model, the lateral stability is ensured by keeping the center of pressure inside the support surface. This is made possible by acting on the frontal joint of the ankle situated on the support side while hip joints put partially the trunk and completely the swing leg back in the sagittal plane.

Gait simulations for walking in straight line and with change of direction confirmed the reliability of our simplified approach. Moreover, the whole of the results indicates also that the quadratic torque performance criterion is sensitive to the non-sagittal movements whereas power based criteria neglect the effects of the low amplitudes movements like those produced by our model in the frontal and transverse planes.

Bibliography

Bessonnet, G., and Rostami, M. (2001), Sagittal gait of a biped robot during the single support phase. Part 2: optimal motion, *Robotica* 19, 241-253.

Channon, P.H., Hopkins, S.H., and Pham, D.T. (1992), Derivation of optimal walking motions for bipedal walking robot, Part 2, *Robotica* 10, 165-172.

Chettibi, T., and Lehtihet, H. E. (2002), A new approach for point to point optimal motion planning problems of robotic manipulators, *6th Biennial Conference on Engineering systems design and analysis, ASME conference*, July 8-11 2002, Istanbul, Turkey.

Chevallerau, C., and Aoustin, Y. (2001), Optimal reference trajectories for walking and running of a biped robot, *Robotica* 19, Part 2, 557-569.

Espiau, B., and Sardain, P. (2000), The Anthropomorphic Biped BIP200, *Proceedings of the International Conference on Robotics and Automation*, San Francisco, CA-USA, April 2000, 3997-4002

Hirai, K., Hirose, M., Haikawa, Y., and Takenaka, T. (1998), The Development of Honda humanoid robot, *Proceedings of the International Conference on Robotics and Automation*, Leuven, Belgium, 1321-1326.

Kaneko, K., Kanehiro, F., Kajita, S., et al. (2004), Humanoid robot HRP-2, *Proceedings of the International Conference on Robotics and Automation (ICRA 2004)*, April 26 - May 1, New Orleans, USA, 1083-1090.

McMahon, T.A. (1984), Mechanics of locomotion, *International Journal of Robotics Research* 3(2), 4-28.

Saidouni, T., and Bessonnet, G. (2002), Gait trajectory optimization using approximation functions, *5th Int. Conference on Climbing and Walking Robots (CLAWAR)*, 25-27 September 2002, Paris, France, 709-716.

Saidouni, T., and Bessonnet, G. (2004), Generating Impactless Gait of a Bipedal Robot, *Proceedings of the 11th World Congress in Mechanism and Machine Science*, 1-4 April, 2004, Tianjin, 1532-1536.

Yamaguchi, J., Nishino, D., and Takanishi, A. (1998), Realization of dynamic biped walking varying joint stiffness using antagonistic driven joints, *Proceedings of the International Conference on Robotics and Automation*, Leuven, Belgium, 2022-2029.

Chapter IV

Control Methods and Systems

The Largest Feedback-Linearizable Subsystem of a Class of Wheeled Robots Moving on an Inclined Plane

Danielle Sami Nasrallah [1†], Jorge Angeles [1†], and Hannah Michalska [2†]
[1] Department of Mechanical Engineering
[2] Department of Electrical and Computer Engineering
[†] Centre for Intelligent Machines
McGill University, Montreal, Canada
{danielle; angeles; michalsk}@cim.mcgill.ca

Abstract. This paper reports on the computation of the largest feedback-linearizable subsystem for a class of wheeled robots: the mobile wheeled pendulums (MWP). The control of MWP-class robots is quite challenging due to the non-actuated central body. Linear control techniques applied to this class of robots are restricted to the velocity level and constrain the motion to the neighborhood of the operating point. Thus, considering nonlinear control techniques is a better alternative. For that, an investigation of the dimension of the largest feedback-linearizable subsystem is essential and remains the first step towards the development of nonlinear controllers. To the best of the authors' knowledge, this is the first attempt to compute the largest feedback linearizable subsystem for mobile wheeled pendulums moving on an inclined plane.

1 Introduction

In this paper the authors compute the dimension of the largest-feedback linearizable subsystem of an anti-tilting outdoor mobile robot, ATOM, moving on an inclined plane. ATOM is composed of three rigid bodies: the central body, a cylinder whose center of mass is offset from its geometric center, and two spherical wheels rotating about the central body. According to its structure, ATOM pertains to the category of Mobile Wheeled Pendulums, MWP.

Many developments in the field of MWP have been reported recently: the US patent behind the Ginger and then the Segway Human Transporter projects (Kamen et al., 1999), JOE, a mobile inverted pendulum (Grasser et al., 2002) and, more recently, Quasimoro, a quasiholonomic mobile robot (Salerno and Angeles, 2004). In the last two references, linear control techniques have been employed for the control of the heading and steering velocities of the robot, by linearizing the system around an equilibrium point, thus removing the need for developing nonlinear control strategies and applying nonlinear state feedback. However, for advanced control objectives, such as controlling the position and orientation of the robot, which can be extremely interesting for a wide range of applications, linear control techniques fail to work, as the states associated with the robot's position and orientation become uncontrollable after linearization. Hence, resorting to nonlinear state and feedback transformations is necessary. In a recent study, Pathak et al. (2005) computed the dimension of the largest feedback-linearizable subsystem in a wheeled pendulum system moving on a horizontal plane. In the present paper a generalization of the foregoing result is presented by allowing the robot to move on an inclined plane. It should be stressed that such a generalization is not trivial as the model pertaining to the motion on an inclined plane is much more complex than that corresponding to the motion on a horizontal plane (Nasrallah et al., 2005). Additionally, the computation presented here employs a parametric frame-invariant formulation which renders it robust and valid for MWP-class robots whether the pendulum is inverted or not. To the best of the authors' knowledge, controllability issues dealing with MWP moving on an inclined plane have not been investigated yet, which constitutes the novelty of the work reported here. The method adopted for the computation of the largest feedback-linearizable subsystem was described in (Marino, 1986), and consists of the computation of a sequence of non-increasing integers, which turns out to be the set of controllability indices of the largest feedback-linearizable subsystem. Finally the results obtained here were employed in order to construct a posture control of ATOM robot based on a multi-loop feedback structure (Nasrallah et al., 2006).

The paper is organized as follows: Section 2 summarizes the notation and definitions that will be used in the course of this work; Section 3 derives briefly the mathematical model of the robot and its state-space formulation; Section 4 introduces preliminary computations that are essential for further steps; and Section 5 computes the dimension of the largest feedback-linearizable subsystem.

2 Notation and Definitions

Given the nonlinear control system

$$\dot{\mathbf{x}} = \mathbf{f}(\mathbf{x}) + \sum_{i=1}^{m} \mathbf{g}_i(\mathbf{x}) u_i \tag{2.1}$$

where the state $\mathbf{x}$ belongs to an open set U of $\mathbb{R}^n$, while $\mathbf{f}$, $\mathbf{g}_1, \ldots, \mathbf{g}_m$ are smooth vector fields defined on U.

In the present work the computation of the largest feedback-linearizable subsystem is based on the method introduced in (Marino, 1986). The set of controllability indices of the largest-feedback linearizable subsystem of (2.1), k_i^*, is given by

$$k_i^* = \text{cardinality}\{n_j \geq i, j \geq 0\}$$

where n_i is a non-increasing sequence given by the codimensions of $\overline{\mathcal{G}}_{i-1}$ in $\mathcal{L}_i$ as follows:

$$n_i = \dim\mathcal{L}_i - \dim\overline{\mathcal{G}}_{i-1}, \quad n_0 = \dim\mathcal{G}_0$$

and $\overline{\mathcal{A}}$ denotes the involutive closure of $\mathcal{A}$, i.e., the smallest involutive distribution containing $\mathcal{A}$. The computation of the distributions $\mathcal{G}_i$ and $\mathcal{L}_i$ is outlined below.

For each point $\mathbf{x}$ of $\mathcal{U}$, the distribution $\mathcal{G}(\mathbf{x})$ is defined as the vector space spanned by the vector fields $\mathbf{g}_1(\mathbf{x}), \ldots, \mathbf{g}_m(\mathbf{x})$, i.e,

$$\mathcal{G}(\mathbf{x}) = \text{span}\{\mathbf{g}_1(\mathbf{x}), \ldots, \mathbf{g}_m(\mathbf{x})\}$$

The distribution $\mathcal{G}_f$ is, in turn,

$$\mathcal{G}_f(\mathbf{x}) = \text{span}\{\mathbf{f} + \mathbf{g} : \mathbf{g} \in \mathcal{G}\}$$

The distribution $\mathcal{G}_i(\mathbf{x})$ is defined, in turn, as

$$\mathcal{G}_i(\mathbf{x}) = \text{span}\{\mathcal{G}_{i-1}, [\mathcal{G}_f, \mathcal{G}_{i-1}]\}, \quad \mathcal{G}_0 = \mathcal{G} \text{ where } [\mathcal{A}, \mathcal{B}] = \text{span}\{[\mathbf{a}, \mathbf{b}] : \mathbf{a} \in \mathcal{A}, \mathbf{b} \in \mathcal{B}\}$$

and $[\mathbf{a}, \mathbf{b}]$ is the Lie product of the vector fields $\mathbf{a}$ and $\mathbf{b}$. Moreover, the distribution $\mathcal{L}_i$ is defined as

$$\mathcal{L}_i(\mathbf{x}) = \text{span}\{\text{ad}_f^i \mathcal{G}, \overline{\mathcal{G}}_{i-1}\}$$

where $\text{ad}_f^i \mathcal{A} = \text{span}\{\text{ad}_f^i \mathbf{a} : \mathbf{a} \in \mathcal{A}\}$ and $\text{ad}_f^i \mathbf{a} = \left[\mathbf{f}, \text{ad}_f^{i-1} \mathbf{a}\right]$, $\text{ad}_f^0 \mathbf{a} = \mathbf{a}$

3 Mathematical Model

As stated above, ATOM is composed of the central body, carrying instruments and control hardware, and two wheels. The spherical shape of the wheels allows the robot to recover its posture after flipping over, thus rendering it anti-tilting. The centers of mass of the wheels coincide with the geometric centers of the sphere, while the center of mass of the central body is offset from its geometric center. The wheels are denoted bodies 1 and 2, while the central body is labeled body 3. The symbols used for the robot model are summarized in Table 1. Figure 1 shows ATOM.

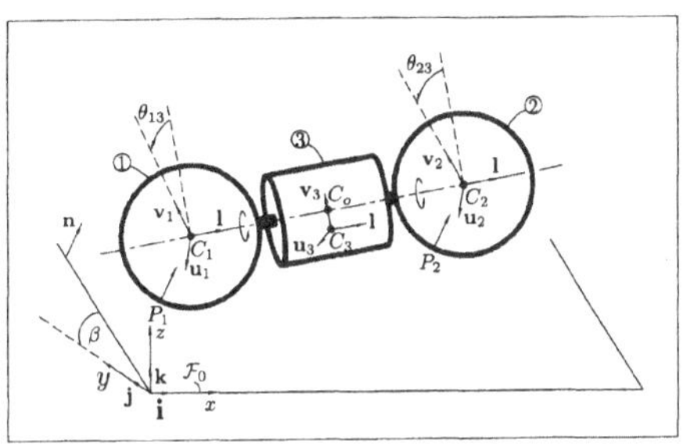

Figure 1. ATOM robot

In a previous work (Nasrallah et al., 2005) the mathematical model of ATOM moving on a general warped surface was developed. The terrain geometry enters the dynamics explicitly via the vectors normal to the ground at the contact points. Moreover, the particular case corresponding to the motion on an inclined plane was also included. Furthermore, in a more recent work (Nasrallah et al., 2006) a model-reduction was achieved in order to facilitate understanding the intrinsic dynamics property of the system. Below we display only the final results. The six-dimensional reduced vector $\mathbf{q}$ describing the posture of the robot is given by $\mathbf{q} = \begin{bmatrix} x_{\mathbf{c}_o} & y_{\mathbf{c}_o} & \mathbf{r}^T & r_0 \end{bmatrix}^T$, where $x_{\mathbf{c}_o}$ and $y_{\mathbf{c}_o}$ denote the $x-$ and $y-$ coordinates of $\mathbf{c}_o$ while r_0 and $\mathbf{r}$ are the Euler-Rodrigues parameters used to describe the orientation of body 3 in the inertial frame. The three-dimensional vector $\mathbf{v}$ of independent velocities is, in turn, $\mathbf{v} = \begin{bmatrix} v_c & \omega_{3p} & \omega_{3l} \end{bmatrix}^T$ where v_c is the heading velocity of the robot, namely, $v_c = (r_w/2)(\dot{\theta}_{13} + \dot{\theta}_{23} + 2\omega_{3l})$. As for ω_{3l}, it represents the projection along the line of (wheel) centers of the angular velocity $\boldsymbol{\omega}_3$ of the central body, given by $\boldsymbol{\omega}_3 = \omega_{3p}\mathbf{n} + \omega_{3l}\mathbf{l}$, while ω_{3p} is the robot steering rate, namely, $\omega_{3p} = (r_w/b)(\dot{\theta}_{13} - \dot{\theta}_{23})$. The nine-dimensional reduced state vector $\mathbf{x}$ thus becomes $\mathbf{x} = \begin{bmatrix} \mathbf{q}^T & \mathbf{v}^T \end{bmatrix}^T$. The input torques τ_1 and τ_2 are transformed as follows $\tau_p = \tau_1 + \tau_2$ and $\tau_m = \tau_1 - \tau_2$.

b	Distance between the wheel centers
$\mathbf{c}_i$	Position vector of the center of mass C_i of the ith body, $i = 1, 2, 3$
$\mathbf{c}_o$	Position vector of the geometric center C_o of the central body
d	Offset between the geometric center and the center of mass of the central body
$\{\mathbf{i}, \mathbf{j}, \mathbf{k}\}$	Right-handed orthogonal triad describing the orientation of the inertial frame $\mathcal{F}_0$
$\mathbf{l}$	Unit vector along the line of wheel centers, directed from C_1 to C_2
m_c	Mass of the central body
m_w	Mass of each wheel
$\mathbf{n}$	Unit vector normal to the inclined plane
r_w	Radius of each wheel
r_0, $\mathbf{r}$	Euler-Rodrigues parameters describing the orientation of the central body
$\{\mathbf{u}_i, \mathbf{l}, \mathbf{v}_i\}$	Right-handed orthogonal triad describing the orientation of the ith body
$\mathbf{v}_3$	Unit vector directed from C_3 to C_o
$\mathbf{I}_c$	Inertia matrix of the central body given by $\mathrm{diag}(I_{cu}, I_{cl}, I_{cv})$
$\mathbf{I}_w$	Inertia matrix of each wheel given by $\mathrm{diag}(I_{wu}, I_{wl}, I_{wv})$
θ_{i3}	Angular displacement of the ith wheel with respect to the central body
τ_i	Input torque applied to the ith wheel
$\boldsymbol{\omega}_i$	Angular velocity vector of the ith body in $\mathcal{F}_0$

Table 1. List of Symbols

The full state-space model of the system is

$$\dot{\mathbf{x}} = \mathbf{f}(\mathbf{x}) + \mathbf{g}_p(r_0, \mathbf{r})\tau_p + \mathbf{g}_m(r_0, \mathbf{r})\tau_m \tag{3.1}$$

where

$$\mathbf{f}(\mathbf{x}) = \begin{bmatrix} v_c\,\mathbf{h}^T\mathbf{i} \\ v_c\,\mathbf{h}^T\mathbf{j} \\ (1/2)(r_0\mathbf{1} - \mathbf{R})(\omega_{3p}\mathbf{n} + \omega_{3l}\mathbf{l}) \\ -(1/2)\mathbf{r}^T(\omega_{3p}\mathbf{n} + \omega_{3l}\mathbf{l}) \\ \mathbf{f}_v \end{bmatrix}, \quad \mathbf{g}_p = \begin{bmatrix} \mathbf{0}_6 \\ \mathbf{g}_{v_p} \end{bmatrix} \text{ and } \mathbf{g}_m = \begin{bmatrix} \mathbf{0}_6 \\ \mathbf{g}_{v_m} \end{bmatrix}$$

with

$$\mathbf{f}_v = \begin{bmatrix} r_w(F_b + F_c + G_b + G_c) \\ (2r_w/b)(F_a + G_a) \\ F_b + G_b \end{bmatrix}, \quad \mathbf{g}_{v_p} = \begin{bmatrix} r_w J_d \\ 0 \\ -J_c \end{bmatrix} \text{ and } \mathbf{g}_{v_m} = \begin{bmatrix} 0 \\ (2r_w/b)J_a \\ 0 \end{bmatrix}$$

Moreover, $\mathbf{h}$ is a unit vector given by $\mathbf{h} = \mathbf{l} \times \mathbf{n}$, while $\mathbf{R}$ is the cross-product matrix (CPM)[1] of vector $\mathbf{r}$, $\mathbf{1}$ is the 3×3 identity matrix, and

$$\begin{aligned} F_a(r_0, \mathbf{r}, \mathbf{v}) &= J_a[-4C_d\omega_{3p}\omega_{3l} + (2/r_w)C_b\omega_{3p}v_c] \\ F_b(r_0, \mathbf{r}, \mathbf{v}) &= (J_c - J_b)(b/r_w)C_b({\omega_{3p}}^2 + {\omega_{3l}}^2) + J_b(b/r_w)C_d{\omega_{3p}}^2 \\ F_c(r_0, \mathbf{r}, \mathbf{v}) &= -J_d(b/r_w)C_b({\omega_{3p}}^2 + {\omega_{3l}}^2) - J_c(b/r_w)C_d{\omega_{3p}}^2 \end{aligned}$$

[1]The CPM of a vector $\mathbf{v} \in \mathbb{R}^3$ is defined, for every $\mathbf{x} \in \mathbb{R}^3$ as well, as $\mathbf{V} = \mathrm{CPM}(\mathbf{v}) = \partial(\mathbf{v} \times \mathbf{x})/\partial\mathbf{x}$

$$G_a(r_0,\mathbf{r}) = -2J_a m_c g d (r_w/b)(\mathbf{v}_3\times\mathbf{n})^T\mathbf{k}$$
$$G_b(r_0,\mathbf{r}) = (J_c - J_b)(2m_w + m_c) g r_w\ \mathbf{h}^T\mathbf{k} + J_b m_c g d\ \mathbf{u}_3^T\mathbf{k}$$
$$G_c(r_0,\mathbf{r}) = -J_c m_c g d\ \mathbf{u}_3^T\mathbf{k} - J_d(2m_w + m_c) g r_w\ \mathbf{h}^T\mathbf{k}$$

$$C_a(r_0,\mathbf{r}) = m_c(r_w d^2/b)\mathbf{h}^T\mathbf{u}_3\ \mathbf{h}^T\mathbf{v}_3 \qquad C_b(r_0,\mathbf{r}) = m_c(r_w^2 d/b)\mathbf{h}^T\mathbf{v}_3$$
$$C_c(r_0,\mathbf{r}) = (r_w/b)(I_{cu} - I_{cv})\mathbf{h}^T\mathbf{u}_3\ \mathbf{h}^T\mathbf{v}_3 \qquad C_d(r_0,\mathbf{r}) = C_a + C_c$$
$$J_a(r_0,\mathbf{r}) = (1/2)/(I_a - I_b + 2I_e) \qquad J_b(r_0,\mathbf{r}) = I_c/(I_c I_d - 2{I_f}^2)$$
$$J_c(r_0,\mathbf{r}) = (I_c - I_f)/(I_c I_d - 2{I_f}^2) \qquad J_d(r_0,\mathbf{r}) = (I_d/2 - I_f)/(I_c I_d - 2{I_f}^2)$$
$$I_a = 2(r_w/b)^2 I_{wu} + I_{wl} + m_w r_w^2 + m_c r_w^2/4 \qquad I_b = -2(r_w/b)^2 I_{wu} + m_c r_w^2/4$$
$$I_c = I_{wl} + m_w r_w^2 + m_c r_w^2/2 \qquad I_d = I_{cl} + m_c d^2$$
$$I_e(r_0,\mathbf{r}) = (r_w/b)^2[(I_{cu} + m_c d^2){\mathbf{h}^T\mathbf{v}_3}^2 + I_{cv}{\mathbf{h}^T\mathbf{u}_3}^2] \qquad I_f(r_0,\mathbf{r}) = m_c(r_w d/2)\mathbf{h}^T\mathbf{u}_3$$

4 Preliminary Computations

By introducing $\boldsymbol{\lambda}$ as $\boldsymbol{\lambda} = \begin{bmatrix} \mathbf{r}^T & r_0 \end{bmatrix}^T$, we have the identities below

$$\frac{\partial(\mathbf{a}^T\mathbf{b})}{\partial\boldsymbol{\lambda}} = \frac{\partial\mathbf{b}}{\partial\boldsymbol{\lambda}}^T\mathbf{a} + \frac{\partial\mathbf{a}}{\partial\boldsymbol{\lambda}}^T\mathbf{b},\quad \frac{\partial(k\mathbf{b})}{\partial\boldsymbol{\lambda}} = k\frac{\partial\mathbf{a}}{\partial\boldsymbol{\lambda}} + \mathbf{a}\left(\frac{\partial k}{\partial\boldsymbol{\lambda}}\right)^T \text{ and } \frac{\partial(\mathbf{a}\times\mathbf{b})}{\partial\boldsymbol{\lambda}} = \mathbf{A}\frac{\partial\mathbf{b}}{\partial\boldsymbol{\lambda}} - \mathbf{B}\frac{\partial\mathbf{a}}{\partial\boldsymbol{\lambda}}$$

where $\mathbf{A}$ and $\mathbf{B}$ are the CPMs of vectors $\mathbf{a}$ and $\mathbf{b}$, respectively. For further computations it will be convenient to introduce the four-dimensional vector $\mathbf{a}_\lambda$, where $\mathbf{a}$ can be $\mathbf{h}$, $\mathbf{l}$, or $\mathbf{n}$, namely,

$$\mathbf{a}_\lambda = \begin{bmatrix} \mathbf{a}^T(r_0\mathbf{1} + \mathbf{R}) & -\mathbf{r}^T\mathbf{a} \end{bmatrix}^T \tag{4.1}$$

Differentiating $\mathbf{a}_\lambda$ with respect to $\boldsymbol{\lambda}$ leads to:

$$\frac{\partial\mathbf{a}_\lambda}{\partial\boldsymbol{\lambda}} = \begin{bmatrix} (r_0\mathbf{1} - \mathbf{R}) \\ -\mathbf{r}^T \end{bmatrix}\frac{\partial\mathbf{a}}{\partial\boldsymbol{\lambda}} + \begin{bmatrix} \mathbf{A} & \mathbf{a} \\ -\mathbf{a}^T & 0 \end{bmatrix} \tag{4.2}$$

Moreover, in the sequel, we will need $\partial\mathbf{f}/\partial\mathbf{x}$, $\partial\mathbf{g}_p/\partial\mathbf{x}$ and $\partial\mathbf{g}_m/\partial\mathbf{x}$ which are evaluated below:

$$\frac{\partial\mathbf{f}}{\partial\mathbf{x}} = \begin{bmatrix} \dfrac{\partial\mathbf{f}}{\partial\mathbf{q}} & \dfrac{\partial\mathbf{f}}{\partial\mathbf{v}} \end{bmatrix},\quad \frac{\partial\mathbf{g}_p}{\partial\mathbf{x}} = \begin{bmatrix} \mathbf{O}_{6\times 9} \\ \dfrac{\partial\mathbf{g}_{vp}}{\partial\mathbf{x}} \end{bmatrix} \text{ and } \frac{\partial\mathbf{g}_m}{\partial\mathbf{x}} = \begin{bmatrix} \mathbf{O}_{6\times 9} \\ \dfrac{\partial\mathbf{g}_{vm}}{\partial\mathbf{x}} \end{bmatrix} \tag{4.3}$$

where

$$\frac{\partial\mathbf{g}_{vp}}{\partial\mathbf{x}} = \begin{bmatrix} \mathbf{O}_{3\times 2} & \dfrac{\partial\mathbf{g}_{vp}}{\partial\boldsymbol{\lambda}} & \mathbf{O}_{3\times 3} \end{bmatrix} \text{ and } \frac{\partial\mathbf{g}_{vm}}{\partial\mathbf{x}} = \begin{bmatrix} \mathbf{O}_{3\times 2} & \dfrac{\partial\mathbf{g}_{vm}}{\partial\boldsymbol{\lambda}} & \mathbf{O}_{3\times 3} \end{bmatrix}$$

with

$$\frac{\partial\mathbf{g}_{vp}}{\partial\boldsymbol{\lambda}} = \begin{bmatrix} r_w\dfrac{\partial J_d}{\partial\boldsymbol{\lambda}} & \mathbf{0}_4 & \dfrac{\partial J_c}{\partial\boldsymbol{\lambda}} \end{bmatrix}^T \text{ and } \frac{\partial\mathbf{g}_{vm}}{\partial\boldsymbol{\lambda}} = \begin{bmatrix} \mathbf{0}_4 & (2r_w/b)\dfrac{\partial J_a}{\partial\boldsymbol{\lambda}} & \mathbf{0}_4 \end{bmatrix}^T$$

Furthermore, the range and nullspace of $\partial\mathbf{l}/\partial\boldsymbol{\lambda}$ and $\partial\mathbf{h}/\partial\boldsymbol{\lambda}$ were derived. For sake of brevity, we display here the final results[2].

$$\mathcal{R}\left(\frac{\partial\mathbf{l}}{\partial\boldsymbol{\lambda}}\right) = \text{span}\{\mathbf{h}\} \text{ and } \mathcal{R}\left(\frac{\partial\mathbf{h}}{\partial\boldsymbol{\lambda}}\right) = \text{span}\{\mathbf{l}\} \tag{4.4}$$

[2]The demonstration can be retrieved from URL http://www.cim.mcgill.ca/~danielle

5 Largest Feedback-Linearizable Subsystem

As stated in Section 2, in order to compute the dimension of the largest feedback-linearizable subsystem, the indices n_i and k_i^* must be determined. Clearly,

$$n_0 = \dim \mathcal{G}_0 = 2, \ \mathcal{G}_0 = \text{span}\{\mathbf{g}_p, \mathbf{g}_m\}$$

and $n_1 = \dim \mathcal{L}_1 - \dim \overline{\mathcal{G}}_0$. For that we need to determine $\mathcal{L}_1$ as well as $\overline{\mathcal{G}}_0$. The distribution $\mathcal{L}_1$ is given by $\mathcal{L}_1 = \text{span}\{\text{ad}_f\mathcal{G}, \overline{\mathcal{G}}_0\} = \text{span}\{[\mathbf{f}, \mathbf{g}_p], [\mathbf{f}, \mathbf{g}_m], \overline{\mathcal{G}}_0\}$ while the computation of $\overline{\mathcal{G}}_0$ requires the Lie product $[\mathbf{g}_p, \mathbf{g}_m]$. By virtue of eqs. (3.1) and (4.3), it is a simple matter to show that $[\mathbf{g}_p, \mathbf{g}_m]$ vanishes, which implies that $\mathcal{G}_0$ is involutive, i.e., $\overline{\mathcal{G}}_0 = \mathcal{G}_0$.
As for the computation of $[\mathbf{f}, \mathbf{g}_p]$ we have $[\mathbf{f}, \mathbf{g}_p] = \mathbf{f}_{p1} - \mathbf{f}_{p2}$ where

$$\mathbf{f}_{p1} = \frac{\partial \mathbf{g}_p}{\partial \mathbf{x}}\mathbf{f} = \begin{bmatrix} \mathbf{0}_6 \\ \mathbf{t}_{p1} \end{bmatrix} \text{ and } \mathbf{t}_{p1} = \frac{1}{2}\frac{\partial \mathbf{g}_{vp}}{\partial \boldsymbol{\lambda}}(\omega_{3p}\mathbf{n}_\lambda + \omega_{3l}\mathbf{l}_\lambda)$$

while

$$\mathbf{f}_{p2} = \frac{\partial \mathbf{f}}{\partial \mathbf{x}}\mathbf{g}_p = \begin{bmatrix} \mathbf{f}_p \\ \mathbf{t}_{p2} \end{bmatrix}, \ \mathbf{f}_p = \begin{bmatrix} r_w J_d \mathbf{h}^T \mathbf{i} \\ r_w J_d \mathbf{h}^T \mathbf{j} \\ -(J_c/2)\mathbf{l}_\lambda \end{bmatrix} \text{ and } \mathbf{t}_{p2} = r_w J_d \frac{\partial \mathbf{f}_v}{\partial v_c} - J_c \frac{\partial \mathbf{f}_v}{\partial \omega_{3l}}$$

Upon applying the same reasoning, we have $[\mathbf{f}, \mathbf{g}_m] = \mathbf{f}_{m1} - \mathbf{f}_{m2}$ where

$$\mathbf{f}_{m1} = \frac{\partial \mathbf{g}_m}{\partial \mathbf{x}}\mathbf{f} = \begin{bmatrix} \mathbf{0}_6 \\ \mathbf{t}_{m1} \end{bmatrix} \text{ and } \mathbf{t}_{m1} = \frac{1}{2}\frac{\partial \mathbf{g}_{vm}}{\partial \boldsymbol{\lambda}}(\omega_{3p}\mathbf{n}_\lambda + \omega_{3l}\mathbf{l}_\lambda)$$

while

$$\mathbf{f}_{m2} = \frac{\partial \mathbf{f}}{\partial \mathbf{x}}\mathbf{g}_m = \begin{bmatrix} \mathbf{f}_m \\ \mathbf{t}_{m2} \end{bmatrix}, \ \mathbf{f}_m = \begin{bmatrix} 0 \\ 0 \\ (r_w/b)J_a\mathbf{n}_\lambda \end{bmatrix} \text{ and } \mathbf{t}_{m2} = (2r_w/b)J_a \frac{\partial \mathbf{f}_v}{\partial \omega_{3p}}$$

Therefore $\mathcal{L}_1 = \text{span}\{\mathbf{f}_{p1} - \mathbf{f}_{p2}, \mathbf{f}_{m1} - \mathbf{f}_{m2}, \mathbf{g}_p, \mathbf{g}_m\}$ and its dimension is obviously four, thereby leading to

$$n_1 = 4 - 2 = 2$$

Now, n_2 is obtained as $n_2 = \dim \mathcal{L}_2 - \dim \overline{\mathcal{G}}_1$. The distribution $\mathcal{G}_1$, after simplification, turns out to be identical to $\mathcal{L}_1$, while the distribution $\mathcal{L}_2$ is computed as:

$$\mathcal{L}_2 = \text{span}\ \{\text{ad}_f^2\mathcal{G}, \overline{\mathcal{G}}_1\} = \text{span}\ \{[\mathbf{f}, [\mathbf{f}, \mathbf{g}_p]], [\mathbf{f}, [\mathbf{f}, \mathbf{g}_m]], \overline{\mathcal{G}}_1\}$$

The construction of $\overline{\mathcal{G}}_1$ proceeds as follows:

$$[\mathbf{g}_p, \mathbf{f}_{p1} - \mathbf{f}_{p2}] = \begin{bmatrix} \mathbf{0}_6 \\ \dfrac{\partial(\mathbf{t}_{p1} - \mathbf{t}_{p2})}{\partial \mathbf{v}}\mathbf{g}_{vp} + \dfrac{\partial \mathbf{g}_{vp}}{\partial \mathbf{x}}\mathbf{f}_{p2} \end{bmatrix}$$

Since $\mathbf{t}_{p1}$ and $\mathbf{t}_{p2}$ are both linear in terms of v_c, ω_{3p}, and ω_{3l}, which are the components of the vector $\mathbf{v}$, it becomes apparent that $\partial(\mathbf{t}_{p1} - \mathbf{t}_{p2})/\partial \mathbf{v}$ does not depend of $\mathbf{v}$.
Moreover, $(\partial \mathbf{g}_{vp}/\partial \mathbf{x})\mathbf{f}_{p2}$ is also independent of $\mathbf{v}$. Therefore, $[\mathbf{g}_p, \mathbf{f}_{p1} - \mathbf{f}_{p2}] = [\ \mathbf{0}_6^T \quad \mathbf{t}_{pp}^T(\boldsymbol{\lambda})$

By the same token we have $[\mathbf{g}_m, \mathbf{f}_{m1} - \mathbf{f}_{m2}] = \begin{bmatrix} \mathbf{0}_6^T & \mathbf{t}_{mm}^T(\boldsymbol{\lambda}) \end{bmatrix}^T$. Further, using the Jacobi identity leads to: $[\mathbf{g}_p, [\mathbf{f}, \mathbf{g}_m]] = [\mathbf{g}_m, [\mathbf{f}, \mathbf{g}_p]] = \begin{bmatrix} \mathbf{0}_6^T & \mathbf{t}_{pm}^T(\boldsymbol{\lambda}) \end{bmatrix}^T$. It is unnecessary to investigate the recursive Lie products of the vectors $[\mathbf{g}_p, \mathbf{f}_{p1} - \mathbf{f}_{p2}]$, $[\mathbf{g}_m, \mathbf{f}_{m1} - \mathbf{f}_{m2}]$, and $[\mathbf{g}_p, \mathbf{f}_{m1} - \mathbf{f}_{m2}]$ because the result vanishes always (the demonstration of this fact is similar to that involving the Lie product of $\mathbf{g}_p$ and $\mathbf{g}_m$). Moreover, since the above three vectors have the same structure as $\mathbf{g}_p$ and $\mathbf{g}_m$, i.e., their first six components are all zero and their last three depend only on $\boldsymbol{\lambda}$, one vector among the three can be chosen, say $[\mathbf{g}_p, \mathbf{f}_{m1} - \mathbf{f}_{m2}]$, can be chosen. The set of the three vectors $\mathbf{g}_p$ and $\mathbf{g}_m$ and $[\mathbf{g}_p, \mathbf{f}_{m1} - \mathbf{f}_{m2}]$ is now linearly independent. From previous computations it follows that:

$$[[\mathbf{f}, \mathbf{g}_p], [\mathbf{f}, \mathbf{g}_m]] = [\mathbf{f}_{p1}, \mathbf{f}_{m1}] - [\mathbf{f}_{p1}, \mathbf{f}_{m2}] - [\mathbf{f}_{p2}, \mathbf{f}_{m1}] + [\mathbf{f}_{p2}, \mathbf{f}_{m2}]$$

and since $[\mathbf{f}_{p1}, \mathbf{f}_{m1}]$, $[\mathbf{f}_{p1}, \mathbf{f}_{m2}]$, and $[\mathbf{f}_{p2}, \mathbf{f}_{m1}]$do not generate any new brackets, we only need to determine the augmentation of $\bar{\mathcal{G}}_1$ by $[\mathbf{f}_{p2}, \mathbf{f}_{m2}]$.

$$[\mathbf{f}_{p2}, \mathbf{f}_{m2}] = \begin{bmatrix} \dfrac{\partial \mathbf{f}_m}{\partial \mathbf{q}} \mathbf{f}_p - \dfrac{\partial \mathbf{f}_p}{\partial \mathbf{q}} \mathbf{f}_m \\ \dfrac{\partial \mathbf{t}_{m2}}{\partial \mathbf{q}} \mathbf{f}_p + \dfrac{\partial \mathbf{t}_{m2}}{\partial \mathbf{v}} \mathbf{t}_{p2} - \dfrac{\partial \mathbf{t}_{p2}}{\partial \mathbf{q}} \mathbf{f}_m - \dfrac{\partial \mathbf{t}_{p2}}{\partial \mathbf{v}} \mathbf{t}_{m2} \end{bmatrix}$$

It is noteworthy that both $\mathbf{t}_{p2}$ and $\mathbf{t}_{m2}$ do not appear in the upper block (the first six components) of $[\mathbf{f}_{p2}, \mathbf{f}_{m2}]$, which is a function of vector $\boldsymbol{\lambda}$ only. Moreover, the lower block (the last three components) can be spanned by the three independent vectors, $\mathbf{g}_p$, $\mathbf{g}_m$ and $[\mathbf{g}_p, [\mathbf{f}_{m1} - \mathbf{f}_{m2}]]$, as noticed above. Thus, in generating further bracketing, there is no need to consider the nine-dimensional vectors $\mathbf{f}_{p2}$ and $\mathbf{f}_{m2}$, as they can be substituted by $\mathbf{f}_p$ and $\mathbf{f}_m$. After simplification we obtain:

$$[\mathbf{f}_p, \mathbf{f}_m] = \begin{bmatrix} \alpha_1 \, \mathbf{h}^T \mathbf{i} + \alpha_2 \, \mathbf{l}^T \mathbf{i} \\ \alpha_1 \, \mathbf{h}^T \mathbf{j} + \alpha_2 \, \mathbf{l}^T \mathbf{j} \\ \alpha_3 \, \mathbf{h}_\lambda + \alpha_4 \, \mathbf{l}_\lambda + \alpha_5 \, \mathbf{n}_\lambda \end{bmatrix}$$

where α_i, for $i = 1 \ldots 5$, are all functions of $\boldsymbol{\lambda}$. It is obvious that $[\mathbf{f}_p, \mathbf{f}_m]$ cannot be expressed as a linear combination of $\mathbf{f}_p$ and $\mathbf{f}_m$; thus, further bracketing must be performed. The first two components of $\mathbf{f}_p$ and $[\mathbf{f}_p, \mathbf{f}_m]$ are clearly independent and do not appear in $\mathbf{h}_\lambda$, $\mathbf{l}_\lambda$ and $\mathbf{n}_\lambda$. It is hence justified to limit further bracketing to the four dimensional vectors obtained using the last four components of $\mathbf{f}_p$ and $[\mathbf{f}_p, \mathbf{f}_m]$, or, equivalently, $\mathbf{n}_\lambda$ and $\alpha_h \mathbf{h}_\lambda + \alpha_l \mathbf{l}_\lambda + \alpha_n \mathbf{n}_\lambda$, where α_h, α_l, and α_n can be considered constant.

$$[\mathbf{n}_\lambda, \alpha_h \mathbf{h}_\lambda + \alpha_l \mathbf{l}_\lambda + \alpha_n \mathbf{n}_\lambda] = \alpha_h(\beta_1 - 2)\mathbf{l}_\lambda + \alpha_l(\beta_2 + 2)\mathbf{h}_\lambda, \ \frac{\partial \mathbf{h}}{\partial \boldsymbol{\lambda}} \mathbf{n}_\lambda = \beta_1 \mathbf{l} \text{ and } \frac{\partial \mathbf{l}}{\partial \boldsymbol{\lambda}} \mathbf{n}_\lambda = \beta_2 \mathbf{h}$$

Next we consider $[\mathbf{n}_\lambda, [\mathbf{n}_\lambda, \alpha_h \mathbf{h}_\lambda + \alpha_l \mathbf{l}_\lambda + \alpha_n \mathbf{n}_\lambda]]$, which constitutes a part of the bracket $[[\mathbf{f}, \mathbf{g}_m], [[\mathbf{f}, \mathbf{g}_m], [[\mathbf{f}, \mathbf{g}_p], [\mathbf{f}, \mathbf{g}_m]]]]$. We obtain, after simplification,

$$[\mathbf{n}_\lambda, [\mathbf{n}_\lambda, \alpha_h \mathbf{h}_\lambda + \alpha_l \mathbf{l}_\lambda + \alpha_n \mathbf{n}_\lambda]] = (\beta_1 - 2)(\beta_2 + 2)(\alpha_h \mathbf{h}_\lambda + \alpha_l \mathbf{l}_\lambda)$$

From now on it is unnecessary to compute further brackets since they will not generate a four-dimensional vector linearly independent from $\mathbf{h}_\lambda$, $\mathbf{l}_\lambda$ and $\mathbf{n}_\lambda$. Therefore, $\overline{\mathcal{G}}_1$ becomes

$$\begin{aligned}\overline{\mathcal{G}}_1 &= \text{span}\,\{\,\mathbf{g}_p, \mathbf{g}_m, [\mathbf{g}_p, [\mathbf{f}, \mathbf{g}_m]]\,, [\mathbf{f}, \mathbf{g}_p]\,, [[\mathbf{f}, \mathbf{g}_p]\,, [\mathbf{f}, \mathbf{g}_m]]\,, [\mathbf{f}, \mathbf{g}_m]\,, [[\mathbf{f}, \mathbf{g}_m]\,, [[\mathbf{f}, \mathbf{g}_p]\,, [\mathbf{f}, \mathbf{g}_m]]]\,,\\ &\qquad [[\mathbf{f}, \mathbf{g}_m]\,, [[\mathbf{f}, \mathbf{g}_m]\,, [[\mathbf{f}, \mathbf{g}_p]\,, [\mathbf{f}, \mathbf{g}_m]]]]\}\\ &= \text{span}\,\{\,\mathbf{e}_1, \mathbf{e}_2, \mathbf{l}_e, \mathbf{h}_e, \mathbf{n}_e, \mathbf{e}_7, \mathbf{e}_8, \mathbf{e}_9\}\end{aligned}$$

where $\mathbf{e}_i$, for $i = 1, 2, 7, 8$, and 9, are the Euclidean basis vectors with the ith entry equal to unity, respectively. The remaining vectors $\mathbf{h}_e$, $\mathbf{l}_e$ and $\mathbf{n}_e$ are each of the form $\mathbf{a}_e = \begin{bmatrix} \mathbf{0}_2^T & \mathbf{a}_\lambda^T & \mathbf{0}_3^T \end{bmatrix}^T$, where $\mathbf{a}_\lambda$ has been defined in eq. (4.1) and $\mathbf{a}$ is equal to $\mathbf{h}$, $\mathbf{l}$, and $\mathbf{n}$, respectively. For the construction of $\mathcal{L}_2$ two additional brackets must be computed, namely, $\text{ad}_f^2\mathbf{g}_p$ and $\text{ad}_f^2\mathbf{g}_m$. Following the same reasoning as above we can prove that these two brackets can be expressed as linear combinations of the vectors fields spanning $\overline{\mathcal{G}}_1$, thereby leading to $\mathcal{L}_2 = \mathcal{G}_1$. By the same token, $n_2 = 0$. Since the sequence of integers n_i is non-increasing, the computation stops at n_2. Hence, $k_1^* = \text{cardinality}\{n_j \geq 1, j \geq 0\} = 2$, $k_2^* = \text{cardinality}\{n_j \geq 2, j \geq 0\} = 2$.

Conclusions and Future Work

MWP-class robots are partially feedback-linearizable with controllability indices $k_1^* = 2$ and $k_2^* = 2$. The implication is that the system outputs, which are to be determined in terms of the control objectives, must be such that their total relative degree must not exceed four. It is noteworthy that the position of the center of mass of the central body with respect to the line of wheel centers has no effect on the computation of the controllability indices conducted above, the result thus obtained being applicable to any MWP-class robot, whether the *pendulum* is inverted or not.

Bibliography

D. L. Kamen, R. R. Ambrogi, R. J. Duggan, R. K. Heinzmann, B. R. Key, A. Skoskiewicz, P. K. Kristal, *Transportation Vehicles and Methods*, US patent 5,971,091, 1999.

F. Grassser, A. D'Arrigo, S. Colombi and A. C. Rufer, *JOE: A Mobile, Inverted Pendulum*, In *IEEE Trans. on Industrial Electronics*, Vol. 49, No. 1, pp. 107-114, 2002.

A. Salerno and J. Angeles, *The Control of Semi-Autonomous Self-Balancing Two-Wheeled Quasiholonomic Mobile Robots*, In *Proc. of the 15th CISM-IFToMM Symposium on Robot Design, Dynamics and Control (RoManSy)*, 2004.

K. Pathak, J. Franch and S. K. Agrawal, *Velocity and Position Control of a Wheeled Inverted Pendulum by Partial Feedback Linearization*, In *IEEE Trans. on Robotics*, Vol. 21, No. 3, pp. 505-513, 2005.

R. Marino, *On the largest feedback linearizable subsystem*, In *Systems and Control Letters*, Vol. 6, pp. 345–351, 1986.

D. S. Nasrallah, J. Angeles and H. Michalska, *Modeling of an Anti-Tilting Outdoor Mobile Robot*, In *ASME Proc. of the 5th International Conference on Mutlibody Systems, Nonlinear Dynamics, and Control*, 2005.

D. S. Nasrallah, J. Angeles and H. Michalska, *Velocity and Orientation Control of an Anti-Tilting Mobile Robot Moving on an Inclined Plane*, to appear in *IEEE Proc. of the International Conference on Robotics and Automation*, 2006.

Leg Thrust Control for Stabilization of Dynamic Gaits in a Quadruped Robot

Joaquin Estremera and Kenneth J. Waldron[1]

Stanford University, Department of Mechanical Engineering, Palo Alto, CA, USA.
[joaquine,kwaldron]@stanford.edu

Abstract. Several parameters must be regulated to control and stabilize running robots: for example, the regulation of the leg touchdown angles and the hip torques has been successfully used in the past to control quadruped machines. However, the regulation of leg thrust can also be an effective technique for controlling the characteristics of the gait (apex height, forward speed, attitude rate) and it can play an important role in gait stabilization. This paper presents the mechanical systems, the models and control strategies employed to regulate leg thrust on the KOLT quadruped robot. Two thrust control methods are proposed and tested experimentally. The closed loop method regulates thrust through the control of the hip liftoff speed, a conceptually simple control strategy that has the property of stabilizing pronk gaits without the need of central feedback. An open loop control method to regulate the energy added in each hop is also presented. Both experiments with the KOLT robot and simulations show the efficacy of the proposed approaches.

1 Introduction

Over the past decades there has been a considerable interest in investigating dynamic gaits for legged robots. However, several major problems have to be solved before high speed locomotion is achieved in a fully operational, self-contained robot. One of the main challenges is the design of mechanical systems able to deliver the energy needed to sustain high speed locomotion during the short and intermittent periods of foot-ground contact. When this constraint is considered jointly with acceptable energetic efficiency and weight, the demands placed on the actuation systems are difficult to fulfill. The systems described in (Brown and Zeglin, 1998; Waldron and Nichol, 2004) represent two attempts to achieve energy efficient locomotion systems.

The stabilization and control of fast dynamic gaits is another challenge that is yet to be solved. In animals, high speed gaits are supposed to be stabilized by the passive dynamics of the musculo-skeletal structure and by simple feed-forward mechanisms, more than by controllers relying on central feedback (Herr and McMahon, 2001). In the case of robots, two main approaches have been proposed to generate stable dynamic gaits. The first approach consists of obtaining the return maps that describe the behavior of the machine to find regions in which the gait is stable. Murphy discovered that the stability of a bound gait could be predicted as a function of the non-dimensional moment of inertia (Raibert, 1986), defined by: $\hat{I} = I/(ml^2)$ where I is

[1] The authors would like to thank Paul Csonka and Surya P. N. Singh for their assistance in this work. The authors also acknowledge the support of the National Science Foundation, Grant No. IIS-0208664 and the Secretaria de Estado de Universidades of the Spanish Ministry of Education and Science.

the moment of inertia of the robot around its pitch axis, m is its mass and l is half the distance between fore and rear hips. The bound gait is passively stable only if $\hat{I} < 1$, a condition that sets a design constraint for bounding robots. For example, in Raibert's quadruped (Raibert, 1990) $\hat{I}$ was estimated to be 0.422 (Neishtadt and Li, 1991), so a stable bounding gait could be obtained without the need of any active control. Berkemeier (1998) derived approximate analytic return maps for the pronk and the bound gaits on planar models running in place. He found that in a pronk gait and for a given non-dimensional inertia, different discrete regions of stability existed depending on the apex height. Another example of this approach is the method used to find a stable cycle for the running robot SCOUT II (Poulakakis *et al.*, 2003). A return map obtained by numerical simulation was employed to find the touchdown angles for the legs that generated a passively stable bounding gait, which was successfully implemented in the robot.

A second strategy consists of actively controlling the attitude of the machine to stabilize the gait. In Raibert's quadruped, trot, pace and pronk gaits were stabilized by applying torques to the hip joints during leg stance (Raibert, 1990), which were computed as a function of the body pitch and roll angles and rates by a PD controller. In (Berkemeier, 1996) another high level controller was proposed to stabilize a pronk gait by independently regulating fore and rear leg thrust. Differential leg thrust was also used by Marehfka (2003) to stabilize bound and gallop gaits. The leg thrusts (and the leg touchdown angles) were computed at the top of flight by a fuzzy controller tuned in simulation to achieve the desired state at the next top of flight. Though the last two methods proved to be effective in simulation they have not been tested in real quadrupeds.

The regulation of leg thrust as an effective way to control and stabilize real running quadruped robots has not been addressed sufficiently in the past. For example, the Raibert quadruped, did not use any method to regulate leg thrust, which was simply pre-adjusted. In the SCOUT II quadruped robot the leg longitudinal thrust cannot be actuated independently, and its single actuator per leg adds energy to the bound gait rotating the leg around its hip joint while a spring provides the required compliance. The robot SCAMPER (Furusho *et al.*, 1995) performed a bound gait by actuating its knee with an electric motor working in speed control mode, a method similar to that presented in section 4.2.

This paper presents ongoing work on the KOLT robot related to leg thrust control and stabilization of simple running gaits. This work is intended to be a step towards complex high speed gaits such as the gallop. The paper is organized as follows: The experimental platform, the KOLT quadruped, is described in section 2. A simple control strategy to stabilize a pronk gait is proposed in section 3. Two methods implemented in the robot to control leg thrust are described in section 4. Experimental and simulation results showing the performance of these methods are presented and discussed in section 5. Finally, section 6 offers some conclusions.

2 Description of the System

The experimental platform employed is the KOLT (Nichol *et al.*, 2004) quadruped (see Figure 1-a). This robot measures 1.8x0.6x0.8 m and weights 75 Kg. The distance between hips is 0.73 m, and the moment of inertia around the pitch axis is 19.1 Kgm^2. Hence, the non-dimensional moment of inertia is $\hat{I} = 1.91$, outside the region in which the bounding gait is passively stable.

The legs are based on a mammal type configuration, and they have 3 active degrees of freedom driven by brushless motors. The hip pitch and roll joints are driven by high torque

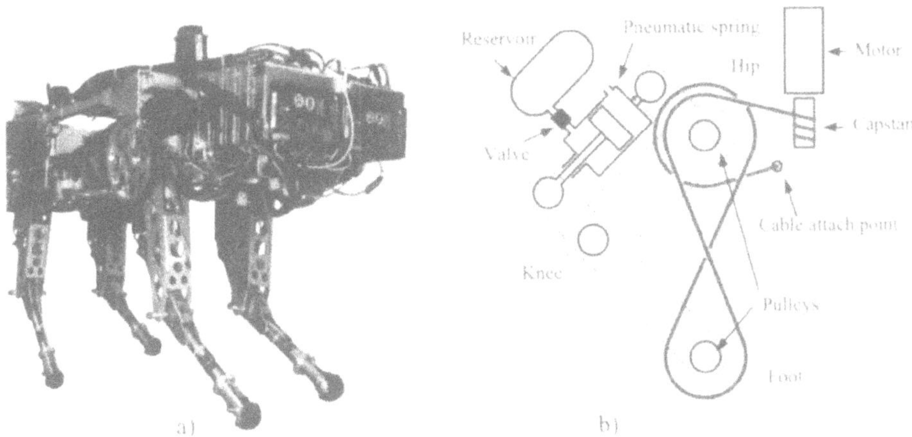

Figure 1. a) The KOLT quadruped robot. b) Schematic of the leg thrusting system.

motors and the reduction ratio is only 5.3:1 to allow a natural dynamic behavior of the quadruped during leg stance. The knee joint is actuated in an agonist-antagonist way by a motor that produces leg flexion and a pneumatic spring which produces leg extension (see Figure 1-b). The shaft of the motor directly drives a capstan on which a cable is wound. The arrangement of this cable around two pulleys located in the thigh joint and near the foot permits the decoupled actuation of the knee joint without interfering in the thigh joint motion (Waldron and Nichol, 2004). During flight the motor winds up the cable causing the flexion of the leg while compressing the pneumatic spring. The pneumatic spring is connected to an air reservoir through a one-way check valve which passively permits flow from the spring to the reservoir and prevents flow in the opposite direction. This valve can be actively opened to allow air to flow back to the spring. The phases and main events during a leg hop are summarized next:

1. During flight the knee motor flexes the knee compressing the spring in order to store pressurized air in the reservoir and to produce ground clearance. The hip motors move the foot to the desired touchdown position.
2. The knee motor unwinds the cable, letting the air spring extend the leg in preparation for landing. The one-way check valve retains the pressurized air in the reservoir.
3. After landing the leg is flexed and the spring is compressed under the weight of the robot. Kinematic and potential energy of the robot are stored in the spring as elastic energy. Hip motors are inactive during stance so the hip acts as a passive joint.
4. The valve is opened so that the pressurized air stored in the reservoir can return to the spring. The leg extends while the elastic energy stored in the spring and reservoir is converted into kinetic and potential energy causing the liftoff of the foot.

Each leg is controlled by an independent microcontroller in charge of executing the low level joint position control as well as generating the leg cycle described above. Feedback is provided by a foot switch activated during ground contact and encoders in the motor shafts. An additional encoder is placed on the knee, as the knee motor and the knee joint do not move synchronously

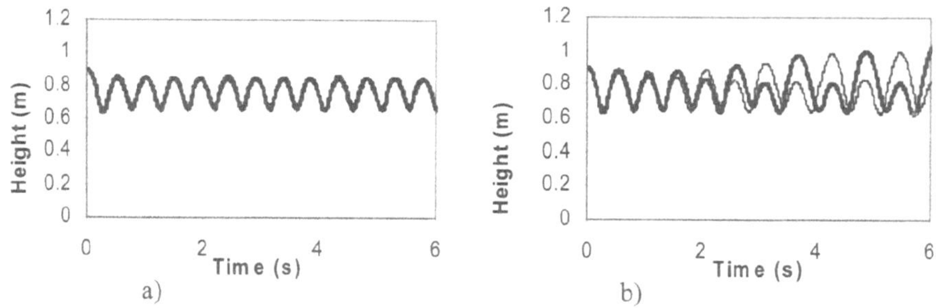

Figure 2. Height of the fore and rear hips (thick and thin lines respectively) in a simulated pronk gait a) employing the liftoff speed controller b) without liftoff speed controller.

during leg stance. A central computer is in charge of coordinating the four leg microcontrollers and estimating the overall state of the robot (attitude, attitude rates, forward speed, etc). A simple Raibert style controller (Raibert, 1990) is employed to produce bound and pronk gaits.

3 Liftoff Speed Controller

A simple strategy was designed to regulate the leg thrust. The hip liftoff speed, defined as the maximum linear speed of the hip related to the foot (in the foot-hip direction), is controlled in closed loop mode during stance. In this way the initial speed of the hip at the beginning of the flight phase can be determined. This approach has several advantages:

1. Liftoff speed is easily measurable in a real robot, while magnitudes such as the impulse or the energy added by the leg should be estimated or measured using more complex sensors.
2. This low level controller can help to stabilize a pronk gait without the need of central feedback. The local feedback loop can make the system less dependent on the initial conditions from hop to hop, and more robust against disturbances.
3. This approach can simplify the complex natural behavior of the running machine, so that simpler high level controllers can make use of it.

This method was tested employing the DynaMechs simulation library (McMillan *et al.*, 1996), and an accurate physical model of the KOLT robot. In simulation the position of the knee is controlled by a PD law, acting as a damped spring. When the knee is maximally compressed during stance the desired position of the knee joint is changed instantly to produce additional thrust. The liftoff speed can be controlled in simulation with arbitrary precision by contracting the leg when the desired foot-hip speed has been reached. Figure 2-a shows the behavior of the robot employing this controller while pronking, while Figure 2-b shows the behavior when this controller is not employed. In both cases the robot is dropped with the body level and thrust was adjusted to produce similar apex heights. Clearly, the pronk and the bound gaits are not naturally stable in the simulated robot, at least for the commanded hop height, but the pronk can be stabilized with the use of this simple controller. Other simple paired-foot gaits such as the trot or the pace can benefit of the same approach.

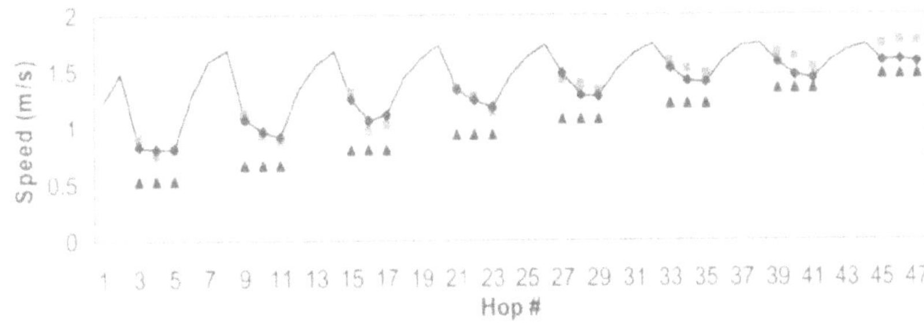

Figure 3. Target liftoff speed (triangles), and liftoff speed estimated from knee encoder (rhombus and solid line) and from time of flight data (gray squares), in 48 consecutive hops.

4 Leg Thrust Control Methods

Two methods have been employed to regulate leg thrust in the real KOLT robot. The first one implements the liftoff speed control described in section 3. The techniques used to estimate and control the leg liftoff speed are described in sections 4.1 and 4.2. The second method controls the amount of energy injected as a function of valve timing and is presented in section 4.3.

4.1 Liftoff Speed Estimation

The liftoff speed is estimated from data provided by the knee encoder. Knee position θ_{knee} is differentitiated and filtered to obtain a linear foot speed $V_{liftoff}$ along the foot-hip direction:

$$V_{liftoff}(t) = \left(\alpha\left(\theta_{knee}(t) - \theta_{knee}(t-1)\right) + (1-\alpha)V_{liftoff}(t-1)\right)L_{shank}\sin(\theta_{knee}(t)/2)/T_s \qquad (1)$$

L_{shank} is the length of the shank and thigh links of the leg and T_s is the sampling frequency. The parameter α of the first order filter was adjusted experimentally to obtain an accurate estimation. In order to do this, the liftoff speed estimate was compared with a second estimate computed from the measured time of flight (assuming a ballistic trajectory of a point mass located in the hip). Figure 3 shows both estimates for different liftoff speeds, after the filter was tuned. The delay introduced by the filter was 2-4 ms for the working sampling period (T_s = 2ms). As a comparison, leg extension during stance lasts about 50ms. The error in the estimation of liftoff speed was about 10%, for liftoff speeds between 0.7 and 1.7 m/s.

4.2 Closed Loop Leg Thrust Control

The first method proposed regulates the liftoff speed with the use of sensory feedback. When the knee has reached its maximum compression point during stance the valve is opened. The knee speed is then monitored during leg extension. When the desired linear liftoff speed has been reached the knee motor is actuated to wind up the cable and flex the leg, interrupting in this way the thrust phase. However, during the stance the cable is not under tension and is considerably slack. Consequently, there is a time lag before the motor can effectively flex the leg and this is the principal difficulty found when controlling the liftoff speed. This time lag depends on the position of the foot at the instant the desired speed is reached.

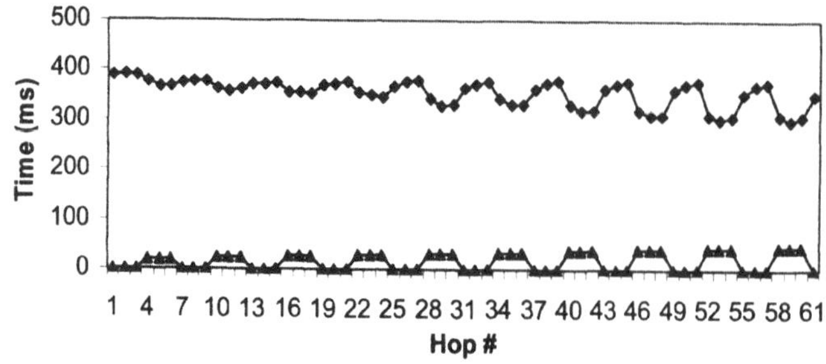

Figure 4. Times of flight (rhombus) obtained when the valve is opened a variable time (triangles) after maximum leg compression, in 61 consecutive hops.

4.3 Open Loop Leg Thrust Control

The second method proposed regulates the leg thrust by varying the instant at which the check valve is opened. Intuitively, the injected energy will be maximum if the valve is opened when the leg is maximally compressed, and any lower amount of energy can be added opening the valve later. Let P_{init} be the initial pressure in the spring and reservoir when the spring is maximally extended, i.e., its length is L_{max}. When the spring is compressed by the motor to a position L_{comp} the pressure ascends to P_{comp} given by:

$$P_{comp} = P_{init}((V_{res} + A_{cyl}L_{\max}) \ / \ (V_{res} + A_{cyl}L_{comp}))^{\gamma} \tag{2}$$

where A_{cyl} is the area of the piston V_{res} is the volume of the reservoir and γ=1.4. When the spring is extended again to length L_{max} in preparation for landing the pressure descends to P_{land}:

$$P_{land} = P_{comp}(L_{comp} \ / \ L_{\max})^{\gamma} \tag{3}$$

The leg is then compressed during landing, and extends again until the valve is opened at spring length L_{valve}, absorbing energy in this process. When the valve is opened, ideally the pressures in the reservoir and the spring equalize instantly. The leg then continues extending until the initial state (L_{max}, P_{init}) is reached, returning energy. The net energy added in this process is given by the sum of both actions:

$$E_{added} = \int_{L=L_{\max}}^{L=L_{valve}} A_{cyl}P_{land} \ (L_{\max} \ / \ L)^{\gamma} \ dL + \int_{L=L_{valve}}^{L=L_{\max}} A_{cyl}P_{init} \ (L_{\max} \ / \ L)^{\gamma} \ dL \tag{4}$$

Substituting (2) and (3) and integrating, the added energy is given by:

$$E_{added} = \frac{A_{cyl}P_{init}}{1-\gamma}\left(L_{comp}^{\gamma}\left(\frac{V_{res} + A_{cyl}L_{max}}{V_{res} + A_{cyl}\ L_{comp}}\right)^{\gamma} - L_{max}^{\gamma}\right)(L_{valve}{}^{1-\gamma} - L_{\max}{}^{1-\gamma}) \tag{5}$$

Thus we can control the added energy modifying the magnitudes L_{comp}, L_{valve} or L_{max}. The first option proved to be more inaccurate in our experimental system, so we adopted the second solution. The third option is equivalent to the method presented in section 4.2.

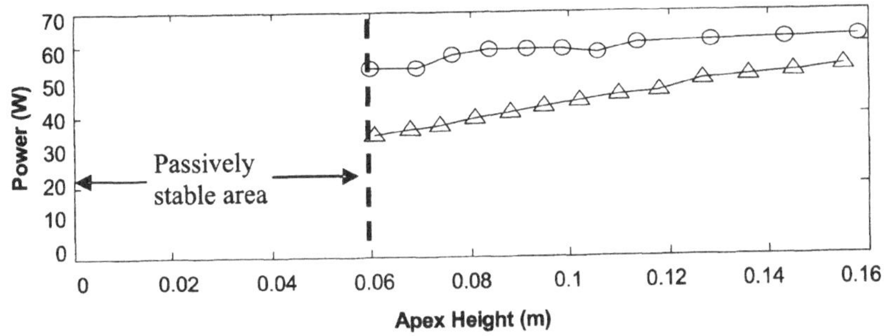

Figure 5. Power employed to maintain a pronk beyond the passively stable region as a function of the apex height, using the closed loop controller (circles), and using an artificial constraint (triangles) in order to stabilize the simulated gait.

5 Experiments and Simulations

5.1 Two Legged Bound

In order to evaluate the performance of the two proposed thrust controllers, the KOLT robot was suspended by the rear part of the body, so that it could hop on its fore legs. To test the closed loop controller different target liftoff speeds were set, in sets of three hops. Sets of hops with maximum thrust were intercalated to avoid excessive energy loss. Figure 3 shows the target liftoff speed and two estimates of the liftoff speed obtained, calculated from knee encoder and time of flight data. This experiment shows that despite the time delay imposed by the mechanical characteristics of the leg, the liftoff speed can be controlled at least partially. The time between the activation of the motor and the foot take-off is below 16 ms. For a typical hip acceleration of 30 m/s^2 that causes an error up to 0.48 m/s, in agreement with the results. Figure 3 shows that in the first hop after the maximum thrust hops the error is higher than in subsequent hops, because the legs are compressed further after landing, producing a larger acceleration, and thus a larger error. Figure 4 depicts the results obtained with the open loop controller. Variable times of flight are obtained varying the valve timing. The times of flight (and hence the liftoff speeds) are roughly proportional to the time from the bottom of flight to the valve opening. The range of liftoff speeds obtained is much shorter than the range obtained with the closed loop controller. This is because the open loop controller can only regulate the additional energy injected in the system, while the closed loop controller can also dissipate energy to obtain shorter hops.

5.2 Four Legged Bound

Two simulations in which the robot performed a four legged bound were made to test the described thrust controllers. In these experiments the four legs worked independently (see section 2) and no leg coordination mechanisms were implemented to force bound or pronk gaits. The pronk gait was passively stable for apex heights lower than 0.06m, and thus the open loop controller was useful only in this region. However, the closed loop controller could stabilize the gait for higher apex heights Figure 5 shows the average power employed to maintain the stable pronk beyond the passive stable region (up to 2.5 times the maximum passively stable apex height). To offer a comparison, Figure 5 also shows the average power used by the open loop controller when a level posture is forced in the simulation to achieve stability. This simulation shows how stability can be achieved at the cost of higher energy expenditure.

6 Conclusion

The regulation of leg thrust is required to control and stabilize running gaits. A simple closed loop controller is proposed here to regulate the leg thrust and to stabilize a pronk gait. Despite its simplicity, this low level controller is able to stabilize an inherently unstable robot without employing any central feedback. Both simulations and experiments are employed to show the validity of this approach. Additionally, an open loop thrust controller and a model of the leg thrusting mechanism of the KOLT robot have been presented. In the future these models and methods can be employed by higher level algorithms to generate complex gaits such as the gallop. However, the effectiveness of these strategies has only been tested in a pronk gait with zero forward speed. Future work includes the extension of these techniques to non-zero forward speed conditions, and integration in higher level controllers to achieve high speed locomotion.

Bibliography

Berkemeier, M. (1996). A Model for Quadrupedal Running-in-Place in the Plane. Proceedings of the 35th FA15 950 Conference on Decision and Control. Kobe, Japan. 3581-3586.

Berkemeier, M. (1998). *Modelling the dynamics of quadrupedal running*. International Journal of Robotics Research. 17:971–985.

Brown, H. and Zeglin, G. (1998). The bow leg hopping robot. *Proceedings of the IEEE International Conference on Robotics and Automation*. Leuven, Belgium. 781-786.

Furusho, J., Sano, A., Sakaguchi, M., Koizumi, E. (1995). Realization of Bounce Gait in a Quadruped Robot with Articular-Joint- Type Legs. *International Conference on Robotics and Automation*. Nagoya, Japan. 697-702.

Herr, H.M. and McMahon, T.A. (2001) A Galloping Horse Model. *The International Journal of Robotics Research*. 20(1): 26-37.

Marehfka, D.W., Orin, D.E., Schmiedeler, J.P., Waldron, K.J. (2003) *Intelligent Control of Quadruped Gallops*. IEEE/ASME Transactions on Mechatronics. 8:446-456.

McMillan, S., Orin, D.E. and McGhee, R.B. (1996). A computational framework for simulation of underwater robotic vehicle systems. *Autonomous Systems*. 3:253–268.

Neishtadt, A., and Li, Z. (1991). *Stability proof of Raibert's four legged hopper in bounding gait.* Technical Report 578. New York state University.

Nichol J.G., Singh S.P., Waldron K.J., Palmer, L.R., and Orin, D.E. (2004). System design of a quadrupedal galloping machine. *International journal of Robotics Research*. 23:1013–1027.

Poulakakis, I., Papadopoulos, E. and Buehler, M. (2003). On the Stable Passive Dynamics of Quadrupedal Running, *Proceedings of the IEEE International Conference on Robotics and Automation*. Taipei, Taiwan. 1368-1373.

Raibert, M.H. (1990). Trotting, pacing and bounding by a quadruped robot. *Journal of biomechanics*. 23:79–98.

Raibert, M.H. (1986) Legged Robots that balance, MIT Press, Cambridge MA.

Waldron, K.J. and Nichol J.G. (2004). Architectural Issues in Running Machines. *Proc. 15th CISM-IFToMM Symposium on Robot Design*, Dynamics and Control. Montreal, Canada.

Vehicle Dynamics of Redundant Mobile Robots with Powered Caster Wheels

Yuan Ping Li* and Teresa Zielinska† and Marcelo H. Ang Jr.* and Wei Lin‡
* National University of Singapore, Faculty of Engineering, Singapore,
e-mail: {liyuanping,mpeangh}@nus.edu.sg
† Warsaw University of Technology, Faculty of Power and Aeronautical Engineering, Poland,
e-mail: teresaz@meil.pw.edu.pl
‡ Singapore Institute of Manufacturing Technology, Singapore,
e-mail: wlin@SIMTech.a-star.edu.sg

Abstract. The dynamic model of redundantly actuated mobile robots with powered caster wheels is derived based on vehicle dynamics. The contact stability problem of wheeled mobile robots is introduced and stable contact condition that characterizes the bounds of contact stability is derived. Sliding mode observer is proposed to estimate the robot velocity using wheel angular velocity and joint torque information. Actuation redundancy of the robot is utilized to satisfy the stable contact condition in trajectory tracking applications.

1 Introduction

More and more important applications of wheeled mobile robots (WMRs) e.g. planetary exploration and mobile manipulation, require accurate dynamic models of WMRs in order to achieve high control performance. Most of the previous literature only consider the linkage dynamics of the rigid bodies on the robot and ignore the dynamic effects of wheel-ground interaction by assuming that the wheels are under pure rolling without slipping motion. However, theory of vehicle dynamics (Bekker, 1956; Wong, 2001), a well established discipline in automobiles dealing with dynamic properties of rolling motion, has revealed that different nonlinear dynamic effects and disturbances will be generated in different wheel-ground interaction conditions. Therefore, this paper discusses the dynamic modelling of WMRs taking into account the wheel-ground interaction. Specifically, dynamic effects caused by the phenomenon of slip is mainly analyzed.

Many WMRs with Powered Caster Wheels (PCWs, also known as offset steerable wheels) have been developed (Wada and Mori, 1996; Yi and Kim, 2002; Li et al., 2005) and even commercialized e.g. the Nomadic XR4000 robot (Holmberg and Khatib, 2000). One important benefit of using PCW is that WMRs with PCWs can generate omnidirectional mobility. However, due to the coupling effects between the rolling and offset steering of the wheel, slip phenomenon is more significant in both longitudinal and lateral directions than other type of wheels. Moreover, most of the WMRs with PCWs use redundant actuators to achieve high payload but cause internal forces between wheels due to the over-constraints in their kinematics. This further magnifies the occurrence of slip phenomenon. These problems are also encountered by us in controlling our omnidirectional mobile robot with 4 PCWs as shown in Figure 1. The robot is redundantly actuated as totally 8 actuators on 4 wheels are used to control the robot.

This paper is an extension of the work (Li et al., 2006). In (Li et al., 2006), dynamic modelling of WMRs with PCWs is improved by considering the effect of wheel-ground interaction. In this paper, we further explore the effects of slip ratio, which is one of the main factors of wheel-ground interaction, on the vehicle dynamics. Moreover, we extend the non-slip conditions derived in (Li et al., 2006) to stable contact condition by considering WMRs locomotion as a multifingered grasping task. Without introducing extra sensors, we apply sliding mode observer to estimate the robot velocity by using the wheel angular velocity and joint torque information. Simulation results are presented to demonstrate the effectiveness of the proposed methods.

Figure 1. Omnidirectional mobile robot with 4 powered caster wheels.

This paper is organized as follows. Section 2 derives the dynamic model of mobile robots with powered caster wheels using vehicle dynamics. Section 3 introduces the contact stability problem for wheeled mobile robots and derives the stable contact condition to characterize the bounds of contact stability. Simulation results of controlling the robot using the vehicle dynamics model and sliding mode observer are demonstrated in Section 4 followed by the conclusion in Section 5.

2 Dynamics Analysis of WMRs with Powered Caster Wheels

2.1 Vehicle Dynamics

Frame assignments and variable definitions of a WMR with n PCWs is shown in Figure 2. The equations of motion of the overall system can be derived using Operational Space Framework (Khatib, 1987) as

$$\Lambda \ddot{P} + \Phi = GF_c \tag{2.1}$$

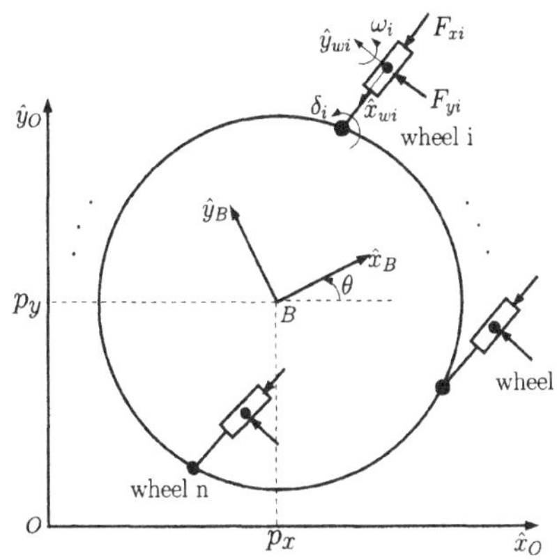

Figure 2. Diagram of a mobile robot with n powered caster wheels.

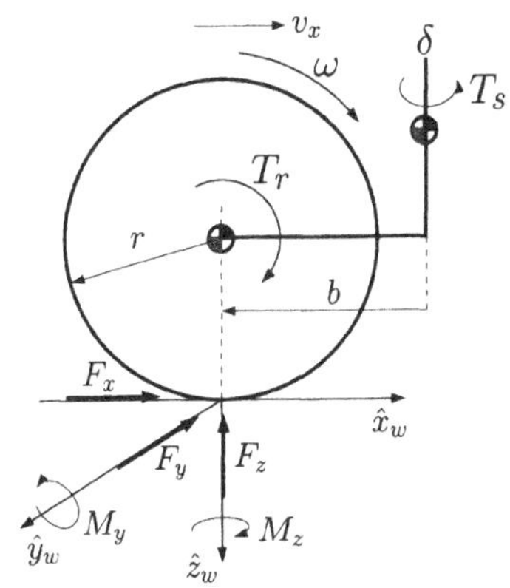

Figure 3. Force equilibrium of a powered caster wheel.

where Λ is the inertial matrix, $P = [p_x, p_y, \theta]^T$ is the operational position and orientation vector of the robot and $\ddot{P}$ is the second derivative of P, Φ is the Coriolis and centrifugal force vector and $F_c = [F_{x1}, F_{y1}, \cdots, F_{xn}, F_{yn}]^T$ is the vector of contact forces. G is the transformation matrix that maps the sum of contact forces into generalized operational forces/torques. Gravitational force is not considered as we assume the robot operates on level ground.

2.2 Wheel Dynamics

Applying the theory of vehicle dynamics (Wong, 2001), the motion and force system of a PCW can be demonstrated with Figure 3. Besides the "normal force" F_z that the ground acts on the robot, the wheel is subjected to longitudinal "tractive force" F_x and "rolling resistance moment" M_y due to the rolling motion. The wheel is also subjected to "lateral force" F_y and "aligning torque" M_z due to the steering motion. The equations of motion of a PCW are given as

$$I_r \dot{\omega} = T_r - M_y - F_x r \tag{2.2}$$

$$I_s \dot{\delta} = T_s - M_z - F_y b \tag{2.3}$$

where I_r and I_s are the inertia tensors of the wheel w.r.t. its rolling and steering axis respectively; ω and δ are the rolling and steering speeds of the wheel respectively; T_r and T_s are the torques of the rolling and steering actuators respectively; r is the radius of the wheel and b is the offset length of the wheel.

Note that the effects of M_y and M_z on the wheel dynamics are usually much smaller than that of the contact forces F_x and F_y, so we will limit our discussion on the contact forces in this paper. Also note that when contact forces F_c are used to relate the vehicle dynamics model (2.1) to the wheel dynamics model (2.2 and 2.3), inertia tensors I_r and I_s should include not only the wheel inertia but also the inertia of the vehicle chassis.

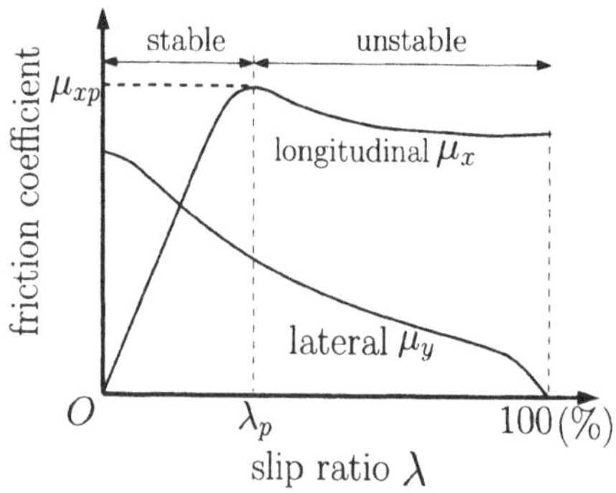

Figure 4. Longitudinal and lateral friction coefficients vary with slip ratio.

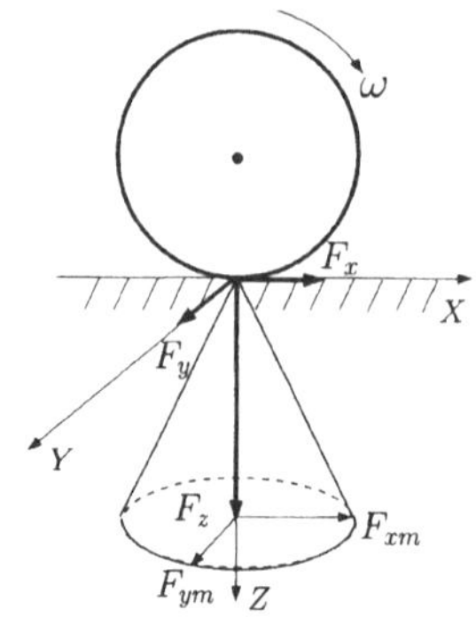

Figure 5. Contact ellipsoid cone of a rolling wheel.

2.3 Relation of Slip Ratio and Contact Force

Contact forces F_x and F_y are usually described in the form of frictional force i.e.

$$F_x = sgn(\omega)\mu_x F_z \hat{x}_w \tag{2.4}$$

$$F_y = -sgn(\delta)\mu_y F_z \hat{y}_w \tag{2.5}$$

where $\hat{x}_w$ and $\hat{y}_w$ are the unit vectors of the wheel frame basis; μ_x and μ_y are defined as longitudinal and lateral friction coefficients which depend mainly on ground conditions and wheel-ground relative motion. The quantity "slip ratio", which characterizes the wheel-ground relative motion, is defined as

$$\lambda = \frac{\omega r - v_x}{max\{\omega r, v_x\}} \times 100\% \tag{2.6}$$

where v_x is the actual longitudinal velocity of the wheel w.r.t. the ground.

Variations of μ_x and μ_y with λ as shown in Figure 4 are important characteristic curves (Wong, 2001) in vehicle dynamics. In the $\mu_x(\lambda)$ curve, μ_x increases with λ until it reaches its peak value μ_{xp} which corresponds to slip ratio λ_p. Further increase of slip ratio beyond λ_p results in rapid decrease of μ_x and the wheel slides or spins on the ground unstably. In the $\mu_y(\lambda)$ curve, μ_y decreases rapidly with increase of λ.

3 Contact Stability of WMRs

WMRs locomotion can be considered as a task where multiple interconnected wheels grasp the ground. This is similar to multifingered grasping tasks where an object is grasped by a multifingered robotic hand. By such an observation, we can consider the "contact stability" problem (Nakamura, 1991) for WMRs as in multifingered grasping tasks. We define contact stability of WMRs locomotion as the ability of the mobile robot to maintain stable rolling contact with the ground when the robot is subjected to disturbing forces. The problem of "contact stability" is especially important for WMRs because the rolling contact between the wheels and the ground is more sensitive to disturbing forces than static contact which is usually the case for multifingered grasping tasks.

3.1 Stable Contact Condition

It is intuitive to specify conditions that characterize the contact stability of WMRs in terms of contact forces. As shown in Figure 4, maximum longitudinal force F_{xm} is reached when slip ratio is λ_p, and maximum lateral force F_{ym} is reached when slip ratio is zero. Evaluating all possible values of slip ratio, an ellipse with F_{xm} and F_{ym} as the major and minor axes can be obtained. Moreover, evaluating the variation of these ellipses with the normal force F_z, an ellipsoid cone, where a point within the cone corresponds to a stable contact state, can be obtained. We call this "contact ellipsoid cone" as shown in Figure 5. Therefore, a condition characterizing the contact stability is that all contact forces need to be within their contact ellipsoid cones.

Let $f_d = [f_{dx}, f_{dy}]^T$ be the unexpected external disturbing force which is bounded by f_d^*, i.e. $\| f_d \| \leq f_d^*$. Thus, a stable contact condition can be derived from the concept of contact ellipsoid cone as

$$\frac{(F_{xi} + f_{dx}^*)^2}{F_{xm}^2(F_{zi})} + \frac{(F_{yi} + f_{dy}^*)^2}{F_{ym}^2(F_{zi})} \leq 1 \tag{3.1}$$

Stable contact condition described by Equation 3.1 is a extension of the non-slip conditions derived in the work (Li et al., 2006).

3.2 Contact Force Distribution

In order to satisfy the stable contact condition 3.1, contact force distribution scheme based on the actuation redundancy of the robot can be utilized. A general solution of the vector of contact forces F_c can be derived from Equation 2.1 as

$$F_c = G^{\dagger}(\Lambda \ddot{P} + \Phi) + (I_{2n\times 2n} - G^{\dagger}G)f_o \tag{3.2}$$

where $I_{2n\times 2n}$ is the identity matrix of order $2n$, $G^{\dagger} = G^T(GG^T)^{-1}$ is the pseudo-inverse of matrix G and $f_o \in R^{2n}$ is an arbitrary vector.

In Equation 3.2, the first term in the right hand side corresponds to the contact force components that generate the resultant operational motion of the robot while the second term corresponds to the nullspace of matrix G that generates no effect on the resultant operational motion of the robot. Vector f_o, which is projected onto the nullspace of matrix G, is determined in a way that the final contact forces F_c computed from Equation 3.2 satisfy the stable contact condition 3.1.

4 Sliding Mode Observer

4.1 Observer Design

The evaluation of slip ratio requires information of wheel angular velocity and vehicle actual velocity. In usual cases, sensors are only available to measure wheel angular velocities and actuator torques. So the vehicle actual velocity needs to be estimated with observers. In a similar application, Unsal and Kachroo (1999) have compared the performances of different nonlinear observers and Sliding Mode Observers was found to provide more satisfactory performance than that of Extended Kalman Filter. The main benefits of sliding mode observers are also well known for their robustness to parametric uncertainty and external disturbances (Slotine and Li, 1991). Therefore, we will adopt sliding mode observer in this paper to estimate the actual velocity of the vehicle.

We first reformulate the system in state space form. We choose state variables as the wheel angular velocity ω and actual velocity of the wheel v_x

$$x = \begin{bmatrix} x_1 \\ x_2 \end{bmatrix} = \begin{bmatrix} \omega \\ v_x \end{bmatrix} \tag{4.1}$$

We can rewrite Equation 2.1,2.2, 2.4 and 2.6 as

$$\begin{aligned} \dot{x}_1 &= f_1 + gu \\ \dot{x}_2 &= f_2 \\ y &= Cx \end{aligned} \tag{4.2}$$

where

$$\begin{aligned} f_1 &= -M_y/I_r - \mu_x(x_1, x_2)F_z r/I_r \\ f_2 &= -\Phi/\Lambda + G\mu_x(x_1, x_2)F_z/\Lambda \\ g &= 1/I_r \\ u &= T_s \\ C &= [1, 0] \end{aligned} \tag{4.3}$$

Now we can define sliding mode observer as follows

$$\begin{aligned} \dot{\hat{x}}_1 &= \hat{f}_1 + \hat{g}u - h_1\tilde{y} - k_1 sgn(\tilde{y}) \\ \dot{\hat{x}}_2 &= \hat{f}_2 - h_2\tilde{y} - k_2 sgn(\tilde{y}) \end{aligned} \tag{4.4}$$

where $\hat{x}$, $\hat{f}$ and $\hat{g}$ are estimations of x, f and g respectively. Measurement error $\tilde{y} \equiv C(\hat{x} - x)$, observed from the structure of the observer, is chosen as the sliding variable.

Readers are referred to (Slotine et al., 1989) for the details of the observer derivation and its analysis as well as techniques to choose the gains h_1, h_2, k_1 and k_2 so that the estimated states finally match their actual values.

4.2 Application: Simulation Results

We simulated the control diagram of Figure 6 in Matlab. This diagram demonstrate a new control scheme that incorporates the vehicle dynamics model, stable contact condition and sliding mode observer as discussed in this paper.

In the simulation, we commanded the system to track a simple straight line trajectory so that the coupling effects between the longitudinal and lateral dynamics (which are not discussed in this paper) are not generated. In order to generate slip phenomenon and other real effects in the simulation, we introduce random disturbing force with range of $(-0.5F_c, 0.5F_c)$ to the system.

The straight line desired trajectory for the simulation implies that no steering action is involved and only the $\mu_x(\lambda)$ relation is required for the computation of slip ratio in the simulation. For convenience, we adopted following simple model which is proposed in (Peng and Tomizuka, 1990)

$$\mu_x = \frac{2\mu_{xp}\lambda_p\lambda}{\lambda_p^2 + \lambda^2} \tag{4.5}$$

Table 1 shows the parameters and gains of the observer used in the simulation. m_r, m_s and m_c are the mass of the wheel disc, steering link and vehicle chassis respectively.

Table 1. simulation parameters

r	0.055 (m)	b	0.020(m)	λ_p	16(%)	μ_{xp}	0.98
m_r	0.65(kg)	m_s	2(kg)	m_c	116(kg)	F_z	310.5(N)
h_1	2	h_2	10	k_1	2	k_2	10

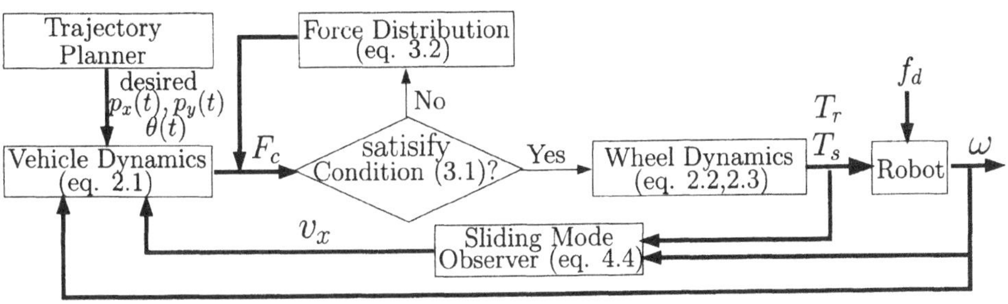

Figure 6. Simulation diagram with sliding mode observer.

Velocity tracking error and estimation error of robot velocity is shown in Figure 7. The tracking error is mainly caused by modelling error and disturbances. Estimation error is relatively small but suffers from chattering. The discussion of chattering elimination is out of the scope of this paper.

Figure 8 shows the actual slip ratio and the estimated slip ratio. It can be noticed that although the slip estimation is effective, the slip ratio is always underestimated by the proposed sliding mode observer.

5 Conclusion

This paper analyzes the vehicle dynamics of wheeled mobile robots with powered caster wheels. Condition that describes the limits of contact stability in terms of contact forces, is derived. Force distribution scheme is proposed to satisfy the stable contact condition. Sliding mode observer is proposed to estimate the system states and its effectiveness is demonstrated by simulation. Future work is to work on the problem of slip based traction control. We also plan to implement the proposed methods on the real robot.

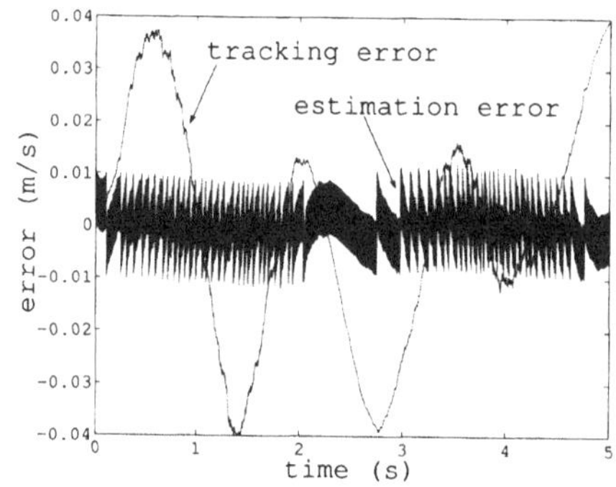

Figure 7. Velocity tracking error and estimation error of robot velocity.

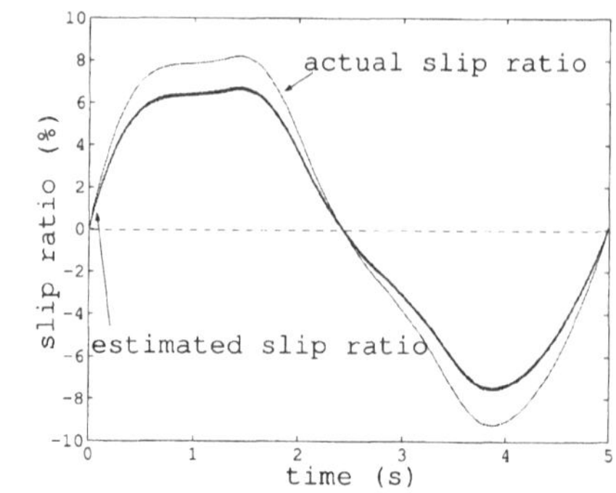

Figure 8. Actual slip ratio and estimated slip ratio.

Bibliography

M.G. Bekker. *Theory of Land Locomotion.* University of Michigan Press, 1956.

R. Holmberg and O. Khatib. Development and control of a holonomic mobile robot for mobile manipulation tasks. *Intl. J. Robotics Research*, 19(11):1066–1074, 2000.

O. Khatib. A unified approach for motion and force control of robot manipulators: The operational space formulation. RA-3(1):43–53, 1987.

Y.P. Li, D.N. Oetomo, Marcelo H. Ang Jr., and C.W. Lim. Torque distribution and slip minimization in an omnidirectional mobile base. *Intl. Conf. Advanced Robotics*, pages 567–572, 2005.

Y.P. Li, T. Zielinska, Marcelo H. Ang Jr., and W. Lin. Wheel-ground interaction modelling and torque distribution for a redundant mobile robot. *IEEE Intl. Conf. Robotics and Automation*, 2006.

Y. Nakamura. *Advanced Robotics: Redundancy and Optimization.* Addison-Wesley Publishing Company, 1991.

H. Peng and M. Tomizuka. Vehicle lateral control for highway automation. *Proc. Amer. Contr. Conf.*, pages 788–794, 1990.

J.-J.E. Slotine, J.K. Hedrick, and E.A. Misawa. On sliding observers for nonlinear systems. *ASME J. Dynamic Syst. Meas. Contr.*, 109:245–252, 1989.

J.-J.E. Slotine and W. Li. *Applied Nonlinear Control.* Prentice Hall, 1991.

Cem Unsal and Pushkin Kachroo. Sliding mode measurement feedback control for antilock braking systems. *IEEE Trans. on Control Systems Technology*, 7(2):271–281, 1999.

M. Wada and S. Mori. Holonomic and omnidirectional vehicle with conventional tires. *Proc. IEEE Intl. Conf. Robotics and Automation*, 4:3671–3676, 1996.

J.Y. Wong. *Theory of Ground Vehicles.* John Wiley and Sons, Inc., 3rd edition, 2001.

B.J. Yi and W.K. Kim. The kinematics for redundantly actuated omni-directional mobile robots. *J. Robotic Systems*, 12(6):255–267, 2002.

Matrix-based Supervisory Controller of Transition-Function Specified Robot Controllers

Niak Wu Koh*, Cezary Zieliński†, Marcelo H. Ang, Jr.* and Ser Yong Lim ‡

* National University of Singapore, Department of Mechanical Engineering
email: (nwkoh, mpeangh)@nus.edu.sg

† Warsaw University of Technology, Faculty of Electronics and Information Technology
email: C.Zielinski@ia.pw.edu.pl

‡ Singapore Institute of Manufacturing Technology
email: sylim@SIMTech.a-star.edu.sg

Abstract. Robot control systems generally require a layer that manages, in a modular manner, the hardware or low level control and a supervisory layer which allows task specification for the execution of a task. This paper proposes the transition-function specified controllers as a lower layer, with an improved matrix-based supervisory controller for the high level operations.

1 Introduction

With a growing popularity in discrete event systems, its usage has been largely favoured by the robotics and manufacturing community especially from the stance of task planning and modelling (Morandin and Kato, 2005; Fourquet et al., 2005). Task planning describes the ability to decompose a task into a set of primitives. Recomposing this primitive set into a cognitively logical sequence, which when executed, enables the task to be carried out autonomously or with user intervention as and when needed.

Whether a primitive has completed its execution is dependent on the current and terminal state of the system. Therefore, a framework that caters to the development of these primitives and the transmission of the system's sensor information is mandatory. Control of a system essentially boils down to the processing of information about the current state of the system and its environment enabling a defined task to be executed.

The remainder of this article is organized as follows: Section 2 introduces the matrix-based controller. Section 3 provides a brief outlook into the transition-function based approach while Section 4 proposes the use of the matrix model as a supervisory controller. Section 5 concludes.

2 Matrix-based Discrete Event Controller

With the use of manufacturing engineering concepts, a powerful rule-based matrix model was developed by Tacconi and Lewis (1997) which has only been implemented to flexible manufacturing systems (see Boghan et al., 2002; Mireles and Lewis, 2001). The pursuit of this research is to employ the discrete event controller to showcase its potential as a tool for the execution of complex robotic tasks. Preliminary work regarding the development and implementation of the matrix model from the modified to its improved state can be viewed in Koh et al. (2005a) and Koh et al. (2005b) respectively.

A task consists of a sequence of jobs and its corresponding resources which, via the use of input signals and a set of conditions, triggers the execution of each job. The entire task is complete once an output signal is present.

With the matrix model, assembly/job sequencing, addition of resources, analyses for deadlock and its avoidance and a dispatching design can be carried out allowing a thorough analysis with a convenient solution to the simulation of the system in the form of a Petri Net (Murata, 1989). The improved matrix discrete event model is described in the subsequent sections.

2.1 Discrete Event Model State Equation

The task can be described by:

$$\bar{x} = F_v \otimes \bar{v}_c \oplus F_r \otimes \bar{r}_c \oplus F_u \otimes \bar{u} \oplus F_D \otimes \bar{u}_D$$
$$\rightarrow \bar{x}_i = F_v(i,j) \otimes \bar{v}_{c_j} \oplus F_r(i,j) \otimes \bar{r}_{c_j} \oplus F_u(i,j) \otimes \bar{u}_j \oplus F_D(i,j) \otimes \bar{u}_{D_j} \tag{2.1}$$

The condition (x) represents the state of the system and the equations shows how it evolves over time.

Table 1. Variable definitions for Equation (2.1)

i,j	[Row, Column]
F_v	Job sequencing matrix
F_r	Resource requirements matrix
F_u	Parts input matrix
F_D	Dispatching matrix

x	Condition
v	Job vector
r	Resource vector
u	Input signal
u_D	Dispatch control vector

As the equations used in the matrix model are logical equations, standard matrix multiplication and addition are replaced by AND/OR algebra and all vectors and matrices are binary. $\otimes$ represents an AND operation and $\oplus$ an OR operation. The over bar is a logical negation and is defined as follows: For any component $a(i)$ of a natural number vector **a**

$$\bar{a} = 0 \quad if \quad a(i) > 0 \qquad \bar{a} = 1 \quad if \quad a(i) \leq 0$$

Each of the matrices are explained as follows:

- F_v determines which relevant job should be completed before condition x is satisfied. When $v_c = 1$, a job is said to be complete.
- F_r determines which relevant resources should be present before condition x is satisfied. When $r_c = 1$, a resource is said to be currently available.

- F_u determines which relevant input signal should be present before condition x is satisfied. When $u = 1$, an input signal is said to be present.
- F_D determines which relevant dispatch signal should be present before condition x is satisfied. When $u_D = 1$, a dispatch signal is said to be present. This matrix is used to determine the priority of the operations when resources are shared.

2.2 Job Start Equation

The start of a job is indicated by the following equation:

$$\begin{aligned} \bar{v}_s &= S_v \otimes \bar{x} \oplus U \otimes \bar{v}_c \\ \rightarrow \bar{v}_{s_i} &= S_v(i,j) \otimes \bar{x}_j \oplus U(i,j) \otimes \bar{v}_{c_j} \qquad v_{s_i} \neq v_{c_j} \end{aligned} \tag{2.2}$$

where S_v is a rectangular job start matrix and U is any user-defined matrix of n x m dimensions (n can equal m). n will assume the number of elements in the $\bar{v}_s$ column vector. Job i starts when $v_{s_i} = 1$. Equation (2.2) can be read as follows: Job i will start iff the relevant conditions (determined by S_v) are satisfied and the relevant job(s) is/are complete. This general representation can be used for a concurrent and dependent operation.

For a sequential operation

$$\bar{v}_s = S_v \otimes \bar{x} \oplus I \otimes \begin{bmatrix} 0 \\ \bar{v}_c \end{bmatrix} \tag{2.3}$$

where I is an n x n identity matrix with n assuming the number of elements in the $\bar{v}_s$ column vector. It is worth noting that the presence of the $\bar{v}_c$ element in the latter part of the equation serves as an added validity check since the status of $\bar{v}_c$ is already present in $\bar{x}$.

By decomposing a complex task into a series of elemental sequences, task primitives, used to represent each subtask, can be assigned to the job start vector v_s.

2.3 Resource Release Equation

The equation indicating the release of a resource is:

$$\begin{aligned} \bar{r}_s &= S_r \otimes \bar{x} \oplus U \otimes \bar{r}_c \\ \rightarrow \bar{r}_{s_i} &= S_r(i,j) \otimes \bar{x}_j \oplus U(i,j) \otimes \bar{r}_{c_j} \end{aligned} \tag{2.4}$$

where S_r is a rectangular resource release matrix and U is any user-defined matrix of n x m dimensions (n can equal m). n will assume the number of elements in the $\bar{r}_s$ column vector. Resource i is released when $r_{s_i} = 1$. Equation (2.4) can be read as follows: Resource r_s can only be released iff the relevant conditions (determined by S_r) are satisfied and the relevant user-defined resource(s) is/are idle.

For a sequential operation

$$\bar{r}_s = S_r \otimes \bar{x} \oplus I \otimes \bar{r}_c \tag{2.5}$$

where I is an n x n identity matrix with n assuming the number of elements in the $\bar{r}_s$ column vector. It is worth noting that the presence of the $\bar{r}_c$ element in the latter part of the equation serves as an added validity check since the status of $\bar{r}_c$ is already present in $\bar{x}$.

2.4 Process Output Equation

Sy determines the set of conditions that need to be satisfied before a product is outputted:

$$y = S_y \otimes x \rightarrow y_i = S_y(i,j) \otimes x_j \tag{2.6}$$

A theoretical insight to the improved matrix-based supervisory controller is discussed in Koh et al. (2005c). Figure 1 depicts the matrix model. The S matrices determine which set of conditions must be satisfied to release the resources required by the starting of a job. Once a job is complete, the F matrices is used to obtain the next set of conditions, based on the system's current state, which will be used by the S matrices.

The matrix-based supervisory controller allows an effective and simple online control by applying a set of discrete event control signals. Together with the Petri net marking transition equation (Murata, 1989), the matrix formulation provides a complete dynamical description that can be used for analysis and computer simulation (Koh et al., 2005b).

3 Behavioural Control Case in a Nutshell

Task planning begins with the decomposition of a complex task into a sequential series of elemental subtasks/jobs/primitives. From the point of view of clarity of a task description that is to be executed by a system, it is useful to group the steps of the commanded evolution of the control subsystem state into sequences which are termed as primitive behaviours (Zielinski, 2005). Once a decomposed cognitive plan of a task is available, Equation (2.2) is used where v_s would represent the primitives.

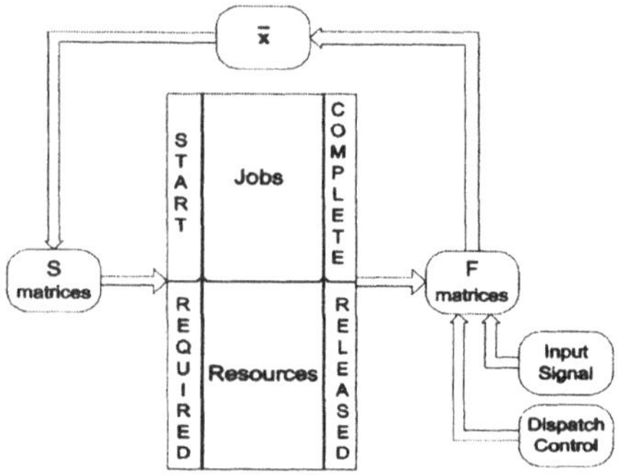

Figure 1. Depiction of the Matrix-based Discrete Event Controller.

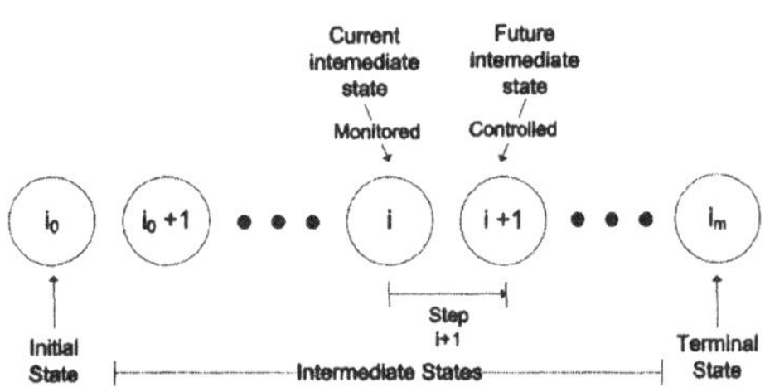

Figure 2. Evolution of the system state during the execution of an instruction/job.

3.1 Behaviour of an Agent

A behaviour can be defined as a sequence of total states

$$ {}^{q}b_j^i = {}^{q}b_j = \{c_j^{i+1}, c_j^{i+2}, ..., c_j^{i+n_s}\} \tag{3.1}$$

where n_s is the number of steps in a behaviour and q denotes a numeric identifier of this reaction, i denotes the current instant and the next considered instant is denoted by $i+1$. Each sequence of states $c_j^{i+1}, c_j^{i+2}, ..., c_j^{i+n_s}$ is generated by one of the functions ${}^{m}f_{c_j}$, i.e., the function defining the primitive behaviour. n_f partial functions are defined as

$$ {}_{y}c_j^{i+1} = {}^{m}f_{c_j}({}_{x}c_j^i) \qquad m = 1, ..., n_f \tag{3.2}$$

whereby the control subsystem uses

$$ {}_{x}c_j^i = \{c_{c_j}^i, {}_{x}c_{e_j}^i, {}_{x}c_{V_j}^i, {}_{x}c_{T_j}^i\} \tag{3.3}$$

to produce

$$ {}_{y}c_j^{i+1} = \{c_{c_j}^{i+1}, {}_{y}c_{e_j}^{i+1}, {}_{y}c_{V_j}^{i+1}, {}_{y}c_{T_j}^{i+1}\} \tag{3.4}$$

and hence the transition functions

$$\begin{aligned} c_{c_j}^{i+1} &= f_{c_{c_j}}(c_{c_j}^i, {}_{x}c_{e_j}^i, {}_{x}c_{V_j}^i, {}_{x}c_{T_j}^i) \\ {}_{y}c_{e_j}^{i+1} &= f_{c_{e_j}}(c_{c_j}^i, {}_{x}c_{e_j}^i, {}_{x}c_{V_j}^i, {}_{x}c_{T_j}^i) \\ {}_{y}c_{V_j}^{i+1} &= f_{c_{V_j}}(c_{c_j}^i, {}_{x}c_{e_j}^i, {}_{x}c_{V_j}^i, {}_{x}c_{T_j}^i) \\ {}_{y}c_{T_j}^{i+1} &= f_{c_{T_j}}(c_{c_j}^i, {}_{x}c_{e_j}^i, {}_{x}c_{V_j}^i, {}_{x}c_{T_j}^i) \end{aligned} \tag{3.5}$$

c_j is the state of the control subsystem of an agent
e_j is the state of the effector of the agent
V_j is the bundle of virtual sensor readings and
T_j is the information from/to the other agents

The state of an agent is thus defined as

$$ s_j = \{c_j, e_j, V_j, T_j\} \tag{3.6}$$

Each program instruction causes a certain change in the state of a system. If we were to associate with an instruction a state-transition pair: *initial state* and *terminal state*, a finite number of discrete event states can be defined using the mapping *sem* as

$$ sem : II \mapsto [S \to S^*] \tag{3.7}$$

where ii is an instruction and II the set of instructions, $ii \in II$.

This formal approach to the behavioural aspect of a robot programming framework was developed by Zielinski (2005)(see Zielinski, 1995 for more details).

In the case of a purely reactive system, the choice of the function ${}^{m}f_{c_j}$ is based on testing predicates ${}^{q}p_{c_j}$, $q = 1, .., n_p$, which take as arguments only the components of ${}_{x}c_{V_j}^i$. In pseudo-code, it can be expressed as

$$ if\ {}^{q}p_{c_j}({}_{x}c_{V_j}^i)\ then\ {}_{y}c_j^{i+1} := {}^{m}f_{c_j}({}_{x}c_j^i) \tag{3.8}$$

Pseudo-code (3.8) represents a single-step behaviour, i.e., $n_s = 1$ (refer to Equation (3.1) for the definition of n_s). For the case of a multi-step behaviour, the pseudo-code is expressed as

$$if\ {}^{q}p_{c_j}({}_{x}c_j^i)\ then\ {}^{q}b_j({}_{x}c_j^i) \tag{3.9}$$

The control program is composed of an endless loop containing a sequence of instructions of form (3.9). Each iteration of the loop contains several control steps i. The required computations, i.e., computation of ${}_{y}c_j^{i+\epsilon}$, $\epsilon = 1,..,n_s$, and the execution of behaviours, i.e., transmissions $e_j \mapsto {}_{x}c_{e_j}^i$, $V_j \mapsto {}_{x}c_{V_j}^i$, $c_{T_j} \mapsto {}_{x}c_{T_j}^i$, are bundled together within ${}^{q}b_{c_j}({}_{x}c_j^i)$, $q = 1,..,n_p$.

4 Sensor Utilization

If notation (3.7) is assumed, the execution of an instruction begins in an initial state, ends in a terminal state and traverses a sequence of intermediate states. Robots are generally controlled by digital computers so the execution of each instruction is subdivided into steps. Each step results in a change of the system state from one intermediate step to the next.

In each intermediate (or current) state, the state of a system can be measured or monitored by sensors. However, the future intermediate states can only be influenced, i.e., controlled. An initial state can be treated as a current intermediate state at the beginning of an instruction execution. A terminal state is the current intermediate state in which the execution of the instruction terminates. As the initial and terminal states are special cases of current intermediate states, both of them can only be monitored (see Figure 2).

The monitoring of the system state is performed by receptors. Raw data obtained from them cannot be utilized directly to monitor or control a system. It has to be transformed into a useful form by data aggregation. In consequence, a virtual sensor reading, V, is obtained.

5 Matrix Model as the Supervisory Controller

Consider a simple task (see Figure 3) where an end-effector moves to two different coordinates (jobs v_{s_1} and v_{s_2}) in cartesian space. With reference to Equation (2.3),

$$\begin{bmatrix} \bar{v}_{s_1} \\ \bar{v}_{s_2} \end{bmatrix} = \begin{bmatrix} 1 & 0 & 0 \\ 0 & 1 & 0 \end{bmatrix} \otimes \begin{bmatrix} \bar{x}_1 \\ \bar{x}_2 \\ \bar{x}_3 \end{bmatrix} \oplus \begin{bmatrix} 1 & 0 \\ 0 & 1 \end{bmatrix} \otimes \begin{bmatrix} 0 \\ \bar{v}_{c_1} \end{bmatrix} \tag{5.1}$$

$$\begin{aligned} \bar{v}_{s_1} &= \bar{x}_1 &\rightarrow\ v_{s_1} &= x_1 \\ \bar{v}_{s_2} &= \bar{x}_2 \oplus \bar{v}_{c_1} &\rightarrow\ v_{s_2} &= x_2 \otimes v_{c_1} \end{aligned}$$

which is read as follows: Job 1 can start iff condition x_1 is satisfied; Job 2 can start iff condition x_2 is satisfied and job 1 is complete.

In concordance with Figure 2, a *Move* instruction can be depicted as shown in Figure 4. With the initial and terminal states (monitored), and by employing the formal

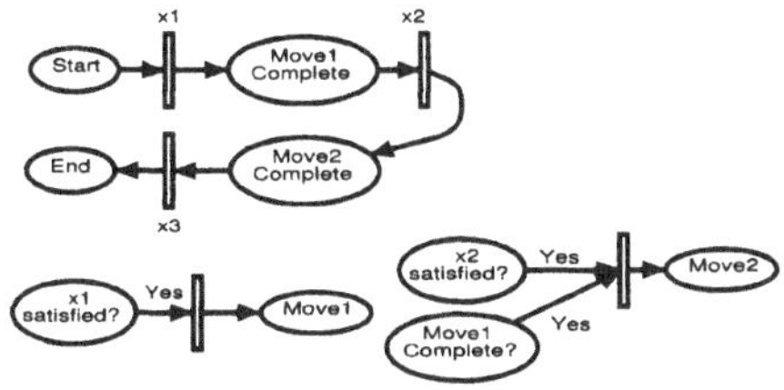

Figure 3. Petri Net representation of a simple move task. Circles/ellipses represent the places and rectangles are the transitions.

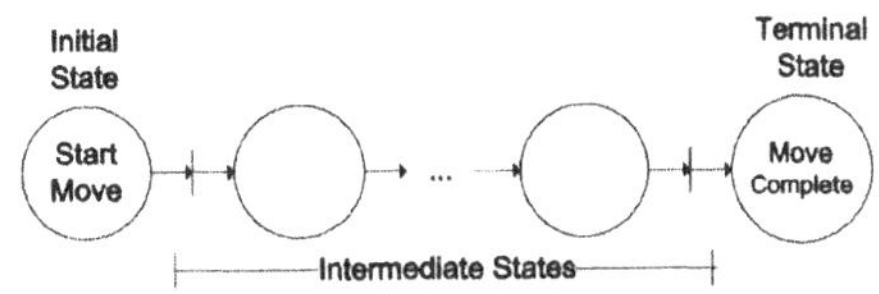

Figure 4. Petri Net representation of a general move instruction.

framework discussed above, the current status of a job or an action can be determined since the aggregated virtual sensor readings, V, are available.

When certain conditions are satisfied and a job can be executed, the supervisor sends a command to the control level (see Figure 5). Once a job is complete, the transition-function specified controller will pass a token, through transmission buffers, T, to the higher level supervisory controller signifying completion of a particular action. Therefore, the supervisory controller *sees* only the tokens providing transparency of the lower level.

Condition, x, quantifies the current state of the system in terms of the availability of resources and the completion of jobs via sensor utilization. Since the state of the system changes with time, x is thus dynamic. Depending on the virtual sensor readings, any one of the vector of conditions can be satisfied which in turn will trigger the start of a job or the release of a resource.

By manipulating the matrix entities in Equation (5.1), the sequence of the jobs and its dependents can be altered without the need to re-structure a program for the execution of a separate task. Viewing each job as a task primitive, the decision to opt for a completely different task is possible simply by choosing the required primitive from a predefined library after task decomposition.

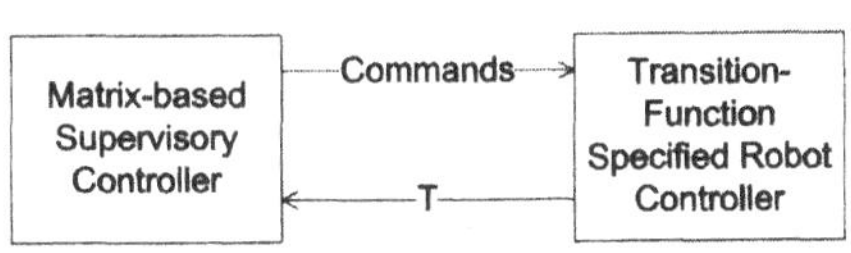

Figure 5. Matrix-based Supervisory Controller with the Transition-Function Specified Controllers. T is the transmission buffer.

6 Conclusion

Robotic tasks in general require a great deal of robustness since the system might encounter some unforeseen problem preventing the task from completion. Information regarding the current and terminal state of a primitive behaviour is essential to the supervisory controller for the execution of a task. By employing the transition-function based approach at the lower level, this necessity is realized.

This framework adopts a transition function-based formalism which introduces rigor into the design and implementation of a single or multi-agent system. By decomposing a large system into modular components, the design and implementation solely by providing the relevant code for the specified functions would suffice for an operational system. Through their public interfaces, these components provide only the data that is necessary for the computation of the control of an agent which is utilized by the transition functions resident in the control subsystem to compute the next state of this subsystem.

Bibliography

S. Boghan, F. L. Lewis, Z. Kovacic, A. Grel, and M. Stajdohar. An implementation of the matrix-based supervisory controller of flexible manufacturing systems. In *IEEE Trans. On Control Systems Technology*, volume 10 of *5*, pages 709–716, September 2002.

J-Y. Fourquet, V. Padois, P. Chiron, and A. Mauratille. Reactive behavior and dynamic sequencing for nonholonomic mobile manipulators. In *IEEE Int. Conference on Computational Intelligence, Robotics and Autonomous Systems*, 13-16 December 2005.

N. W. Koh, M. H. Ang Jr, and S. Y. Lim. Implementation of a matrix-based discrete event controller for robotic tasks. In *IEEE Asian Conf. for Industrial Automation and Robotics*, 11-13 May 2005a.

N. W. Koh, M. H. Ang Jr, and S. Y. Lim. Robotic tasks employing an improved matrix-based discrete event controller. In *Int. Symp. on Collaborative Research in Applied Science*, 7-9 October 2005b.

N. W. Koh, M. H. Ang Jr, and S. Y. Lim. A theoretical insight to the improved matrix-based supervisory controller. In *IEEE Int. Conf. on Computational Intelligence, Robotics and Autonomous Systems*, 13-16 December 2005c.

Jr J. Mireles and F. L. Lewis. Intelligent material handling: Development and implementation of a matrix-based discrete-event controller. In *IEEE Trans. on Industrial Electronics*, volume 48 of *6*, pages 1087–1097, December 2001.

Jr O. Morandin and Edilson R.R. Kato. Virtual petri nets as a modular modeling method for planning and control tasks of fms. In *Int. J. Computer Integrated Manufacturing*, volume 18, pages 100–106. Federal University of Sao Carlos, Taylor and Francis Group, March-May 2005.

T. Murata. Petri nets: Properties, analysis and applications. In *Proceedings of IEEE*, volume 77, pages 542–580, April 1989.

D. A. Tacconi and F. L. Lewis. A new matrix model for discrete event systems: Application to simulation. In *IEEE Control Systems*, volume 17 of *5*, pages 62–71, October 1997.

C. Zielinski. Robot programming methods. DSc thesis, Politechnika Warszawska, 1995.

C. Zielinski. Formal approach to the design of robot programming frameworks: The behavioural control case. In *Bulletin of the Polish Academy of Sciences*, volume 53, pages 1–11, 2005.

Control Architecture for Sensor-Based Two-Handed Manipulation*

Wojciech Szynkiewicz†, Cezary Zieliński†, Witold Czajewski‡ and Tomasz Winiarski†

† Warsaw University of Technology (WUT), Faculty of Electronics and Information Technology, Institute of Control and Computation Engineering, ul. Nowowiejska 15/19, 00–665 Warsaw, Poland, e-mail: [W.Szynkiewicz,C.Zielinski,T.Winiarski]@ia.pw.edu.pl

‡ WUT, Faculty of Electrical Engineering, Institute of Control and Industrial Electronics, ul. Koszykowa 75, 00-662 Warsaw, Poland, e-mail: W.Czajewski@isep.pw.edu.pl

Abstract. In this paper we propose a sensor-based control architecture for two-handed robotic manipulation in a partially structured environment. We deal with two major aspects of sensor-based manipulation: visual localisation and recognition of objects and utilization of force and torque measurements. We also take into account that service robots should be endowed with reasoning capabilities about motions that have to executed and that high level operations must be translated into trajectories to be executed by the manipulators. As a benchmark task we chose the manipulation of Rubik's cube.

1 Introduction

The development of service/personal robots functioning in a human-oriented environment is a major scientific and technological challenge. One of the challenging problems for service robots is grasping and manipulating objects in a semi-structured environment, where object properties are not known *a priori* and sensing is prone to errors. The resulting uncertainty in the relationship between the object and robotic hand makes it difficult to control contact forces and establish a successful grasp or accurately position the object. Thus sensors are essential for dexterous manipulation in a semi-structured or unstructured domains.

In our everyday lives we perform many operations in which our two hands cooperate to move, rotate, and manipulate diverse objects. Our research is focused on understanding the nature of two handed cooperative manipulation and enhancing the manipulative skills of the dual-arm service robot. Two cooperative hands, if properly utilized, are capable of grasping and manipulating a much larger class of objects, including long, flexible, irregularly shaped, or complex objects (e.g., Rubik's cube). While many solutions for single-handed manipulation can be easily adopted, the whole potential of two cooperative

*This work was supported by Polish Ministry of Science and Information Technology grant: 4 T11A 003 25.

hands cannot be fully utilized without a deeper understanding of their unique characteristics. Many studies have been devoted to single-handed manipulation, for overview see (Okamura et al., 2000). However, two-handed manipulation has attracted less attention, especially in an uncalibrated workspace. There are a few works on a two-handed robotic manipulation (see for example (Zöllner et al., 2004; Zhang et al., 1999)), but the literature pertaining to the analysis of bimanual operations performed by humans is quite extensive, especially in the field of a Human-Computer Interaction e.g. (Guiard, 1987; Hinckley et al., 1998). In general, two-handed manipulation can be classified into uncoordinated and coordinated tasks (Guiard, 1987). The latter can be further subdivided into symmetric and asymmetric ones. In this paper we are focused on the symmetric coordinated tasks, and we propose the solution of the Rubik's cube puzzle as an example of such a task. To implement two-handed manipulation utilizing vision, force and tactile sensing, the `MRROC++` robot programming framework is used (Zieliński et al., 2001; Zieliński, 2002).

2 Sensor-Based Two-Handed Manipulation

In this particular case we are interested in a symmetric coordinated manipulation in which both hands are manipulating the same object, thus creating a closed kinematic chain (Szynkiewicz, 2003). The task that the system has to execute can be divided int two phases: acquiring the cube and manipulating it. For the execution of the first phase visual servoing (Kragic and Christensen, 2003) combined with stiffness control are utilized, while in the second phase Task Frame Formalism (TFF) (Mason, 1981; Khatib, 1987) (i.e., selection of directions for position or generalized force control) is used. Efficient realization of TFF requires the development of sensor-based manipulation primitives which robustly perform task-relevant functions and can be easily composed into sequences or complex nets, which are solutions to more complex manipulation tasks. The manipulation primitives constitute a unified programming interface between the controller and high-level task planner. To describe these primitives we use TFF, which is easily implemented within the `MRROC++` programming framework.

The task of solving Rubik's cube needs several sensor-based manipulation primitives such as:

- approaching the cube while avoiding collisions by using visual information,
- grasping the cube using force/torque measurements for stiffness control (at a later stage we shall include eye-in-hand and tactile sensors mounted in the jaws to improve grasping capabilities),
- sensorless motions, to present the cube to the global camera for identifying the cube's initial state,
- turning the faces of the cube while avoiding jamming using information from force/torque sensor (later also tactile sensors) for implementing TFF.

Figure 1 presents the geometrical structure of the two-arm manipulation system, and coordinate frames F_i attached to the appropriate components of the system together with the distribution of sensors. However, the coordinate frames attached to sensors are not shown for reasons of clarity. Homogenous transformation matrices $T_{i,j} \in \mathbb{SE}(3)$ (where $\mathbb{SE}(3)$ is a special Euclidean group) are mappings between the appropriate coordinate

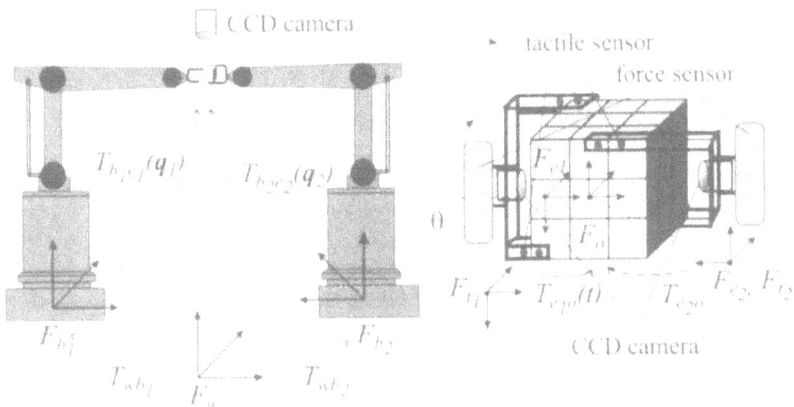

Figure 1. Two-handed robotic system.

frames. To describe the task for each hand two task coordinate frames $F_{t_i}, i = 1, 2$ are introduced as shown in Figure 1. These frames are used to plan manipulation primitives such as approach trajectories for both hands, grasp an regrasp operations, and hand movements to turn the cube faces.

Given a view of the scene, the robot should be able to recognize the cube and localize it in the robot workspace. As a result of the visual localisation the position and orientation of the coordinate frame F_o attached to the cube with respect to (w.r.t) world coordinate frame F_w is calculated as described in section 3.

From the high-level task planning system, i.e. Rubik's Cube solver, a sequence of the twists of the faces is obtained. We define any 90, 180, and 270 degree twist of a face as one move. We never twist the same face twice in a row, because the same result can be obtained with a single twist of the face. Hence we have 18 different possible moves, three for each of the six cube faces. To solve Rubik's Cube we use Iterative-Deepening-A* (IDA*) algorithm (Russell and Norvig, 1995) with heuristics based on pattern databases (Korf, 1997). The task planning system contained in the coordinating process transforms the solution obtained from Rubik's cube solver into a proper sequence of manipulation primitives. In the `MRROC++` framework these primitives are implemented as motion generators, which are used by the `Move` instructions (Zieliński, 2001).

3 Visual Identification of the Rubik's Cube

The process of visual identification of the cube consists of three tasks: 3-D localisation, visual guidance, configuration identification (Czajewski and Staniak, 2004). The 3-D localisation of the cube is based on a projective transformation that maps points from one plane to another. If an object of interest can be represented by n planar points X_i in homogeneous coordinates, $i = 1, \ldots, n$, the relationship between the points X_i and points x_i in an image of the same object is given by:

$$x_i = PX_i$$

where P is a matrix representing the homography. Given at least four correspondences, the matrix P can be computed. Since the matrix P is predetermined in the camera calibration phase, we can now invert the process and solve the above equation for Xi. In our case one plane is one of the cube's faces and the other plane is the camera's sensor. The resulting transformation lets us calculate the 3-D position and orientation of the cube with respect to the camera (and the gripper – with hand-eye coordination). However, in order to determine the transformation precisely we have to accurately find at least 4 known planar points in the image. This is the most difficult part of the entire process. In the first phase of image processing, we are looking for a number of adjacent quadrilateral and convex regions of almost homogeneous colour. The search for 7 particular colours of interest (one for each side of the cube and black for grooves separating the tiles) is done in the HSV colour space. It must be preceded, however, by a precise camera colour calibration. Naturally, the image around the cube is also classified by the colour matching algorithm, but it is very unlikely that random spots will form groups of convex quadrilaterals. When the cube's tiles are recognized, they must be assigned to respective sides of the cube and their relative position on the most distinctive side must be determined (otherwise the 3-D localisation of the cube is impossible). This task is relatively easy if all of the tiles are properly identified, however in most cases some of the tiles are missing and the cube identification algorithm is not always successful. In the final step, the 3-D transformation is calculated based on the biggest planar polygon found on the cube (see Fig. 2). The precision of position and orientation estimation depends primarily on the camera/object distance, but also on the image quality and the accuracy of the tiles identification. For close distances the positional uncertainty is roughly 1%. The angular uncertainty varies form single to several degrees. The second task – visually guiding the gripper – is based on exactly the same principle. The main difference is that the in-gripper camera is positioned almost perpendicularly to the cube, hence only one side of the cube is visible and the identification process is much easier. In the close range (10-15cm) the position and orientation estimation precision is 1mm and up to 5 degrees respectively (usually less). In the last task of the entire process, the cube's configuration is identified so that the initial conditions for the cube solving algorithm are known. In this stage every tile of the cube is recognized, both in terms of its colour and location on the cube. The same colour and shape matching algorithm is used, but this time each face of the cube is presented to the camera sequentially by either of the two manipulators. This makes the identification of tiles fairly easy.

4 Control Architecture

The structure of the controller produced by using the MRROC++ framework for the purpose of executing the two-handed manipulation task is presented in Figure 3. The user's task is coordinated by a single process called the Master Process MP. It is assumed that in general there can be any number of effectors in the system (e.g.: manipulators), but in our case we have just two. From the point of view of the executed task MP is the coordinator

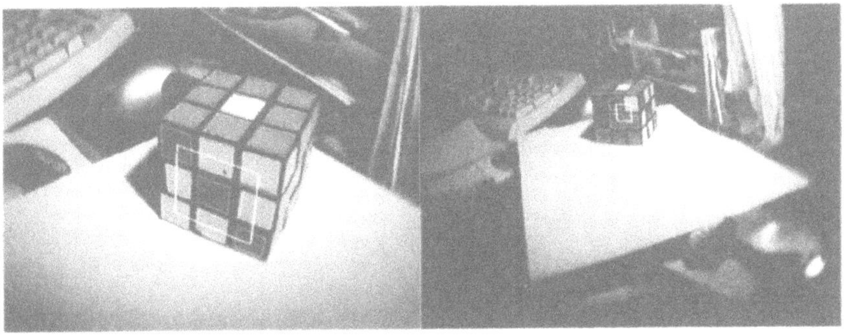

Figure 2. The Rubik's Cube: all visible tiles properly identified (left); few properly identified tiles (right) – the most distinctive planar polygon is marked bold white.

of all effectors present in the system. It is responsible for trajectory generation in multi-effector systems, where the effectors cooperate tightly – as is the case in the presented system. In our case the MP is responsible both for producing the plan of the motions of the faces of the cube and subsequently the trajectory generation for both manipulators. This trajectory can be treated as a crude reference trajectory for both arms. At a later stage this trajectory is modified by taking into account the force readings.

Each effector has two processes controlling it: Effector Control Process ECP and Effector Driver Process EDP. The first one is responsible for the execution of the user's task dedicated to this effector (in our case the task is defined by the MP– it is defined by the reference trajectory that is to be executed by the manipulator), and the other for direct control of this effector. The EDP is responsible for direct and inverse kinematics computations as well as for both position and force servocontrol. The EDP has been divided into several threads: EDP_MASTER (communication with the ECP including interpretation of its commands), EDP_TRANS (designation of force and position control directions), EDP_SERVO (motor position servocontrol) and EDP_FORCE (force measurements).

Figure 4 presents the structure of the position-force controller. The figure contains two parts: the macrostep execution part and the step execution part. The macrostep is a pose increment as obtained from the motion generator, which computes it on the basis of the information obtained by the MP or ECP from the VSP serving the active camera (either the overhead or one of the in-gripper cameras). The ECP sends the `SET ARM POSE_FORCE_TORQUE` command (M) to the EDP_MASTER thread of the EDP. This command contains the definition of the macrostep that must be executed. The definition specifies: `pos_xyz_rot_xyz` and `force_xyz_torque_xyz` vectors, number of servo sampling periods (motion steps) defining the duration of the macrostep, `selection_vector` and `local_reference_frame`. The `local_reference_frame` is represented by the ${}^{G}_{L}\mathbf{T}$ matrix relating the local to the global reference frame. In the following the left superscript denotes the reference frame (L – local, G – global) of the associated vector quantity. The right subscript in parenthesis is the control step number, counted from the beginning of the macrostep execution.

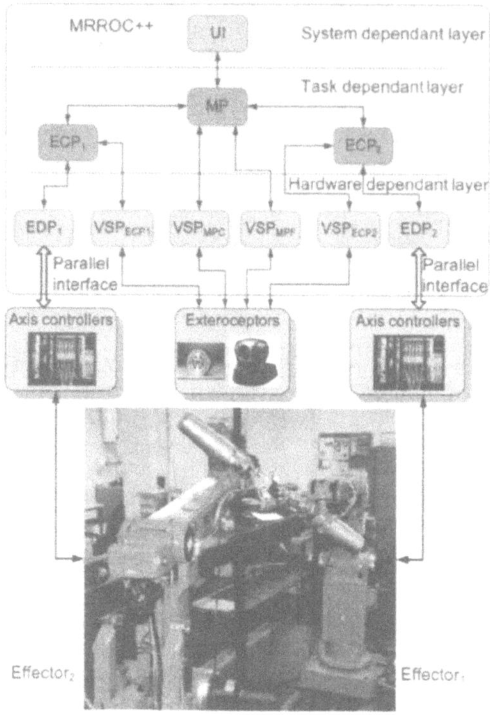

Figure 3. MRROC++ based controller for two-arm system.

The `pos_xyz_rot_xyz` vector defines $\Delta\,^{L}\mathbf{P}_{p}$, the end-effector pose increment for the current macrostep. This increment is subdivided by the StGen block into pose increments for each step within the macrostep $\Delta\,^{L}\mathbf{P}_{p(k)}$:

$$\Delta\,^{L}\mathbf{P}_{p(k)} = \frac{\Delta\,^{L}\mathbf{P}_{p}}{n} \tag{4.1}$$

where n is the number of steps in the macrostep, and k is the current step number.

In each step the force/torque error vector $^{L}\mathbf{F}_{e(k)}$ is the difference between the desired force/torque vector $^{L}\mathbf{F}_{d}$ (defined by `force_xyz_torque_xyz`) and the measured force/torque vector $^{L}\mathbf{F}_{c(k)}$:

$$^{L}\mathbf{F}_{e(k)} = {}^{L}\mathbf{F}_{d} - {}^{L}\mathbf{F}_{c(k)} \tag{4.2}$$

The vector $^{L}\Delta\mathbf{P}_{f(k)}$ is obtained from the force/torque control law:

$$\Delta\,^{L}\mathbf{P}_{f(k)} = K\,^{L}\mathbf{F}_{e(k)} \tag{4.3}$$

where K is the viscosity.

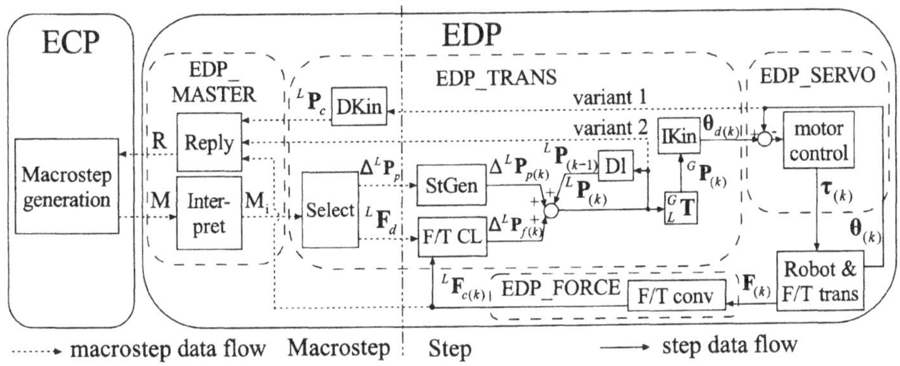

Figure 4. Structure of the position-force controller.

The end-effector pose setpoint for the current step of the macrostep execution is described by the vector ${}^{L}\mathbf{P}_{(k)}$:

$$ {}^{L}\mathbf{P}_{(k)} = {}^{L}\mathbf{P}_{(k-1)} + \Delta\, {}^{L}\mathbf{P}_{p(k)} + \Delta\, {}^{L}\mathbf{P}_{f(k)} \tag{4.4} $$

It is transformed into the global reference frame:

$$ {}^{G}\mathbf{P}_{(k)} = {}^{G}_{L}\mathbf{T}\, {}^{L}\mathbf{P}_{(k)} \tag{4.5} $$

The new motor position vector $\theta_{d(k)}$ is the solution of the inverse kinematic problem for ${}^{G}\mathbf{P}_{(k)}$. The vector $\theta_{d(k)}$ is a part of the command sent to the motor controllers implemented in the EDP_SERVO. Motor controllers force the motors to move in the direction of the setpoint by delivering the vector $\tau_{(k)}$ consisting of the values of PWM duty cycles. The end-effector pose that is delivered to the ECP, for the purpose of high-level feedback, can originate either directly with the encoder measurements that are subsequently transformed by the procedure solving the direct kinematics problem (variant 1) or it is simply the rewritten set value for step n (variant 2). The EDP_FORCE thread transforms the raw force/torque measurements vector $\mathbf{F}_{(k)}$ into the vector ${}^{L}\mathbf{F}_{c(k)}$.

5 Conclusions

This is a report of the currently ongoing research aspiring to propose a general procedure for designing service robots. Unlike their industrial counterparts, service robots must perform their operations relying heavily on sensing, reacting and planning, rather than precision of motions. Thus, our research is focused on developing diverse sensing capabilities, especially visual and tactile (i.e., force), that dominate in humans, who operate in the same environments as service robots. Furthermore we have concentrated on endowing the robot with the ability of quick reaction to sudden changes in the environment (thus low level force control), and the ability to reason about objects in the environment (for the time being this is mimicked by producing the plan of motions of Rubik's cube faces – this task requires the translation of symbolic motions of Rubik's cube faces into trajectories of motions of the two manipulators, including grasping and regrasping operations). Finally, we have developed methods for integrating all of the above components into a consistent and modular system with a clear structure (for that purpose we use as an implementation tool the MRROC++ robot programming framework, which in turn had been formally specified.This paper proposes the general system structure and shows how force control and visual processing is being fused together.

Bibliography

W. Czajewski and M. Staniak. Visual identification of the rubik's cube in human environment. Przegląd Elektrotechniczny, *(4):311–314, 2004.*

Y. Guiard. Assymetric division of labor in skilled bimanual action: The kinematic chain as a model. Journal of Motor Behavior, *19(4):486–517, 1987.*

K. Hinckley, R. Pausch D. Proffitt, and N. Kassell. Two-handed virtual manipulation. ACM Transactions on Computer-Human Interaction, *5(3):260–302, 1998.*

O. Khatib. A unified approach for motion and force control of robot manipulators: The operational space formulation. Int. J. of Robotics and Automation, *3(1):43–53, 1987.*

R. Korf. Finding optimal solutions to rubik's cube using pattern databases. In Proc. of the National Conf. on Artificial Intelligence (AAAI-97), *pages 700–705. July 1997.*

D. Kragic and H. I. Christensen. Robust visual servoing. Int. J. of Robotics Research, *22(10-11):923–939, 2003.*

M. Mason. Compliance and force control for computer controlled manipulators. IEEE Transactions on Systems, Man, and Cybernetics, *(11):418–432, 1981.*

A.M. Okamura, N. Smaby, and M.R. Cutkosky. An overview of dexterous manipulation. In Proc. of the IEEE Int. Conf. on Robotics and Automation, *pages 255–262, 2000.*

S. Russell and P. Norvig. Artificial Intelligence: A Modern Approach. *Prentice Hall, Upper Saddle River, N.J., 1995.*

W. Szynkiewicz. Motion planning for multi-robot systems with closed kinematic chains. In Proc. of the 9th IEEE Int. Conf. on Methods and Models in Automation and Robotics, Międzyzdroje, *pages 779–786, Aug. 25-28 2003.*

J. Zhang, Y. von Collani, and A. Knoll. Interactive assembly by a two-arm-robot agent. Robotics and Autonomous Systems, *29(1):91–100, 1999.*

C. Zieliński. By how much should a general purpose programming language be extended to become a multi-robot system programming language? Advanced Robotics, *15(1): 71–95, 2001.*

C. Zieliński. Motion generators in MRROC++ based robot controller. In 14th CISM–IFToMM Symposium on Robotics, Ro.Man.Sy'02, Udine, Italy, CISM Courses and Lectures no.438, *pages 299–306. Springer, July 1–4 2002.*

C. Zieliński, W. Szynkiewicz, K. Mianowski, and K. Nazarczuk. Mechatronic design of open-structure multi-robot controllers. Mechatronics, *11(8):987–1000, Nov 2001.*

R. Zöllner, T. Asfour, and R. Dillmann. Programing by demonstration: Dual-arm manipulation tasks for humanoid tobots. In Proc. of the IEEE/RSJ Int. Conf. on Intelligent Robots and Systems (IROS2004), Sendai, Japan, *pages 479–484, Sep. 28–Oct. 2 2004.*

Chapter V

Humanoids

The mechanical improvements of the anthropomorphic flutist robot WF-4RII to increase the sound clarity and to enhance the interactivity with humans

Jorge Solis [1], Kei Suefuji [2], Koichi Taniguchi [2], and Atsuo Takanishi [1,3]

[1] Waseda University, Department of Mechanical Engineering, Tokyo, Japan
[2] Waseda University, Graduate School of Science and Engineering, Tokyo, Japan
[3] Waseda University, Humanoid Robotics Institute, Tokyo, Japan
e-mail: solis@kurenai.waseda.jp, takanisi@waseda.jp

Abstract. The development of the anthropomorphic flutist robot, at Waseda University, has focused on imitating the human flute playing by mechanically reproducing the organs involved during such activity. Our research aims in understanding several aspects of the human flute playing: clarifying the human motor control while performing skillful activities, enabling robots to express ideas and feelings in musical terms and proposing new ways of interaction between the human and the robot. In this paper the new version of the flutist robot, the Waseda Flutist Robot No. 4 Refined II, is presented. The improvements of the mechanical system in order to improve the sound clarity of the robot's performance and to enhance the interaction with humans are described. A set of experiments were carried out to verify the effectiveness of the mechanical components. As a result, the robot is able of performing a musical score with more clarity and furthermore, the robot can interact with humans more natural using the vision system.

1 Introduction

The relation between the human and the music has a long history dating from the antiquity, during which poetry, dancing and music were inseparable and constitute an important mean of communication of everyday life. During the golden era of automata, 17th and 18th centuries, the music also served as a tool for understanding how the human organism is able of performing high skilful activities. In particular, some researchers were interested in studying how the human is able of performing musical instruments, such as the flute. For that reason, Vaucanson developed a mechanical flutist player in order to understand the breathing mechanism. Vaucanson presented "The Flute Player" to the Academy of Science in 1738 (Doyon, 1966). For this occasion he wrote a lengthy report carefully describing how his flutist can play exactly like an alive person (Figure 1).

*A part of this research was done at the Humanoid Robotics Institute (HRI), Waseda University. This research was supported (in part) by a Gifu-in-Aid for the WABOT-HOUSE Project by Gifu Prefecture. A part of this research has been also supported by the Japan Society for the Promotion of Science (JSPS). We would like to express our thanks to Mr. Kunimitsu Wakamatsu for his valuable advices.

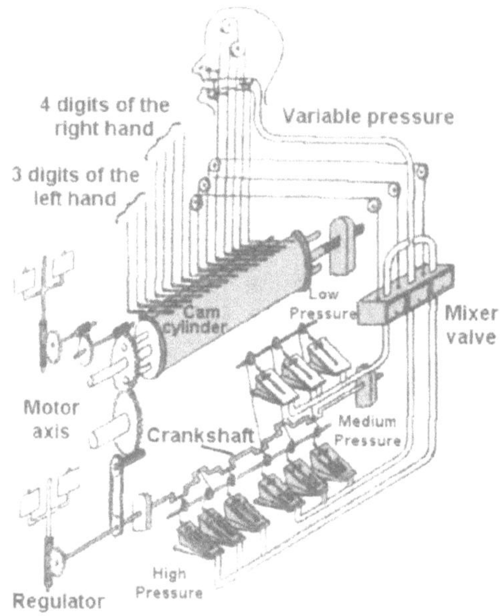

Figure 1. Diagram of the Vaucanson's flute-player mechanism (Doyon, 1966).

At the present time, the advances in computer technology, electronics, artificial intelligence, speech synthesis and understanding and vision processing have led to breakthrough in robotic technology. In fact, nowadays, several researchers have developed the humanoid robots which roughly emulated the physical dynamics and mechanical dexterity of the human body as a mean for understanding better the human motor control (Fukui et al., 2005; Ogura et al., 2005; etc.).

In 1984, the first attempt of developing a musical robot was done by Waseda University with the WABOT-2 which was able of playing a concert organ. Then, in 1985, the WASUBOT built also by Waseda, could read a musical score and play a repertoire of 16 tunes on a keyboard instrument. Kato et al. (1973) argued that the artistic activity such as playing a keyboard instrument would require human-like intelligence and dexterity. In 1989, the University of Electro-Communication developed the Musician Robot (MUBOT), which can automatically play a violin or cello (Kajitani, 1989). The MUBOT was developed with the premise that music should be played by robot without remodeling the musical instruments at all. The MUBOT was developed for the following purposes: a study in robot engineering, as an entertainment robot and as a performance simulator for the study of musical engineering. Sobh et al. (2003) has also been carried out research on developing robot musicians; however, their approach is based on remodeling the musical instrument to open new domains of research for musical engineering. The robot musician band is composed by the following instruments: drum, percussion, electric guitar, violin, cello, and several kinds of guitars which are automatically controlled to perform a score.

Alford et al. (1999) has developed a musician robot for playing the Theremin that is an electronic musical instrument played without physical contact.

Since 1990, the research of the Waseda flutist robot has been focused on understanding the human flute playing mechanism by reproducing the organs involved in such a task. The human flute playing mechanism is characterized for a highly skillful interaction and coordination of several organs that enables humans to express ideas, feelings, etc. Thus, our research also aims in understanding the human motor control, proposing different applications where the robot can interact with humans (playing a performance in a duet (Okuma et al., 2003) and teaching basic skills to beginners (Solis et al., 2004). In this paper, we are presenting the newest Waseda Flutist Robot No.4 Refined II which has improved the sound clarity of the musical performance by re-designing the lung system to reduce the mechanical noise, the flute mounting systems to increase the stiffness of the flute holding, and finally the eye system was added in order to enhance the interactivity between the robot and human. A set of experiments are presented and discussed to verify the improvements.

2 The Waseda Flutist Robot No.4 Refined II

2.1 Overview

In this year, the Waseda Flutist Robot No. 4 Refined II (WF-4RII) is presented where two main improvements were done: the clarity of the sound produced by the robot (improvements of the lung and flute mounting systems) and the enhancement of the interaction with humans (addition of the eyes system). As a result, this new version has a total of 43-DOFs (Figure 2): lips (5-DOFs), the neck (4 DOFs), the fingers (12 DOFs), the arms (14-DOFs), the lung and air valve (3-DOFs), the throat (1-DOF), the tongue (1-DOF), and the eyes (3-DOFs). Each of the organs has been designed as similar as possible to the human size.

2.2 Mechanical System

The lips, neck, arms, tongue, neck, finger, and throat systems of the WF-4RII had not presented any modification respect to the previous version WF-4R (Solis et al., 2005). The lips system has a total of 5-DOFs to control with high accuracy the width, thickness, angle and length of the air beam. The upper lip has a total of 3-DOFs (inside, outside, forward) while the lower lip has 2-DOFs (forward and inside). Each of the degree-of-freedom of the lip system is actuated by a unit composed by a DC motor and a ball screw connected by a timing belt. The neck system has a total of 4-DOFs (upper and lower pitch, yaw and roll) to expand the range of positioning and attitude control of the lips. A harmonic drive was used to assure the positioning control. The arms system has been implemented similar to the human with a total of 7-DOFs: the upper arm has 4-DOFs (serial mechanism), the forearm has 2-DOFs (parallel mechanism), and the wrist with 1-DOF. The tonguing system has been implemented inside a case with acrylic plastic, similar to the oral cavity of the human. A rod fixed on the axis of the rotation pushes the cam follower of the tongue, mounted on the slide table, to convert circular motion into linear motion to imitate the double tonguing mechanism. The fingers system is

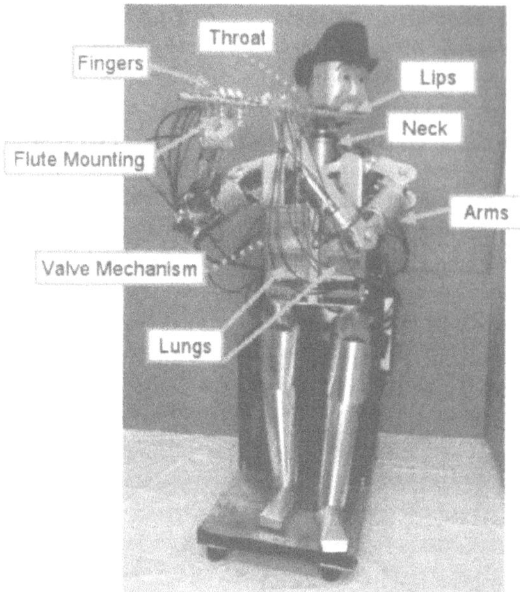

Figure 2. The Anthropomorphic Flutist Robot No. 4 Refined II.

composed by 12-DOFs: left hand with 5-DOFs and the right hand with 7-DOFs. Each finger consists of a link mechanism actuated by DC servomotors with pulley and wire. The throat system is only used to produce vibrations to the air beam by using a coil motor which presses directly a rubber tube where the air stream flows from the lungs to the oral cavity.

The lung and flute mounting systems were modified in order to improve the clarity of the sound produced by the robot. The lung system on the WF-4R was designed using two piston cylinders and actuated using a DC servomotor with a motor belt driving and a ball screw linear guide. This mechanical design produced a high noise during the breathing performance. Therefore, the new lung system was designed using two vane mechanisms (Figure 3a), each of them with 1-DOF, actuated directly by an AC motor to increase the control accuracy and the inhalation speed. Furthermore, an air valve mechanism (with 1-DOF) has been designed to control the opening and closing of the air flow valve of each vane mechanism (Figure 3b). The flute mounting section of WF-4R was composed by a pipe joint which could be freely adjusted but the position reproducibility during the calibration procedure and the stiffness of this mechanism was quite low. Therefore, the new mounting system was designed using a fixed component attached to the flute which locks it with a chunk device located on the palm of the hand (Figure 4). Therefore, the stiffness of the attachment was increased as well as the reproducibility of the attaching positioning.

Finally, the eyes system was developed and it is composed by 3-DOFs corresponding

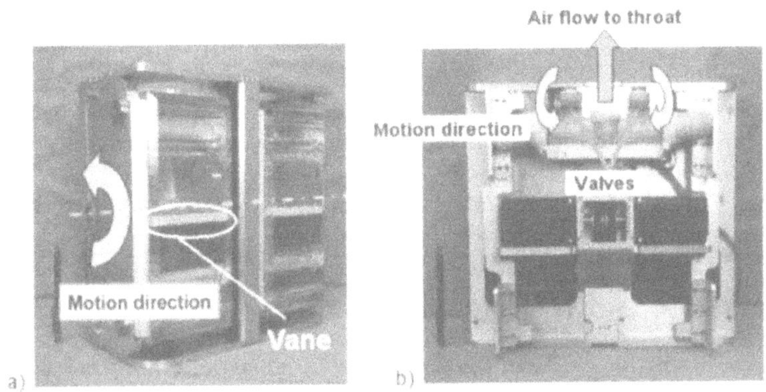

Figure 3. a) New lung's system; b) Valve mechanism system.

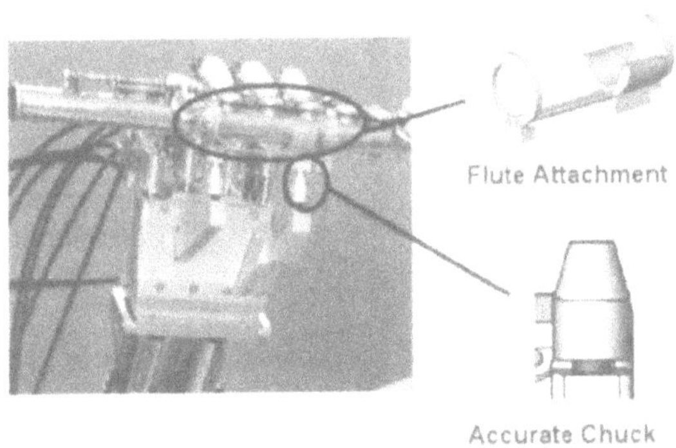

Figure 4. Flute mounting system of WF-4RII.

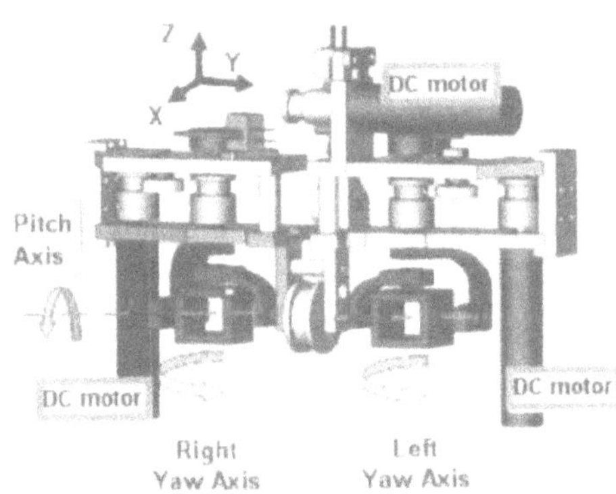

Figure 5. Detail of the eye system of WF-4RII.

to the yaw (horizontal displacement) for each eye and the pitch (vertical displacement) which is common for both eyes (Figure 5). A tendon driven gimbal mechanism was selected to reduce the mechanical size. The range motion of the eye mechanism has been designed nearly similar to the human (from -30° to 30° for both pitch and roll). The hardware of the vision system is composed by a video board (Picolo Pro2 from Eurosys) mounted on a PC (Pentium 4 3.2GHz) with Windows XP. The video board is then connected to the CCD cameras attached to the eyes system; where the frame rate for each camera is 30 frames per second.

3 Experiments and Results

In order to verify if the clarity of the sound was improved, the lung and flute mounting systems were tested. At first, we have measured the mechanical noise produced by the lung system during the breathing process and then, we measure the stiffness of the flute mounting during the flute performance. In the case of the lung system, we have recorded the WF-4R and WF-4RII performing the E5 note two times; where an inhalation phase was programmed between both notes. In Figure 6a, the level of noise (in dB) of both versions is presented. Between 1 *sec* and 2 *sec* of the recording performance, the level of noise of the WF-4RII was reduced considerably (around 6dB) compared with the previous version. Additionally, the effectiveness of the air valve system was tested. Basically, the rate of the air flow which is taken out from both lungs was measured. As a result, the air flow rate was increased to 4.0 $10^{-3}m^3$ in the WF-4RII compared with the 3.5 $10^{-3}m^3$ on the WF-4R. The air flow rate is strongly related to the stability and clarity of the sound.

In the case of the flute mounting system, we measured the force acting on the lips while the flute is held and the keys of the flute were pressed and released with the fingers to simulate a musical performance. As a result, in Figure 6b, the force measured on the new flute mounting system remained stable (without variations during the performance) and lower respect to the previous version. This is because there was no gap between the flute and the lips which increased considerably the stiffness of the flute holding and therefore, a high control positioning accuracy can been achieved to produce the desired note.

Finally, to verify the effectives of the eyes system in order to enhance the interactivity with humans, a face tracking system was developed using the API's provided by the Open Source Computer Vision Library (OpenCV). The main tasks of the vision system are detecting the presence of human faces when approaching to the robot and maintaining visual eye contact with the partner during the performance. The face tracking algorithm was implemented based on the frontal face detection proposed by Viola and Jones (2001) and the continuously adaptive mean-shift algorithm proposed by Bradski (1998). Therefore, the coordinates x and y (which corresponds to the region where the face is located from the tracking algorithm) are transformed from the camera reference (*pixels*) to the motor reference (*pitch* and *yaw*) by taking into account the range of motion of the eye mechanism. Furthermore, to assure a smooth motion trajectory, a linear interpolation was used. Finally; such data is then sent to the PC Controller to update the corresponding command positions of the eyes. We verified the eyes system response by asking a person to move first just along the horizontal direction; and then to move only along the vertical direction. The system response along the pitch (vertical) and yaw (horizontal) directions are shown in Figure 7. The input reference is obtained from the proposed tracking algorithm. The movement of the user along the horizontal direction corresponds to the left part of the dashed line, while the right part belongs to the vertical motion of the person.

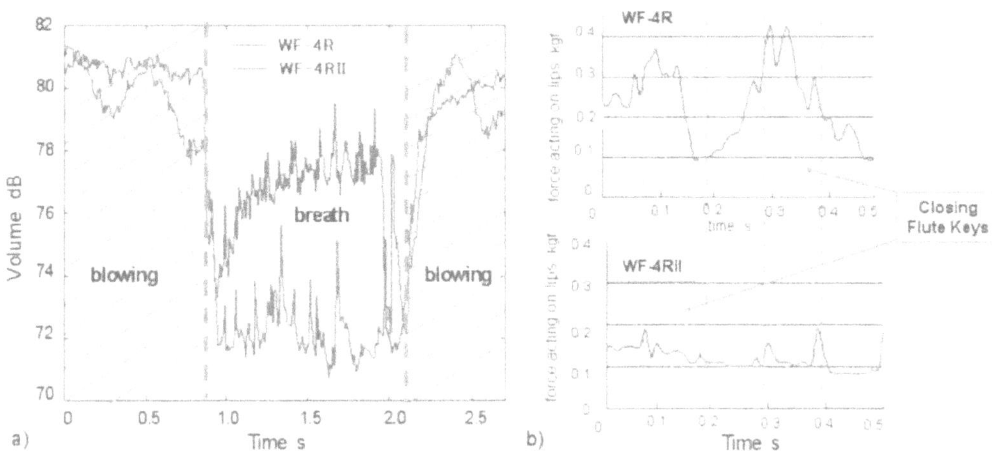

Figure 6. a) Mechanical noise response of the lungs; b)Force acting on the lips while using the flute mounting system.

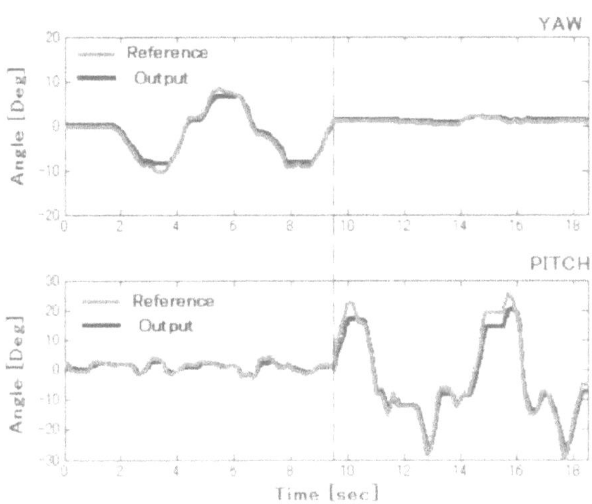

Figure 7. Eye's system response.

4 Conclusions

In this paper; the mechanical improvements, done on the Waseda Flutist Robot No. 4 Refined II, to enhance its performance and the interactivity with humans were presented. As a result; the new lung system reduced the mechanical noise during the breathing process, the valve mechanism increased the air flow mass coming out from the lungs, the new flute mounting increased the stiffness of the flute holding and finally the eyes system responded smoothly while tracking a human face.

Bibliography

A. Alford, S. Northup, K. Kamakura, K.W. Chan, and J. Barile. Music playing robot. In *Proceedings of the 1999 International Conference on Field and Service Robots*, pages 174–178, 1999.

G.T. Bradski. Computer vision face tracking as a component of a perceptual user interface. In *Proceedings of the 1998 Workshop on Applications of Computer Vision*, pages 214–219, 1998.

A. Doyon. *Jacques Vaucanson: mecanicien de genie*. PUF, 1966.

K. Fukui, K. Nishikawa, S. Ikeo, E. Shintaku, K. Takada, H. Takanobu, M. Honda, and A. Takanishi. Development of a talking robot with vocal cords and lips having human-like biological structures. In *Proceedings of the IEEE International Conference on Intelligent Robotic Systems*, pages 2526–2531, 2005.

M. Kajitani. Development of musician robots. *Journal of Robotics and Mechatronics*, 1: 254–255, 1989.

I. Kato, S. Ohteru, H. Kobayashi, K. Shirai, and A. Uchiyama. Information-power machine with senses and limbs. In *Proceedings of the CIS-IFTiMM Symp. Theory and Practice of Robots and Manipulators*, pages 12–24, 1973.

Y. Ogura, T. Kataoka, T. Aikawa, K. Shimomura, H.O. Lim, and A. Takanishi. Evaluation of various walking patterns of biped humanoid robots. In *Proceedings of the IEEE International Conference on Robotics and Automation*, pages 605–610, 2005.

I. Okuma, K. Chida, S. Isoda, Y. Saisu, A. Takanishi, and K. Wakamatsu. Time-scale performance control of new anthropomorphic flutist robot. In *Proceedings of the 21st Annual Conference of the Robotics Society of Japan*, page 3H15, 2003.

T.M. Sobh, B. Wang, and K.W. Coble. Experimental robot musicians. *Journal of Intelligent and Robotic Systems*, 38:197–212, 2003.

J. Solis, M. Bergamasco, S. Isoda, K. Chida, and A. Takanishi. The anthropomorphic flutist robot wf-4 teaching flute playing to beginner students. In *Proceedings of the IEEE International Conference on Robots and Automation*, pages 146–151, 2004.

J. Solis, M. Bergamasco, S. Isoda, K. Chida, and A. Takanishi. The anthropomorphic flutist robot wf-4r: from mechanical to perceptual improvements. In *Proceedings of the IEEE International Conference on Intelligent Systems*, pages 1929–1934, 2005.

P. Viola and M.J. Jones. Rapid object detection using a boosted cascade of simple features. In *Proceedings of the IEEE International Conference on Computer Vision and Pattern Recognition*, pages 511–518, 2001.

Mechanical Design of Emotion Expression Humanoid Robot WE-4RII

Kazuko Itoh [1,2], Hiroyasu Miwa [3,4,5], Massimiliano Zecca [2,4], Hideaki Takanobu [2,5,6], Stefano Roccella [7], Maria Chiara Carrozza [2,7], Paolo Dario [2,7] and Atsuo Takanishi [1,2,4,5,8]

[1] Department of Mechanical Engineering, Waseda University, Tokyo, Japan
[2] RoboCasa, Tokyo, Japan
[3] Digital Human Research Center, National Institute of Advanced Industrial Science and Technology (AIST), Tokyo, Japan
[4] Institute for Biomedical Engineering, ASMeW, Waseda University, Tokyo, Japan
[5] Humanoid Robotics Institute (HRI), Waseda University, Tokyo, Japan
[6] Department of Mechanical Systems Engineering, Kogakuin University, Tokyo, Japan
[7] ARTS Lab, Scuola Superiore Sant'Anna, Pontedera, Italy
[8] Advanced Research Institute for Science and Engineering, Waseda University, Tokyo, Japan
e-mail: itoh@suou.waseda.jp (K. Itoh), takanisi@waseda.jp (A. Takanishi)

Abstract. A Personal Robot is expected to become popular in the future. It is required to be active in joint work and community life with humans. Therefore, we have been developing new mechanisms and functions for a humanoid robot that can express emotions and communicate naturally with humans. In this paper, we present the mechanical design of the Emotion Expression Humanoid Robot WE-4RII, which was developed by integrating the Humanoid Robot Hands RCH-1 into the previous version WE-4R*. The robot has four of the five human senses for detecting external stimuli: visual, tactile, auditory and olfactory, and 59-DOFs for expressing motion and emotions. It is capable of expressing seven basic emotional patterns.

1 Introduction

Robots have become indispensable for human life. Most popular robots are of the industrial kind with various functions, such as assembly and conveyance. However, we hope that personal robots which are active in joint work and community life with humans will become popular in the future. Such personal robots must adapt to partners and the environment and to communicate naturally with humans. Therefore, new mechanisms and functions are developed in order to realize the natural communication by expressing the emotions, behaviors and personality in a human-like manner.

Research is being conducted into communication robots in the field of robotics. Breazeal and

* We would like to thank the Humanoid Robotics Institute (HRI) consortium, General Directorate for Cultural Promotion and Cooperation of the Italian Ministry of Foreign Affairs, ARTS Lab, NTT Docomo, SolidWorks Corp. for their support. In addition, this research was supported by the Ministry of Education, Science, Sports and Culture, Grant-in-Aid for Young Scientists (B) (No. 17700211) and a Grant-in-Aid for the WABOT-HOUSE Project by Gifu Prefecture.

Scassellati (1999) have developed an expressive robotic creature that expresses facial expressions using its eyes, eyelids, eyebrows and mouth. It communicates with humans using visual information from CCD cameras. Kobayashi and Hara (1996) have developed a head robot that uses the fourteen Action Units of Ekman (see Ekman and Friesen, 1978). It can express six basic facial expressions as quickly as human using 24-DOFs with air compressors for actuators. It can also recognize human facial expressions using CCD cameras and reciprocate the same facial expressions back.

We have been developing the humanoid robots from viewpoints of both robotics and psychology in order to realize human-like motion. The Human-like Head Robot WE-3 (Waseda Eye No.3) series have been developed as a mechanical model since 1995 (Miwa et al., 2002). So far, we have achieved coordinated head-eye motion with V.O.R. (Vestibular-Ocular Reflex), depth perception using the angle of convergence between the two eyes, adjustment to the brightness of an object with the eyelids and four sensations, visual, auditory, cutaneous and olfactory sensations. In addition, we produced emotional expressions using not only the face, but also the upper body by the Emotion Expression Humanoid Robot WE-4 (Waseda Eye No.4) series with the waist, 9-DOFs emotion expression humanoids arms (Itoh et al., 2004) and humanoid robot hands RCH-1 (Robo Casa Hand No.1) (Zecca et al., 2006). Moreover, a mental model was developed for humanoid robots based on psychology. We introduced the mental space with three independent parameters, the mood, the second order equations of emotion, the robot personality, the need model, consciousness model, behavior model and memory model (Itoh et al., 2005) into the mental model. In this paper, we describe the mechanical features of the latest Emotion Expression Humanoid Robot WE-4II (Waseda Eye No.4 Refined II).

2 Mechanical Hardware

Figure 1 presents the hardware overview of Emotion Expression Humanoid Robot WE-4RII. The robot has 59 DOFs for expressing motions and emotions as shown in Table 1, and four of the five human senses for detecting external stimuli: visual, auditory, olfactory and tactile. The height and weight of its upper body are 970[mm] and 59.3[kg], respectively. The hand is the Humanoid

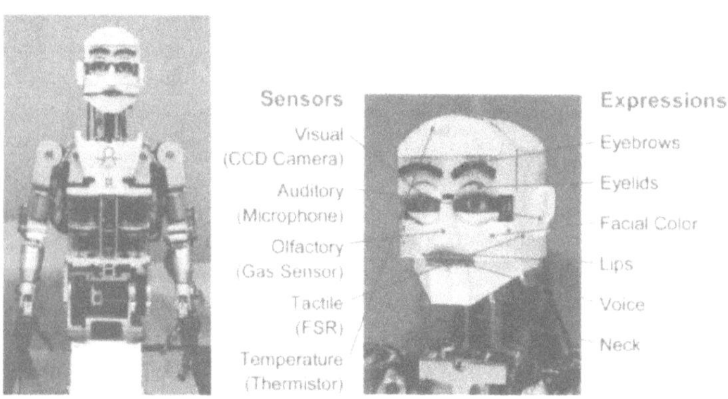

(a) Whole view (b) Head part

Figure 1. Emotion Expression Humanoid Robot WE-4RII.

Table 1. DOFs configuration of WE-4RII.

Part	DOF
Neck	4
Eyes	3
Eyelids	6
Eyebrows	8
Lips	4
Jaw	1
Lungs	1
Waist	2
Arms	18
Hands	12
Total	59

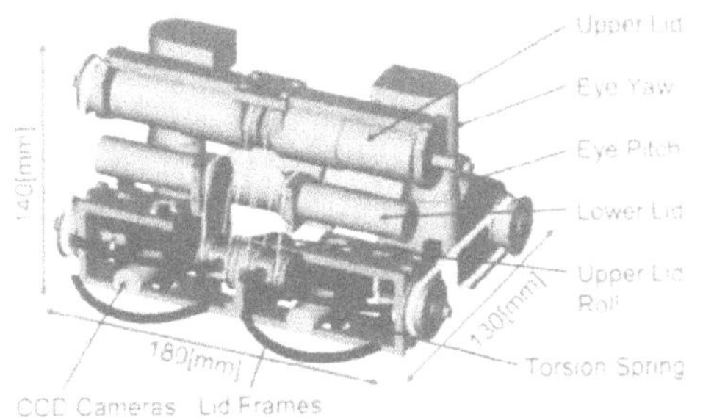

Figure 2. Compact eye unit mechanism.

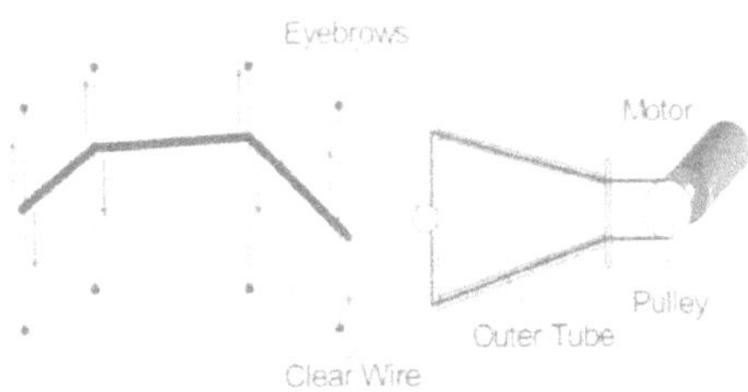

Figure 3. Eyebrows mechanism.

Robot Hand RCH-1 (RoboCasa Hand No.1) which is designed and developed by ARTS Lab, Scuola Superiore Sant'Anna in order to express not only emotional expressions but also active behavior.

2.1 Facial Mechanism

For miniaturization of the head, the Eye Unit was developed as shown in Figure 2. The weight in the Eye Unit was reduced by adopting both a belt driven mechanism and a wire driven mechanism, and the motor positions were optimized. Since, the eyeball pitch axis motion mechanically synchronizes the opening and closing of upper eyelid motion, the coordinated eyeball-eyelids motion is controlled by the hardware much easily.

In this Eye Unit, the eyeballs have 1-DOF for the pitch axis and 2-DOFs for the yaw axis. The pitch axis is driven by a belt driven mechanism with a DC motor and a harmonic drive system. The yaw axes are driven by a tendon driven mechanism with a DC motor and a torsion spring. The maximum angular velocity of the eyeballs is similar to that of a human at 600[deg/s].

The eyelid consists of the upper eyelid and the lower eyelid. The upper eyelid has 1-DOF for opening and closing and 1-DOF for rotating. The lower eyelid has 1-DOF for opening and closing. The opening and closing motion of the upper and lower eyelids is driven by a tendon driven mechanism using a DC motor, while the rotating motion of the upper eyelid is driven by a direct driven mechanism with a DC motor. The maximum angular velocity of opening and closing eyelids is 900 [deg/s]. The robot can blink within 0.3[s], which is as fast as a human can.

In addition, WE-4RII expresses its facial expression using its eyebrows, eyelids, lips, jaw, facial color and voice. The eyebrows consist of flexible sponges, and each eyebrow has 4-DOFs as shown in Figure 3. And, spindle-shaped springs were used for the lips. Their shape changes by pulling in 4 directions, and they are opened and closed by the jaw motion. The facial color was realized by using red and blue EL (Electro Luminescence) sheets, which were applied to the cheeks. For the voice system, the robot has a small speaker set in the jaw. The voice is a synthetic voice made by LaLaVoice 2001 (TOSHIBA Corporation).

2.2 Neck, Waist and Lungs

Figures 4, 5 and 6 show the neck, waist and lungs mechanism, respectively. The neck has 4-DOFs, which are the upper pitch, the lower pitch, the roll and the yaw axes. The lower pitch is

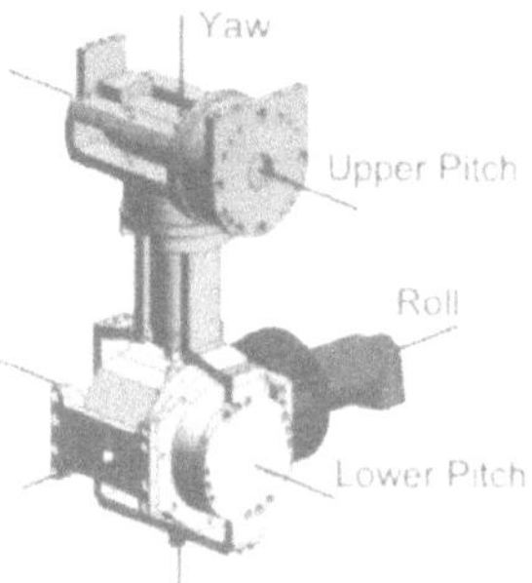

Figure 4. Neck mechanism.

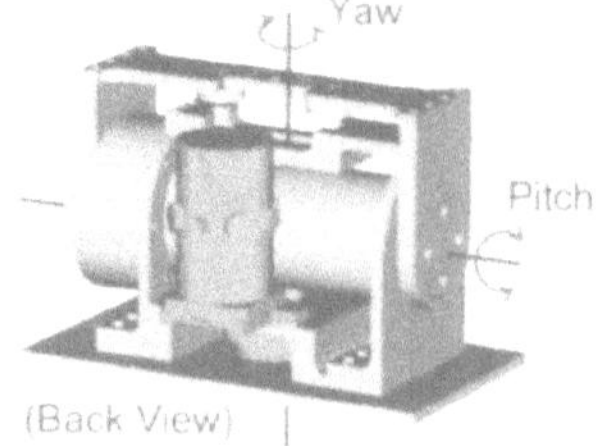

Figure 5. Waist mechanism.

Figure 6. Lungs mechanism.

driven by a direct driven mechanism with an AC motor and a harmonic drive system. Other axes are driven by a direct driven mechanism with a harmonic drive systems' DC motor. The maximum angular velocity of each axis is similar to that of a human at 160[deg/s].

The waist has 2-DOFs, the pitch and yaw axes. The pitch axis is driven by a direct driven mechanism with a harmonic drive systems' DC motor, and the yaw axis is driven by a belt driven mechanism with a DC motor and a harmonic drive system. The robot pursues a target using not only coordinated head-eye motion, but also coordinated waist-head-eye motion with V.O.R.

In addition, the robot has lungs in its chest. The lungs consist of 2 sets of a cylinder and a piston, a ball screw and a DC motor. WE-4RII can display breathing motion, which expresses more emotional motion in addition to breathing air for the olfactory sensation.

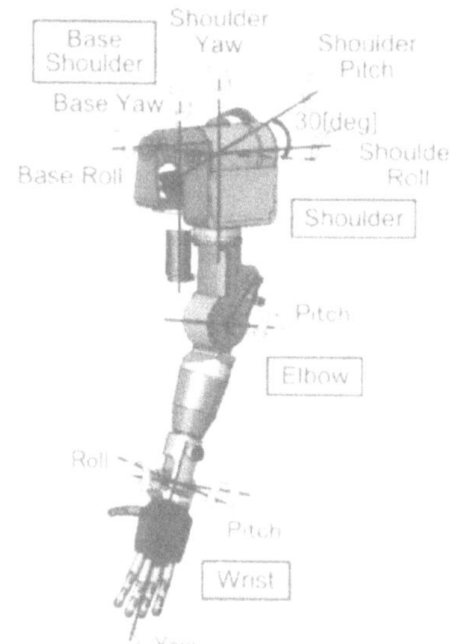

Figure 7. Arm configuration.

2.3 9-DOFs Emotion Expression Humanoid Arm

The 9-DOFs Emotion Expression Humanoid Arm is shown in Figure 7. It consists of the 2-DOFs base shoulder, the 3-DOFs shoulder, the 1-DOF elbow and the 3-DOFs wrist. The yaw axis of the base shoulder and 3 axes of the shoulder are driven by a direct driven mechanism with DC motors and harmonic drive systems. Since the roll of the base shoulder requires large torque, a direct driven mechanism with an AC motor and a harmonic drive system is adopted. The pitch axis of the shoulder is leant at a 30-degree angle to avoid a singular point in a common use range. The elbow adopts a belt driven mechanism, in which the output axis of a motor is connected to a harmonic drive system by a timing belt. The wrist joints of the previous version WE-4R were driven by DC motors with planetary gear system. The hand motion of WE-4R wasn't stable since

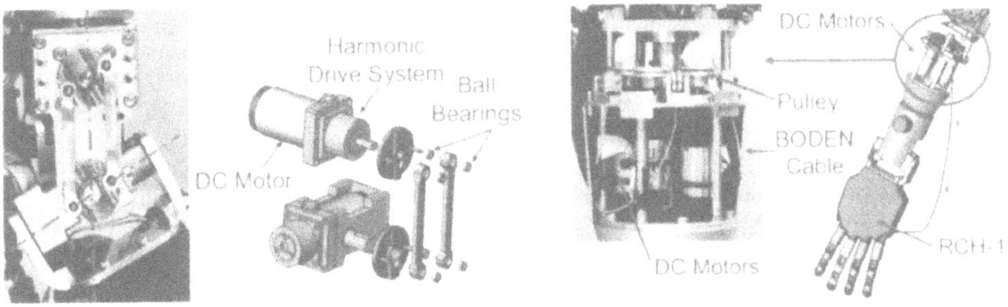

(a) Wrist mechanism (b) Actuator system of RCH-1

Figure 8. Improvement of arm mechanism.

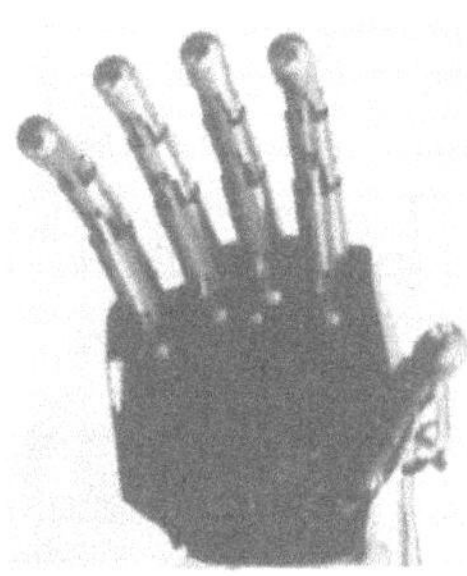

Figure 9. RCH-1.

Table 2. Sensors on WE-4RII.

Part	Sensation		Device	Quantity
Head	Visual		CCD Camera	2
	Auditory		Microphone	2
	Cutaneous	Tactile	FSR	26
		Temperature	Thermistor	1
		Weight	Current Sensor	2
	Olfactory		Semiconductor Gas Sensor	4
Hand	Cutaneous	Tactile	Contact Sensor	16
		Tactile	FSR	4
		Force	3D Force Sensor	2

there was too much backlash. Moreover, it was necessary to redesign the forearms for mounting the finger's motors. Therefore, we changed the gear system to small harmonic drive systems in order to reduce the backlash and miniaturize the wrist mechanism. Especially, we designed the link mechanism as shown in Figure 8(a) for the pitch axis of the wrist. The link mechanism transmits the motor power with the two links, which were supported by four ball bearings to reduce the slant caused between inner and outer rim.

2.4 Humanoid Robot Hand RCH-1

Figure 9 shows the humanoid robot hand RCH-1 (Zecca et al., 2006). RCH-1 is designed and developed by ARTS Lab, Scuola Superiore Sant'Anna. RCH-1 has five fingers and six active DOFs and 10 passive DOFs. Each finger has 1-DOF for extension and flexion movement. Moreover, the thumb has one more DOF for abduction and adduction movement. Extension and flexion motions are driven by single cable actuated by a DC motor. RCH-1 can grasp an object following the object shape by special mechanisms. Therefore, it is not necessary to independently control all joints. Abduction and adduction motion are driven by direct driven mechanism with a DC motor mounted inside the palm.

In order to integrate RCH-1 into WE-4R, the actuation system for extension and flexion of RCH-1 has been mounted inside the forearm of WE-4RII as shown in Figure 8(b), thus mimicking the position of *flexor digitorum* and *extensor digitorum* in the human forearm. The fingers and motors are connected with thin wire in BODEN CABLE, which is an outer tube for endoscopes.

3 Fore Sensations of WE-4RII

WE-4RII has visual, auditory, olfactory and tactile sensors on its head, and tactile sensors on its hands as shown in Table 2. Regarding visual sensor, WE-4RII has two color CCD cameras (CS6550, Tokyo Electronic Industry Co. Inc.) in its eyes. WE-4RII calculates the gravity and area of the targets. The robot can recognize any color as the targets and it can recognize eight targets at the same time. It also can recognize the distance from its eyes to the targets by the angle of convergence between the two eyes. If there are multiple targets in the robot's view, WE-4RII autonomously selects a suitable color as the target according to the situation. With regard to the auditory sensor, WE-4RII has condenser microphones in each ear. It can localize the sound direction from its loudness in a 3D space. For olfactory sensation, we set four semiconductor gas sensors (SB-19, SB-30, SB-AQ1A and SB-E32, FIC Inc.) in WE-4RII's nose. The robot can quickly distinguish the smells of alcohol, ammonia and cigarette smoke. In addition, we used FSRs (model 406, Interlink Electronics, Inc.) as the tactile sensor and set them on the cheeks, forehead, top and side of the head of WE-4RII. The robot can recognize the difference in touching behaviors such as "push," "hit" and "stroke".

Meanwhile, WE-4RII has on/off contact sensors and FSRs on RCH-1. The on/off contact sensors, which are film shaped switch, detect contact with the objects for grasping. FSRs set on the dorsum of RCH-1 are used for interaction with a human.

4 Total System Configuration

We control WE-4RII by three computers (PC/AT compatible) connected by Ethernet. PC1 (Pentium4 3.0 [GHz], Windows XP) captures the visual images from CCD cameras. It calculates gravity and brightness of the targets, and sends them to PC2. PC2 (Pentium4 2.66 [GHz], Windows XP) obtains and analyzes the outputs from the olfactory and cutaneous sensors and the sounds from microphones by using 12 [bit] A/D boards and a soundboard, respectively. In addition, PC2 determines the robot's mental state and controls DC motors excepting RCH-1. The control data of RCH-1 is sent to PC3. PC3 (Pentium III 1.0[GHz], Windows 2000) obtains and analyzes the sensor data of RCH-1 and sends it to PC2. And, PC3 controls DC motors on RCH-1.

5 Emotional Expressions

5.1 Definition of Emotional Patterns

WE-4RII can express its emotion by using the upper body motion: the facial expression, arms, hands, waist and neck motion. First, the seven facial patterns of "Happiness", "Anger", "Disgust", "Fear", "Sadness", "Surprise", and "Neutral" were defined by using the Six Basic Facial Expressions of Ekman (Ekman and Friesen, 1978). Next, we made 13 patterns of the upper body for each emotional expression by using the relation between the emotion and the posture

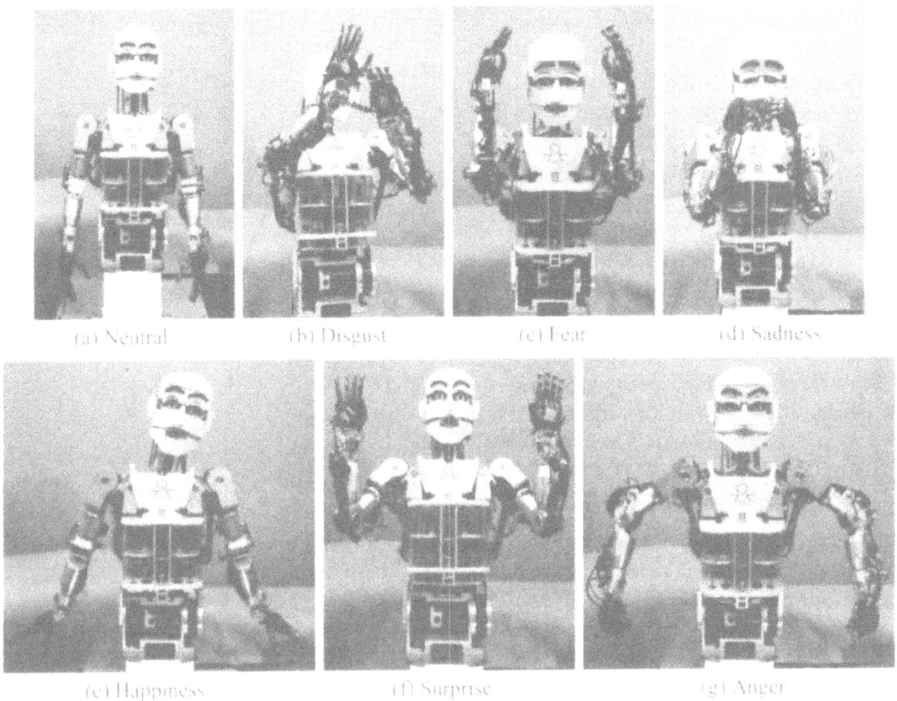

Figure 10. Emotional expressions of WE-4RII.

proposed by Hama et al. (2001). We defined one pattern with the highest recognition rate in pre-experimental evaluation as the emotional pattern. In addition, both the posture and the motion velocity were controlled in order to realize the effective emotional expression, since the motion velocity is as important as the posture for emotional expressions. For example, WE-4RII moves its body quickly and slowly for surprise and sadness emotional expressions, respectively. We defined the motion velocities by comparing the various motion velocities. Figure 10 shows the emotional expressions exhibited by WE-4RII.

5.2 Experimental Evaluation

We evaluated how the emotions of WE-4RII are transmitted by its facial expressions and upper bodily expressions. First, we showed 18 subjects (averaged age: 21) the movies of the six basic emotional expressions exhibited by the previous version WE-4R and WE-4RII. In the movies, the motion patterns of WE-4R and WE-4RII were the same except for the hand motion. Next, the subjects chose an emotion that they thought the robot expressed. We examined the correct recognition rates of those emotional expressions. Finally, we compared the recognition rates of WE-4R and WE-4RII. The experimental results are presented in Figure 11.

As a result, the WE-4RII's recognition rate of "Happiness" was 5.5 [points] higher than WE-4R. All subjects correctly recognized the "Surprised", "Sadness", "Anger" and "Disgust" emotional expressions of WE-4RII. However, the recognition rate of "Fear" was 5.5 [points] lower than WE-4R's rate since the some subjects thought the "Fear" emotional expression as "Disgust" emotional expression. In total, the averaged recognition rate of all emotional expressions of WE-4RII was 2.8 [points] higher than the WE-4R's averaged recognition rate. Therefore, we considered that the integration of the humanoid robot hands was effective for the improvement of emotional expressions. However, the emotional expressions in Fig.4 are Japanese patterns. We need to deal with cultural differences.

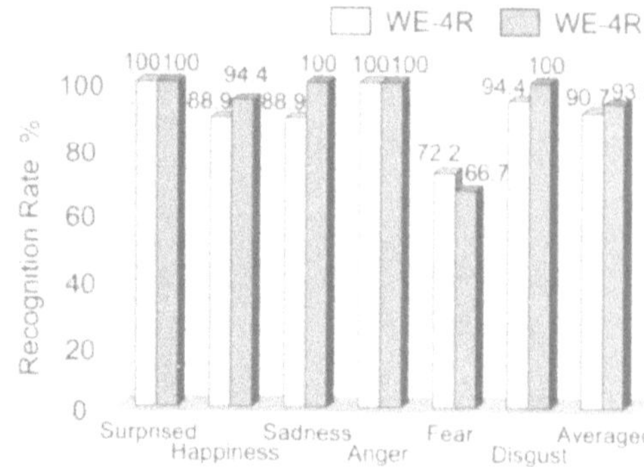

Figure 11. Results of the experimental evaluation.

6 Conclusions and Future Work

In this paper, we presented the Emotion Expression Humanoid Robot WE-4RII developed by integrating the Humanoid Robot Hands RCH-1 into the previous version WE-4R. We confirmed that WE-4RII can effectively express its emotions.

In the future, we would like to increase the emotional expressions and behaviors and to develop a model for generation of most suitable emotional expression and behavior according to the situation. In addition, a method for objective evaluation of robots has not been proposed yet. It is very important to measure the psychical effect of robots on humans in real time and with high reliability by objectively evaluating robots. Therefore, we would like to develop new evaluation system for the humanoid robots.

Bibliography

Breazeal, C., and Scassellati, B. (1999). How to build robots that make friends and influence people. *Proceedings of the 1999 IEEE/RSJ International Conference on Intelligent Robots and Systems*, 858-863.

Ekman, P., and Friesen, W. V. (1978). *Facial Action Coding System*. Consulting Psychologists Press Inc.

Hama, H., et al. (2001). *Kanjo Shinrigaku heno Syotai (in Japanese)*. Saiensu. 162-170.

Itoh, K., Miwa, H., Takanobu, H., and Takanishi, A. (2004). Mechanical Design and Motion Control of Emotion Expression Humanoid Robot WE-4R. *Proceeding of the 15th CISM-IFToMM Symposium on Robot Design, Dynamics, and Control*, ROM04-14.

Itoh, K., et al. (2005). Application of Neural Network to Humanoid Robots –Development of Co-Associative Memory Model. *Neural Networks*, Vol.18, No.5-6, 666-673.

Kobayashi, H., and Hara, F. (1996). Real Time Dynamic Control of 6 Basic Facial Expressions on Face Robot. *Journal of the Robotics Society of Japan*, Vol.14, No.5, 677-685.

Miwa, H., Takanobu, H., and Takanishi, A. (2002). Human-Like Head Robot WE-4RV for Emotional Human-Robot Interaction. *ROMANSY 14 –THEORY AND PRACTICE OF ROBOTS AND MANIPULATORS*, 519-526.

Zecca, M., et al. (2006). From the Human Hand to the Humanoid Hand: Development of RoboCasa Hand #1. to be submitted to 16th CISM-IFToMM Symposium on Robot Design, Dynamics, and Control.

The Mechatronic Design of a Human-like Robot Head

Karsten Berns[1] Carsten Hillenbrand[1] and Krzystof Mianowski[2]

[1] Robotics Lab, University of Kaiserslautern, Germany, e-mail: berns@informatik.uni-kl.de
[2] Institute of Aircraft Engineering and Applied Mechanics, Warsaw University of Technology, Poland

Abstract. In this paper the design concept as well as the constructed human-like robot head is introduced. Main goal of the head design is the support of an adequate human machine interaction. Therefore, our robots head should be able to support non-verbally communication based on facial expressions but should also have the skills to observe the intention of a human operator. Based on experience done with a graphical simulation system, the artificial skin movement is examined to express specific facial expressions. These results lead to the basic for the mechanical head design. In the paper also the neck design and the new eye construction is presented. At the end of the paper the sensor system and the control architecture of the head is introduced.

1 Introduction

Worldwide, several research projects focus on the development of humanoid robots. Especially for the head design there is an ongoing discussion if it should look like a human head or if a more technical optimized head construction (Breazeal, 2003, Esau et al. 2003, Takanobu et al., 2002) should be developed. The advantage of a technical head is, that there is no restriction according to the design parameters like head size or shape. This fact reduces the effort for mechanical construction. On the other hand, if realistic facial expressions should be used to support communication between a robot and a person, human likeness could increase the performance of the system. The aim of the humanoid head project of the University of Kaiserslautern is to develop both a very complex robot head able to simulate the facial expressions of humans while perceiving its environment by a sensor system (stereo-camera system, artificial nose, several microphones, ..) similar to the senses of a human. On the other hand, the robot head should look like a human head to examine, if its performance due to non-verbally communication will be higher compared to technical heads.

The starting point of our research was the human-like mask, which was designed by the company Clostermann Design Ettlingen, Germany (see figure 1). In the following first our head design is introduced. The first task for the head is the facial expression of emotions. Therefore, based on Ekmans Facial Action Coding System (Ekman, 1978) adequate action units are selected for the emotions that shall be expressible by the head. This is examined with the help of our graphical simulation system, which will be introduced. The results obtained from this are used as guidelines for the head and the eye construction. The neck pose supports also the expressions of emotions. On the other hand the neck similar that a human head is necessary for the detection of a human operator and its environment. Therefore, a neck was constructed based on design parameters of a human neck, which will be described. In parallel to the mechanical design the computer architecture and the sensor system are developed. In section 4 a short introduction to our sensor system and the implemented control architecture is given.

2 The graphical simulation system for the support of the design process

In (Ekman, 1978) a set of action units is introduced which are mainly used for the expression of emotions by means of facial expression. Each of these units consists of one or several muscles which are connected to specific parts of the skin. Ekman has shown that an activation of the muscles of a specific set of action units lead to an expression of the basic emotions like joy, fear, sadness, surprise or disgust. For example, the emotion 'disgust' is expressed if the action units 'brow lowerer', 'lip corner depressor' and 'chin raiser' are activated at the same time. Also action units exist for the neck and the eye movement to support the expression of specific emotions. If for the skin movements each of the action units is implemented by one motor and only the minimum number of units is selected for the expression of emotions, at least 30 motors must be installed in a robot head. For the design of a robot head able to express emotion like humans, it is first necessary to determine an adequate skin movement.

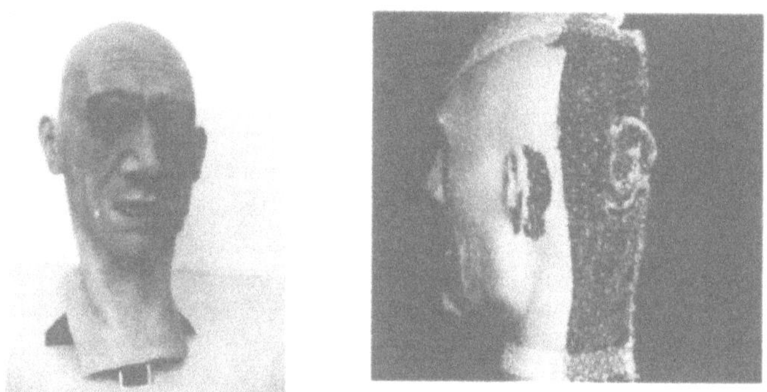

Figure 1: The human-like mask, which was the starting point for our head construction (left). Simulation of the head with the artificial skin (right)

To find a solution for this problem a simulation of the artificial skin was implemented together with the Computer Graphics Group (Prof. Hagen) of the University of Kaiserslautern. Starting with a 3D laser scan (using the Minolta-Digitizer Vi900) of the artificial head covered with a silicon skin, a triangular mesh with about 1,6 Mio triangles was generated and reduced by standard mesh simplification algorithms to about 100000 triangles. Then, the resulting mesh (see Figure 1, right) was modelled as a spring/damper system, which describes the behavior of the silicon skin when external forces are applied. Based on this simulation of the artificial skin, several tests were performed to find out where and how the artificial skin can be moved in order to simulate an action unit. After the realisation of Ekman's action units in the simulation system, tests with about 60 students are performed to optimise the position and the direction of the motion of the action units. With these results an optimal placement for the action units are performed which yield to the final mechanical design concept.

3 The mechanical construction

The mechanics of the head consists of three parts the basic unit (cranial bone) including the lower jam, the neck, and the artificial eye construction. Beside the eye construction, which is built at the moment in our mechanical workshop, all mechanical components are installed in our head. In the following a short introduction of the basic unit and the eye construction is given. The main focus of the section is the design aspect and the implementation of the neck.

3.1 Basic unit and eye design

In the basic unit 10 metal plates, which can be moved via wires, are glued on the silicon skin. The plate areas as well as its fixing positions on the skin and the direction of its movement are optimized in a simulation system mentioned above (see Figure 2, left). As actuators, 10 servomotors are used to pull and push the wires. Additionally, a servo-motor is used to raise and lower the lower jaw.

The requirements for the eye construction are strongly related to the human eyes. The eye has 3 DOF one degree for moving up and down (+-40°), one for left and right (+-45°), and an additional one for the eyelashes (30°). The eye design was done in a way that 2 dragonfly cameras could be included. The total weight of the eye design including the camera is less than 200g. The max. angular velocity of the eye axis are 200°/s. The angular velocity of the eyelashes is about 120°/s.

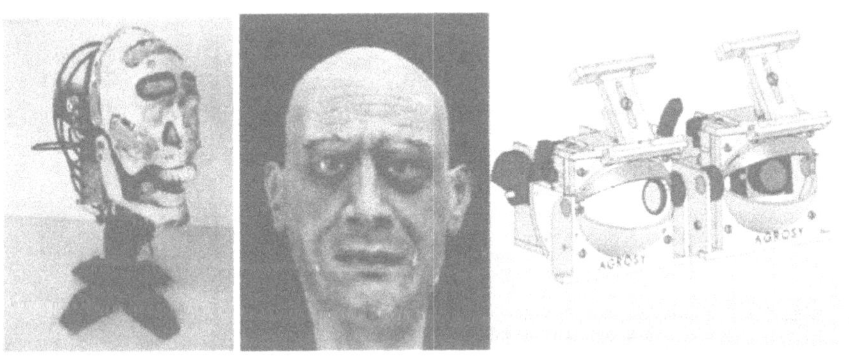

Figure 2. Mechanical construction of the humanoid robot head including 11 servo motors for the skin and mouth movements (left), the artificial skin glued on the basic unit (middle), and the new construction of the artificial eyes for stereo image processing (right)

3.2 The neck design

Analysis

Biological neck contains bones, joints, tendons, ligaments, muscles and nerves. Bones and muscles of the neck region of human body are shown in Figure 3. It is one of the most flexible regions of the spine, which consists of vertebrae, seven shock-absorbing discs, muscles, and vertebral ligaments to hold them in place. The uppermost cervical disc connects the top of the spinal column to the base of the skull (for further information see (Berme, 1990, Morecki, 1980).

The basic control task of the neck in nature is to keep its pose in a "neutral" position whenever possible. The neck region of the spine treated as a bio-mechanism consists of the links and joints is shown in Figure 4. Two upper joints of this fragment have one degree of freedom (DOF) each, so they can be treated as kinematic pairs of fifth class. The next six pairs have three DOF's each, and so they can be treated as kinemtic pairs of third class. Of course the angular range of motion of each pair is rather small, but because of long serial chain (see Figure 4) the global number of passive degrees of freedom is high, while the ranges of resulted motion are high. In Figure 5 and Figure 6 the ranges of resulted motions of the neck region of the man are shown. Due to the view in Figure 5, the ranges of inclination motion from the side could be estimated as +/- 45°. Form Figure 6 the ranges of in the frontal plane can be estimated as +/- 50°.

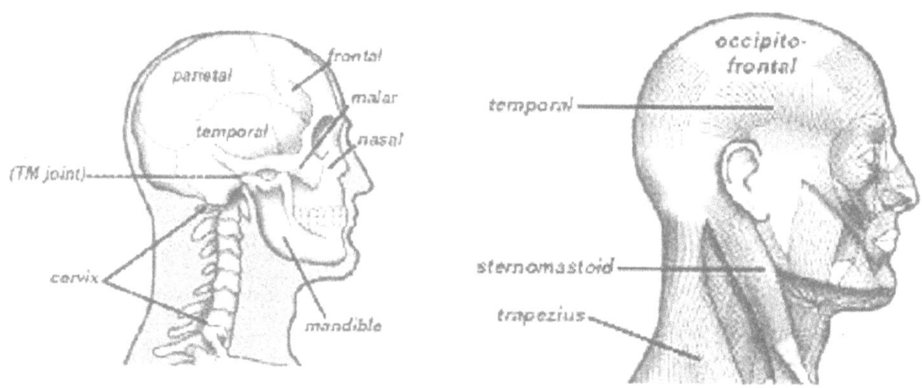

Figure 3. Bones (left) and muscles (right) of the human neck.

Figure 4. Links and joints of the human neck (left)and its kinematic structure (right)

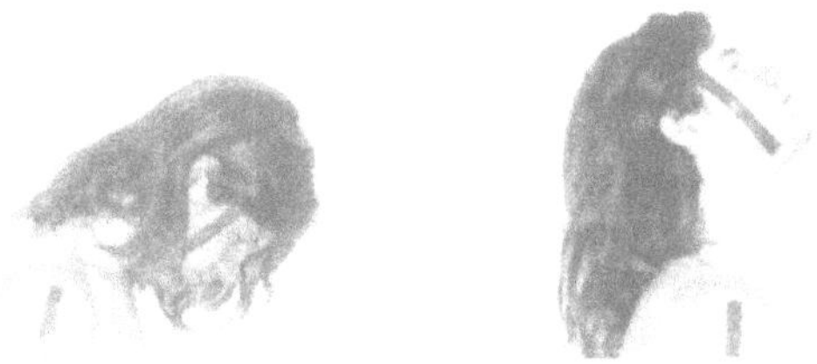

Figure 5. Ranges of angular motion of the neck shown from the side

Figure 6. Ranges of angular motion of the neck shown from the front.

Mechanical neck design

As one requirement of the neck construction it was assumed, that the global motion of the neck will be realize by the kinematic chain with only three degrees of motion. It should be stressed, that the possible poses of the artificial neck should be appropriate to support the human machine interaction. For the design of the neck basic characteristics of the geometry, kinematics and dynamics of a human neck is considered.

From the analysis of the neck (Figure 5 and Figure 6) with 3 DOF of motion a ball joint could be an adequate solution. Unfortunately, it is very hard to design appropriate driving system for such a solution. So, it has been decided, that for kinematic functions of the natural neck serial chain similar to Cardan joint solution will be applied (see Figure 7). The first degree of freedom is the rotation over vertical axis. The range of this rotation for artificial neck was assumed as +/- 90°. The second degree of freedom is the inclination of the neck over horizontal axis in the side plane. The range of this motion was assumed as +/- 30°. The third degree of freedom is the inclination of the neck in frontal plane. It is rotating around the axis which is moving accordingly to the second degree of freedom. The range of this motion was assumed as +/- 30°. The design of the neck was done with the CAD program ProEngineer (see Figure 7).

The rule of the design structure was simply similar to the nature one: keep the artificial neck in a "neutral" position whenever possible. This problem was solved in a way, that spring elements (not shown in Figure 7) has been applied for balancing the gravity forces and moments. The springs are stretched between special bridges mounted on the ring outside the central Cardan joint and previous/second links. For the driving system Faulhaber electric rotational motors with gears are selected. As result a mechatronic construction was designed, which is very compact and which simulated in its dimension and the degree of motion the "natural" neck in an appropriate way. For the compensation of external gravitational torque of the neck, the system of spring has been applied. It is shown in Figure 8a. The characteristic of gravitational external torque, additional torque produced by springs and the sum of them is shown in Figure 8b). One can see, that the resultant torque not exceed 0.2Nm in the range of angle –30°÷30°.

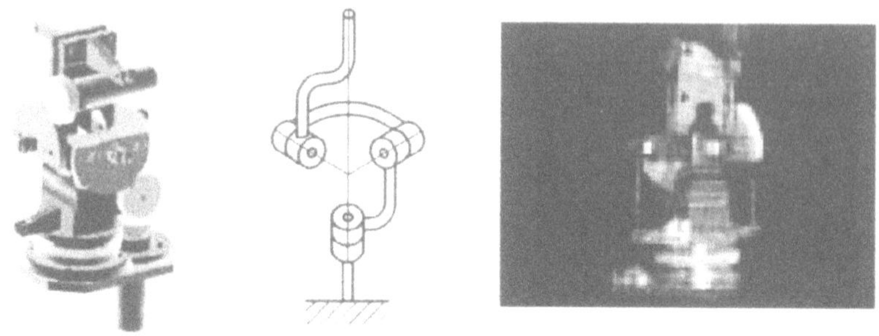

Figure 7. Artificial neck. Left the Kinematic scheme of the neck and the ProEngineer design, right the mechanical construction

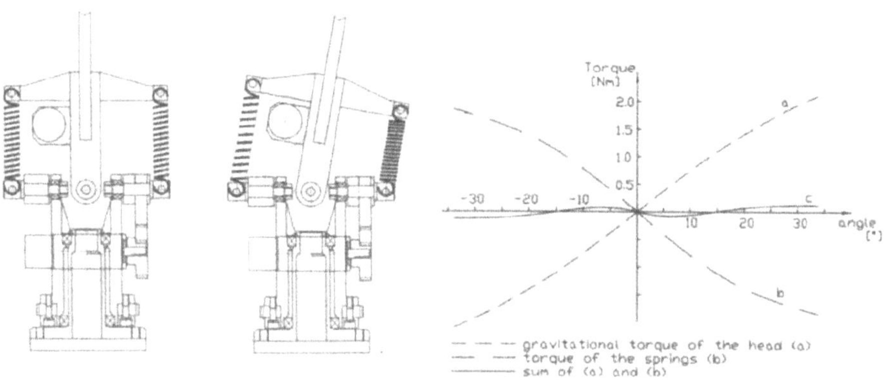

a) b)

Figure. 8. Spring compensations of gravitational torque of artificial neck:
a) general view of the neck with compensational springs,
b) external characteristic of the torque for control

4 Sensor system and control architecture

Sensor system

Beside encoders fixed on the DC motors the neck and the eye movements, an inertial system are integrated in the head which measures the angular velocity and the acceleration in all 3 DOFs. This gives an estimation of the pose of the head. Also two microphones (fixed in the ears) and a loudspeaker are included in the head. The main sensor system for the interaction with human is the stereo-vision system, which consist of two dragonfly cameras. Several experiments have been performed like the detection of a human head (see also (Braun et al., 2005). At the moment the integration of the stereo camera system in the eye design of the head is under development.

Control architecture

The control of the servo and DC motors as well as the determination of the pose from the inertial system is done with a DSP (Motorola 56F803) connected to a CPLD (Altera EPM 70 128). In total 5 of these computing units are installed in the head one for the inertial system, two for the DC motors for eye control, one for the 3 DC motors of the neck and one of the 11 servo motors which move the skin. These computing units are connected via CAN-bus to an embedded PC (see Figure 9). The two microphones and the loudspeaker are connected to the sound card of the embedded PC. The cameras, which are included in the eye construction, use the firewire IEEE 1394 input channel of the embedded PC. The calculation of movements of the different facial expressions is done on a Linux-PC. The behaviour based control is implemented with the help of the Modular Controller Architecture (MCA). MCA is a modular, network transparent and realtime capable C/C++ framework for controlling robots (see Scholl et al. 2000 for details). MCA is conceptually based on modules and edges between them. Modules may be organized in module groups, which allow the implementation of hierarchical structures.

5 Summary and outlook

In this paper a human-like robot head is introduced, which will be used to interact with humans. One focus of the present research is how the facial expressions of a human being can be transferred to a robot head. Based on a simulation of the silicon skin of the head and the implementation of Ekman´s action units facial expressions were simulated. The simulation was also used to reduce the number of actuators. Based on these results the head was constructed. The neck construction was also done based on the geometry and the degree of motion of a human head

Present research is focusing on the implementation of behaviours for the control of the action units in the eye, artificial skin and the neck. In addition we have started to set up and emotional architecture and design a concept for the modelling of a human in the environment of the robot head.

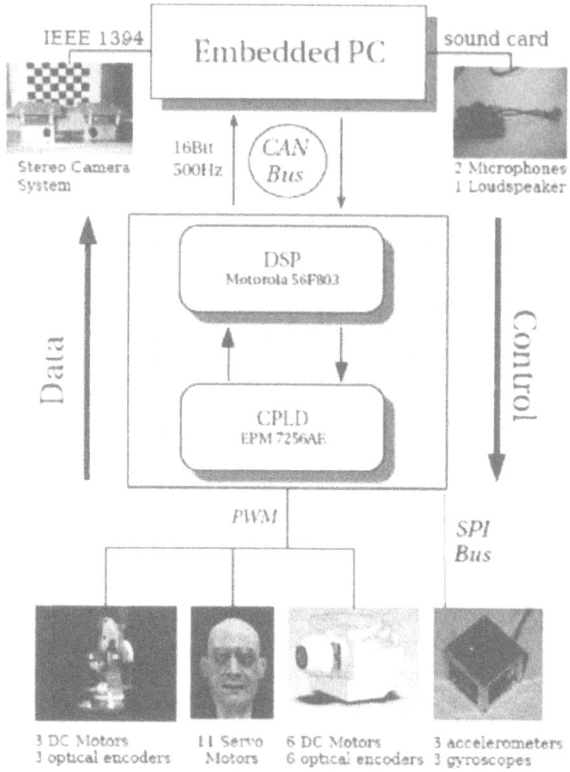

Figure 9. The control architecture applied to the humanoid robot head

Bibliography

Berme N., Capozzo A. (1990) Biomechanics of human movement, Bertec Corporation, Worthington, Ohio, USA

Berns, K., Braun, T. (2005) Design concept of a human like robot head, *Humanoids 2005*, Tsukuba, Japan

Braun, T., Szentpetery, K., Berns, K. (2005). Detecting and following humans with a mobile robot. In *Proceedings of the EOS Conference On Industrial Imaging and Machine Vision.*

Breazeal, C. (2003). Emotion and sociable humanoid robots. *Int. J. Hum.-Comput. Stud.*, 59(1-2):119–155.

Ekman, P, and Friesen, W. (1978). Facial Action Coding System. *Consulting psychologist Press, Inc.*

Esau, N., Kleinjohann, B., Kleinjohann, Stichling, D. (2003) Mexi -machine with emotionally extended intelligence: A software architecture for behavior based handling of emotions and drives, In *Proceedings of the 3rd International Conference on Hybrid and Intelligent Systems (HIS'03)*

Morecki, A.(1980) *Biomechanics of motion*, Proc. Of Int. Centre for Mechanical Sciences, Springer Verlag

Scholl, K.U., Kepplin, V., Albiez, J., Dillmann, R. (2000) Developing robot prototypes with an expandable modular controller architecture. In *Proceedings of the International Conference on Intelligent Autonomous*. Systems, Venedig, pp. 67–74.

Takanobu, H., Takanishi, A., Miwa, H. (2002). Development of human-like head robots for modelling human mind and emotional human robot interaction. *IARP International workshop on Humanoid and human Friendly Robotics*, pages 104–109.

Human Equilibrium Control Principles Implemented into a Biped Humanoid Robot

T. Mergner[1], F. Huethe[1], C. Maurer[1] and C. Ament[2]

[1] University of Freiburg, Neurology, Germany, e-mail: mergner@uni-freiburg.de
[2] University of Freiburg, Microsystems Engineering, Germany, e-mail: ament@imtek.de

Abstract. We describe how human equilibrium control principles, which were derived from neurophysiological experiments, can be implemented in a biped humanoid robot ('PostuRob'). Stance control in humans uses sensorbased feedback which involves mainly three sensors that measure: (1) body motion in space (vestibular system), (2) body motion with respect to the feet (ankle angle proprioception), and (3) torque between body and feet (ankle torque proprioception). The sensor signals are not used directly for feedback, but instead are used to internally reconstruct the physical stimuli in the outside world by means of sensor fusions. These reconstructions yield internal estimates of the external force fields such as gravity, the external contact forces (e.g. push), and support surface motion. The estimates are fed into an ankle angle proprioceptive feedback loop. By way of the signal fusions the robot's control adjusts to the external stimulus situations, a fact that allows for low loop gain and robust control. This control differs from that of the biped robots described in the literature. They tend to use a global stability measure such as the COP (centre of pressure) or ZMP (zero moment point) and this control does not adjust to the external stimulus situation. Our PostuRob is used in a biorobotics approach to better understand stance control deficits of neurological patients and the effects of therapy and rehabilitation.

1 Introduction

Can we nowadays claim that we understand the control principles of human upright stance well enough to use them as blueprints for constructing an equilibrium control system of a biped robot? Given this, can we use such a robot in medical research for better understanding impaired stance control in patients and the effects of therapy and rehabilitation? Finally, in which way does the presumed human control principles differ from the engineering ones currently used in robotics? We refer to these three issues, which emerged from our current research, as 'demonstrator aspect', 'biorobotics aspect', and 'bionics aspect'.

Demonstrator aspect. Biological research nowadays uses more and more dynamic computer models to formalize and simulate identified sensorimotor functions and this even when they contain non-linearities (example in Maurer et al. 2005). This also applies to the research of human control of upright stance (van der Kooij et al. 2001; Peterka 2002; Mergner et al. 2003). These models may well serve as blueprints for constructing corresponding machines. But why should we implement them into machines instead of relying on the computer simulations? One reason is that successful implementation of the models into real world media adds credit to their validity (the demonstrator aspect). For our colleagues, the functioning of our robot (Fig. 1) has indeed a much higher explanatory and persuasive power than our computer models had in the past.

Biorobotics aspect. Another reason to implement the biological models into robots is that they may be used in biological research (biorobotics). There has recently been a extended debate on this approach (Webb 2001; target article plus comments). The following aspect, however, has not been sufficiently addressed. Imagine the investigation of a neurological patient on a motion platform in a posturographic laboratory with the help of various devices for measuring kinematic and kinetic parameters. The complexity of this situation poses problems when we try to mimic it in computer simulations for comparison of simulated and experimental data, because the many simplifications and formalizations required for the modelling move us away from reality. The problem may partially be resolved by applying a new method, the so-called *'hardware in the loop (HIL)' simulation.* In this approach only the control mechanism under consideration is implemented as a software model on an 'Embedded PC' (real-time; under control of a host PC), while all other components remain hardware. In our example, this would mean that we can use the laboratory and its recording equipment as before, replace the human body with its sensors and actuators by a humanoid robot, whose "brain" (Embedded PC) contains a software model for human stance control (Fig. 2). We therefore hold that our medical research can well take advantage from biorobotics.

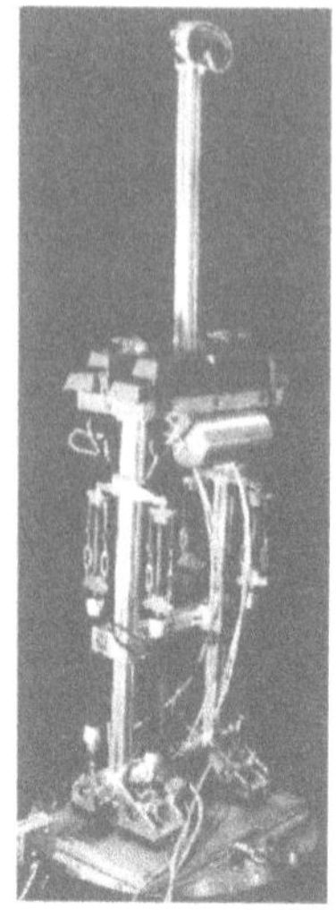

Figure 1: Photograph of 'PostuRob' on motion platform. Its aluminium skeleton consists of two rigid legs fixed to a pelvic girdle and a spine ('body'). Centre of mass is mainly represented by two plumb weights on pelvis. Each leg carries a front and back pneumatic 'muscle' with 'tendons' (springs) to move the body with respect to the foot about the 'ankle joint'.

Bionics aspect. As to the differences in the control principles between humans and previous robots, one may predict that robot technology may profit from the use of biological principles (bionics). This notion intuitively draws on several basic differences, such as that the control strategy of humans have developed during million years of phylogenesis, while the engineering sciences are rather recent, that the rather compliant human tissues is advantageous over the metallic materials the robots are currently built from, and so forth. But we will try to answer the question more specifically.

A difficulty in this comparison may arise from the fact that there exist considerable differences among the equilibrium control principles currently used in biped robots. However, it appears common to most of them to use sensorbased feedback, calculating from the sensor inputs a global stability measure such as the COP (centre of pressure) and ZMP (zero moment point) etc. and to use this to keep the robots equilibrium within stability limits (example, Popovic et al. 2005). This principle leaves open which of the external stimuli is causing a given perturbation. In contrast, the human control principle reconstructs from different sensors the external stimuli and uses these estimates for feedback control, as we will show.

In the following, we first describe the most relevant sensors used by humans for their control of dynamic behaviour in space and give technical equivalents. Then, we describe the principles that humans use to fuse the sensor signals for their self-motion perception and posture control. Furthermore, we explain how we implemented these principles into the robot and mention first results. Finally, we report on its possible applications in medicine and point out difference as compared to the classical engineering way to implement such a control task.

2 Sensors and sensor fusion principles

From physiological and clinical work, it is known that the following sensor systems play an outstanding role in the control of human dynamic behaviour in space:

(1) Gravito-inertial "body-in-space" sensor - Vestibular system. Anatomically, this system is located in the inner ear of the skull and consists of two parts, the macular (or otolith) organs and the semicircular canals. Their functions are those of a 3D accelerometer and a 3D gyrometer, respectively. The fact that humans are able to distinguish between head tilt and linear acceleration despite the equivalence principle has been explained by a '*canal-otolith interaction*' in the central nervous system, CNS (Mergner and Glasauer 1999). This interaction represents a sensor fusion that yields three sets of signals of body motion: (a) 3D rotational velocity, (b) 2D angle with respect to gravitational vector, and (c) 3D linear acceleration. The vestibular signals, unlike the other sensor signals, appear to show pronounced slow spontaneous fluctuations (Mergner et al. 2001) reminiscent of 1/f noise.

(2) Joint angle sensor (kinematic proprioception). The position sense of the limbs is known to involve muscle spindle receptors as well as receptors in the skin and in the capsules of the joints. Humans combine these inputs in the CNS to a joint angle signal and, for special purposes, a joint angular velocity signal. A simple technical equivalent would be a goniometer.

(3) Joint torque sensor (kinetic proprioception). This measure appears to involve Golgi tendon organs in the muscle tendons (Duysens et al. 2000). During upright stance, humans also use somatosensory foot sole pressure receptors deep in the foot arch to ob-

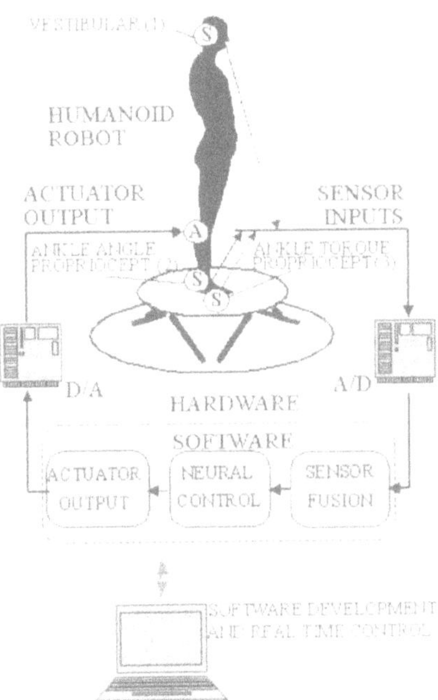

Figure 2: HIL simulation with actuator (A) and sensors (S)

tain a measure of the Centre Of Pressure (COP) (Maurer et al. 2000) which upon force closure between foot and support can be used as an equivalent of ankle joint torque (van der Kooij et al. 2005). In the robot, one can derive a measure of joint torque from force sensors under forefeet and heels of the robot's feet and/or from force sensors in the muscle mountings.

There are certainly many more sensor systems involved in human equilibrium control, for instance visual sensors. The three suffice, however, to explain the control principles. Their dynamic properties need not be considered here, unless stated otherwise.

Sensor fusions. They are considered in biology often under the view of redundancy effects where a given performance profits, e.g. in terms of temporal and spatial accuracy, from involving two or more sensors rather than one. Therefore it is important to note that each of the above three sensors provide distinct aspects of the physical stimuli, with redundancy occurring only in certain stimulus conditions (e.g., only if the foot is stationary in space is a given body-in-space angle equal to the body-to-foot angle). Using different sensors, appropriate sensor fusions may resolve the problem of *underdetermination of a physical stimulus by the signals of a single sensor.* We learned these lessons in psychophysical experiments of human self-motion perception (overview, Mergner 2002). A general rule drawn from these experiments was that humans in their perception tend to reconstruct from several sensory sources the underlying physical stimulus.

Our psychophysical studies suggested that these sensor fusions play a role also in sensorimotor control (Mergner et al. 1997). When we presented our subjects with vestibular as well as neck and leg proprioceptive stimuli during head, trunk, and leg rotations, we observed that they use proprioception in two ways: (1) for their perception of intersegmental motion, and (2) to perform with its help coordinate transformations of the vestibularly derived space reference from its origin (vestibular organ) in the head via the trunk to the feet and their support. Applied to postural control of upright stance, this could mean that (i) ankle angle proprioception is used in a feedback loop for body-on-foot(support) stabilization, and (ii) a vestibularly derived signal of support-in-space motion is fed into this loop and upgrades it into a body-in-space control.

Human perception during dynamic behaviour in space deals with the kinematic aspects of this behaviour, while the kinetic aspects (dynamics) tend to remain subconscious (Mergner 2002). Yet, with the general sensor fusion rule (internal reconstruction of external stimuli) we were able to include the kinetic aspects into a concept on how humans control their upright stance and we performed preliminary tests (Mergner et al. 2003). Care was taken to include all relevant external stimuli (i.e. body support surface motion, force fields such a gravity, and contact forces) as they would occur, for instance, during the following scenario: "The waiter on the ship brought me a glass of beer. Without looking at it, he balanced it across the swaying deck through a pushing crowd."

Human posture control studies. Our concept was finally tested in experiments. To this end, we simplified the 'waiter on the ship' scenario in that we restricted all stimuli to the sagittal (anterior-posterior) rotational plane and ascertained that, with small external stimuli, all rotations occurred primarily in the ankle joint. Subjects stood on a motion platform that was used for the support motion stimulus and, furthermore, pull stimuli were applied to their bodies by means of cable winches. These simplifications allowed us to model the scenario as an 'inverted pendulum'. It will be described together with the other parts of the model in the next section. Comparing simulated with experimental Bode plot data in a parameter identification procedure, model parameters were determined and, furthermore, changes in sensor weights across stimulus conditions were quantified. The results are published elsewhere (Maurer et al. 2005). They clearly show that human equilibrium control can be described by a sensorbased feedback model. This enabled us to take our model as a blueprint for constructing the robot.

3 The blueprint

The *'inverted pendulum' model* (Fig. 3) consists of two parts, a body (head, trunk, and legs) and the foot and its support. Both parts are inter-connected by the ankle joint. Primary body position is upright and foot position level, with the body weight pressing the foot firmly on its support (i.e. a motion platform; Fig. 1). Ankle torque T_{ank} in Fig. 3 results from muscles (actuators, assumed to be ideal, not shown) that produce torque and from passive torque (due to viscous-elastic muscle and elastic tendon properties; box BIOM, for biomechanics). It determines the COP. The torque affects body-in-space position *BS* depending on body inertia (box 'BODY INERTIA'). *BS* excursion in turn tends to lead to a COM (centre of mass) gravitational torque (T_{grav}; via box GRAV, for gravity), as does an external pull stimulus on the body ('External Torque', T_{ext}). Body-to-foot angle *BF* depends on *BS* and foot-in-space angle *FS* in the form of $BF = BS - FS$ (*FS* stems from tilt of the support surface and thus serves as tilt stimulus input). Details of these and the following parts of the model are given elsewhere (Maurer et al. 2005).

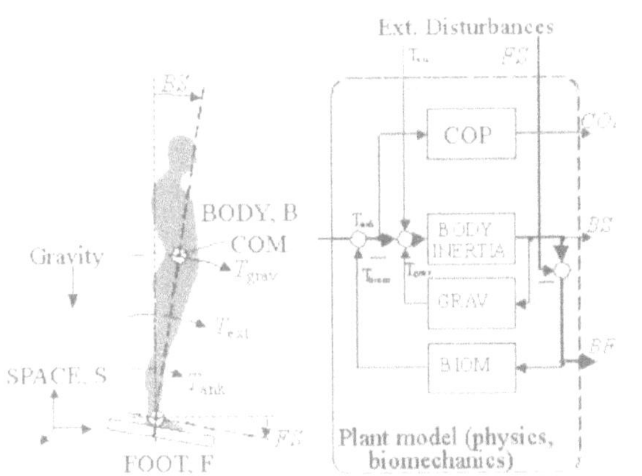

Figure 3: Plant model (arrows indicate direction sign of torques)

The mechanical part is combined in Fig. 4 with the sensor part (S), sensor fusion part (F), and controller part (C) of the model. The output signals of the mechanical part (COP, BS, BF) are used as inputs of the *three sensors* described above (ankle TORQUE SENSOR, body-In-SPACE SENSOR, and ankle ANGLE SENSOR, respectively). Their outputs provide measures of T_{ank} (τ_{ank}), BS (*bs* and first derivative $b^{\cdot}s$), and BF (bf, $b^{\cdot}f$).

Sensor fusions. The external mechanical stimuli are given by the support surface motion (FS), the contact force (T_{ext}), and gravity (T_{grav}). Their reconstruction is shown in part F of the model (Fig. 4) in terms of the following internal estimates:

Estimate of support surface tilt FS, *fs*. The internal estimate *fs* is derived in the form of $f^{\cdot}s = b^{\cdot}s - b^{\cdot}f$ (a coordinate transformation of the space reference to the support surface), with $f^{\cdot}s$ then being fed through a detection threshold and a velocity-to-position integration. Subsequently, the *fs* signal is fed into an ankle angle proprioceptive feedback loop, which serves body-on-foot stabilization ('local loop') and is upgraded into a body-in-space stabilization by *fs* + *bf* = *bs'''*. Note that *bs'''* is determined by the ANGLE SENSOR signal *bf* when FS= 0° (stationary platform, *fs*= 0°), while it becomes mainly determined by the In-SPACE SENSOR during platform motion (*bf* and *b'f* contributions to *bs'''* then tend to cancel each other). Note furthermore that not only motion or inclination of the support surface may lead to a *fs* feedback contribution, but also compliance and unevenness of the support surface.

Estimate of the COM's gravitational torque T_{grav}, τ_{grav}. The estimate of τ_{grav} is proportional to, and hence derived from the vestibular body-in-space signal *bs* (box aG2 contains transforming and gain factor). Since here the head is fixed on the trunk, no coordinate transformation of the in-space signal from the head to the COM in the trunk is required. τ_{grav} is passed through a threshold and then back transformed (box b) into an equivalent of a body-in-space signal, *bs'*.

Estimate of the external torque T_{ext}, τ_{ext}. The estimate τ_{ext} is obtained by combining τ_{ank} (output of TORQUE SENSOR) with *bs*-derived predictions of the COM's gravitational torque and the actively produced torque (in box 'Torque*'; see Maurer et al. 2005 for details and low-pass filtering of the constituents, not shown). After passing it through a threshold, τ_{ext} is transformed into a bs equivalent, *bs''*. Note that *bs''* adds to the feedback only to the extent that an external contact force is applied. This may not only be a push or pull stimulus, but also an extra load on the body or a change in load distribution. Note furthermore that the gravitational term in box 'Torque*' is negative so that the contribution of *bs* via τ_{ext} tends to cancel that via τ_{grav}. Thus, *bs* from the In-SPACE SENSOR contributes to the feedback only if the TORQUE SENSOR is dismissed (together with box 'Torque*').

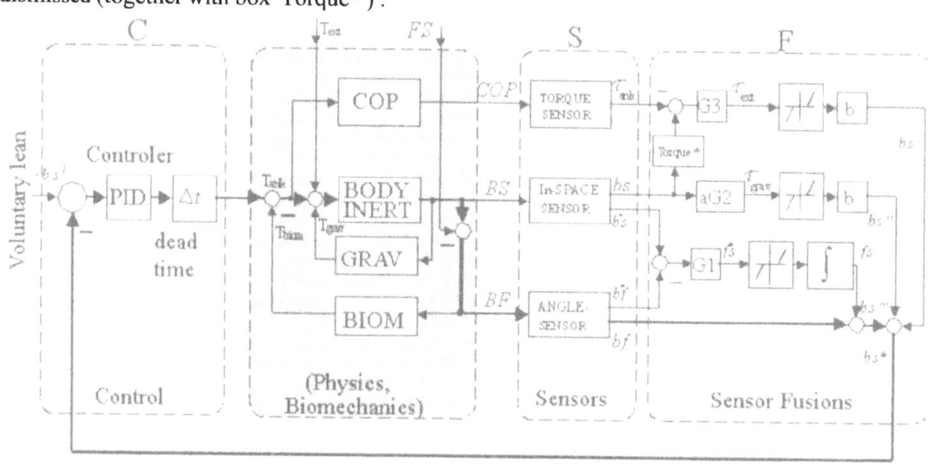

Figure 4: Blue print in the form of a block diagram ('local loop' in bold)

The *bs'*, *bs''*, and *bs'''* signals are combined to one feedback signal *bs**, which is subtracted in the Control part of the system from the voluntary command signal *bs'* (Fig. 4). The controller (box PID, for P, proportional, I, integrative, D, differential factors) is adjusted to produce an adequate corrective torque. The various delay times in the system are represented by a single dead time element (box Δt; taken from human data to be 150 ms). As mentioned before, the control is implemented in Simulink/Matlab on an Embedded PC under control of a host PC (Fig. 2).

We refrain here from explaining the role of velocity and position thresholds in the sensor fusions; they were observed in our human posture studies, represent central mechanisms rather than sensor properties, and are involved in sensory re-weighting mechanisms (Maurer et al. 2005).

A description of the robot's hardware (height, 1.9 m; weight, 100 kg) with actuators (pneumatic muscles) and sensors as well as the relevant model parameters are given elsewhere (online:

www.neurologie.uniklinik-freiburg.de/forschung/sensorfusion/PostuRob/).

At this web site, we also give first demonstrations of the robot's performance which show that its balancing behaviour resembles that of our human subjects.

4 Discussion

We take up the framework of Introduction, but slightly change its order.

Demonstrator aspect and bionics aspect. We show that equilibrium control principles of human upright stance, which we derived from biological studies, can successfully be implemented in the equilibrium control of a biped humanoid robot. Not surprisingly, the robot shows postural responses to external stimuli that resemble those of humans (similar gain and phase curves in Bode diagrams). Furthermorc, the robot successfully performs 'voluntary' body lean movements while simultaneously compensating for support tilt and contact force stimuli, a task whose performance challenges even human subjects. Which features make the human control principles different from those currently used in biped robots? As mentioned before, the latter use for feedback control a measure of the robot's stability, which they extract from the sensor signals. In the human control, in contrast, fusions of signals from different sensors are used to reconstruct the physical stimuli that are causing the sensor signals. Thus, it regains information that is lost by the *underdetermination of physical stimuli when characterising them by single sensor signals*. The achieved estimates of the physical stimuli are then used by means of the feedback to create the appropriate countermeasures to these stimuli. By way or the estimates and their fusion in the feedback the control system adjusts to the stimuli and their composition (see above, estimate of T_{ext}) so that the overall loop gain can be low, a fact which yields a 'soft' and robust control that tolerates a relatively long dead time of the loop and sensor inaccuracies. Furthermore, active participation of the In-SPACE SENSOR in the control is largely limited to situations with support tilt (see above, estimate *fs*). In fact, it is this sensor which is the most inaccurate one among the three sensors used. The internal stimulus representation create a meta-level that in future applications may be suited for incorporating communicated, predicted, and stored knowledge about the stimuli into the estimates. Finally, upon failure of sensors and or sensor fusion components, the system's function is only partially impaired.

We hold that the robot's equilibrium control can be extended to the other planes of space including the translational ones. Note that we have not yet implemented an internal estimate of linear support acceleration explicitly, but implicitly the system in its present form is already coping successfully with this stimulus. The reason is that the stimulus leads by means of body inertia to a torque that becomes represented in the signal of the TORQUE SENSOR and the τ_{ext} estimate. We hold furthermore that the control is modular and therefore will allow us to add more body segments and joints and that it can incorporate further sensors as well as cognition.

Biorobotics aspect. As mentioned in Introduction, research of complex biological systems is performed more and more in a hypothesis-driven way with a back and forth between experiments and model simulations. We conceive that the HIL methods with a robot as hardware can help to overcome the problem arising from the fact that the formalisations and simplifications in modelling tend to make the model unrealistic. We expect that this approach will allow us to explore into human sensorimotor control functions and to develop concepts as to the malfunctions underlying sensorimotor deficits in patients and to therapies. An additional aspect is that the approach may help to construct neural prostheses and orthoses (exoskeletons in the form of shells around body segments, equipped with motors that guide, perform, or assist segment movements). They may be designed to aid patients with motor problems (e.g. paraplegic patients) or with sensory deficits (e.g. loss of sensors). *The technical realization of such orthoses would be rather straight forward, because parts of the robot and the HIL system can be directly transformed into the design.* We hold that this 'human' control design has a much better chance to gain patients' compliance than a pure technical one.

ACKNOWLEDGMENTS: Supported by DFG Me 715/5-3, 715/6-1.

Bibliography

Duysens J, Clarac F, Cruse H (2000) Load-regulating mechanisms in gait and posture: comparative aspects. *Physiol Rev* 80:83-133

Maurer C, Mergner T, Bolha B, Hlavacka F (2000) Vestibular, visual, and somatosensory contributions to human control of upright stance. *Neursci Let* 281:99-102

Maurer C, Mergner T, Peterka RJ (2005) Multisensory control of human upright stance. *Exp Brain Res* (DOI 10.1007/s00221-005-0256.y)

Mergner T (2002) The Matryoshka Dolls principle in human dynamic behavior in space - A theory of linked references for multisensory perception and control of action. *Curr Psychol Cogn* 21:129-212

Mergner T, Glasauer S (1999) A simple model of vestibular canal-otolith signal fusion. *Ann New York Acad Sci* 871:430-434

Mergner T, Huber W, Becker W (1997) Vestibular-neck interaction and transformations of sensory coordinates. *J Vestib Res* 7:119-135

Mergner T, Maurer C, Peterka RJ (2003) A multisensory posture control model of human upright stance. In:Prablanc C, Pélisson D, Rossetti Y (eds) Neural control of space coding and action production. *Progr Brain Res* 142:189-201

Mergner T, Nasios G, Maurer C, Becker W (2001) Visual object localization in space. Interaction of retinal, eye position, vestibular and neck proprioceptive information. *Exp Brain Res* 141:33-51

Peterka RJ (2002) Sensorimotor integration in human postural control. *J Neurophysiol* 88:1097-1118

Popovic MB, Goswami A, Herr H (2005) Ground reference points in legged locomotion: definitions, biological trajectories and control implications. *Int J Robotics Res* 24:1013-1032

van der Kooij H, Jacobs R, Koopman B, van der Helm F (2001) An adaptive model of sensory integration in a dynamic environment applied to human stance control. *Biol Cybern* 84:103-115

van der Kooij H, van Asseldonk E, van der Helm FC (2005). Comparison of different methods to identify and quantify balance control. *J Neurosci Methods* 145:175-203

Webb B (2001) Can robots make good models of biological behaviour? *Behav Brain Sci* 24:1033-1094

Development of a New Humanoid Robot to Realize Various Walking Pattern Using Waist Motions*

Yu Ogura[1], Hiroyuki Aikawa[1], Kazushi Shimomura[1], Hideki Kondo[1], Akitoshi Morishima[1], Hun-ok Lim[2,3], and Atsuo Takanishi[1,3]

[1] Waseda University, Graduate School of Science and Engineering, Japan, e-mail: *uogura@ruri.waseda.jp*
[2] Kanagawa University, Department of Mechanical Engineering, Japan, e-mail: *holim@ieee.org*
[3] Waseda University, Humanoid Robotics Institute, Japan, e-mail: *takanisi@waseda.jp*

Abstract. A new humanoid robot capable of various motions is proposed in this paper, which has a 2-DOF pelvis and 2-DOF trunk. A new upper body mechanism is mounted on a leg machine, WABIAN2/LL, and a full scale humanoid robot, WABIAN-2, is constructed in this study. Its trunk is designed to bend forwards, backwards and sideways and rotate in combination with a 2-DOF pelvis. Also, the upper body is designed in way that makes it able to use a walk-assist machine. Basic walking experiments with and without a walk-assist machine are conducted, and the effectiveness of the mechanism of WABIAN-2 is confirmed.

1 Introduction

Many research groups have been studying biped humanoid robots in order to realize robots that can coexist with humans and perform a variety of tasks. For example, a research group from HONDA has developed humanoid robots—P2, P3, and ASIMO (see Sakagami et al., 2002). The Japanese National Institute of Advanced Industrial Science and Technology (AIST) and Kawada Industries, Inc. have developed HRP-2P (see Kaneko and Fujiwara et al., 2002, 2003). The University of Tokyo developed H6 and H7 (see Nishiwaki et al., 2000, 2002), and the Technical University of Munich developed Johnnie (see Löffler et al., 1999). Waseda University developed the WABIAN series that realized humanoid walking with a 3-DOF trunk motion and a 3-axis zero moment point (ZMP) compensation using the trunk (see Lim et al., 1999, 2002). Korea Advanced Institute of Science and Technology (KAIST) also developed a 41-DOF humanoid robot— KHR-2 (see Kim et. al, 2005). These human-size biped robots achieved stable and dynamic walking.

Meanwhile, if these humanoid robots can use rehabilitation or welfare instruments like a walk-assist machine as shown Fig. 1(a), if these humanoid robots can use rehabilitation or welfare instruments, as shown in Fig. 1(b), they could help in testing such instruments quantitatively. Therefore, we—Waseda University research group—have proposed the use of a

* This research was supported in part by a Grant-in-Aid for the WABOT-HOUSE Project by the Gifu Prefecture, NEDO (the New Energy and Industrial Technology Development Organization) and the JSPS Research Fellowships for Young Scientists. The authors would also like to express our gratitude to SolidWorks Japan K.K. for their support.

biped humanoid robot as a human motion simulator. The main advantages of the simulator can be considered to be as follows: (1) The measurement of the angle and the torque required at each joint can be measured easily and quantitatively as compared to the corresponding values in the case of a human measurement. (2) Experiments using such robots can help identify leg defects of a human from an engineering point of view. (3) A robot can replace humans as experimental subjects in various dangerous situations: experiments involving the possibility of falling, tests with incomplete prototype instruments, simulations of paralytic walks with temporarily locked joints.

Such experiments require a humanoid robot with mechanisms and a control system that enables it to closely replicate a human. However, humans have more redundant DOFs than conventional biped humanoid robots; this feature enables them to achieve various motions. Therefore, the assumption of a DOF configuration that is necessary to reproduce such motions is one of the very important issues in the development of a humanoid robot.

In advance of this study, we already have developed a new biped walking robot named WABIAN-2/LL (WAseda BIpedal humANoid-2 Lower Limb, Fig.2). Moreover, we have developed an algorithm that enables the robot to stretch its knees in steady walking avoiding singularity by using waist motion, and carried out stretch walking experiment by using this robot (see Ogura et al. 2004). WABIAN-2/LL without upper limb originally developed as a lower limb system for a humanoid type robot WABIAN-2(WAseda BIpedal humANoid-2).

In this paper, a new humanoid robot—WABIAN-2— with two 7-DOF legs, a 2-DOF waist, a 2-DOF trunk, and two 7-DOF arms is proposed. First, a DOF configuration based on researches from the viewpoint of clinical medicine is described. Next the mechanisms and control system of the robot are presented. In the development of the robot, new design principle for a robot which can use walking assist machine is set as first goal of this study. The effectiveness of the development of the humanoid robot is confirmed by walking simulation.

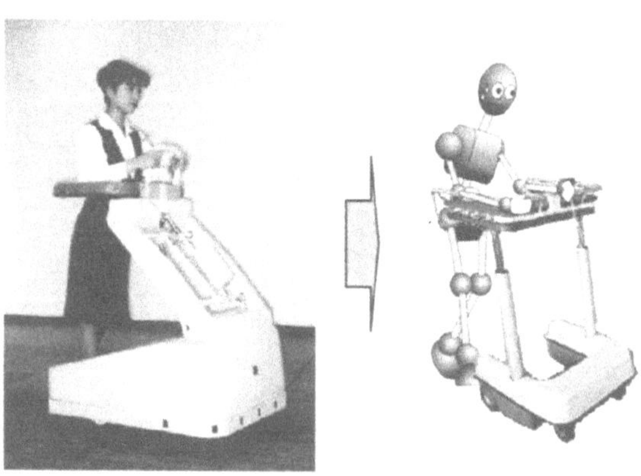

(a) by human (b) by robot

Figure 1. Walk-assist machine

Figure 2. WABIAN-2 LL

2 Mechanical Design of WABIAN-2

2.1 Overview

Figure 3 presents the DOF configuration of the new humanoid robot. Wabian-2 has two 7-DOF legs, a 2-DOF waist, a 2-DOF trunk, two 7-DOF arms, a 3-DOF neck and two 3-DOF hands. In this study, the initial pose of the robot is defined as standing straight, and rotational direction of each joint is defined by using inertial coordinate system fixed on the ground as shown in Fig. 3.

Frameworks of WABIAN-2 are mainly made of duralumin in order to realize antithetical concepts; light weight, high stiffness and wide movable range. Each actuator system of joint consists of a DC motor, a Harmonic drive gear, a lug belt and two pulleys. This double speed reduction mechanism allows high reduction ratio, and also a joint axis to be set apart from a motor axis. This mechanism provides designs for a human-like joint mechanism without a big projection. In this paper, we mainly focus on the development of the waist, trunk and arms. Figure 4 presents a photo of WABIAN-2. The details are described as follow.

2.2 Waist mechanism

Klopsteg et al. (1963) have proposed the result of the gate analysis of humans. Figure 5 shows the pelvis and the knee motion plotted in the steady walking phase. The data is based on experimental results of 8 people walking motion who do not have physical handicaps. In the result, human's pelvis motion in steady walking is observed in frontal plane (defined as roll motion in this study) and horizontal plane (defined as yaw motion). Waist motion in sagittal plane (defined as pitch motion) is seldom observed. According to this a humanoid robot which can perform walks similar to human should be able to move its hips in roll and yaw axes. These hip movements have to be independent in its trunk position.

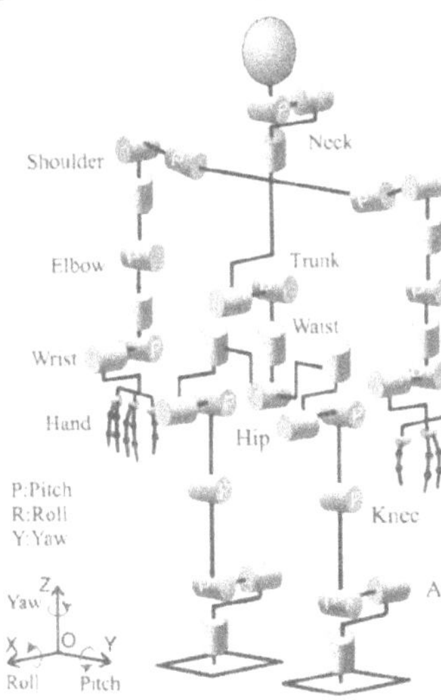

Figure 3 DOF configuration

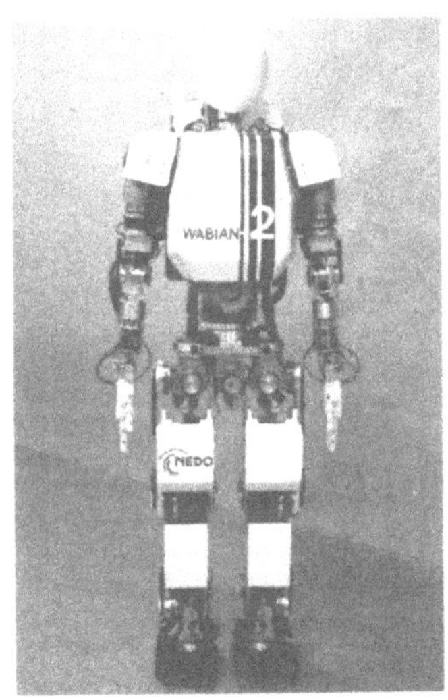

Figure 4. Photo of WABIAN-2

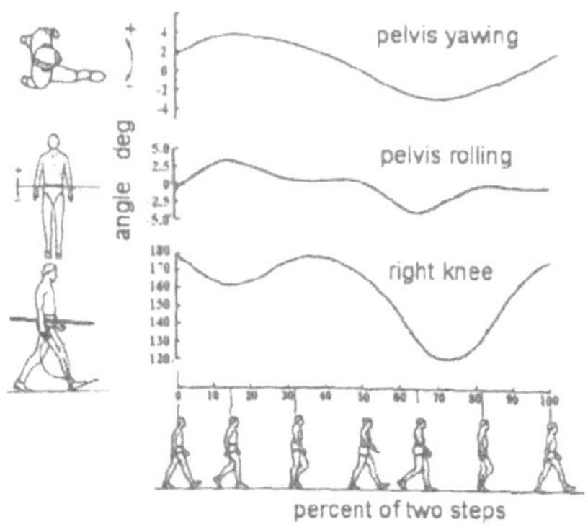

Figure 5 Human's pelvis and knee motion [10]

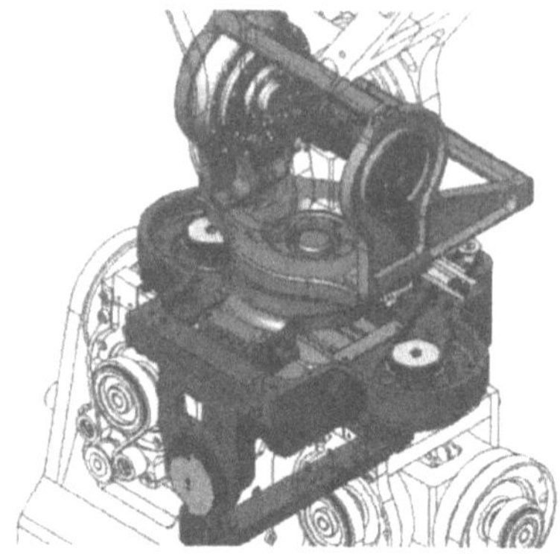

Figure 6. Waist mechanism

Figure 6 shows the 2-DOF waist system. 2-DOF waist combination of a roll and a yaw joint is attached on the middle between the hip joints. The roll axis and yaw axis should be perpendicular to each other, and crossing the middle point between the two hip joints. This will result in minimizing the displacement of the trunk by waist motion and simplifying the kinematics calculation.

2.3 Trunk

Humans can move their trunk independent of the pelvic motion. The Japanese Association of Rehabilitation Medicine and the Japanese Orthopaedic Association have established the basic rules of representation and measurement methods for the range of motion (ROM) (see Nakamura and Saito 1992). The general concept of ROM does not always imply joints or articulation. However, it serves as a useful guide in the development of a humanoid robot. Table 1 lists the rules and results for trunk motions (thoracic and lumbar spines). These ROM measurements for an average human were either carried out in a sitting position or with instruments that held the pelvis in initial position. This table shows that humans are capable of three dimensional motions at least in their trunk. These trunk motions are essential for a human motion simulator to mimic various human motions. For example, humans bend their trunks in order to rise from a sitting position, walk with a limp swinging their trunk to maintain the balance of the entire body, or twist their trunks to walk fast.

Table 1 ROM of thoracic and lumber spines

Direction of movement	Flexion	Extension	Rotation		Lateral bending	
			Left	Right	Left	Right
ROM	45	30	40	40	50	50
Figures	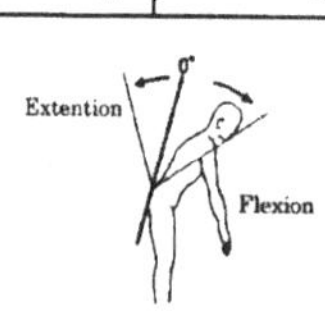		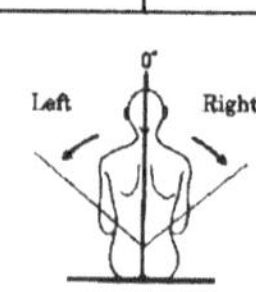		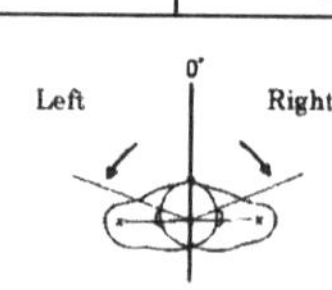	

Figure 7 shows the 2-DOF trunk system. 2-DOF trunk having a pitch and a roll joint is assembled over the waist. In addition, the yaw joint mounted on the waist is able to be used as yaw rotation for both the hips and the trunk. This DOF configuration of waist and trunk gives substantial 3-DOF trunk motions mentioned in Table 1.

2.4 Arms

The arm of WABIAN-2 has 7-DOF. Figure 8 shows the 3D-CAD. The arms were designed in such a way that can support the robot balance while it is walking. It includes three actuators for fingers that can bend like human's hand fingers. Moreover, the arms were designed to hold the robot weight while it leans on a walking assist machine.

Since the robot could lean on a walking assist machine, most of the robot weight will become on both the elbow. In order to determine the suitable angles of the arms some software simulation were conducted. In this simulation, the robot leans on a walking assist machine using its forearms. The elbow angle is 15deg from a posture bend at right angle, when the arm supports a half weight of the robot (30kg, the two arms support a whole weight of the robot (60kg)).

3 EXPERIMENTS

3.1 Walking Experiments

Walking experiments with two walking styles were carried out on a horizontal flat plane by using WABIAN-2. First, conventional walking style which is the previous walking with a constant waist height and bending knees at all times. Second, stretch walking in which walking pattern generated using predetermined knee trajectories to imitate a human gait. The experiments were done with a step cycle of 0.96s/step, a step height of 0.03m and five different step lengths from 0.00m to 0.20m with 0.05m difference. The results showed that stable walking with all step lengths can be performed. Figure 9 shows a scene of the walking experiment with a step length of 0.20m/step.

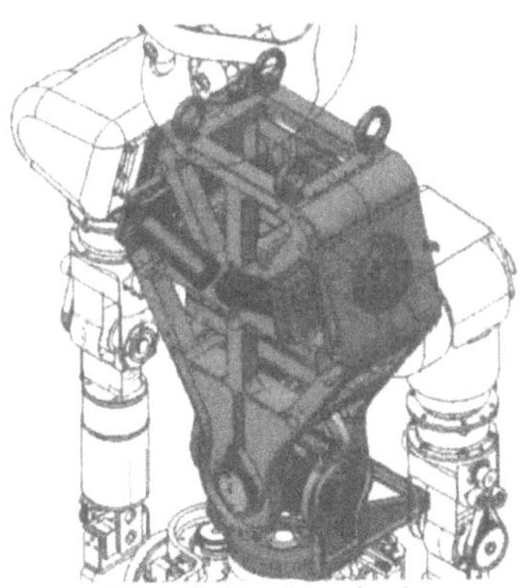

Figure 7 Trunk mechanism

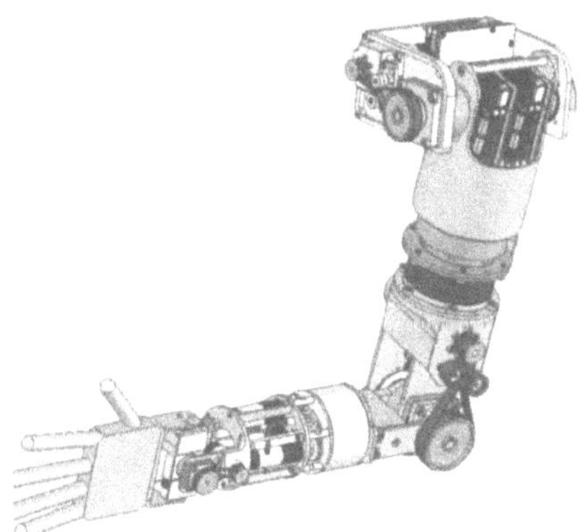

Figure 8 Arm mechanism

To compare two style walks with a human walk, measurements of vertical force applied to 6-axes force sensor on foot were carried out in the experiments of the step length of 0.20m. Figure 10 shows the results of the right foot in each walking cycle in steady walking. A measurement of a human was carried out by using Force Plates. In the case of the conventional walking, substantially constant vertical forces are applied in the support phase. On the other hand, the result of the stretch walking shows double peaks of vertical forces as like as the measurements of human's.

3.2 Walking Experiment with a walk-assist machine

In order to verify the effectiveness of the control method for forearms and hardware of the humanoid robot, assisted-walking experiments with a walk-assist machine were carried out. The experiments were conducted with a step cycle of 0.96 s/step, a step height of 0.03 m, a step length of 0.20 m, and three different heights of arm rests of the walk-assist machine that were varied every 0.05 m from 0.85 m to 0.95 m. The walking styles employed in these experiments are conventional with constant waist height and knees bent throughout the entire walking phase.

The results showed that stable walking with all step lengths is possible. Figure 11 shows a picture of this walking experiment with an arm rest height of 0.90 m/step. Figure 12(a) and (b) show the graphs of experimental data with regard to the reaction forces applied to the right arm and the right foot in the experiments. Furthermore Figure 13 shows the experimental results of the energy consumptions of knee joint actuators in the experiments.

Figure 12 and 13 show that the forces applied to the right arm are increased as the armrest of the walk- assist machine is set lower. On the other hand, the forces applied to the foot are decreased in this case. These results imply that the adjustment of the height of the armrest can adjust load distribution between the upper limb and lower limb.

(a) Conventional walking (b) Stretch walking

Figure 9 Walking experiments

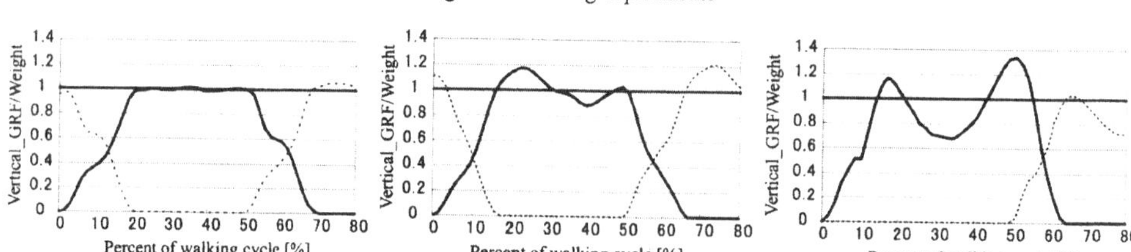

(a)Conventional walking (b) Stretch walking (c) A human

Figure 10 Comparison of ground reaction forces

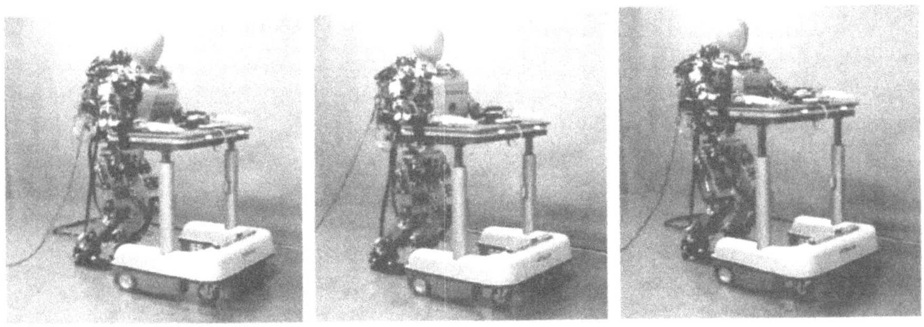

(a) Height of armrest: 0.85m (b) Height of armrest: 0.90m (c) Height of armrest: 0.95m

Figure 11 Walking Experiments with a walk-assist machine

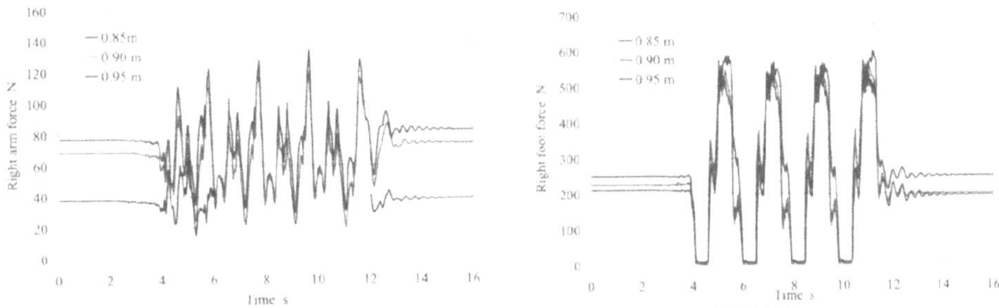

(a) Right forearm (b) Right leg

Figure 12. Reaction forces data

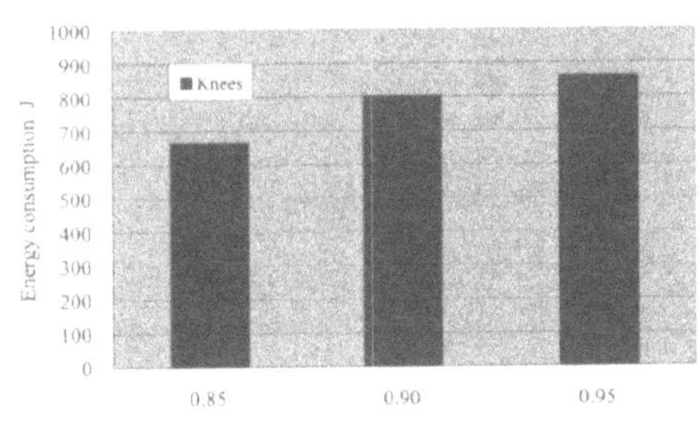

Figure 13 Comparison of energy consumptions of knee joints

In addition, some clinicians frequently commented the following: In a general clinic, the initial position of the armrests is set according to the height of the user's elbow; however, if the user has no handicap, it is set higher than the initial position. If the user has lower-limb disability such as a person who requires a cane to walk, the height of the armrest is decreased. These results show that WABIAN-2 can provide quantitative data, which supports qualitative clinical opinions.

4 Conclusion and future work

This paper describes development of a new humanoid robot that can be as a hardware simulator for human motion. The humanoid robot has 7-DOF legs, a 2-DOF waist, a 2-DOF trunk, and 7-

DOF arms. In the development of the robot, new design principle for a robot which can use walking assist machine is proposed. Moreover a method for redundant trunk motion to control forearms is proposed. Mechanical design and effectiveness of the control method of the robot are confirmed by several kinds of experiments.

In the near future, it is necessary to utilize walking parameters, such as foot trajectory, ZMP trajectory, and so on, to get an efficient walking motion. Also, a hardware simulator system capable of being applied to the evaluation of welfare machines or robots shall be proposed. Moreover the effectiveness of this proposal shall be verified through many experiments.

Bibliography

Sakagami, Y., Watanabe, R., Aoyama, C., Matsunaga, S., Higaki, N. and Fujimura, K. (2002). The intelligent ASIMO: System overview and integration. In *Proc. 2002 IEEE/RSJ Int. Conference on Intelligent Robots and Systems*, 2478-2483.

Kaneko, K., Kanehiro, F., Kajita, S., Yokoyama, K., Akachi, K., Kawasaki, T., Ota, S. and Isozumi, T. (2002). Design of Prototype Humanoid Robotics Platform for HRP. In *Proc. 2002 IEEE/RSJ Int. Conference on Intelligent Robots and Systems*, 2431-2436.

Fujiwara, K., Kanehiro, F., Kajita, S., Yokoi, K., Saito, H., Harada, K., Kaneko, K. and Hirukawa, H. (2003). The First Human-size Humanoid that can Fall Over Safely and Stand-up Again. In *Proc. 2003 IEEE/RSJ Int. Conference on Intelligent Robots and Systems*, 1920-1926.

Nishiwaki, K., Sugihara, T., Kagami, S., Kanehiro, F., Inaba, M. and Inoue, H. (2000). Design and Development of Research Platform for Perception-Action Integration in Humanoid Robot: H6. In *Proc. 2000 IEEE/RSJ Int. Conference on Intelligent Robots and Systems*, 1559-1564.

Nishiwaki, K., Kagami, S., Kuniyoshi, Y., Inaba, M. and Inoue, H. (2002). Online Generation of Humanoid Walking Motion based on a Fast Generation Method of Motion Pattern that Follows Desired ZMP. In *Proc. 2002 IEEE/RSJ Int. Conference on Intelligent Robots and Systems*, 2684-2689.

Löffler, K., Gienger, M. and Pfeiffer, F. (1999). Sensor and Control Design of a Dynamically Stable Biped Robot. *In Proc. 1999 IEEE Int. Conference on Robotics and Automation*, 484-490.

Lim, H., Ishii, A. and Takanishi, A. (1999). Motion Pattern Generation for Emotion Expression. In *Proc. 1999 Int. Symposium on Humanoid Robots*, 36-41.

Lim, H., Kaneshima, Y. and Takanishi, A. (2002) Online Walking Pattern Generation for Biped Humanoid with Trunk. In *Proc. 2002 IEEE Int. Conference on Robotics and Automation, Washington*, 3111-3116.

Kim, J., Park, I., Lee, J., Kim, M., Cho, B. and Oh, J. (2005) System Design and Dynamic Walking of Humanoid Robot KHR-2. In *Proc. 2005 IEEE Int. Conference on Robotics and Automation*, 1443-1448.

Ogura, Y., Aikawa, H., Lim, H. and Takanishi, A. (2004) Development of a Human-like Walking Robot Having Two 7-DOF Legs and a 2-DOF Waist. In *Proc. 2004 IEEE Int. Conference on Robotics and Automation*, 134-139.

Ogura, Y., Kataoka, T., Shimomura, K., Lim, H. and Takanishi, A. (2004). A Novel Method of Biped Walking Pattern Generation with Predetermined Knee joint Motion. In *Proc. 2004 IEEE/RSJ Int. Conference on Intelligent Robots and Systems*, 2831-2836.

Klopsteg, P. E. and Wilson, P. D. et al. (1963). *Human Limbs and Their Substitutes*, New York Hafner.

Nakamura, R. and Saito, H. (1992). *Kiso Undougaku (Fundamental Kinesiology)* . Ishiyaku Publishers, Inc. 4th ed.

From the Human Hand to a Humanoid Hand: Biologically-Inspired Approach for the Development of RoboCasa Hand #1

Massimiliano Zecca*[1,2,3], Stefano Roccella[3], Giovanni Cappiello[3], Kazuko Ito[4], Kazutaka Imanishi[4], Hiroyasu Miwa[1,5,6], M. Chiara Carrozza[2,3], Paolo Dario[2,3], Atsuo Takanishi[1,2,4,5,7]

[1]Consolidated Research Institute for Advanced Science and Medical Care, Waseda University, Tokyo, Japan, email: zecca@aoni.waseda.jp; takanisi@waseda.jp
[2]RoboCasa, Waseda University, Tokyo, Japan
[3]ARTS Lab, Scuola Superiore Sant'Anna, Pisa, Italy
[4]Department of Mechanical Engineering, Waseda University, Tokyo, Japan
[5]Humanoid Robotics Institute (HRI), Waseda University, Tokyo, Japan
[6]Digital Human Research Centre, National Institute of Advanced Industrial Science and Technology (AIST), Tokyo, Japan
[7]Advanced Research Institute for Science and Engineering, Waseda University, Tokyo, Japan

Abstract. In a society getting older year by year, Robot technology (RT) is expected to play an important role. In order to achieve this objective, the new generation of personal robots should be capable of a natural communication with humans by expressing human-like emotion. In this sense, the hands play a fundamental role in communication, because they have grasping, sensing and emotional expression ability. This paper presents the recent results of the collaboration between the Takanishi Lab of Waseda University, Tokyo, Japan, and the Arts Lab of Scuola Superiore Sant'Anna, Pisa, Italy, and RoboCasa in a biologically-inspired approach for the development of a new humanoid hand. In particular, the grasping and gestural capabilities of the novel anthropomorphic hand for humanoid robotics RCH-1 (RoboCasa Hand No.1) are presented.

*The authors would like to express their thanks to Okino Industries LTD, OSADA ELECTRIC CO. LTD, SHARP CORPORATION, Sony Corporation, Tomy Company LTD and ZMP INC. for their financial support to the Humanoid Robotics Institute; to the Italian Ministry of Foreign Affairs, General Directorate for Cultural Promotion and Cooperation, for its support to RoboCasa; to thank ARTS Lab, NTT Docomo, SolidWorks Corp., the Advanced Research Institute for Science and Engineering of Waseda University, Prof. H. Kimura for their supports to the research. Partial support has been provided by a Grant-in-Aid for the WABOT-HOUSE Project by Gifu Prefecture; by Waseda University Grant for Special Research Projects (No. 266740); by JSPS for the Post-Doc fellowship received by Dr. Zecca.

1 Introduction

The average age of most industrialized countries is rising fast due to an increased expectation of life and a contemporary reduced child birth rate (NIPS (2003); OECD (2004)). While today there are about 2.8 workers per retiree, this figure is estimated to fall to 1.4 in 2050. Thus, there is considerable expectation that in the next years there will be a growing need for greater medical and nursing care services. One possible solution to support this aging society could be given by the next-generation Robot Technology (RT). These new robots should be capable of a smooth and natural interaction and communication with the human society and its environment, in particular for home and personal assistance of elderly and/or handicapped persons (Takanishi (2004)). Next generation robots, then, should be able not only to work together with humans, but also to express human-like emotions.

In this sense, the aspect and the functionalities of the hand assume a major role. The hand, in fact, not only is a fundamental organ for exploration, manipulation and interaction with objects and other persons (Table 1; Chao (1989); Kapandji (1982)), but it is also a basic tool for communication and expression of emotions through gestures (Kendon (2004)).

This paper describes the realization of RoboCasa Hand No. 1 (RCH-1), recently developed at RoboCasa, the Italy-Japan Joint Research Laboratory on Humanoid & Personal Robotics between the Humanoid Research Institute of Waseda University, Tokyo, Japan, and the ARTS Lab of Scuola Superiore Sant'Anna, Pisa, Italy.

2 The model: the human hand

The model for the realization of an artificial hand is of course the human hand, which is a truly remarkable instrument capable of realizing very complex and useful tasks using an extremely efficient combination of mechanisms, sensing, actuation and control functions (Chao (1989); Kapandji (1982)). The human hand is not only an effective tool but also an ideal instrument to acquire information from the external environment thanks to a huge number of sensors (Table 1). Moreover, it is capable of expressing emotions and feelings through gesture. MacKenzie and Iberall (1994) list about three hundred tasks that hands perform related to the activities of daily living. Therefore the replication of its mechanical capabilities and of its sensory–motor functionalities is a continuous challenge for scientists and engineers.

As a general criterion, artificial hands could be divided into two main categories:

1. prosthetic hands;
2. robotic hands.

Prosthetic devices are usually simple gripper (OttoBock (2001)) with one or two DOFs, with high reliability and robustness, but with poor grasping capabilities. Due to the lack of DOFs, such devices are characterized by a low grasping functionality, because they do not allow adequate encirclement of objects. Conversely, robotic hands could achieve better performance in mimicking human capabilities, but they are complex devices requiring large and bulky controllers (Bekey et al. (1990); Bicchi (2000); Borst

Natural Hand performance	
Number of DoFs	22
Wrist mobility	2+1 DoFs
Total volume	≃50 cc
Weight	400 g
Type of Grasps	Power Grasps, Precision Grasps
Force of power grasp	>500 N (age 20–25); >300 N (age 70–75)
Two fingers force	>100 N
Tapping force	1–4 N
Max. tapping frequency	4.5/sec.
Range of flexion	100°, depending on the joint
Max. duration of grasp	Var. with energy
Number of sensors	≃ 17'000
Proprioceptive sensing:	Position; Movement; Force
Exteroceptive sensing:	Acceleration; Force; Pain; Pressure; Temperature
Proportional Control	Ability to regulate force and velocity according to the type of grasp, the object, etc.
Stability	The grasp is stable against incipient slip or external load
Possible flexions	Limited only by muscular fatigue

Table 1. Main characteristics of the human hand (adapted from Zecca (2003)).

	Size (norm.)	# of fingers	DoFs	Actuators (Type)	Control	# of sensors	Weight (gr)	Force (N)	Speed (sec)
Human hand	1	5	22	38 (I+E)	E	≃ 17'000	≃ 400	>300	0.25
Ottobock SUVA	1	3	1	1 (E)	I	2	600	≃ 100	<1
MARCUS Hand	≃ 1.1	3	2	2 (I)	I	3	–	–	–
Southampton hand	≃ 1	5	6	6 (E)	E	–	400	38	≃ 5
Hokkaido hand	>1	5	7	7 (E)	E	–	125	–	–
Karlsruhe	≃ 1	5	17	17 (E)	E	–	≃ 120	12	0.1
NTU Hand	≃ 1	5	17	17 (E)	E	35	1570	–	–
Utah/MIT	≃ 2	4	16	32 (E)	E	16	–	31.8	–
Belgrade/USC	≃ 1.1	4	4	4 (E)	E	23+4	–	–	–
Stanford/JPL Hand	≃ 1.2	3	9	12 (E)	E	–	1100	-45	–
DLR Hand II	≃ 1.5	4	13	13 (E)	E	64	320	30	–
Robonaut Hand	≃ 1.5	5	12 + 2	14 (E)	E	43 +	–	–	–
RoboCasa Hand #1	≃ 1	5	16	6 (I+E)	E	24	350	≃ 40	0.25

Table 2. Comparison among several types of artificial hands (modified after Zecca (2003)). The size of each hand is normalized respect to the adult human hand. In the Actuator type and in the Control rows, '*I*' means '*internal*', and '*E*' means '*external*'. *Speed* indicates the time required for a full closing and opening.

et al. (2003); Jacobsen et al. (1984); Salisbury and Craig (1982)). Table 2 presents a comparison among several types of artificial hands.

RCH-1 aims at filling the gap between these two main classes of artificial hands, by realizing a device as close as possible to the human hands in terms of dimensions and performance without increasing too much the complexity of the system.

3 RoboCasa Hand #1 (RCH-1)

RCH-1 should mimic as close as possible the performance of the natural hand summarized in Table 1. In addition, further requirements on the specifications for grasping and gesture should be fulfilled.

Grasping capabilities Grasping capabilities are required for a better interaction with humans and with the surrounding environment, especially considering a robot for home and personal assistance. At least the following grasping types are needed:

- **cylindrical grasping**: e.g. to grasp a can, or a small bottle, etc.;
- **spherical grasping**: to grasp small balls, or fruits like orange, apple, and so on;
- **tip pinch**: (thumb and index, or thumb and middle finger) for small objects like candies, pens, etc.;
- **lateral grasping**: for keys, paper, name cards, credit cards, and so on.

Besides the single hand grasping of small objects, the robot should also be capable of two hands grasping of large objects (i.e.: ball, big toys, puppets, etc). Additional functionalities could be the measurement of the hardness of the grasped object, and the recognition of different surfaces.

Gestural capabilities A "gesture" is a form of non-verbal communication made with a part of the body and in particular with the hand, and used instead of verbal communication (or in combination with it). The language of gesture is rich in ways for individuals to express contempt, hostility, or approval towards others. Most people use gestures and body language in addition to words when they speak; some ethnic groups and languages use them more than others do, and the amount of such gesturing that is considered culturally acceptable varies from one location to the next. Thus, the robot should be able to point at objects, waving (calling people), close the hand (fist), shake the hand with its partners, closing mouth when yawn, and some other simple signs like goodbye, ok, good, peace sign, counting, and so on. The robot should also be capable to adapt to change in the working environment, as the same gesture can be interpreted in a different way according to the person and the context. More details and examples on world-wide known gestures can be found in Kendon (2004).

3.1 Realization of the first prototype

According to the above mentioned requirements, two new hands (left and right) have been designed and fabricated at ARTS lab. These new hands have then been integrated into the Emotion Expression Humanoid Robot WE-4R (Waseda Eye #4 Refined: Miwa et al. (2002, 2003)). The result of the integration, named WE-4RII, is shown in Figure 1(a).

Each new hand has in total 16 Degrees of Freedom (DOF) or 6 Degrees of Motion (DOM): 1 DOM/3DOFs for flexion/ extension of each finger, plus one 1DOM/1DOF for thumb adduction/abduction. Each finger is underactuated (Figure 1(b)), and its movement is driven by a single cable actuated by a motor placed in the forearm (Maxon RE-max17 4.5W 216012), thus mimicking the human muscular structure (Figure 2). The

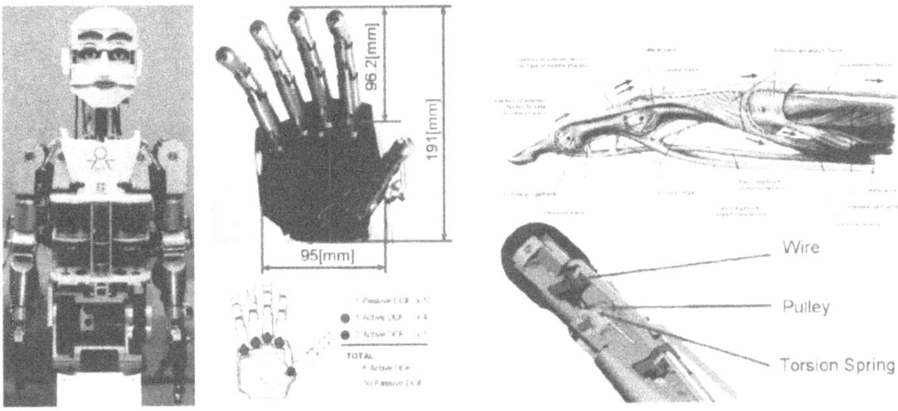

(a) Left: Emotion Expression Humanoid Robot WE-4RII; upper right: RCH-1 with its main dimensions; lower right: active and passive DOFs.

(b) Top: anatomy of the human finger; bottom: schematic view of the finger of RCH-1.

Figure 1. Overview of RCH-1 and details of the finger design.

flexion of the three phalanxes is provided by a single cable connected to the actuator in the forearm, while the extension is obtained trough torsion springs. The motor for thumb ad/abduction is integrated into the palm (Faulhaber 1016M006G) and it is placed in a homologous position as the *abductor pollicis* in the human hand (Figure 3).

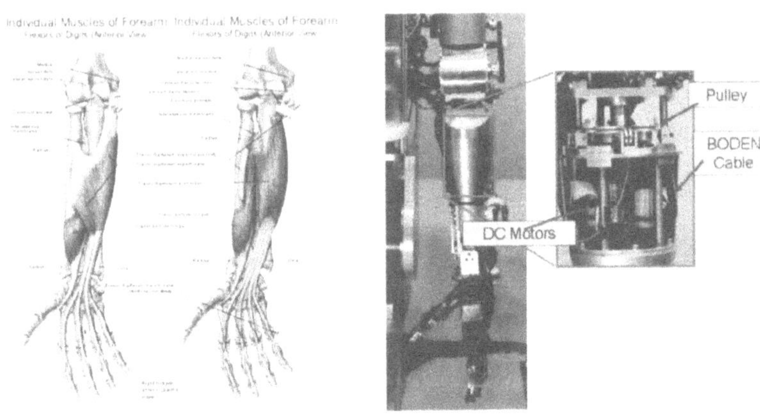

(a) Human anatomy

(b) WE-4RII's forearm

Figure 2. Left: superficial and deep muscles in the human forearm; Right: view of the forearm of WE-4RII, and detailed view of the finger motor block.

RCH-1 has the same dimensions of a average adult. Each finger is 92.2 mm long,

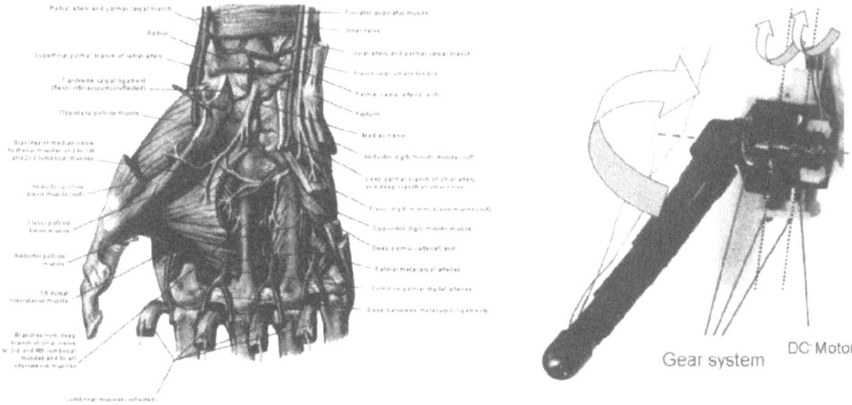

(a) Human anatomy (b) RCH-1's finger

Figure 3. Left: palmar view of the human hand; Right: details of the abduction-adduction mechanism of RCH-1. The motor is placed in the same position as the *abductor pollicis*.

with a diameter of 14 mm; the palm is 95 mm in width and 45 mm in thickness; the overall length of the hand is 192.2 mm; the total weight of the hand is about 320 gr without silicon cover. A more detailed description of the RCH-1 and its integration into the WE-4RII can be found in Zecca et al. (2004).

4 Advanced behaviors for WE-4RII

The ability of WE-4R to show emotions have been tested in several occasions. However, it lacked the capability to interact with objects and people. Thanks to RCH-1, instead, WE-4RII is capable of several additional behaviors. Some examples are shown in Figure 4 (handshake; victory), Figure 5 (dumbbell; maracas), and Figure 6 (throwing; bye-bye).

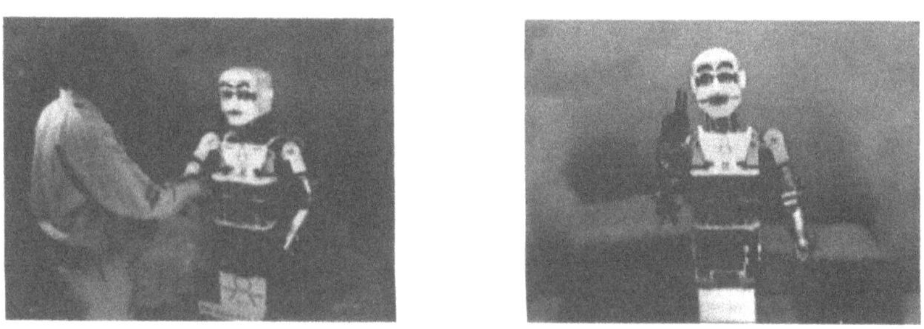

(a) Handshake (b) Victory!

Figure 4. Advanced behaviors for WE-4RII - Part I.

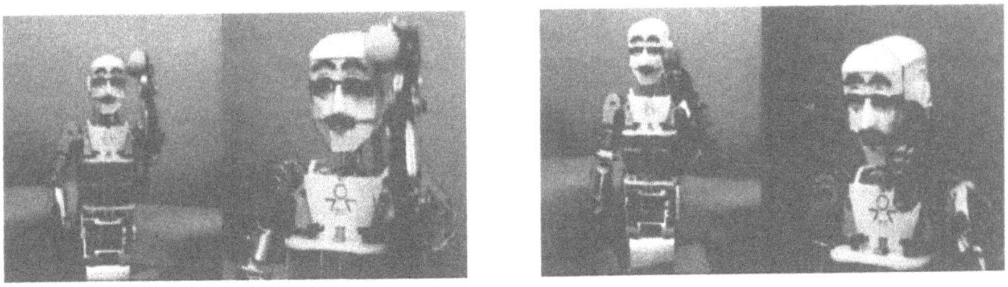

(a) Doing exercise with a dumbbell (b) Playing maracas

Figure 5. Advanced behaviors for WE-4RII - Part II.

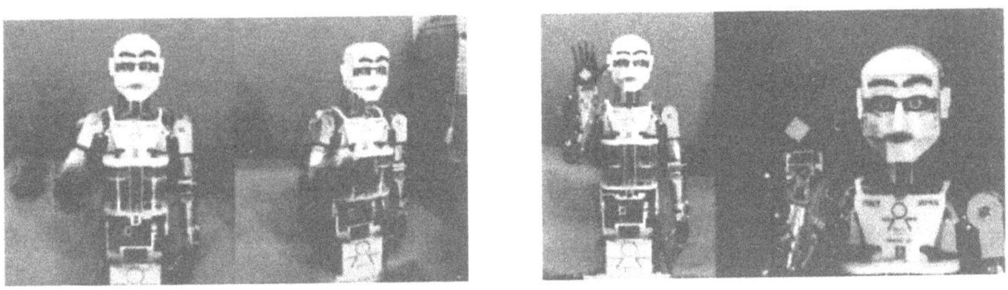

(a) Throwing an object (b) Saying "bye-bye" and waving the hand

Figure 6. Advanced behaviors for WE-4RII - Part III.

5 Results and Discussion

In a society that is getting older year by year, there is a considerable expectations that robot technology could provide a substantial help to the population. However, in order to achieve a smooth and natural integration of the robots, emotions play a major role. A key aspect in successfully emotion expression is the presence of a human-like hand, capable not only of grasping but also of signs and gestures.

In this paper, a biologically inspired approach to the development of a new anthropomorphic hand for humanoid robots, named RCH-1, has been presented. The flexion of the five finger is in fact obtained with the replication of the architecture of the *flexor digitorum superficialis* and of the *flexor digitorum profundus*, by integrating 5 DC motors in the forearm (Figure 2). The extension of the fingers, however, is obtained by using torsion springs integrated into the structures of the phalanxes (Figure 1(a)). The abduction-adduction mechanism of the thumb replicates the function of *abductor pollicis*, *adductor pollicis longus* and *brevis* (Figure 3). However in RCH-1 these functions are obtained with one motor only.

RCH-1 has been successfully integrated into the Emotional Expression Humanoid Robot WE-4RII developed in Takanishi-lab of Waseda University in the framework of the RoboCasa project, co-designing also the forearm and the control system. The effectiveness of the new hands for the interaction with the environment and with the users has been proved through several experiments. With RCH-1, moreover, several additional behaviors became possible.

Although RCH-1 represents a significant advancement of the state of the art, it still has several drawbacks. For example, the structure of the fingers is slightly too fragile, and sometimes the mechanical stops (which should prevent the phalanxes to bend backwards) break. Another limitation is the absence of a human-like skin, in terms of both compliance and sensory system. Furthermore, the current shape of the palm (i.e. flat) slightly reduces the grasping capabilities. The current activities in RoboCasa aim at solving these problems.

Bibliography

G.A. Bekey, R. Tomovic, and I. Zeljkovic. *Dextrous Robot Hands*, chapter Control Architecture for the Belgrade/USC Hand, pages 136–149. NY: SpringerVerlag, 1990.

A. Bicchi. Hands for dexterous manipulation and robust grasping: a difficult road toward simplicity. *IEEE Transactions on Robotics and Automation*, 16(6):652–662, 2000.

C. Borst, M. Fischer, S. Haidacher, H. Liu, and G Hirzinger. DLR hand II: experiments and experience with an anthropomorphic hand. In *International Conference on Robotics and Automation*, volume 1, pages 702–707, 2003.

E. Chao. *Biomechanics of the Hand.* World Scientific, 1989.

S. C. Jacobsen, J.E. Wood, D.F. Knutti, and K.B. Biggers. The Utah/MIT hand: work in progress. *The International Journal of Robotics Research*, 3(4):21–50, 1984.

I. A. Kapandji. *The Physiology of the Joints, Vol. 1, Upper Limb.* Edinburgh: Churchill Livingstone, 1982.

Adam Kendon. *Gesture: Visible Action as Utterance.* Cambridge University Press, 2004.

C.L. MacKenzie and T. Iberall. *The grasping hand.* Amsterdam: North Hollan, 1994.

H. Miwa, T. Okuchi, H. Takanobu, and A. Takanishi. Development of a New Human-like Head Robot WE-4. In *IROS2002*, volume III, pages 2443–2448, 2002.

H. Miwa, H. Takanobu T. Okuchi, K. Itoh, and A. Takanishi. A new mental model for humanoid robots for human-friendly communication - introduction of learning system, mood vector and second order equations of emotion. In *IEEE International Conference on Robotics & Automation*, pages 3588–3593, 2003.

NIPS. Population statistics of japan 2003. Technical report, National Institute of Population and Social Security Research, Hibiya, Chiyoda-ku, Tokyo, Japan, 2003.

OECD. Population statistics. Technical report, Organisation for Economic Co-operation and Development, `http://www.oecd.org`, 2004.

OttoBock. *SUVA Hand technical manual.* Otto Bock GmbH, Duderstadt, DE, 2001.

J. K. Salisbury and J. J. Craig. Articulated hands: force control and kinematic issue. *The International Journal of Robotics Research*, 1(1):4–17, 1982.

A. Takanishi. World robot declaration. International Robot Fair 2004, 2004.

M. Zecca. *On the development of a cybernetic prosthetic hand.* PhD thesis, Scuola Superiore Sant'Anna, Pisa, Italy, 2003.

M. Zecca, S. Roccella, M.C. Carrozza, H. Miwa, K. Itoh, G. Cappiello, J.J. Cabibihan, M. Matsumoto, H. Takanobu, P. Dario, and A. Takanishi. On the development of the emotion expression humanoid robot WE-4RII with RCH-1. In *2004 Int. Conf. on Humanoid Robots*, volume 1, pages 235–252, 2004.

Chapter VI

Biology and Robotics – Specialized Tools and Methods

Towards Realistic Surgical Simulation: Biomechanics of Needle Insertion into the Brain

Tonmoy Dutta-Roy [1] Adam Wittek[1,3] Zeike Taylor[1] Kiyoyuki Chinzei[2] Toshikatsu Washio[2] and Karol Miller[1,3]

[1] Intelligent Systems for Medicine Laboratory, The University of Western Australia, Crawley/Perth, Australia, {tduttar, adwit, ztaylor, kmiller}@mech.uwa.edu.au
[2] Institute for Human Science and Biomedical Engineering, Surgical Assist Technology Group, AIST, Tsukuba, Japan, {k.chinzei, washio.t}@aist.go.jp

Abstract. Robotic surgery has been recognised in the past few years to have immense potential. Understanding the biomechanics of surgical procedures is central to robotic surgery. Needle insertion into soft tissues is a common surgical procedure but the biomechanics of needle insertion is poorly understood. We developed a computational biomechanical model to understand the mechanics of needle insertion into the brain tissue. The brain tissue is considered as a single phase continuum undergoing finite deformations. A non-linear constitutive model of the brain tissue is used. Precise geometrical model of the brain is obtained from MRI images and the brain mesh is created. The brain computational model is verified by comparing with previously published experimental results for porcine brain indentation. The reaction forces acting on the needle during insertion are obtained using fully non-linear, explicit Finite Element procedures in time domain.

1 Introduction

Advantages of surgical robots and manipulators are well recognised in clinical and technical communities. They include precision, accuracy and potential for remote surgery. Precision and accuracy are key factors when performing the most common neurosurgical procedure: needle insertion into the brain.

Planning needle insertion is seemingly simple: a needle is aimed at an anatomical target whose location can be determined from pre-operative radiological images (*e.g.*, magnetic resonance images or MRI). In reality, reaching the target requires advanced surgical skills. This is because of the change of target position caused by brain tissue deformation under the load imposed by the needle. Thus, trajectory planning for the manipulators performing a surgical procedure, such as needle insertion into the brain, must include information about current, i.e. intraoperative organ deformation. Although such information can be obtained using intraoperative MRI, currently tens of seconds are required to produce and analyse a new set of images. One possible method of dealing with these delays could be predicting organ deformation

[3] The study has been supported by grants provided by the University of Western Australia (UWA) Grant Number 12104012 under the UWA Research Grant Scheme and ARC Linkage International Grant Number LX0560460.

by means of computer simulations. When organ deformation can be predicted and the tissue mechanical properties are known, interaction force between the end-effecter (i.e. needle or other surgical tool) and the organ surface during surgery can also be computed. Thus, the simulation can be used to predict reaction forces acting on the needle as well as to provide realistic force and tactile feedback for virtual reality surgical simulators.

Realistic simulations cannot be done without thorough understanding of the biomechanics of needle insertion. Yet, most of the previous studies on needle insertion have been focused on haptics, visual feedback and surgery planning. With the exception of Simone and Okamura (2002), Maurin et al. (2004) and Washio and Chinzei (2004), only a few studies attempted to understand the biomechanics of needle insertion.

DiMaio and Salcudean (2003) developed interactive needle insertion simulators for haptic and visual feedback. Alterovitz et al. (2005) dealt with the simulation and planning of needle insertion for radioactive seed implantation for robotic surgery in prostate brachytherapy. Nienhuys (2003) used non-linear finite deformation analysis to predict deformation of soft objects subjected to cutting. Heverly et al. (2005) used a fracture mechanics approach to model needle insertion.

Despite experimental evidence that the brain and other soft tissues exhibit non-linear constitutive behaviour and the obvious fact that during needle insertion procedure tissue undergoes large deformation, studies conducted by Simone et al. (2002) and O'Leary et al. (2003) rely on linear stress-strain relationship and assume infinitesimal deformations when calculating organ deformation and the needle reaction forces. In contrast, our work employs finite deformation procedures and a non-linear constitutive model of the brain tissue to calculate the forces acting on the needle during insertion thus giving higher reliability for the calculated forces.

Subsequent parts of the paper will deal with establishing biomechanics of needle insertion (Section 2), a computational model for insertion simulation (Section 3), and application of this model (Section 4).

2 Biomechanics of needle insertion: experimental data

The mechanics of needle insertion into very soft organs have been determined by Washio and Chinzei (2004) in their experiments on porcine liver and spleen. We have based our work on their findings. The results by Washio and Chinzei (2004) are summarized in Figure 1. At point 1 in Figure 1, the needle touches the organ surface and organ deformation starts. This is accompanied by an increase in force acting on the needle. As the needle moves forward further, the force continues to increase since the membrane (i.e. pia matter in case of the brain) surrounding the organ stretches and the organ deforms. Consequently, elastic energy is stored in the membrane. At point 2 in Figure 1, the membrane ruptures accompanied by a drop in force acting on the needle tip. The elastic energy stored in the membrane is released. This is followed by "springback" of both the organ (brain) and the membrane (pia). After springback, the needle proceeds further into the organ. Therefore, based on results by Washio and Chinzei (2004), we divided needle insertion into four steps:

1. Maximum deformation of the brain and the pia just before failure of pia
2. Failure of the pia
3. Springback of the pia and the brain
4. Further insertion of the needle into the brain.

In this paper, we deal only with step 1 as steps 2 to 4 would require extensive experimental study to determine the failure mechanism/mechanisms of the pia and brain tissues. Subsequent sections of this paper describe the biomechanical and computational models for the simulation of step 1 (till point 2 in Figure 1).

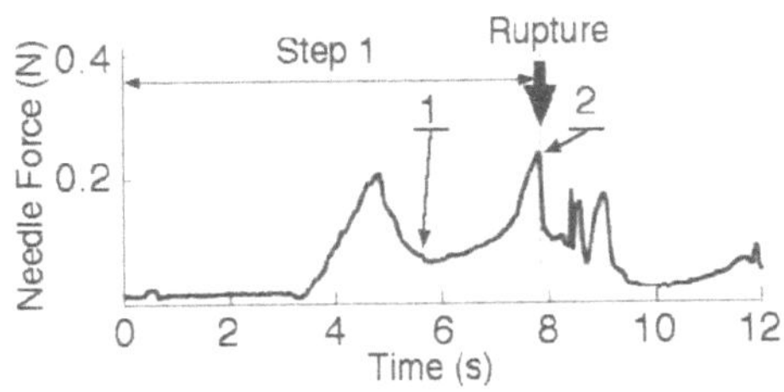

Figure 1 Needle force for constant insertion speed (5 mm/sec). Adapted from Washio and Chinzei (2004).The first peak can be attributed to cardiac and/or pulmonary motions.

3 Biomechanical Model

3.1 Physical Model

Brain: The effects of cerebrospinal fluid flow in the brain were disregarded as the time scale of this flow is many times larger than that of the surgical procedures. Consequently, the brain tissue was treated as a single phase continuum. During needle insertion, the brain tissue undergoes large deformations. Therefore, unlike in the previous studies discussed in Section 1, hyper-viscoelastic constitutive model according to Miller and Chinzei (2002) was used for the brain tissue. This model is described using the following formulae:

$$W = \frac{2}{\alpha^2} \int_0^t [\mu(t-\tau) \frac{d}{d\tau}(\lambda_1^\alpha + \lambda_2^\alpha + \lambda_3^\alpha - 3)] d\tau \quad (1)$$

$$\mu = \mu_0 [1 - \sum_{k=1}^{n} g_k (1 - e^{-\frac{t}{\tau_k}})] \quad (2)$$

where W is the potential function, λ_i are the principal stretches, μ_0 is the instantaneous shear modulus in undeformed state, τ_k are the characteristic times, g_k are the relaxation coefficients and α is the material coefficient which can assume any real value without any restrictions.

Constants for the brain tissue constitutive model were taken from the study by Miller and Chinzei (2002) (Table 1).

Table 1: Material constants for the constitutive model of the brain tissue

Instantaneous Response	μ_0=842 Pa	
	Characteristic Time	Relaxation Coefficients
k=1	τ_1=0.5 sec	g_1= 0.450
k=2	τ_2=50 sec	g_2= 0.365

Pia matter: The pia matter is a thin tissue membrane surrounding the brain. During needle insertion, it can be subjected to very large deformations. Therefore, we used a hyperelastic material model proposed by Weiss et al. (1996) and Puso et al. (1998) to model the pia. The model parameters were determined using Poisson's ratio of 0.49 and Young's modulus of 1 MPa – a value derived from the studies on brain injury by Takhounts et al. (2003) and Zhang et al. (2003).

Skull: The skull has been considered rigid for this study.

Loading: As in the present study we are focusing on the biomechanics of needle insertion into the brain rather than the effects of specific type of needle or needle tip shape on the forces acting on the needle, the needle was not directly modelled. The brain model was loaded through the prescribed motion of four nodes that would be in contact with the needle during the actual insertion (Figure 2). The sum of forces at these nodes was a close approximation of the reaction force between the brain surface and the needle.

Following the previous experiments by Washio and Chinzei (2004) on the needle insertion into very soft organs, a constant velocity of 5 mm/s was used when prescribing the nodal motion. The velocity was applied for 2.4 s, which corresponds to needle insertion depth of 12 mm. This depth was selected as preliminary experiments conducted by the Surgical Assist Technology Group of National Institute for Advanced Industrial Science and Technology (AIST) in Tsukuba, Japan suggests that when inserting a needle with conical tip of diameter 0.5 mm into the porcine brain, the pia failure occurs for insertion depth of around 10-12 mm.

Boundary Conditions: Following the study by Miller et al. (2000), it was assumed that the bottom surface of the brain does not move. Therefore, all nodes on this surface were fixed. Nodes on the brain top surface, on other hand, were constrained only by surface-to-surface contact between the brain and the skull. This contact was defined with a 0.75 mm gap between the brain and the skull. Gap thickness was taken from the study by Miller et al. (2000).

3.2 Continuum Mechanics of Brain Tissue

The brain tissue biomechanics has been detailed by Miller (2002). In this section, only basic ideas are summarised.

The mathematical model of the deformation behaviour of the brain as a single-phase continuum consists of standard solid-mechanics partial differential equations of equilibrium (or dynamics), boundary and initial conditions, and a constitutive model for the brain tissue. During needle insertion, parts of the brain undergo large deformations. Thus, complete non-linear equations must be used, and not, as in some previous studies, equations of linear elasticity.

In the present study, the stresses and strains were measured with respect to the deformed configuration, i.e. Cauchy stress and Almansi strain were used. Thus, the equations of equilibrium are given by:

$$\tau^{ij}_{,i} + \rho F_i = 0 \tag{3}$$

where, τ is Cauchy stress, ρ is mass density and F_i is body force per unit mass in direction i. The comma indicates covariant differentiation with time with respect to deformed configuration and repeated index summation convention has been used.

3.3 Computational Model

Finite element solver: The model was computed using *LS-DYNA* (Livermore Software Corporation, Livermore California, USA) non-linear finite element code. This code uses explicit time integration and accounts for both the constitutive and geometric (i.e. finite deformation formulations are used) nonlinearities.

Brain: The brain geometry for the model was taken from a set of magnetic resonance images (MRI) of the porcine brain obtained by Surgical Assist Technology Group of AIST in Tsukuba, Japan. When conducting MRI scans, the brain was placed in a paper clay cast simulating the skull. MRI images were processed using *SLICER* software (http://slicer.org) to obtain the brain cross-sections. From the sections, 3-D finite element mesh of the porcine brain consisting of 19706 hexahedrons (8 node brick elements) was built using *Hypermesh* (Altair Engineering, USA) pre-processing software (Figure 2). Rationale for using the hexahedron element was that they are known to be the most suitable for explicit non-linear computation of incompressible/nearly incompressible continua, such as the brain, as they do not exhibit volumetric locking (Hughes, 2000).

In order to accurately model the needle-brain interactions, the element size when discretising the brain should not exceed the needle diameter. As the diameter was 0.55 mm, over million elements would be required to discretise the entire brain. This would result in extremely long computation times. Therefore, only "a graft" of dimensions 6x6x12 mm in direct neighbourhood of the needle insertion point was discretised using elements of size 0.5 mm. Elements of size of up to 3.5 mm were used for the remaining part of the brain (Figure 2). Tied contact interface (Hughes et al, 1976) was applied to connect the graft with remaining part of the model.

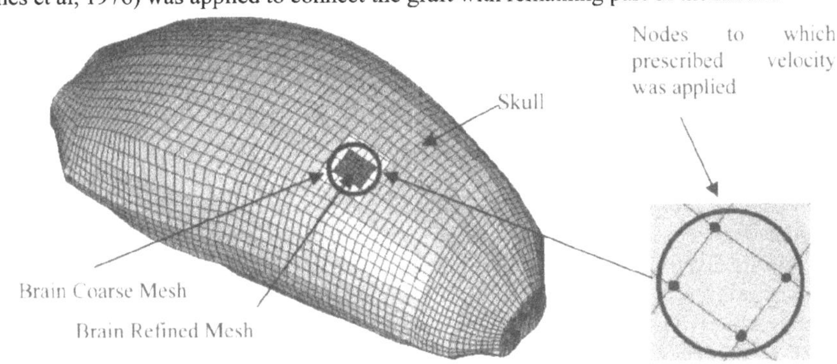

Figure 2 Brain model for needle insertion simulation.

Pia matter: The pia mater surrounding the brain was discretised using 4567 Belytschko-Tsay shell elements (Belytschko et al., 1984) with one integration point across thickness, i.e. these elements had no bending stiffness and behaved as a membrane. Pia thickness was designated as 0.10 mm – a value consistent with the experimental results of Aimedieu et al. (2004).

3.4 Model Verification

No quantitative experimental data on the forces acting on the needle during its insertion into the brain have been obtained so far. Therefore, to evaluate our model in terms of constitutive modelling of the brain and pia tissues, we applied it to simulate the experiments on porcine brain indentation conducted by Miller et al. (2000).

The physical and constitutive models of the brain and pia including the boundary conditions and the contact definitions remained the same as those described for needle insertion in Sections 3.1 and 3.2. The needle was replaced by a cylindrical indenter of diameter 10 mm. The indenter was considered rigid as its Young's modulus is many orders of magnitude greater than that of the brain. The brain was loaded by the prescribed motion of the indenter (Figure 3a). Following experiments by Miller et al. (2000), a constant indenter velocity of 1 mm/s was used.

Figure 3b compares the experimental force-displacement curve obtained by Miller et al. (2000) and that obtained from our simulations. At indentation depth of 3.5 mm, the force predicted by our model is only around 27% lower than the experimental one. This should be regarded as a good agreement taking into account large variability of the mechanical properties of biological tissues.

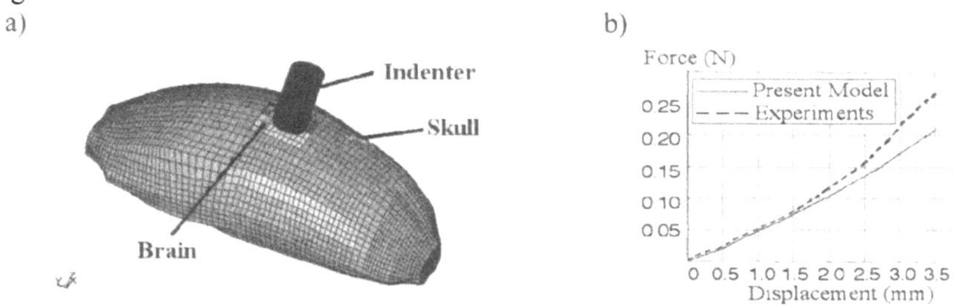

Figure 3 Porcine brain indentation: a) Computational model; b) Comparison of force-displacement curve obtained using this model with experimental results of Miller et al. (2000).

4 Results of Needle Insertion Simulation

Figures 4a and 4b summarise the force-displacement and the force-time curves obtained from our computational model of needle insertion. Due to lack of relevant experimental data, we were not able to perform complete validation of these curves in terms of magnitude of force between the brain and needle. However, qualitative validation can be done by comparing them with the results obtained by Miller et al. (2000) when indenting the porcine brain and those obtained by Washio and Chinzei (2004) when inserting the needle into porcine liver.

The general behaviour of the force-displacement curve obtained from our model agrees well with that obtained by Miller et al. (2000) in their experiments on brain indentation. Differences in the force magnitudes between these two curves can be attributed to two factors: 1) Needle insertion depth in our simulations (12 mm) was much larger than the indentation depth (around 3 mm) in experiments by Miller et al. (2000); 2) The cross-section area of the indenter (78.5 mm^2) was much larger than the needle insertion area in our model (0.237 mm^2).

When only the needle insertion phase preceding the rupture of the membrane surrounding the soft organ (i.e. pia in case of brain) is taken into account, general behaviour of the needle insertion force-time history computed using our model is consistent with that of the needle tip force obtained by Washio and Chinzei (2004) in their experiments on the needle insertion into the porcine liver (compare Figures 1 and 4b).

Although the curve obtained by Washio and Chinzei (2004) exhibits bimodal shape, present authors conclude that the first peak is of relatively small magnitude and can be attributed to the cardiac and pulmonary motions (experiments were conducted *in vivo*).

Overall deformation of the model under the needle at an insertion depth of 12 mm is summarised in Figure 5. It indicates large localised deformation which is very similar to the one observed in the photographs obtained during preliminary experiments on the needle insertion into the brain conducted at AIST, Tsukuba.

It should be also noted that Figures 4 and 5 are clear evidence that no model based on assumptions that the deformations are infinitesimal and the stress-strain relationship is linear can produce reliable results when applied to the simulation of needle insertion into the brain. The deformations experienced by the brain during insertion are far too large to justify these assumptions.

Figure 4 Simulation of needle insertion (needle speed is 5 mm/sec): a) Force-Displacement relationship; b) Force-Time relationship.

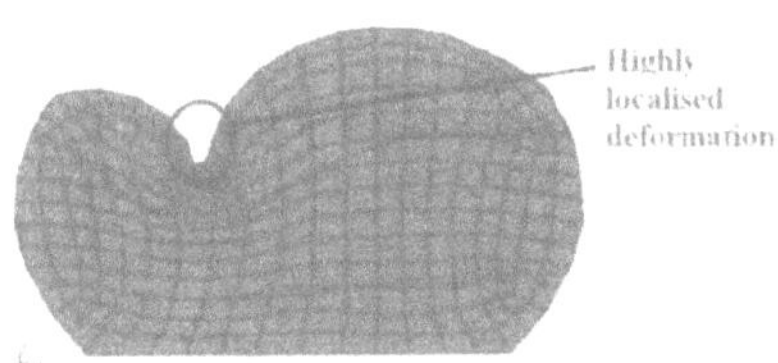

Figure 5 Needle insertion simulation: model cross-section at 12 mm insertion depth.

5 Conclusions and Discussion

The present study indicates that the explicit finite element analysis using non-linear procedures is a feasible method for computing the brain deformation during needle insertion and the reaction forces acting on the needle during insertion. General behaviour of the calculated needle force-time history was consistent with that obtained by Washio and Chinzei (2004). Due to the lack of relevant experimental data at present, we were not able to validate our results in terms of the magnitude of calculated needle force. However, we have confidence in our constitutive model and their constants as well as in the finite element procedures. When these models and procedures were applied to simulate the experiments of brain indentation by Miller et al. (2000), good agreement between the computed and experimental results was obtained (Figure 3b).

The present analysis focused on the needle insertion phase preceding rupture of the membrane (i.e. pia in case of brain) surrounding the soft organ. Therefore, the natural direction for further studies would be to experimentally determine the failure mechanism/mechanisms of the pia and implement a new material model for the simulation of this mechanism.

Bibliography

Aimedieu P. and Grebe R. (2004) Tensile Strength of Cranial Pia Matter: Preliminary Results. *Journal of Neurosurgery* 100: 111-114.

Alterovitz R., Goldberg K., Okamura A. (2005) Planning for Steerable Bevel-tip Needle Insertion Through 2D Soft Tissue with Obstacles. In *'Proceedings of IEEE International Conference on Robotics and Automation (ICRA)'* pp. 1652-1657.

Belytschko T., Lin J., Tsay C.S. (1984) Explicit Algorithms for Non-Linear Dynamics of Shells. *Computational Methods in Applied Mechanical Engineering* 42: 225-251.

DiMaio S.P. and Salcudean S.E. (2003) Needle Insertion Modelling and Simulation. *IEEE Transactions on Robotics and Automation: Special Issue on Medical Robotics* 19: 864-875.

Heverly M., Dupont P., Triedman J. (2005) Trajectory Optimization for Dynamic Needle Insertion. In *'Proceedings of 2005 IEEE International Conference on Robotics and Automation'*. Barcelona, Spain.

Hughes T.J.R., Taylor R.L., Sackman J.L., Curnier A.C., Kanoknukulchai W. (1976) A Finite Element Method for a Class of Contact-Impact Problems. *Journal of Computational Methods in Applied Mechanical Engineering* 8: 249-276.

Hughes T.J.R. (2000) 'The Finite Element Method: Linear Static and Dynamic Finite Element Analysis'. New York: Dover Publications Inc., 207-209.

Miller K., Chinzei K., Orssengo G, Bednarz P (2000) Mechanical Properties of Brain Tissue in-vivo: Experiment and Computer Simulation. *Journal of Biomechanics* 33: 1369-1376.

Miller K. (2002) 'Biomechanics of Brain for Computer Integrated Surgery.' Publishing House of Warsaw University of Technology: Warsaw.

Miller K. and Chinzei K. (2002) Mechanical Properties of Brain Tissue in Tension. *Journal of Biomechanics* 35: 483-490.

Nienhuys H.W. (2003) Cutting in Deformable Objects. University of Utrecht.

O'Leary M.D., Simone C., Washio T., Yoshinaka K., Okamura A.M. (2003) Robotic Needle Insertion: Effects of Friction and Needle Geometry. *In 'Proceedings of 2003 IEEE International Conference on Robotics and Automation'*. Taipei, Taiwan pp. 1774-1780.

Puso M.A., Weiss J.A. (1998) Finite Element Implementation of Anisotropic Quasilinear Viscoelasticity Using a Discrete Spectrum Approach. *ASME Journal of Biomechanical Engineering* 120: 62-70.

Simone C., Okamura A.M. (2002) Modelling of Needle Insertion Forces for Robot-Assisted Percutaneous Therapy. In *'Proceedings of the 2002 IEEE International Conference on Robotics and Automation'*. Washington, DC, USA.

Takhounts E.G., Eppinger R.H., Campbell J.Q., Tannous R.E., Power E.D., Shook S.S. (2003) On the development of the SIMON finite element head model. *Stapp Crash Car Journal* 47: 107-133.

Washio T. and Chinzei K. (2004) Needle Force Sensor, Robust and Sensitive Detection of the Instant of Needle Puncture. In *'Proceedings of 7th International Conference on Medical Image Computing and Computer-Assisted Intervention - MICCAI 2004'*. Saint-Malo, France pp. 113-120.

Weiss J.A., Maker B.N., Govindjee S. (1996) Finite Element Implementation of Incompressible, Transversely Isotropic Hyperelasticity. *Computational Methods in Applied Mechanical Engineering* 135: 107-128.

Zhang L., Bae J., Hardy N.W., Monson K.L., Goldsmith W., Yang H.K., King A.I. (2002) Computational Study of the Contribution of the Brain Vasculature on the Dynamic Response of the Brain. *Stapp Crash Car Journal* 46: 145-163

Robotic System for Femoral & Tibial Osteotomies Assistance

Kaddour Bouazza-Marouf[1], Ian Browbank[1], Jim R Hewit[2], Alan P Slade[2], Stuart I Brown[2], Roger Phillips[3], James Ward[3], Amr MMA Mohsen[3,4], and Kevin P Sherman[3,4]

[1] School of Mechanical and Manufacturing Engineering, Loughborough University, Loughborough, UK, e-mail: k.bouazza-marouf@lboro.ac.uk
[2] University of Dundee, Dundee, UK.
[3] University of Hull, Hull, UK.
[4] Department of Traumatology and Orthopaedics, Royal Hull Hospitals, Hull, UK

Abstract. Typical osteotomy procedures around the knee joint involve the removal of a wedged shaped section of bone from either the proximal tibia or distal femur in order to correct a deformity and alter the alignment of the lower extremities. Unfortunately, the subjective nature of the preoperative planning and free-hand surgical techniques currently employed can lead to significant outcome variability and high complication rates. A prototype computer-robotic system, KneeCAS-1 (Knee Computer-Assisted System), has been developed to assist such osteotomy procedures around the knee joint. It consists of a custom-built six degree-of-freedom " fail-hard" surgical robot, with an instrumented saw driver unit, and a computer based planning system. It is intended to perform automated cutting (sawing) of the osteotomy planes under surgeon supervision.

1 Introduction

Osteotomy procedures around the knee joint, involve the cutting, realignment and subsequent stabilisation of either the tibia or the femur. These procedures are carried out in order to correct a lower limb deformity, correct a misaligned fracture, or to relieve the pain associated with chronic osteoarthritis. An example of closing wedge distal femoral osteotomy, for correction of a valgus deformity, is shown in Figure 1. This procedure involves a skin incision and soft tissue retraction on the inner (medial) side and at the lower (distal) end of the femur. This is then followed by a horizontal distal cut, approximately 5 cm above the knee joint line, and a proximal cut angled according to the desired angle of correction. The wedge-shaped piece of bone, with the base of the wedge on the inner (medial) side of the femur, is then removed and followed by closure and internal fixation of the osteotomy site. In practice, the wedge removal process can be performed using either a hand-held oscillating saw, an osteotome or a combination of both. Unfortunately, the subjective nature of the preoperative planning and free-hand surgical techniques currently employed can lead to significant outcome variability and high complication rates (see, e.g., Insall et al., 1984, Berman et al., 1991, Coventry et al., 1993, and Rinonapoli et al., 1998, Mathews et al., 1998). The considerable scope for error associated with the free-hand wedge removal process gives the most grounds for concern (see Insall, 1993). Postoperative results display significant outcome variability, with both over-correction and under-correction being observed, in relation to the accuracy of knee joint re-alignment. As a result, complications such as delayed/non-union of the osteotomy site, intra-articular fractures, or even the recurrence of deformity and pain in the longer-term, are not uncommon. Hence, the surgical aim of this project is to improve the planning, implementation and outcome of osteotomy procedures.

The use of a fully powered autonomous robot to perform (hip joint) osteotomy procedures has previously been investigated by Moctezuma et al. (1994). A commercially available clean room robot (PUMA 560) was used. However, Moctezuma et al. (1997) acknowledged that an industrial robot cannot be used in a clinical environment due to the high security standards. It is for this reason that we decided to use a custom-built robotic system.

A prototype computer-robotic system has been developed. It is intended to perform automated cutting of the osteotomy planes under surgeon supervision, and it is aimed particularly at correcting complex multi-plane deformities that are difficult with conventional techniques. The computer-robotic system which has been developed consists of a custom-built "fail-safe" six degree-of-freedom surgical robot, with an instrumented saw driver unit, and a computer based planning system.

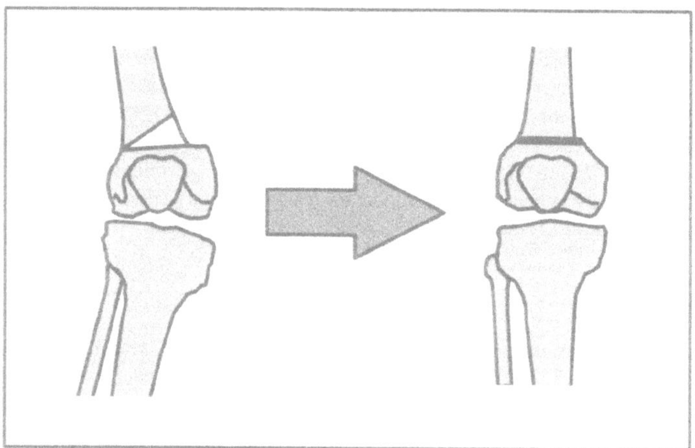

Figure 1. Closing wedge distal femoral osteotomy for correction of valgus deformity. (Right leg shown)

2 Custom-Built Osteotomy Robot

A Custom-Built 6 degrees-of-freedom (DOF) robot has been developed as part of this research to address the issue of safety at the outset. The large working volume and reach associated with industrial robots create significant problems in relation to the intraoperative safety of the patient and the surgical team. A more direct development route, involving the design and manufacture of a custom-built robotic manipulator has therefore been followed in relation to the Osteotomy Robot. On the grounds of safety and ease-of-use, the saw positioning process performed by the robotic system is broken into two distinct phases: "coarse" passive and "fine" active alignment. As depicted schematically in Figure 2, the KneeCAS-1 prototype is therefore a two component system consisting of an active saw-positioning robot mounted on a braked passive positioning system. The floor-standing passive system allows the surgical team to manually locate the active robot at the operative site, and to subsequently maintain this pose throughout the surgical procedure. The six degrees-of-freedom (DoF) active robot is then used to align the saw unit with the pre-planned cutting trajectories. The adoption of this "semi-active" alignment process significantly reduces the working envelope of the active robot, thereby allowing the use of a lighter more compact design, whilst at the same time avoiding the large active motions associated with anthropomorphic robots.

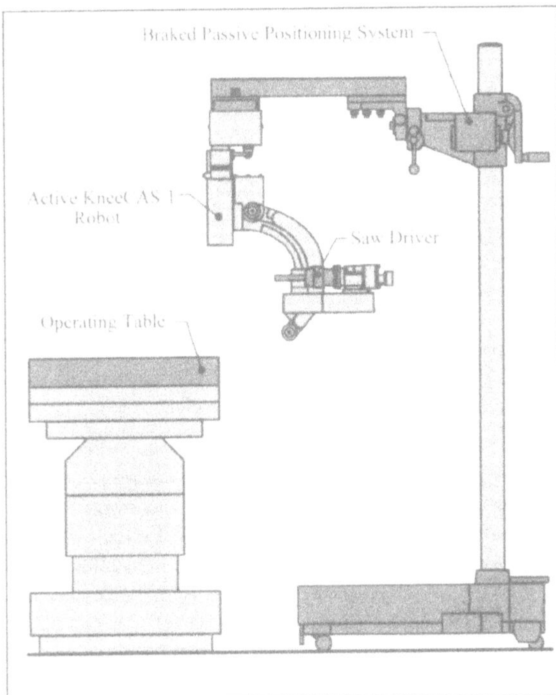

Figure 2. Schematic diagram showing the configuration of the KneeCAS-1 robotic system.

The active manipulator of the custom-built robot has 3 prismatic joints (horizontal, vertical and feed actuators) and 3 revolute joints (Pan, tilt and roll actuators) as shown in Figure 3. The robot has been designed to fail-hard, i.e. should the power to the joints fail for whatever reason all the joints will remain in their current position. The actuator mechanisms for all the joints, except for the roll joint, have been designed to be non-backdriveable, and a specifically designed electromechanical brake with zero-backlash has been used for the roll joint.

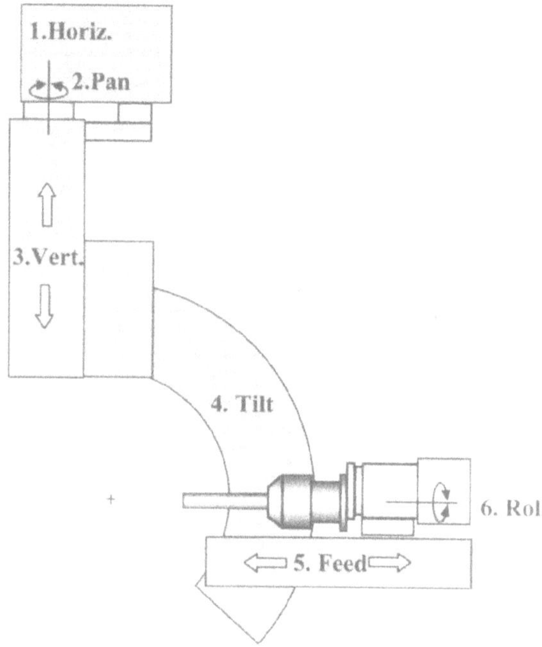

Figure 3. Kinematic configuration of the active manipulator

The chosen kinematic configuration of the active saw-positioning robot ensures that the rotation axes are located in the optimum positions with respect to the femur/tibia, thereby significantly reducing the necessary linear motions and the overall size/weight of the robot. For example, the tilt rotation, which controls the angle-of-attack of the saw blades into the femur/tibia, is implemented by mounting the feed unit on a circular motion guide (arc-shaped rail). When the centre of the arc is correctly positioned, using the passive "coarse" alignment and the active robot's linear degrees-of-freedom, the saw can therefore be safely rotated around a point within the femur/tibia. Similarly, the *pan* rotation, which controls the magnitude of the bone wedge angle, is located in the vicinity of the apex of the wedge. The *roll* rotation has been allocated as the robot's sixth DoF, thus allowing rotation of the saw blades without having to rotate large sections of the robot. The complete robotic manipulator, with the active KneeCAS-1 robot attached to the passive positioning system, is shown in Figure 4.

An instrumented "smart" saw driver unit has been custom-built to provide enhanced bone cutting performance and to minimise bone cutting induced vibrations. It was decided to develop a 'smart' saw because existing traditional sagittal saws do not have the necessary instrumentation, the cutting profile is not optimum and the induced vibration is not desirable. As the blade approaches the end of the cut, the corners of the blade for the existing saws would inevitably protrude beyond the edge of the bone into the surrounding soft tissues, therefore a continuous change in orientation would be necessary for such saws to follow the bone contour. Ideally, the shape of the locus described by the blade would be configured in a small circle, not in a shallow arc like the sagittal saw, allowing the blade to cut right to the edge of the bone with a minimum of protrusion. Another disadvantage of the sagittal saw, for robotic applications is that the single bladed design generates substantial vibration and jerkiness in use. Moctezuma *et al.* (1997) eliminated this by fitting two blades, operating 180° out of phase with each other, thus balancing vibratory and cutting forces. Therefore, in order to achieve the desired fit between the saw blade and the bone, and to provide a balanced and vibration free end-effector, a double bladed design, with blades driven in circular (or elliptical) orbit was developed. Also, to ensure that the saw is not fed beyond the edge of the bone into the surrounding soft tissues and to minimise the risk of thermal osteonecrosis, the saw feed force is measured and the saw blades are equipped with thermocouples which measure the temperature at the cutting face. The feed rate of the robot's end-effector can be adjusted according to cutting conditions, as measured by the "smart" saw.

The robot motion control is currently performed using two levels of control. High level control is performed using a Personal Computer, and the low level control is performed using a custom-built PIC microcontroller system with the necessary power-electronics. The microcontroller controls and monitors the motion of the robot drive units.

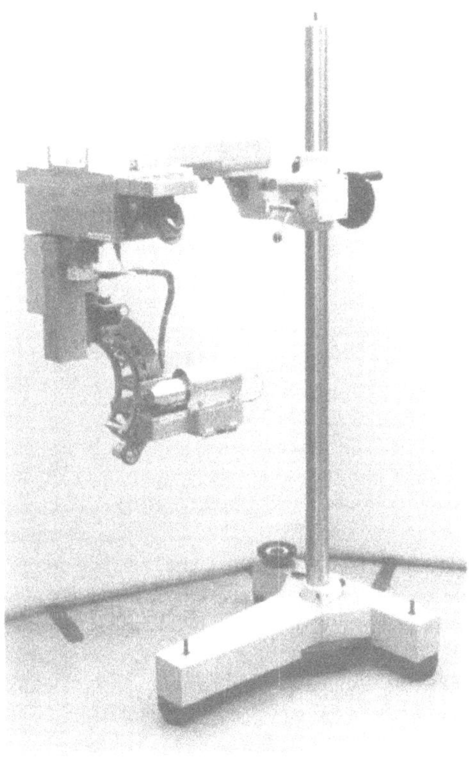

Figure 4. The prototype KneeCAS-1 robotic system.
(Active robot supported by a braked passive positioning system.)

3 Planning and Deployment

The overall robotic-assisted osteotomy procedure involves a planning and implemention stages. This entails a pre-operative surgical procedure planning, datuming of the robot to the bone intra-operatively, robot control and bone cutting, and monitoring of the robot end-effector and the bone cutting process.

The planning of the surgical operation is carried out first, before the surgical procedure takes place, using 3-D models of bone obtained from CT scans, or using a combination of pre-operative planar x-ray films. The objective of the planning stage is to allow the surgeon to interactively place and adjust the osteotomy wedge, and to simulate and visualise the outcome of the operation. The final position and orientation of the wedge is stored and used in the operating theatre to produce the required saw cuts, i.e. to produce the desired trajectories of the end-effector.

In the operating theatre, robotic assistance is provided as follows. The patient's leg is flexed to approximately a ninety degree angle at the knee and secured to an immobilisation device. X-ray images of the patient's leg are then acquired using a fluoroscopic image-intensifier. This is necessary to provide sufficient information for datuming the bone to pre-operative images, and thus producing the required saw cuts which are downloaded to the robot controller. The active robot is attached to the passive positioning system, with both the "smart" saw driver unit and the necessary sterile draping in place, and depending upon which leg is to be operated on, and whether a distal femoral (medial side incision/access) or a proximal tibial (lateral side incision/access) osteotomy is to be performed, the robotic system is moved to the appropriate side of the operating table. The active robot is then coarsely aligned with the patient's anatomy, using the degrees-of-freedom of the passive positioning system. The location of the active robot will then be checked (as described below) and a series of small computer-controlled active motions performed to correct any remaining alignment errors to ensure that the robot's arc-based tilt rotation is correctly aligned with the leg before the robotic-assisted bone cutting process is instigated.

The registration of the robot end-effector (i.e. the saw) to the bone and the provision of the cutting trajectories are provided using a "computer based planning system". The intra-operative registration is established via the use of a real-time opto-electronic position measurement system which is able to track infra-red LEDs mounted on rigid bodies attached to the robot, patient/immobilisation device, and to the intra-operative C-arm fluoroscopy unit. A surface-based registration technique, which employs anatomical landmarks to perform the matching process, is used to register preoperative and intra-operative images/bone models.

The computer based planning system produces a set of transformations for the desired positions and orientations of the saw. Inverse kinematics is then performed to provide the necessary link motions and move the saw to the desired pose which is independently checked using the optical tracking system. Any errors are therefore automatically compensated for. Once the saw is positioned along the desired cut, sawing is initiated by the surgeon, and the position of the saw with respect to the bone is continuously monitored. The sawing/cutting force is also continuously measured and used for both safety (to ensure that the saw does not damage the soft tissue) and to confirm/adjust the saw position with respect to the bone. For safety reasons, it was decided that the feed drive is fully retracted after each saw cut, thus ensuring safe alignment with the next cutting trajectory.

The deployment of the saw is carried out as follows. The registration and planning protocol PC transmits via RS232 serial communication (with an error checking protocol) the current and desired positions of the saw, with respect to a fixed (registration) frame of reference, to Robot PC. These positions are in the form of 4x4 transformation matrices. The two saw position transformation matrices are then utilised by the Robot protocol to compute the desired robot joint positions, using the robot inverse kinematics. The PC then transmits the desired joint positions, via RS232 serial communication, to the robot driver (PIC microcontroller) which drives the joints in closed loop control and monitors all the switches continuously. Also, the protocol within the PIC microcontroller has been configured so that the status of the switches and the readings from the joint encoders are continuously transmitted to the Robot PC, which performs a high level position monitoring of the robot. The saw position, as computed from the robot joint readings, is also transmitted to the planning protocol PC as and when requested (automatically).

A deadman's handle is provided to give complete control of the robot and saw to the surgeon. No motion of the robot can occur if the switch on the deadman's handle is not pressed.

4 System Performance and Evaluation

Laboratory tests have been carried out with KneeCAS-1 during the integration phase. The two main tasks that have been evaluated include the integration of the different components of the overall system and the accuracy of simulated surgical procedures. Such accuracy would depend on the accuracy and the optical tracking and guidance system, and on the accuracy of the robot (which is affected by the load, the accuracy of the Denavit-Hartenberg parameters and the position of the end-effector within the workspace). The accuracy is also affected by any datuming error between the frame of reference of the optical tracker infra-red LEDs holder (attached to the roll unit), and the roll unit frame of reference. However, the iterative approach we have adopted for the deployment of the saw guaranteed accurate results despite such errors, i.e. a robust deployment of the saw. Following every move of the saw to a desired position, the actual position of the saw is measured by the optical tracking system and compared to the desired position. If there is an error greater than 1mm both the actual and desired saw positions are sent to the robot controller to reduce this error. This is done by generating a new move of the saw based on the latter measurements. This has proved very effective, resulting in positional accuracy of saw blade tip within 0.7 mm, and angular error within 1°. These errors, which are the combined errors of the complete system, are within the performance specifications, of ± 1mm and $\pm 1^{\circ}$, specified by the collaborating surgeons at the start of the project. Also, the preliminary test results have shown that the surface finish of bone cuts produced with the "smart" saw driver are at least as good as those obtained with conventional oscillating saws.

5 Discussion

The shortcomings and scope for error associated with existing osteotomy techniques make the postoperative results somewhat unpredictable, and there is evidence that the long-term success of an osteotomy procedure is directly related to surgical precision. A number of computer-assisted surgery applications have therefore been developed with the aim of improving upon the results obtained using existing surgical techniques. However, computer-assisted surgery systems all suffer from an inherent limitation, in that the invasive stage of the surgical intervention is still performed by hand, thereby leaving scope for human error. Automation of the bone cutting process, via robotic-assistance, has therefore been advocated during the current investigation with the aim of maximising intraoperative precision. A prototype computer and robotic assisted surgery system has been developed, as part of this research, to aid orthopaedic surgeons in performing both more accurate planning and implementation of osteotomy procedures around the knee. The integration of the different components of the overall system and the accuracy of simulated surgical procedures have been successfully investigated in laboratory tests.

Because we opted to develop a compact custom-built robotic system, it has been possible to address the intraoperative safety problem from the outset of the development process. A fail-safe system architecture has therefore been developed, which initially supports optimal passive positioning (guided by an optoelectronic position measurement system) of an active robot in relation to the patient's anatomy. The task-specific kinematic configuration of this active robot then ensures that the subsequent trajectory alignment motions are mechanically constrained to "safe" areas of the operating room. The robot's fail-safe brakes can then be engaged, thus ensuring a single degree-of-freedom system during the enhanced bone cutting process. Although this KneeCAS-1 system is still at an early stage of development with respect to a fully operating room compatible system, a preliminary evaluation has demonstrated that the system's robotic components satisfy the relevant clinical and engineering performance requirements.

Bibliography

Berman, A.T., Bosacco, S.J., Kirshner, S. and Avolio, A. Factors influencing long-term results in high tibial osteotomy. *Clinical Orthopaedics and Related research*, 1991, 272, 192-198.

Coventry, M.B., Ilstrup, D.M. and Wallrichs, S.L. Proximal tibial osteotomy: A critical long-term study of eighty-seven cases. *Journal of Bone & Joint Surgery*, 1993, 75A(2), 196-201.

Insall, J.N., Joseph, D.M. and Msika, C. High tibial osteotomy for varus gonarthrosis: A long-term follow-up study. *Journal of Bone & Joint Surgery*, 1984, 66A(7), 1040-1048.

Insall, J.N. Osteotomy. In: Insall, J.N., Windsor, R.E., Scott, W.N., Kelly, M.A. and Aglietti, P. (eds.), *Surgery of the Knee*, Churchill Livingstone, New York, 1993, pp. 635-676.

Mathews, J., Cobb, A.G., Richardson, S. and Bentley, G. Distal femoral osteotomy for lateral compartment osteoarthritis of the knee. *Orthopaedics*, 1998, 21(4), 437-440.

Moctezuma, J.L., Gossé, F. and Schultz, H.-J. A computer and robotic aided surgery system for accomplishing osteotomies. *Proceedings of the 1st International Symposium on Medical Robotics and Computer Assisted Surgery. (MRCAS'94)*, Pittsburgh, Pennsylvania, 22-24 Sept. 1994, 31-35.

Moctezuma, J.L., Schuster, D., Gosse, F. and Schulz, H.J. New oscillating saw for robotic aided surgery. *Proceedings of the Institution of Mechanical Engineers, Part H: Journal of Engineering in Medicine*, 1997, 211(4), 301-308.

Rinonapoli, E., Mancini, G.B., Corvaglia, A. and Musiello, S. Tibial osteotomy for varus gonarthrosis: A 10- to 21-year follow-up study. *Clinical Orthopaedics and Related Research*, 1998, 353, 185-193.

WL-16RII: Prototype of Biped Walking Wheelchair

Yusuke Sugahara [1], Kenji Hashimoto [2], Hiroyuki Sunazuka [2], Masamiki Kawase [2], Akihiro Ohta [2], Chiaki Tanaka [2], Hun-ok Lim [3,5] and Atsuo Takanishi [2,4,5]

[1] Graduate School of Science and Engineering, Waseda University, Tokyo, Japan, E-mail: sugahara@suou.waseda.jp
[2] Graduate School of Science and Engineering, Waseda University, Tokyo, Japan
[3] Department of Mechanical Engineering, Kanagawa University, Kanagawa, Japan
[4] School of Science and Engineering, Waseda University, Tokyo, Japan
[5] Humanoid Robotics Institute, Waseda University, Tokyo, Japan

Abstract. This paper describes the development of the prototype of the biped walking wheelchair. This robot WL-16RII consists of two Stewart Platform type legs and waist with passenger seat. The walking control method based on ZMP criteria and virtual compliance control for this robot were also developed. Through various experiments, the effectiveness of developed hardware and control method was confirmed.

1 Introduction

From the viewpoint that the expected locomotion form for the robots is ultimately biped walking, authors have been studying on the mechanism and control of biped walking robots for around 35 years. To date, however, most humanoid robots have employed wheels and crawlers for their locomotion mechanism. This is not only because biped robots are essentially unstable; another problem is that they are difficult to treat as independent locomotion modules.

Our ultimate research aim is to develop a bipedal robot as a leg module that is sufficient for practical use as a multi-purpose locomotor for robotic systems (Figure 1). We aim to provide a bipedal robot with only lower-limbs and a waist that can walk independently; allowing users to then build the upper body according to their needs. This biped locomotor would be applicable to the welfare field as a walking wheelchair or as a walking support machine that is able to walk up and down stairs carrying or assisting a human.

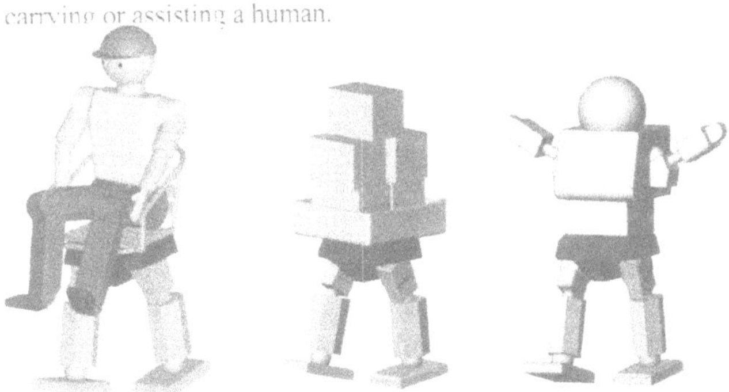

Figure 1. Applications of multi-purpose biped mobile base.

Various researches into walking robots for practical use were carried out by some researchers, as well as other researches into those of the biped walking type. Regarding locomotion modules capable of human transport, researches on the "Walking Chair" (Takeda, et al., 2001) or locomotion module called "My Agent" (Kamada, 1994) have also been carried out. In addition and remaining in the welfare industry field, the electric wheelchair named "iBOT", capable of ascending and descending stairs and slopes was released by Independence Technology L. L. C. (2003), although it employed wheels for its locomotive mechanism.

Some of these researches are done with the concept of reducing DOF (Takita, et al., 2000 and Ota, et al., 2003) while other researches focused only on ascending and descending stairs (Konuma, 2001 and 2003). Furthermore, Takeda (2001) and Kamada (1994) did not focus on dynamic biped walking. Although such methods appear advantageous in terms of cost reduction and ease, their ultimate goal is not versatility. Although "iBOT" is going to be completed as one suitable style of locomotion system on which to ride, its objective does not include operation in any narrow environment with obstacles scattered around such as a Japanese traditional house. Moreover, there are certain problems related to its mechanism, such as requiring a rigid stair rail to ascend and descend stairs, and its inability to move sideways.

Therefore, we developed some biped locomotors, WL-15 (Waseda Leg - No. 15), WL-16, WL-16R and WL-16RII, which have 6-DOF parallel mechanism legs. This robot consists of two legs and a waist. Using this robot, we studied the way to apply the biped walking robot to a mobile base. In 2003, using WL-16, a world-first dynamic biped walking carrying an adult woman and a dynamic biped walking carrying a 60 kg adult man were realized (2004), earlier than the release of the Toyota Motor Corporation's Toyota Personal Robot (2004). In addition, biped walking up and down stairs carrying a human was also realized in 2005.

In this paper, our previous steps toward the realization of biped walking wheelchair are described. Our steps include the development of hardware, control method and various experiments of the newest prototype WL-16RII (Waseda Leg - No. 16 Refined II) shown in Figure 2.

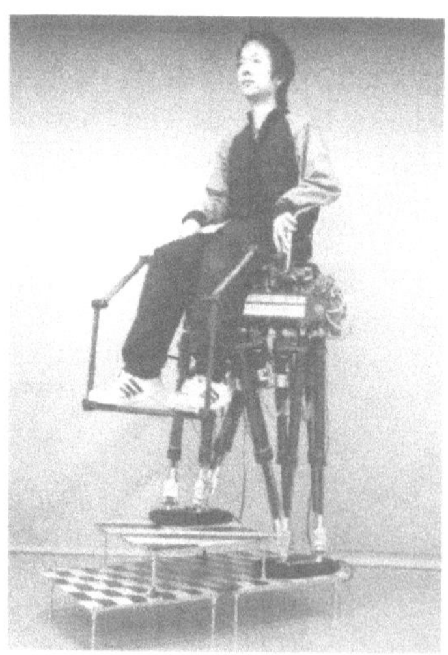

Figure 2. Waseda Leg - No. 16 Refined II.

2 Hardware of Prototype Waseda Leg - No. 16 Refined II

2.1 Basic Design

As stated before, this biped machine was developed to be as a leg module that is sufficient for practical use as a multi-purpose locomotor of the robot system. Therefore, this robot should be possible to carry heavy load. To fulfill this requirement, the DOF configuration of the leg mechanism consists of Stewart Platform as shown in Figure 3.

The weight of WL-16RII is 62 kg including 5 kg battery weight. Each leg mechanism has 6 linear actuators and passive joints at the both side of each linear actuators. Each linear actuator consists of 150W DC servo motor and ball screw, and has negative operation electromagnetic brake, so this robot can hold its posture without power supply. The control computer is arranged at the rear of the pelvis, and the batteries, DC servo drivers and a body angle detector are arranged inside the pelvis. A passenger seat is on the pelvis.

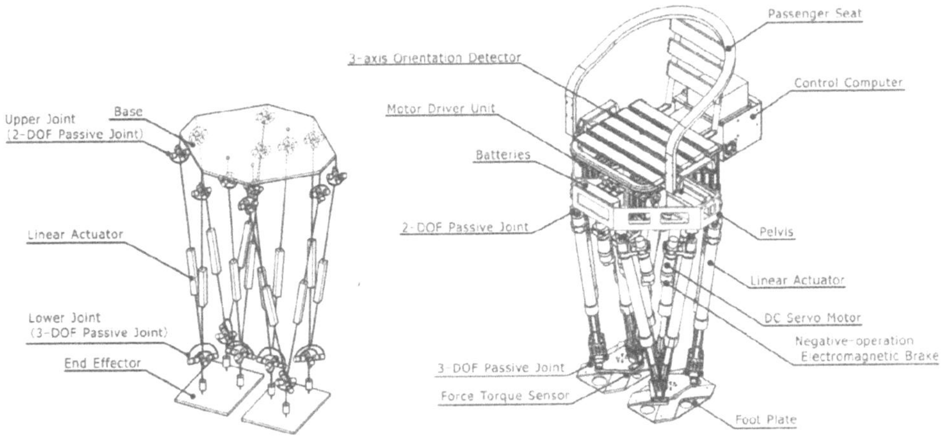

Figure 3. DOF configuration and isometric view of WL-16RII.

2.2 Support Torque Reduction Mechanism

On the design specification of the linear actuator of WL-16RII, the maximum payload is 60 kg. It is not too small, but it cannot be said that it is enough to apply this robot as biped wheelchair.

It is mainly because the torque of DC servo motor is not enough to carry heavier load. So the easiest way to increase the maximum payload is to use other higher power actuators, or to increase the torque using reduction gears with a higher reduction ratio. However, the former leads to being more expensive to produce and too large in human-living environment, and the latter is disadvantageous in walking at higher speed.

Instead, research on robot arms provides a way of reducing the necessary torque of the actuator by compensating for the robot's own weight mechanically. So we focused on the fact that most of the force acting on the foot during walking with a heavy load was to support the robot's own weight, and developed the Support Torque Reduction Mechanism not to compensate for all of the robot's weight mechanically, but to reduce torques needed to support the robot's own weight using a simpler mechanism.

Support Torque Reduction Mechanism consists of two compression gas springs with different spring forces that have a lock function, and are attached to each leg (Figure 4). These mechanisms lock and unlock each spring according to the support / swing phase, and supports mainly vertical force.

During the support phase, the gas spring with a large reaction force is unlocked and activated, and the other is locked. Contrariwise, during the swing phase, the gas spring with a small reaction force is unlocked and the other is locked. When the supporting state changes, the effects of the two gas springs are switched by pushing down and releasing the switch pin of each gas spring using solenoids.

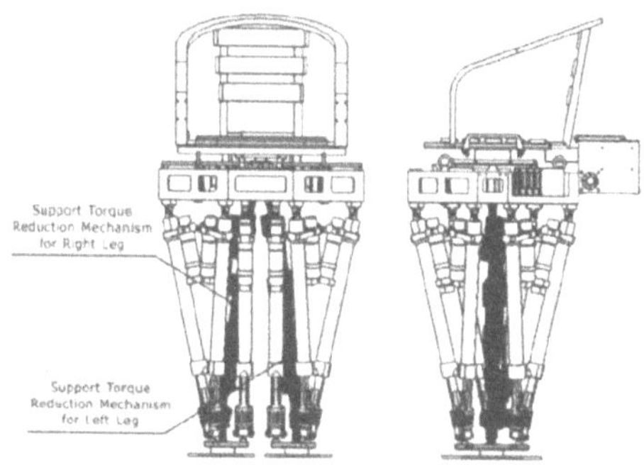

Figure 4. Assembly drawing of WL-16R with support torque reduction mechanism.

3 Walking Control Method

Since WL-16RII is also biped robot that has 6 DOF legs, the walking control problem is basically same as that of existing biped robots that includes humanoid robots. As the control method of this robot, model based walking control method based on ZMP criteria is used.

The walking control method of this robot consists of two main parts: first is the walking pattern generation before walking as the feed forward control of biped walking system, and second is the real-time stabilization control during walking as the feed back control.

3.1 Walking Pattern Generation

This algorithm consists of the following main parts:

1. Modeling of the robot
2. Derivation of the ZMP equations
3. Computation of approximate waist motion
4. Computation of strict waist motion by iteratively computing the approximate waist motion

Let the walking system be assumed as follows:

(1) The robot is a system of particles.
(2) The floor for walking is solid and can not be moved by any force or moment.

(3) A Cartesian coordinate system is determined as shown in Figure 5. The X-axis and Y-axis form a plane which is the same as that of the floor.
(4) The contact region between the foot and the floor is a set of contact points.
(5) The coefficient of friction for rotation around the X, Y and Z-axes is zero at the contact point.

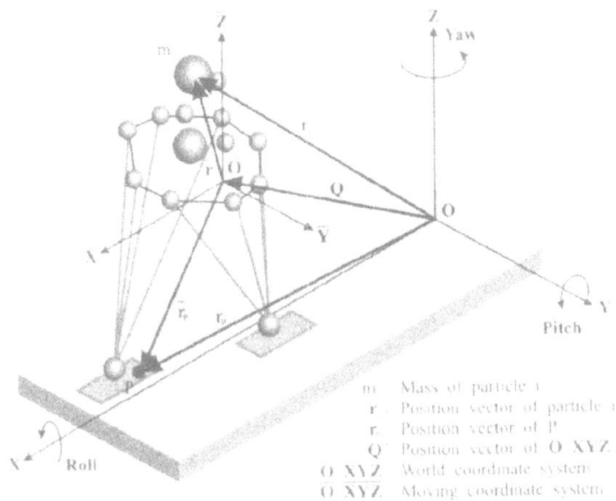

Figure 5. Definition of coordinate system and vectors.

First, we define an approximation model of the waist and the position vectors like Figure 5. The moment balance around point P on the floor can be expressed as below:

$$\sum_{i}^{all\_particles} m_i(\mathbf{r}_i - \mathbf{r}_p) \times (\ddot{\mathbf{r}}_i + \mathbf{G}) + \mathbf{T} = \mathbf{0} \tag{1}$$

If the point P is defined as ZMP, T = 0. We denote the position vector of P as $\mathbf{P}_{ZMP}(x_{ZMP}, y_{ZMP}, 0)$. To consider the relative motion of each part, a moving coordinate is established on the waist of the robot parallel to the fixed coordinate O-XYZ.

$\mathbf{Q}$ (x_q, y_q, z_q) is the position vector of the origin of from the origin of O. Using the moving coordinate frame, Equation (1) can be modified as follows:

$$\sum_{i}^{all\_particles} m_i(\bar{\mathbf{r}}_i - \bar{\mathbf{r}}_{zmp}) \times \{\ddot{\bar{\mathbf{r}}}_i + \ddot{\mathbf{Q}} - \mathbf{G} + \dot{\bar{\boldsymbol{\omega}}} \times \bar{\mathbf{r}}_i + 2\bar{\boldsymbol{\omega}} \times \dot{\bar{\mathbf{r}}}_i + \bar{\boldsymbol{\omega}} \times (\bar{\boldsymbol{\omega}} \times \bar{\mathbf{r}}_i)\} = \mathbf{0} \tag{2}$$

where $\bar{\mathbf{r}}_{zmp}$ is the position vector of ZMP with respect to the O. $\bar{\boldsymbol{\omega}}$ is the angular velocity vector of the origin of $\overline{W} - \overline{XYZ}$.

Assuming that a moving coordinate does not rotate, this equation is expanded into Equations (3) and (4) by putting the terms representing the moment generated by the lower-limb particles on the right-hand side as known parameters, named M_x and M_y respectively.

$$m_W(\bar{z}_W-\bar{z}_{zmp})(\ddot{\bar{x}}_W+\ddot{\bar{x}}_q-g_x)-m_W(\bar{x}_W-\bar{x}_{zmp})(\ddot{\bar{z}}_W+\ddot{\bar{z}}_q-g_z)=-M_y \tag{3}$$

$$m_W(\bar{y}_W-\bar{y}_{zmp})(\ddot{\bar{z}}_w+\ddot{\bar{z}}_q-g_z)-m_W(\bar{z}_W-\bar{z}_{zmp})(\ddot{\bar{y}}_W+\ddot{\bar{y}}_q-g_y)=-M_x \tag{4}$$

These equations are interferential and non-linear because each equation has the same variable $\bar{z}_w$. Thus, by assuming that the waist particles move vertically, the equations can be decoupled and linearized with the right side including the known clause of the left side being newly replaced with M^*_x and M^*_y, as follows.

$$m_W(\bar{z}_W-\bar{z}_{ZMP})\ddot{\bar{x}}_W-m_W(-\bar{g}_z)\bar{x}_W=-M^*_y \tag{5}$$

$$-m_W(\bar{z}_W-\bar{z}_{ZMP})\ddot{\bar{y}}_W+m_W(-\bar{g}_z)\bar{y}_W=-M^*_x \tag{6}$$

In these equations, M^*_x and M^*_y are known because they are derived from the lower-limb's motion and the time trajectory of ZMP. In the case of steady walking, M^*_x and M^*_y are periodic functions because each particle of the lower-limbs and the time trajectory of ZMP move periodically for the moving coordinate $\overline{W}-\overline{XYZ}$. Thus, each equation can be represented as a Fourier series. By comparing the Fourier transform coefficients from both sides of each equation, we can easily acquire the approximate periodic solution for the waist motion.

In order to obtain the strict solutions of the body, the approximate solutions are subtracted into Equation (2) and ZMP is computed. Then, the strict motion of the body is obtained using an iteration method.

3.2 Virtual Compliance Control

Since our walking pattern generation algorithm outputs the stable walking pattern generated offline beforehand, robot becomes unstable by the landing impact according to modeling errors such as the deflection of the structure. There is concern that this landing impact would be fatal when the robot walks with some conditions which makes large deflection of structures and response error of the motor easily, such as the walking with heavy payload or carrying a human. Therefore, using the virtual compliance control, WL-16RII moderates the landing impact and stabilizes walking with such conditions.

4 Experiments

Using developed robot, various experiments has been conducted (Figure 6).

Figure (a) in Figure 6 shows the dynamic biped walking experiment carrying 50 kg payload, and Figure (b) shows the dynamic biped walking experiment carrying 60 kg adult man. In the both of these experiments, step length was 0.1 m/step and walking cycle was 0.96 s/step. By using developed hardware and control method, these experiments were successfully completed.

Figure (c) in Figure 6 shows the dynamic biped walking experiment carrying 94 kg adult man using support torque reduction mechanism. The walking cycle was 0.96 s/step and step length was 0.1 m/step. By using developed mechanism, this experiment was succeeded.

Figure (d) in Figure 6 shows the biped walking experiment ascending and descending stairs carrying 60 kg adult man. The rise of the stair was 200 mm and the walking cycle was 5.76 s/step. Although Stewart Platform has generally narrow workspace, by twisting the waist according to the foot position, this experiment was also succeeded.

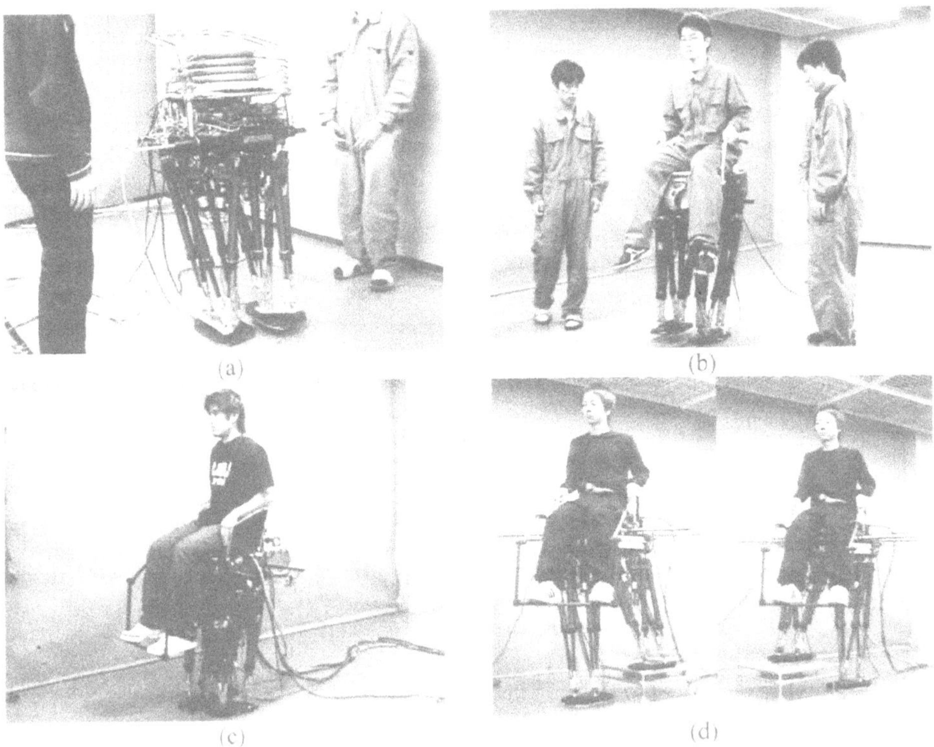

Figure 6. Walking Experiments. (a) Walking experiment carrying 50 kg load, (b) Walking experiment carrying 60 kg adult man, (c) Walking experiment carrying 94 kg adult man using support torque reduction mechanism, (d) Walking experiment ascending and descending 200 mm stair carrying 60 kg adult man.

5 Conclusions and Future Work

As the prototype of biped walking wheelchair, WL-16RII that consists of two Stewart Platform type legs and waist with passenger seat was developed. Support torque reduction mechanism that reduces the required motor torque to carry heavy load was also developed. The walking control method based on ZMP criteria and virtual compliance control method for this robot were also developed. Through various experiments, the effectiveness of developed hardware and control method was confirmed.

Our next goal is to conduct various experiments and improve the hardware and control method toward the practical application.

Acknowledgment

This study was conducted as part of the humanoid project at the Humanoid Robotics Institute, Waseda University, and was done in collaboration with TMSUK Co., Ltd supported by Fukuoka City, SANYO Electric Soft Energy Co., Ltd. and Solid Works Japan K.K. This study was supported in part by the Ministry of Education, Science, Sports and Culture, Grant-in-Aid for Young Scientists (B), 16760208, 2004. It was also supported in part by the Robotics Industry Development Council. This study was supported in part by the 21st Century Center of Excellence Program, Japan Society for the Promotion of Science. The authors would like to express sincere thanks to them for their financial and technical support. The authors would like to thank the city employees and residents of Fukuoka City for their help in conducting experiments.

Bibliography

A. Takanishi, et al. (1985). The realization of dynamic walking by the biped walking robot. *Proceedings of the IEEE ICRA '85*, 459–466.

A. Takanishi, et al. (1990). A control method for dynamic walking under unknown external force. *Proceedings of the IEEE/RSJ IROS '90*, 795–801.

J. Yamaguchi, A. Takanishi and I. Kato. (1994). Development of a Biped Walking Robot Adapting to a Horizontally Uneven Surface. *Proceedings of the IEEE/RSJ IROS '94.*

H. O. Lim, Y. Sugahara and A. Takanishi. (2002). Stabilization walking control for human-like biped robots. *Proceedings of the CISM-IFToMM Symp. on Theory and Practice of Robots and Manipulators*, 471–480.

K. Takita, et al. (2000). Fundamental mechanism of dinosaur-like robot TITRUS-II utilizing coupled drive. *Proceedings of the IEEE/RSJ IROS 2000*, 1670–1675.

Y. Ota, T. Tamaki, K. Yoneda and S. Hirose. (2003). Development of walking manipulator with versatile locomotion. *Proceedings of the IEEE ICRA 2003*, 477–483.

Y. Konuma and S. Hirose. (2001). Development of the stair-climbing biped robot "Zero Walker-1. *Proceedings of the 19th Annual Conf. of the RSJ*, 851–852.

Y. Konuma and S. Hirose. (2003). Development of 2 types of leg-wheel vehicle with the function of stable stair-climbing. *Proceedings of the 8th RSJ/JSME/SICE Robotics Symposia*, 160–167.

Y. Takeda, M. Higuchi and H. Funabashi. (2001). Development of a walking chair (Fundamental investigations for realizing a practical walking chair). *Proceedings of the CLAWAR2001*, 1037–1044.

T. Kamada. (1994). My Agent: A Practical Personal Assistant. *Proceedings of the JSME ROBOMEC '94*, 1107–1112.

Independence Technology L. L. C. (2003). http://www.independencenow.com/ibot/index.html.

Y. Sugahara, et al. (2003). Control and Experiments of a Multi-purpose Bipedal Locomotor with Parallel Mechanism. *Proceedings of the IEEE ICRA 2003*, 4342–4347.

Y. Sugahara, et al. (2004). Realization of Dynamic Walking by a Biped Locomotor Carrying a Human. *Proceedings of the 9th RSJ/JSME/SICE Robotics Symposia*, 96–101.

Y. Sugahara, et al. (2004). Realization of Dynamic Human-Carrying Walking by a Biped Locomotor. *Proceedings of the IEEE ICRA 2004*, 3055–3060.

Y. Sugahara, et al. (2004). Support Torque Reduction Mechanism for Biped Locomotor with Parallel Mechanism. *Proceedings of the IEEE/RSJ IROS 2004*, 3213–3218.

Toyota Motor Corporation. (2004). http://www.toyota.co.jp/en/special/robot/index.html.

Toyota Motor Corporation. (2004). http://www.toyota.co.jp/en/news/04/1203_1d.html.

Y. Sugahara, et al. (2005). Walking Up and Down Stairs Carrying a Human by a Biped Locomotor with Parallel Mechanism. *Proceedings of the IEEE/RSJ IROS 2005*, 3425–3430.

Prosthesis Design by Robotic Approaches Part 1: Metabolical Cost

Werner Schiehlen* and Marko Ackermann†
University of Stuttgart, Institute of Engineering and Computational Mechanics, Germany,
* e-mail: schiehlen@itm.uni-stuttgart.de
† e-mail: ackermann@itm.uni-stuttgart.de

1 Introduction

The design of devices or therapies to improve gait quality of persons presenting gait disorders requires the definition of an appropriate optimizing criterion. Minetti and Alexander (1997) pointed out that humans are minimizing the metabolical cost which is different from the mechanical cost. While earlier work by Alexander (1992) was based on the minimization of the mechanical cost or work, respectively, in a more recent paper (Alexander, 2002) the metabolical cost is minimized. In particular, the improvement of the gait comfort of disabled persons will depend on the metabolical cost since the natural limit cycle does no longer fit to their skeletal bones involved in the locomotion. This paper reviews some fundamentals of metabolical cost, discusses the musculoskeletal dynamics arising from the use of Hill-type muscle models and presents a recently proposed phenomenological metabolic cost model by Umberger et al. (2003).

2 Some Fundamentals of Metabolical Cost

The biological actuators in animal locomotion are the skeletal muscles. The function of the muscle and the origin of its 'fuel' is described by Wasserman et al. (1999) as analogous to a machine: "Skeletal muscle may be considered to be a machine that is fueled by the chemical energy of substrates derived from ingested food and stored as carbohydrates and lipids in the body". This energy is commonly called metabolic energy cost or simply metabolic cost, respectively, and it is believed to play an important role in human locomotion. For instance, Ralston (1976) observed that people walk at speeds that minimize metabolic cost of transport. This criterion is therefore important to understand and study human locomotion, in particular with respect to disabled humans using prosthetic devices.

Metabolical energy cost is a measurement of effort required from the person to perform a given activity. For this reason, many studies assess the performance of prosthetic and orthotic devices and the success of therapies oriented to patients presenting gait disorders by experimental measurements of the metabolical energy expenditure (Schmalz et al., 2002, MacFarlane et al., 1997 and Hsu et al., 2000). Estimations of this quantity can be obtained using, for example, metabolic gas analysis systems (see Macfarlane, 2001).

In order to assess the metabolical cost, or the effort in the sense of energy consumption, without the need of experiments, two basic approaches were used in the literature. In the first approach computational models of the skeletal system are used and the estimations of the energy expenditure are assumed to be proportional to the external mechanical work or to some quantity dependent on the time histories of the moments at the joints. For instance, Zlatnik (1998) designs an intelligently controlled transfemoral prosthesis using a performance criteria based on the moment at the hip to quantify the hip effort. The main advantage of this approach is its simplicity. The joint moments and powers can be calculated by inverse dynamics for known kinematics.

This approach, however, can lead to erroneous estimations of the metabolical cost, since the correspondent work at the joints is poorly correlated to the metabolical energy being demanded (Martin et al., 1993) In other words, muscles cannot be assumed to perform work with constant efficiency. Furthermore, muscles consume metabolical energy during isometric contractions, too, i.e., when no external mechanical work is observed.

The use of models of the musculoskeletal system associated with models of heat production in the muscles can overcome the deficiency of the first approach and deliver more realistic estimations of metabolical energy cost. Indeed, Anderson and Pandy (2001) reproduced metabolical cost and kinematic patterns of human normal walking, using a large-scale model of the musculoskeletal system associated with a model of muscle heat production, by means of a dynamic optimization procedure that minimized metabolical cost per unit of distance traveled.

3 Musculoskeletal System

As pointed out in the previous section, the mechanical cost is not the true cost felt by humans due to the related biomechanical processes. Therefore, the skeletal models with ideal actuators will be complemented by muscle models.

3.1 Biological Actuators and Underdetermination

The skeletal system of humans is actuated by muscles. Since many muscles serve each joint of the skeletal system, muscle forces cannot be directly computed from joint moments. This is the known problem of underdetermination in biomechanics (Nigg and Herzog, 1999). In order to solve this problem optimization procedures are used.

The equations of motion of a multibody system with tree topology can be written as f ordinary differential equations in minimal form

$$\boldsymbol{M}(\boldsymbol{y})\ddot{\boldsymbol{y}} + \boldsymbol{k}(\boldsymbol{y}, \dot{\boldsymbol{y}}) = \boldsymbol{q}(\boldsymbol{y}) + \boldsymbol{B}^T \boldsymbol{u}\,, \tag{3.1}$$

where $\boldsymbol{M}$ is the $f \times f$ symmetric, positive definite mass matrix, $\boldsymbol{y}$ is the $f \times 1$-vector of generalized coordinates, $\boldsymbol{k}$ is the $f \times 1$-vector of generalized Coriolis forces and $\boldsymbol{q}$ is the $f \times 1$-vector of generalized applied forces. The matrix $\boldsymbol{B}$ is the $k \times f$ control input matrix and $\boldsymbol{u}$ is the $k \times 1$-vector of control torques in the joints.

Equation (3.1) describes the dynamics of the skeletal system. The joint moments $\boldsymbol{u}$ are generated by the muscles and by other structures spanning the joints, like the ligaments. Splitting the vector of joint moments into the vector of moments generated by muscles and the vector of moments generated by other structures yields:

$$\boldsymbol{M}(\boldsymbol{y})\ddot{\boldsymbol{y}} + \boldsymbol{k}(\boldsymbol{y},\dot{\boldsymbol{y}}) = \boldsymbol{q}^*(\boldsymbol{y}) + \boldsymbol{R}(\boldsymbol{y})\boldsymbol{f}^m , \tag{3.2}$$

$$\boldsymbol{R}(\boldsymbol{y}) = \boldsymbol{B}^T \boldsymbol{A}_m(\boldsymbol{y}) \tag{3.3}$$

where $\boldsymbol{q}^*$ is the $f \times 1$-vector of generalized forces other than the ones caused by the muscles, $\boldsymbol{f}^m$ is the $m \times 1$-vector of muscle forces, $\boldsymbol{A}_m$ is the $k \times m$-matrix describing the points of action of the muscles and $\boldsymbol{R}$ is the $f \times m$-matrix that transforms the muscle forces into generalized forces. Due to the underdetermination it yields $m \geq k$. The vector $\boldsymbol{q}^*$ includes the vector of ground reaction forces for both feet $\boldsymbol{f}^{gr}$ if the contact between feet and ground is modelled as elastic or if the ground reaction forces are known.

If the kinematics $\boldsymbol{y}(t)$ of the movement is known, Eq. (3.1) can be solved for $\boldsymbol{u}$, since the number of degrees of freedom of the mechanical system f is greater or equal to the number of unknown joint moments k, except for the double stance phase when the contacts are modeled as constraints. In this case a kinematic loop arise.

On the other hand, the number of muscles m is always greater than the number of degrees of freedom of the skeletal system f, what leads to the mentioned underdetermination. In this case an optimization problem is formulated, where the equations of motion constrain the minimization of an objective function. Moreover, many cost functions were proposed to solve the distribution problem in biomechanics. Crowninshield and Brand (1981), for instance, proposed a cost function based on the minimization of muscle fatigue. A typical formulation is as follows: find $\boldsymbol{f}^m$ that minimizes the objective function J, submitted to the equality constraint given by the equations of motion in the form $\boldsymbol{R}\boldsymbol{f}^m = \boldsymbol{b}$ and to the inequality constraint $\boldsymbol{f}^m \geq 0$, because muscles cannot push. Rewriting Eq. (3.2) results in $\boldsymbol{b} = \boldsymbol{b}(\boldsymbol{y},\dot{\boldsymbol{y}},\ddot{\boldsymbol{y}})$. The static optimization problem has then to be solved for each discrete time step considered.

Such a procedure does not consider the dynamics of force production, which will be explained further on, what can lead to results for the muscle forces that are not possible in reality. Furthermore, this approach does not allow to estimate metabolical cost.

3.2 Muscle Models and Musculoskeletal Dynamics

Skeletal muscles are composed by the muscle fibers and the tendon. There exist many different muscle models and levels of complexity, refer to Winters (1990) and Zajac (1989) for extensive reviews. In the field of movement simulation, however, the three-component Hill-type model has been most frequently used. This model consists of a contractile element (CE), a nonlinear parallel elastic element (PE), representing the stiffness of the structures in parallel with the muscle fibers, and a nonlinear series elastic element (SE), representing basically the stiffness of the tendon, see Fig. 1. The force is generated in the CE. The pennation angle of a muscle, given by the angle the muscle fibers do with the line of force application of the muscle, is modeled by an angle α, see figure.

Some variations of these models are also used, where a damping element is added in parallel to the CE and an additional elastic element is added in series to the CE,

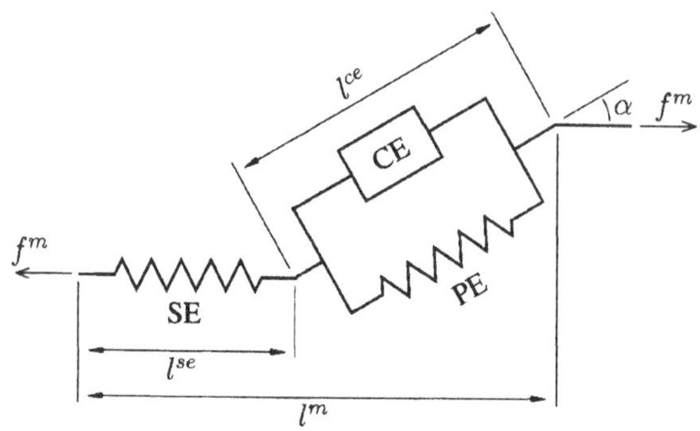

Figure 1. Hill-type muscle model with Contractile Element (CE), a parallel elastic element (PE) and a series elastic element (SE).

representing the stiffness of tissue in series with the muscle fibers. A simplification of these models is composed by a CE in series with a series elastic element and an elastic element in parallel to both. This permits that the parallel elastic element be lumped into the stiffness of a joint, that incorporates the stiffness of the parallel elastic elements of all the muscles and the stiffness of all the other elastic structures spanning the joint. Expressions for this joint stiffnesses can be found, for example, in Riener and Edrich (1999).

The process of force generation is composed by many steps from the neural excitation to the contraction dynamics, refer to Zahalak (1990) for a detailed explanation. In a gross way, the neural action potential, received from the neural system, at the neuromuscular junction triggers a sequence of biochemical processes that result in muscle contraction. This complex process is modeled, as schematically represented in Fig. 2, by two independent processes, the activation dynamics and the contraction dynamics.

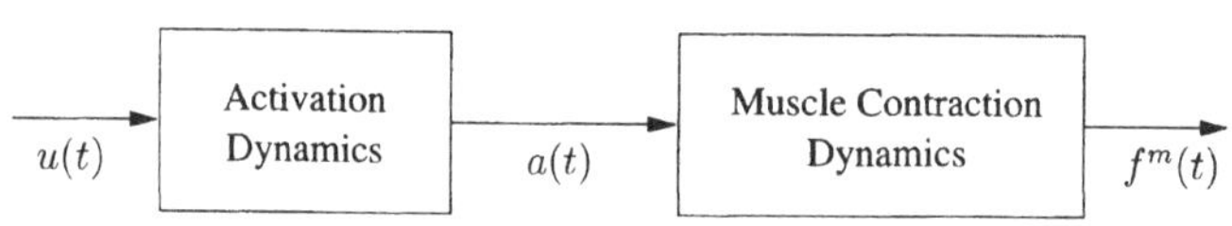

Figure 2. Muscle activation and contraction dynamics.

The activation $a(t)$ represents an internal muscle tissue state, which is associated with the contractile process. The process that leads to the muscle activation $a(t)$ from the neural excitation $u(t)$ is modeled by a first order differential equation that simulates the time delay caused by the biochemical processes involved. This differential equation can be, for example, of the form presented in Nagano and Gerritsen (2001):

$$\dot{a}(t) = (u(t) - a(t))(c_1 u(t) + c_2) \ , \tag{3.4}$$

where c_1 and c_2 are constants, $u(t)$ is the neural excitation $0 \leq u \leq 1$ and $a(t)$ is the muscle activation state $0 \leq a \leq 1$.

The contraction dynamics models the force applied by the muscle on the skeletal system as a function of the muscle tissue state $a(t)$. The force generated by the CE, f^{ce}, is a function of the activation a, its current length l^{ce}, and its current contraction velocity v^{ce}. This relation, the force-length-velocity relation, (3.5), is an intrinsic property of the muscle fibers and incorporates the so called force-length and force-velocity relations,

$$\frac{f^{ce}}{F_{iso}} = f_1\left(a, \frac{v^{ce}}{L_{opt}V_{max}}, \frac{l^{ce}}{L_{opt}}\right) . \tag{3.5}$$

The force-length-velocity curve as well as the tendon force-length relation can be scaled by a few parameters, refer for example to Zajac (1989), namely, the maximal muscle isometric force, F_{iso}, the maximal CE shortening velocity, V_{max}, and the muscle CE optimal length, L_{opt}. Scaling considering the muscle fiber composition is also frequently used.

From Eq. (3.5) and according to Fig. 1 by assuming a pennation angle $\alpha \approx$ constant the contraction dynamics for the muscle model considered can be expressed as:

$$\dot{f}^m = \dot{f}^m(a, v^m, l^m, f^m) \ , \tag{3.6}$$

This dynamics depends on the total muscle length and shortening velocity, l^m and v^m, which are directly computed from the skeletal system generalized coordinates, $\boldsymbol{y}$, and their derivatives, $\dot{\boldsymbol{y}}$. The skeletal system dynamics, described by Eq. (3.2), and the muscle contraction dynamics, described by Eq. (3.6), are, therefore, coupled. In Fig. 3 the dynamics of the whole musculoskeletal system is shown and the mentioned coupling is visualized. In the Fig. 3, $\boldsymbol{u}$ is the $m \times 1$ vector of muscle neural excitations; $\boldsymbol{a}$ is the $m \times 1$ vector of muscle activations; $\boldsymbol{f}^m$, $\boldsymbol{l}^m$ and $\boldsymbol{v}^m$ are the $m \times 1$ vectors of muscle forces, muscle lengths and muscle shortening velocities, respectively.

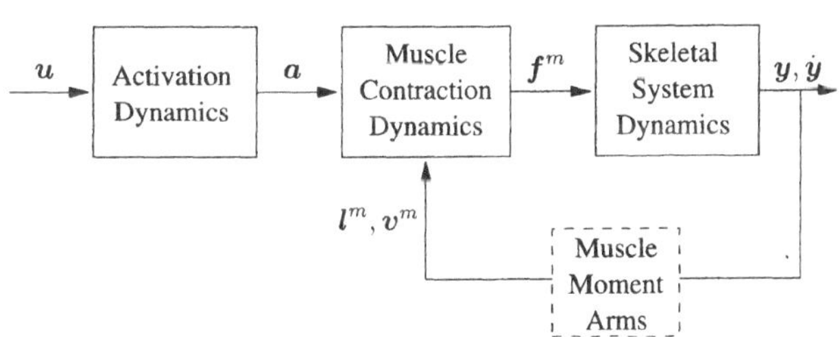

Figure 3. Scheme of the musculoskeletal system dynamics.

4 Metabolic Cost Model

Some fundamentals of metabolic energy cost were presented above. This section briefly introduces models of muscle energy expenditure available in the recent literature, pointing out some of their advantages and drawbacks.

The most simple approach comprises the computation of the net joint mechanical energy or some function of the net joint moments and rotational velocities. The main advantage of this method is its simplicity. However, the energy expenditure estimations obtained are not really related to the actual metabolical cost as discussed above.

Minetti and Alexander (1997) proposed an elegant method to estimate metabolical cost from joint net moments and angular velocities, based on properties of the underlying muscles. Their model was sufficient to identify basic principles correctly, but it is not able to predict quantitatively the locomotion behavior and the metabolical energy cost as pointed out by the authors and underlined by a strong underestimation of the metabolical cost of transport.

Therefore, in order to assess more accurate estimations of muscle energy expenditure, more detailed models are required, taking the process of force generation of each muscle involved into account. Quantitative predictions of energy expenditure could be obtained by models based on the cross-bridge theory by Huxley (1957) since in these models the mechanical processes are directly related to the energetic ones, see also Zahalak (1990). However, these models require many parameters whose values are uncertain. Furthermore, because of their complexity, these models would be difficult to implement in a large-scale musculoskeletal model.

A compromise is achieved with phenomenological muscle energy expenditure models. These models have the advantage of being used in conjunction with the widely employed Hill-type muscle models, see Fig. 1. Furthermore, these empirical models are able to deliver relatively accurate estimations of metabolical energy cost, although the contraction processes are not directly related to the energetic ones. Indeed, Umberger et al. (2003) and Bhargava et al. (2004) proposed independently phenomenological muscle energy consumption models, that were able to predict reliable characteristics of human normal walking in simulation results obtained by dynamic optimizations that minimized metabolical cost of transport.

These models subdivide the total metabolical cost rate, $\dot{E}$, into the muscle CE work rate or power, $\dot{W}$, and the muscle CE heat production rate, $\dot{H}$, $\dot{E} = \dot{H} + \dot{W}$. The work rate is simply the mechanical work performed by the muscle CE,

$$\dot{W} = -f^{ce}v^{ce} , \tag{4.1}$$

resulting in a positive rate if v^{ce} is negative due to shortening. The complete algebraic expression for the muscle heat rate is presented in Umberger et al. (2003) depending on four muscle CE states and five muscle parameters $\dot{H} = \dot{H}(u, a, v^{ce}, l^{ce}, F_{iso}, L_{opt}, ft, PCSA, width)$. The muscle heat rate, described by this expression, is a function of the muscle neural excitation, u, the muscle activation, a, the muscle CE shortening velocity, v^{ce}, the muscle CE length, l^{ce}, and a set of muscle-dependent parameters comprising the muscle maximal isometric force, F_{iso}, the muscle CE optimal length, L_{opt}, the muscle fiber composition in percentage of fast twitch fibers, ft, the muscle physiological cross-sectional area, $PCSA$, and an additional parameter introduced to adjust the force-length relation for each muscle, $width$, see Gerritsen et al. (1998). The total metabolical energy cost rate for a muscle is, therefore, written as

$$\dot{E} = \dot{E}(u, a, v^{ce}, l^{ce}, f^{ce}, \boldsymbol{p}) , \qquad (4.2)$$

where the muscle parameters are summarized in the vector $\boldsymbol{p}$. This expression delivers a more accurate estimation of the metabolical cost rate during a numerical simulation of human motion. In this way, optimal motion and muscle force distribution patterns, and optimal parameters for assistive devices in the sense of energy expenditure can be searched.

5 Conclusion

In the paper, musculoskeletal systems are considered to design devices to improve gait quality of persons presenting musculoskeletal disorders. Since the gait cannot be characterized properly by the mechanical energy consumed, the fundamentals of metabolic cost are reviewed. Then, biological actuators and their underdetermination as well as reliable muscle models are discussed. Models for metabolic cost are introduced, which deliver a more accurate estimation of the effort required to perform a given human motion. In part 2 of this paper, see Ackermann and Schiehlen (2006), the metabolical cost is evaluated for a lower limb system.

Bibliography

M. Ackermann and W. Schiehlen. Prosthesis design by robotic approaches, part 2: Optimization approach. In T. Zielinska, editor, *ROMANSY 16*, Wien, 2006. Springer. to be published.

R. McN Alexander. A model of bipedal locomotion on compliant legs. *Philosophical Transactions: Biological Sciences*, 338(1284):189–198, 1992. URL

R. McNeill Alexander. Energetics and optimization of human walking and running. *American Journal of Human Biology*, 14(5):641–648, 2002.

F. C. Anderson and M. G. Pandy. Dynamic optimization of human walking. *Journal of Biomechanical Engineering*, 123(5):381–390, October 2001.

L. J. Bhargava, M. G. Pandy, and F. C. Anderson. A phenomenological model for estimating metabolic energy consumption in muscle contraction. *Journal of Biomechanics*, 37(1):81–88, January 2004.

Roy D. Crowninshield and Richard A. Brand. Physiologically based criterion of muscle force prediction in locomotionl. *Journal of Biomechanics*, 14(11):793–801, 1981.

K. G. M. Gerritsen, A. J. van den Bogert, M. Hulliger, and R. F. Zernicke. Intrinsic muscle properties facilitate locomotor control – a computer simulation study. *Motor Control*, 2:206–220, 1998.

M.-J. Hsu, D. H. Nielsen, J. Yack, D. G. Shurr, and S.-J. Lin. Physiological comparisons of physically active persons with transtibial amputation using static and dynamic prostheses versus persons with nonpathological gait during multiple-speed walking. *Journal of Prosthetics & Orthotics*, 12(2):60–67, 2000.

A. Huxley. Muscle structure and theories of contraction. *Progress in Biophysics and Biophysical Chemistry*, 7:255–318, 1957.

D. J. Macfarlane. Automated metabolic gas analysis systems. *Sports Medicine*, 31(12): 841–861, 2001.

P. A. MacFarlane, D. H. Nielson, D. G. Shurr, K. G. Meier, R. Clark, J. Kerns, M. Moreno, and B. Ryan. Transfemoral amputee physiological requirements: comparisons between sach foot and flex-foot walking. *Journal of Prosthetics & Orthotics*, 9(4):138–143, 1997.

P.E. Martin, G. D. Heise, and D. W. Morgan. Interrelationships between mechanical power, energy transfers, and walking and running economy. *Medicine & Science in Sports & Exercise*, 25:508–515, 1993.

A. E. Minetti and R. McN. Alexander. A theory of metabolic cost for bipedal gaits. *Journal of Theoretical Biology*, 186:467–476, 1997.

A. Nagano and K. G. M. Gerritsen. Effects of neuromuscular stregth training on vertical jumping performance - a computer simulation study. *Journal of Applied Biomechanics*, 17(2):113–128, 2001.

Benno M. Nigg and Walter Herzog, editors. *Biomechanics of the Musculo-Skeletal System*. John Wiley & Sons, 2dn edition, 1999.

H. J. Ralston. Energetics of human walking. In R. M. Herman et al., editor, *Neural Control of Locomotion*, pages 77–98. Plenum Press, New York, 1976.

R. Riener and T. Edrich. Identification of passive elastic joint moments in the lower extremities. *Journal of Biomechanics*, 32(5):539–544, May 1999.

T. Schmalz, S. Blumentritt, and R. Jarasch. Energy expenditure and biomechanical characteristics of lower limb amputee gait: The influence of prosthetic alignment and different prosthetic components. *Gait & Posture*, 16(3):255–263, december 2002.

B. R. Umberger, K. G. M. Gerritsen, and P. E. Martin. A model of human muscle energy expenditure. *Computer Methods in Biomechanics & Biomedical Engineering*, 6(2):99–111, 2003.

Karlman Wasserman, James E. Hansen, Darryl Y. Sue, Richard Casaburi, and Brian J. Whipp. *Principles of exercise testing and interpretation*. Lippincott Willians & Wilkins, 3rd edition, 1999.

Jack M. Winters. Hill-based muscle models: a system engineering perpective. In Jack M. Winters and Savio L-Y. Woo, editors, *Multiple Muscle Systems*, chapter 5, pages 69–93. Springer-Verlag, 1990.

G. I. Zahalak. Modeling muscle mechanics (and energetics). In Jack M. Winters and Savio L-Y. Woo, editors, *Multiple Muscle Systems*, chapter 1, pages 1–23. Springer-Verlag, 1990.

F. E. Zajac. Muscle and tendon: properties, models, scaling, and application to biomechanics and motor control. *CRC Critical Reviews in Biomedical Engineering*, 19(4): 359–411, 1989.

D. Zlatnik. *Intelligently controlled above knee prosthesis*. Ph.D. thesis nr. 12814, Swiss Federal Institute of Technology, Zürich, 1998.

Prosthesis Design by Robotic Approaches Part 2: Optimization Approach

Marko Ackermann* and Werner Schiehlen†

University of Stuttgart, Institute of Engineering and Computational Mechanics, Germany,
* e-mail: ackermann@itm.uni-stuttgart.de
† e-mail: schiehlen@itm.uni-stuttgart.de

1 Introduction

Robotic approaches are proposed for the design of devices or therapies to improve gait quality of persons presenting gait disorders. The inverse dynamics robotic approach for skeletal models as discussed by Schiehlen and Ackermann (2005) will be extended and applied to musculoskeletal system of disabled. An optimization approach is proposed and discussed that permits the computation of muscle forces, activations and neural excitations by minimizing a more accurate estimation of the metabolical cost. In part 1 of this paper, see Schiehlen and Ackermann (2006), the modeling of muscles and the evaluation of the corresponding metabolical cost is reported in detail. Finally, an example of the lower limb is used to illustrate the procedure of combining an optimal skeletal design with a minimum of metabolical cost.

2 Proposed Optimization Approach

Static optimization approaches to solve the muscle force distribution problem for each time step considered (e.g. Crowninshield and Brand, 1981) can lead to nonphysiological muscle force histories, because they do not consider the muscle activation and contraction dynamics. Furthermore, this approach does not permit accurate estimations of metabolical cost rates and cannot be used when the objective function depends on the time history of the variables. For example, this approach cannot be implemented to optimize muscle forces when the goal is to minimize metabolical cost of transport, which is accepted to be the intrinsic goal in human walking.

A solution to consider such kind of objective function is to perform a dynamic optimization. This approach was successfully implemented recently for the case in which the objective function is the metabolical energy cost per unit of distance traveled (Anderson and Pandy, 2001, Bhargava et al., 2004 and Umberger et al., 2003). These studies could mimic human normal walking patterns like kinematics, walking optimal velocity and metabolic energy cost reasonably. However, the dynamic optimization of such large-scale musculoskeletal models is extremely costly in terms of computational effort, since the differential equations that describe the musculoskeletal dynamics have to be forward integrated several times. This drawback prevents this method of being used widely among biomechanicists. Moreover, the addition of equality constraints or of further terms to the objective function like optimal tracking to solve the problem for known or measured arbitrary movements would probably further increase the computational effort required.

Thelen et al. (2003) proposes an algorithm to solve the problem of muscle force distribution for known movement kinematics, based on a control algorithm that tracks the kinematics of a measured movement. This method is much faster than the conventional dynamic optimization approaches, because it requires only one forward integration of the state equations. This method efficiently enforces the dynamics of the system, but it still needs a static optimization approach for solving the muscle force distribution problem for each time step. Therefore, in opposition to the dynamic optimization approach, the use of the total metabolic cost as objective function is not possible.

We propose a method based on a large-scale static optimization algorithm, that minimizes the total metabolic cost for a given activity and is subject to constraint equations given by the equations of motion and by the admissible values of the neural excitations and of the muscle activations. This method permits to calculate the time histories of the optimal muscle forces in the sense of metabolic energy expenditure. Moreover, this method is computationally more efficient than the dynamic optimization since the differential equations are not integrated. This method resembles a method used in robotics to solve the optimum control problem to optimize robot trajectories by a state parametrization, what avoids the need for integration of the equations of motion, see for example Stryk (1998).

The method requires the discretization of the motion in $n-1$ constant time intervals, resulting in the time vector $\boldsymbol{T} = [t_1 \cdots t_j \cdots t_n]$, where $t_j - t_{j-1} = \Delta t$. The optimization variables are the muscle forces for all time steps and muscles considered. They are ordered in a $mn \times 1$ vector, called vector of global muscle forces,

$$\boldsymbol{F}^m = [\boldsymbol{f}_1^{mT} \quad \boldsymbol{f}_2^{mT} \quad \cdots \quad \boldsymbol{f}_j^{mT} \quad \cdots \quad \boldsymbol{y}_n^{mT}]^T ,$$

where, m is the number of muscles considered and $\boldsymbol{f}_j^m$ is the $m \times 1$ vector of muscle forces at the time instant t_j. The objective function can be, for example, the metabolical cost of transport, computed from the time integral of the metabolical energy cost rate for all m muscles involved, divided by the distance traveled in a complete cycle, ΔS, of the movement,

$$J = \frac{1}{\Delta S} \sum_{i=1}^{m} \left(\int_{t_1}^{t_n} \dot{E}(u_i(t), a_i(t), v_i^{ce}(t), l_i^{ce}(t), f_i^{ce}(t), \boldsymbol{p}_i) dt \right) \tag{2.1}$$

where the index i denotes a quantity related to the muscle i, u is the muscle neural excitation, a is the muscle activation, v^{ce} is the muscle contractile element (CE) shortening velocity, l^{ce} is the muscle CE length, and $\boldsymbol{p}$ is a vector that contains muscle-specific parameters, refer to Umberger et al. (2003) or Schiehlen and Ackermann (2005).

Later on, the computation of $\dot{E}$ from the vector of muscle forces $\boldsymbol{F}^m$ will be explained. The integral is solved numerically using the values of $\dot{E}$ in the discrete time instants t_j, for each function evaluation during the optimization.

The vectors of muscle forces, $\boldsymbol{f}_j^m$, contained in the vector $\boldsymbol{F}^m$, have to satisfy the equality constraint equations given by the equations of motion for all time instants t_j considered,

$$\boldsymbol{M}(\boldsymbol{y}_j)\ddot{\boldsymbol{y}}_j + \boldsymbol{k}(\boldsymbol{y}_j, \dot{\boldsymbol{y}}_j) = \boldsymbol{q}^*(\boldsymbol{y}_j) + \boldsymbol{R}(\boldsymbol{y}_j)\boldsymbol{f}_j^m \tag{2.2}$$

where $\boldsymbol{y}_j$ is the $f \times 1$-vector of generalized coordinates at the time instant t_j, $\boldsymbol{q}^*$ is the $f \times 1$-vector of generalized forces other than the ones caused by the muscles, $\boldsymbol{R}$ is the $f \times m$-matrix that transforms the muscle forces into generalized forces. Due to the underdetermination it yields $m > f$. The vector $\boldsymbol{q}^*$ includes the vector of ground reaction forces for both feet $\boldsymbol{f}^{gr}$ if the contact between feet and ground is modelled as elastic or if the ground reaction forces are known.

Since the kinematics of the movement is known, the only unknowns are the muscle forces, $\boldsymbol{f}_j^m$. Rearranging Eq. (2.2) yields

$$\boldsymbol{A}_{eq\,j} \cdot \boldsymbol{f}_j^m = \boldsymbol{b}_j \ , \tag{2.3}$$

where $\boldsymbol{A}_{eq\,j} = \boldsymbol{R}(\boldsymbol{y}_j)$ and $\boldsymbol{b}_j = \boldsymbol{M}(\boldsymbol{y}_j)\,\ddot{\boldsymbol{y}}_j + \boldsymbol{k}(\boldsymbol{y}_j, \dot{\boldsymbol{y}}_j) - \boldsymbol{q}^*(\boldsymbol{y}_j)$. Writing all the constraint equations given by the equations of motion in a single matrix equation yields:

$$\boldsymbol{A}_{eq} \cdot \boldsymbol{F}^m = \boldsymbol{b} \ , \tag{2.4}$$

where $\boldsymbol{A}_{eq}$ is a $fn \times mn$ "block diagonal" matrix and $\boldsymbol{b}$ is a $fn \times 1$ vector, constructed as:

$$\boldsymbol{A}_{eq} = diag\{\boldsymbol{A}_{eq\,1} \quad \boldsymbol{A}_{eq\,2} \quad \cdots \quad \boldsymbol{A}_{eq\,j} \quad \cdots \quad \boldsymbol{A}_{eq\,n}\} \ , \tag{2.5}$$

$$\boldsymbol{b} = [\boldsymbol{b}_1^T \quad \boldsymbol{b}_2^T \quad \cdots \quad \boldsymbol{b}_j^T \quad \cdots \quad \boldsymbol{b}_n^T]^T . \tag{2.6}$$

In addition, the vector of global muscle forces, $\boldsymbol{F}^m$, has to satisfy the inequality constraint,

$$\boldsymbol{F}^m \geq \boldsymbol{0} \ , \tag{2.7}$$

because muscles cannot push, and the nonlinear equality constraints given by the values that the neural excitations, u_{ij}, and muscle activations, a_{ij} are allowed to assume, for all m muscles and n time instants

$$0 \leq u_{ij} \leq 1 \ , \quad 0 \leq a_{ij} \leq 1 \ , \quad i = 1 \ldots m \ , \quad j = 1 \ldots n \ , \tag{2.8}$$

Additional equality constraints can be imposed for periodic motions to assure that the initial and final neural excitations, muscle activations and muscle forces are equal.

Therefore, the static optimization problem can be formulated as follows: find the global vector of muscle forces, $\boldsymbol{F}^m$, that minimizes the objective function J, for example the metabolical costs of transport, Eq. (2.1), subject to the linear equality constraints given by the equations of motion, Eq. (2.4), the inequality constraints for positive muscle forces, Eq. (2.7), the inequality constraints for the neural excitations and muscle activations, Eq. (2.8), and, if the motion is periodic, the equality constraints that assure the same initial and final values for the neural excitations, muscle activations and muscle forces.

In order to implement the optimization approach proposed, it is necessary to compute, for each function evaluation, i.e. for each time the optimization algorithm enters a new $\boldsymbol{F}^m$, the values of the neural excitation, u, of the muscle activation, a, and of the total metabolic cost rate, $\dot{E}$, for all muscles and all time steps, from the time histories of the muscle forces, contained in $\boldsymbol{F}^m$, and from the known kinematics, $\boldsymbol{y}$ and $\dot{\boldsymbol{y}}$. The computational algorithm is explained below.

The first step is to compute the total muscle length, l^m, and total muscle shortening velocity, v^m, from the generalized coordinates and their derivatives, $\boldsymbol{y}$ and $\dot{\boldsymbol{y}}$, as:

$$l^m = l^m(\boldsymbol{y}) , \quad v^m = v^m(\boldsymbol{y}, \dot{\boldsymbol{y}}) . \tag{2.9}$$

These relations can be obtained by the positions of the muscle origins and muscle insertions in the skeleton, see (Delp, 1990).

The second step is to numerically differentiate the time history of each muscle force, $f^m(t)$, obtaining $\dot{f}^m(t)$. From the time histories of the muscle forces and their time derivatives it is possible to compute v^{ce}, l^{ce} and f^{ce} of the contractile elements of Hill-type muscle models, refer to Schiehlen and Ackermann (2005). The muscle activation is obtained by solving the force-length-velocity relation equation of the muscle as (Schiehlen and Ackermann, 2005)

$$f^{ce} - f^{ce}(a, v^{ce}, l^{ce}, \boldsymbol{p}) = 0 . \tag{2.10}$$

For some models of the muscle force-length-velocity relation, a can be written explicitly as function of v^{ce}, l^{ce} and f^{ce}. For other models, this is not possible and a has to be computed numerically using Eq. (2.10).

The neural excitation, u, is computed by solving the first order differential equation used to model the activation dynamics, see for example Nagano and Gerritsen (2001), for $a(t)$ and $\dot{a}(t)$ known, which leads to a quadratic polynomial in u as

$$(c_1) \cdot u^2(t) + (c_2 - a(t)c_1) \cdot u(t) - (a(t)c_2 + \dot{a}(t)) = 0 , \tag{2.11}$$

where c_1 and c_2 are constants and $\dot{a}(t)$ is obtained by numerical differentiation of $a(t)$.

The last quantity needed to implement the method, for the case in which the total metabolical cost is of interest or is used in the objective function, is the metabolical cost rate, $\dot{E}$, which can be easily computed by using the expression proposed by Umberger et al. (2003) , since, for all muscles and time steps, u, a, v^{ce}, l^{ce} and f^{ce} are now known.

3 Lower Limb System

In this section, the method proposed for estimating the metabolical cost and to solve the muscle force distribution problem is used for a musculoskeletal model of the lower limb of a human. The kinematical data of the normal walking of a test person as well as the ground reaction forces were measured in the gait analysis laboratory of the Institut für Sportwissenschaft at the University of Stuttgart and were used as input to the simulation. This data is documented by Ackermann and Gros (2005).

The model of the lower limb skeletal system adopted is composed by three rigid bodies, the thigh, the shank and the foot. The movement is performed in the sagittal plane and is described by five generalized coordinates, the angle α, describing the rotation of the thigh, the angle β, describing the knee flexion, the angle γ for the ankle plantar flexion, the horizontal position of the hip joint, x_{hip}, and the vertical position of the hip joint, z_{hip}, as schematically shown in Fig. 1b. The pelvis and trunk are assumed to remain in the vertical position during the movement, what is reasonable for normal walking. The masses, center of mass locations, and the mass moment of inertia in the sagittal plane are obtained using the tables in Winter (2005), as functions of the subject's body mass, height, thigh length and shank length.

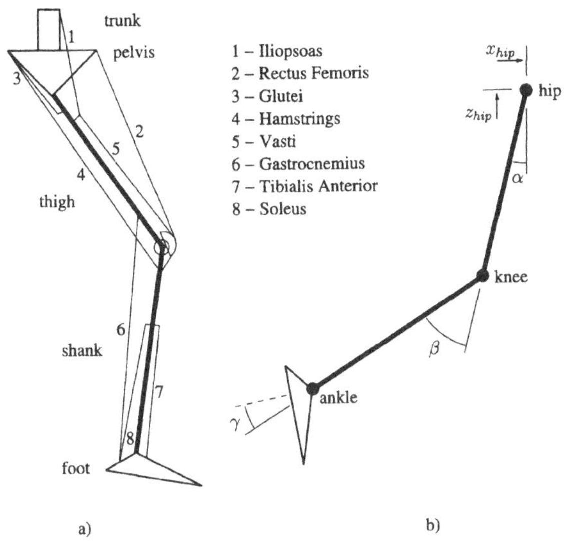

Figure 1. Musculoskeletal model of the lower limb: a) eight muscles and b) generalized coordinates.

The eight muscle units considered in this analysis are shown in Fig. 1. The muscles were modeled by Hill-type muscle models composed by a contractile element CE and a series elastic element SE. In this model all the structures in parallel to the CE and the SE are represented by a total passive moment around the joints, which includes additionally the passive moments generated by all other passive structures around the joints, like ligaments, too. The formula for the passive moments around the hip, knee and ankle are functions of α, β and γ as proposed by Riener and Edrich (1999). A linear damping is added to the knee and hip joints and their values are the approximate average values obtained by pendulum experiments performed by Stein et al. (1996), too. The muscle parameters are selected from Gerritsen et al. (1998), Nagano and Gerritsen (2001), Umberger et al. (2003) and Menegaldo et al. (2003). The phenomenological model for the muscle metabolical cost rate is adopted from Umberger et al. (2003).

The whole walking cycle measured has a duration of 1.14 s. This period is divided into 57 time intervals or 58 time steps, $n = 58$, respectively, which are the reference time steps used. The motion, $\boldsymbol{y}(t)$, and the ground reaction forces were measured in the gait analysis laboratory and $\dot{\boldsymbol{y}}(t)$, and $\ddot{\boldsymbol{y}}(t)$ are obtained by numerical differentiation of $\boldsymbol{y}(t)$.

The approach described in the previous section, uses as objective function the total metabolic cost of transport, Eq. (2.1), and as initial guess for the optimization the result obtained by the static optimization for each time step of Crowninshield and Brand, 1981. The optimization problem is solved using the MATLAB routine *fmincon*. For this example the muscle force-velocity relation is not implemented during the optimization, only the force-length is used. The reason is that, for the model used (presented in Umberger et al., 2003), Eq. (2.10) cannot be explicitly formulated for a if the force-velocity relation is considered. Therefore Eq. (2.10) would have to be solved numerically and the computational effort for the whole optimization would increase considerably. However, neglecting the force-velocity relation will have a small influence on the results, since, for walking, the velocities are low. Furthermore, the metabolical cost were recomputed for the obtained optimal forces using the force-velocity relation, and the resulting difference is approximately 5 %.

The total metabolical cost computed using the metabolical cost of transport as objective function is 215 J. This optimization needed about 20 hours to converge, mainly due to the high number of variables to be optimized, which were $8 \cdot 58 = 464$ (number of muscles $\times$ number of time steps). The computed optimal muscle activation and muscle force histories for the 8 muscles considered, respectively are shown in Fig. 2. Because the motion measured in the experiment was not perfectly periodic, the constraints to assure that final states equal to initial states for periodic movements were weakened. Consequently, the muscle activations and muscle forces at the beginning and at the end of the simulation, are not exactly the same. Please, note that the convergence to a global minima cannot be guaranteed for any deterministic optimization algorithm. Different local minima for this problem were achieved when using different initial guesses. However, the minima that was found by using the results of the static optimization of Crowninshield and Brand (1981) as initial guess for the optimization proposed did achieve lower values for the objective function than the ones obtained for any other initial guess for all the experiments performed.

Although the computational effort to achieve convergence for the approach proposed can be considered high, this method is much more efficient than performing dynamic optimization. Indeed, Anderson and Pandy (2001) needed 10.000 hours of CPU time for their dynamic optimization problem involving a model of the whole body with 54 muscles. Their optimization had 810 control variables to be optimized (54 muscles $\times$ 15 control nodes), while in the example shown in this paper, 464 optimization variables (8 muscles $\times$ 58 time steps) were used, less muscles (8 instead of 54) but more control nodes (58 time steps instead of 15). The method proposed can be considered, therefore, as a viable alternative to dynamic optimization of large-scale musculoskeletal systems if the motion is known or measured. This method permits, for example, to obtain estimations of total metabolical cost and optimal muscle forces for a given body motion efficiently, when the performance criteria adopted by the body is assumed to be the minimization

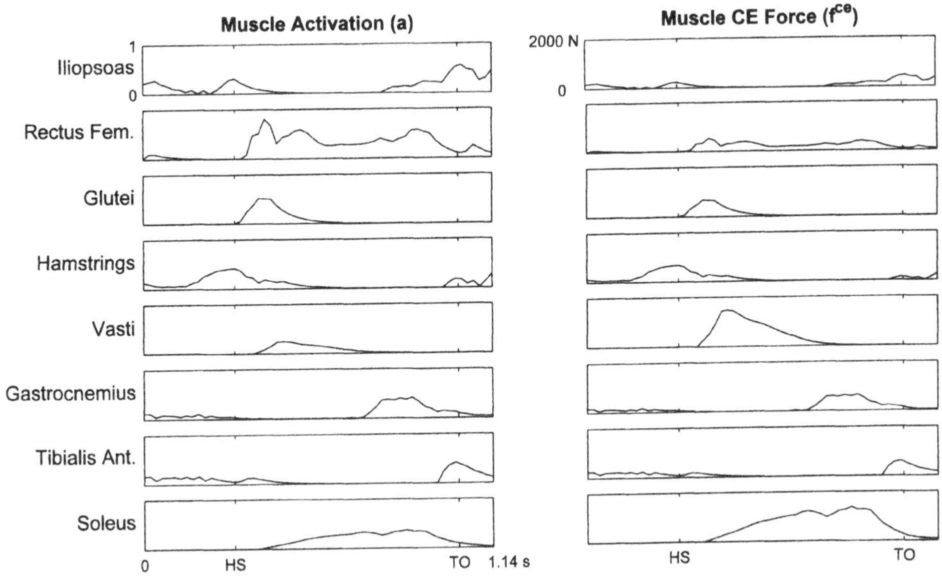

Figure 2. Trajectories of muscle activations and CE forces for the proposed optimization approach. The 'heel strike' is denoted by HS and the 'toe off' by TO.

of the metabolical cost of transport.

4 Conclusion

In this paper musculoskeletal systems are considered to design devices to improve gait quality of persons presenting musculoskeletal disorders. The total metabolical cost is an appropriate design criteria for walking, since for human walking it can be assumed that the body minimizes this quantity.

Recently proposed phenomenological models to accurately estimate energy expenditure based on the dynamics of Hill-type muscle models are used together with a dynamic optimization approach to determine optimal muscle forces and metabolical cost. The dynamic optimization is, however, extremely costly and requires large supercomputing facilities. Therefore, a less demanding global static optimization algorithm is proposed, which minimizes the metabolical cost subject to constraints given by the equations of motion, the admissible values of the neural excitations and the muscle activations. This approach is described in detail. As an example a lower limb system with eight muscles is presented. It turns out that the proposed method works well.

Bibliography

M. Ackermann and H. Gros. Measurements of human gaits. Zwischenbericht ZB-144, Institute B of Mechanics - University of Stuttgart, Stuttgart, 2005.

F. C. Anderson and M. G. Pandy. Dynamic optimization of human walking. *Journal of Biomechanical Engineering*, 123(5):381–390, October 2001.

L. J. Bhargava, M. G. Pandy, and F. C. Anderson. A phenomenological model for estimating metabolic energy consumption in muscle contraction. *Journal of Biomechanics*, 37(1):81–88, January 2004.

Roy D. Crowninshield and Richard A. Brand. Physiologically based criterion of muscle force prediction in locomotionl. *Journal of Biomechanics*, 14(11):793–801, 1981.

K. G. M. Gerritsen, A. J. van den Bogert, M. Hulliger, and R. F. Zernicke. Intrinsic muscle properties facilitate locomotor control – a computer simulation study. *Motor Control*, 2:206–220, 1998.

L. L. Menegaldo, A. T. Fleury, and H. I. Weber. Biomechanical modeling and optimal control of human posture. *Journal of Biomechanics*, 36(11):1701–1712, November 2003.

A. Nagano and K. G. M. Gerritsen. Effects of neuromuscular stregth training on vertical jumping performance - a computer simulation study. *Journal of Applied Biomechanics*, 17(2):113–128, 2001.

R. Riener and T. Edrich. Identification of passive elastic joint moments in the lower extremities. *Journal of Biomechanics*, 32(5):539–544, May 1999.

W. Schiehlen and M. Ackermann. Estimation of metabolical costs for human locomotion. In *Proceedings of IDETC/CIE2005*, volume 1, Long Beach, USA, 2005.

W. Schiehlen and M. Ackermann. Prosthesis design by robotic approaches, part 1: Metabolical cost. In T. Zielinska, editor, *ROMANSY 16*, Wien, 2006. Springer. to be published.

R. B. Stein, M. K. Lebiedowska, D. B. Popovic, A. Scheiner, and H. J. Chizeck. Estimating mechanical parameters of leg segments in individuals with and without physical disabilities. *IEEE Transactions on Rehabilitation Engineering*, 4(3):201–211, 1996.

O. von Stryk. Optimal control of multibody systems in minimal coordinates. *Zeitschrift für Angewandte Mathematik und Mechanik*, (3):1117–1120, 1998.

Darryl G. Thelen, Frank C. Anderson, and Scott L. Delp. Generating dynamic simulations of movement using computed muscle control. *Journal of Biomechanics*, 36: 321–328, 2003.

B. R. Umberger, K. G. M. Gerritsen, and P. E. Martin. A model of human muscle energy expenditure. *Computer Methods in Biomechanics & Biomedical Engineering*, 6(2):99–111, 2003.

David A. Winter. *Biomechanics and Motor Control of Human Movement.* John Wiley & Sons, 3rd*edition*, 2005.

Chapter VII

Innovative technologies in robotics

Sound Source Detection System for Control of an Autonomous Mobile Robot, a Behaviour-Based Approach

Johanna Lantto
Helsinki University of Technology, Automation Technology Laboratory, Espoo, Finland
e-mail: llantto@cc.hut.fi
Technische Universität Kaiserslautern, Arbeitsgruppe Robotersysteme,Kaiserslautern, Germany

Abstract. This paper describes a first implementation of an two-microphone system for sound source localization using the Modular Controller Architecture framework. The system was developed to be a part of an autonomous mobile robot and its behaviour-based control architecture. The goal was to enable the robot to react to sound stimulus and lay groundwork for environment perception using sound signals and sound recognition. The emphasis on the development of the system was on reacting to speech signals. The results of testing the system in stationary experiments, as well as in a real environment where the robot was able to move, are presented and discussed.

1 Introduction

The project described in this paper was the first research done with audio signals in the Arbeitsgruppe. The goal was to produce a working implementation of the direction calculation and connect it to the robot's control architecture to form a platform for later research. The ways of detecting a sound source, based on the so–called Jeffress model of hearing, have been implemented numerous times in the field of robotics. What is special in this system is the integration of the sound source detection system onboard an autonomous mobile robot, Marvin(AG Robotersysteme , 2005), using the behaviour–based control approach, and the Modular Controller Architecture software framework. Though the system was developed with human speech as the primary stimulus, emphasis was not on speech recognition or state of the art signal processing, since they are not in the group's research domain.

2 Modular Controller Architecture framework and behaviour-based control

All research robots of the Arbeitsgruppe Robotersysteme use a behaviour- based control architecture, which is implemented using the Modular Controller Architecture software framework. According to MCA (2005), MCA is "a modular, network transparent and realtime capable C/C++ framework for controlling robots and other kind of hardware. The main plattform is Linux/RTLinux [...]." The core idea of MCA is the division of the entire software into modules with standardized input/output interfaces, to enable code reuse and easily extendable architectures. The functionality of the modules rests in their Control and Sense functions, which receive, process and propagate control and sensory data. The modules are connected with data transporting edges and organized into a hierarchical structure(MCA , 2005). The groups can be connected with other groups and modules, thus forming parts that are the highest level of the control hierarchy. A part excecutes the Control and Sense functions of its modules in a continuous loop. Different parts manage different sections of the robot, like a vision system or motion control. Control can be distributed over several PC:s by connecting parts with TCP/IP interpart edges. The idea of behaviour-based control is based on the subsumption architecture of Brooks (1985), where a control problem is solved by dividing it into behaviours instead of function-based modules. The approach used in this work, which is presented in Proetsch (2004), considers behaviours as uniform modules with defined interfaces. Each behaviour has an activity function a, a target rating function r and a transfer function F. An behaviour receives an input vector $e^{\rightarrow}$, an inhibition i and an activation ι, and generates an output vector $u^{\rightarrow}$. The output vector describes the influence of the behaviour on its environment and other behaviours. The robot also produces an activity value a, which provides information about how vigorously the behaviour is performing its functionality. The target rating is an indication of contentment, that is how far the behaviour is from reaching its intended target. The transfer function determines the behaviour's output vector based on input values and internal representations. Behaviours are classified into three types: Target-based behaviours, progress-based behaviours and activation behaviours.

3 Microphone system and algorithms

The physical system used in this work consisted of two ordinary electret microfones connected via external amplifier circuits to the line-in of an PC. The microphones were attached to Marvin-robot, so that the system turned with the robot. The original aim was to build a 4-microphone system, but technical problems prevented this. The sound capture was implemented using the Advanced Linux Sound Architecture interfaces presented in ALSA project (2005). For the source localization, two different algorithms were implemented: the correlation algorithm and the Dual delay-line(DDL) algorithm. Both are based on the Jeffress model of human binaural hearing (Jeffress , 1948). The essential idea of the Jeffress model is that the location of a sound source in respect to the auditory system is represented by the interaural time difference, or the delay between the arrival of the signal to the left and right ears, or microphones, respectively. The concept is explained in more detail in Liu et al. (2000) . The Jeffress model can be implemented with signal cross-correlation in the time domain, or with the dual delay-line algorithm in the frequency domain. The dual delay-line approach is presented in Liu et al. (2000), where it is used as a basis of multiple sound source localization, and implemented for one sound source in Calmes (2002). The goal was to implement both algorithms, so that performance could be compared.

3.1 Correlation algorithm

The basis of the correlation algorithm is the cross correlation of the signals sampled from the left and right microphones of the system. The discrete form of the running crosscorrelation is

$$\psi(\tau, n) = \sum_{i=0}^{n} r(i)l(i - \tau) \tag{3.1}$$

where τ is the delay and i the index of the current sample in the array of samples. The values of the correlation vector represent the correlation on different values of the delay, so the maximum value of the delay vector corresponds to the moment, when same sound has reached the both microphones. The arrival angle of the sound, in relation to the microphone system, can be obtained with the formula

$$\alpha = \arcsin(\frac{ct_s}{f_s b}), \tag{3.2}$$

where c is the speed of sound (m/s), t_s the location of the maximum correlation in samples, f_s is the sampling rate (Hz) and b is the distance between the microphones (m).

3.2 Dual delay–line algorithm

In dual delay-line implementation of the Jeffress model, the signals are delayed in the frequency domain instead of the time domain. Transformation into frequency domain makes it possible to analyze the spectra of incoming sounds and opens many possibilities for digital signal processing. First, the discrete signals $x_{Ln}(k)$ and $x_{Rn}(k)$ from the left and right microphones are decomposed into $X_{Ln}(m)$ and $X_{Rn}(m)$, m=1,. . . ,M/2, with a M-point Fourier transform. The transformed signals are then fed into a pair of delay lines, and the delay values, at which the two signals are identical, are detected. According to the time shifting theorem of the Fourier transform, the signals at delay value τ_i in the delay line are

$$X_{Ln}^{(i)}(m) = X_{Ln}(m)e^{-j2\pi f_m \tau_i}, \tag{3.3}$$

$$X_{Rn}^{(i)}(m) = X_{Rn}(m)e^{-j2\pi f_m \tau_{I-i+1}}, \quad i = 1, \ldots, I; m = 1, ..., M/2 \tag{3.4}$$

(Calmes , 2002) where n is the index of the currently processed signal timeframe, X_{Ln} and X_{Rn} are the transformed signals, m is the index of the M-point Fourier transform and i the index of the delay in the delay line The frequency $fm = mf_s/M$ for all m. The delay values are

$$\tau_i = \frac{b}{2c} \sin\left(\frac{i-1}{I-1}\pi - \frac{\pi}{2}\right), \quad i = 1, \ldots, I \tag{3.5}$$

where b is the intermicrophone distance and c the speed of sound (in m/s). The angle, where the sound comes to the system, is

$$\alpha_i = 90 - \frac{i-1}{I-1}180, \quad i = 1, \ldots, I \tag{3.6}$$

(in degrees). (Liu et al. , 2000), (Calmes , 2002) When the signal has been delayed on all frequencies and delay values, the point in the delay line, where the difference of signals $\Delta X_n^{(i)}(m)$ is minimal on all frequencies, can be detected.

$$\Delta X_n^{(i)}(m) = \mid X_{Ln}^{(i)}(m) - \Delta X_{Rn}^{(i)}(m) \mid, \quad i = 1, \ldots, I; m = 1, \ldots, M/2 \qquad (3.7)$$

The 3-dimensional difference pattern was in our approach first integrated (averaged) over the frequency spectrum

$$H_n(i) = \frac{2 \sum_{m=1}^{M/2} \Delta X_n^{(i)}(m)}{M}, \quad i = 1, \ldots, I \qquad (3.8)$$

In (Calmes , 2002), also the coincidence patterns of previous timeframes are used in current source estimation, but this approach actually produced worse results for us because of the bad sound quality caused by inadequate hardware. The values of $H_n(i)$ are normalized to 0..1 and inverted to produce the final representation of signal difference on each delay value,

$$Loc_n(i) = \left(1 - \frac{H_n(i) - \min}{\max - \min}\right)^x, \quad i = 1, \ldots, I \qquad (3.9)$$

The peaks of $Loc_n(i)$ correspond to the sound sources. The peaks are enhanced by the factor x, here x=6. Max and min are the maximum and minimum of $H_n(i)$(Calmes , 2002)

3.3 Implementation within MCA

In keeping with the modular approach, the sound source localization is divided into several modules with standardized interfaces. The sound capture, detection of interesting sounds and source localization are all encapsulated in one part, AlsaPart. The output of this part is the detected sound source location (the angle of the source in respect to the microphone system). The internal structure of the AlsaPart depends on the algorithm used. It always contains a module that performs the actual sound capture using ALSA. The sound has to be sampled in finite periods, continuous capture is not possible. The captured samples are forwarded to other parts through a so–called Blackboard object, since transfer of arrays of values is not possible using the inter–module edges. This sets some limits on the implementation of the source detection algorithms. AlsaPart also always contains a level filtering module that compares the volume of the captured sound to the volume of the background noise to determine if other sounds than the background noise are present. The source detection calculations are performed only if the captured sound volume exceeds this threshold.The background noise volume is observed during the first few sound capture cycles. If the correlation algorithm is used, the AlsaPart contains only a direction calculation module in addition to those mentioned earlier. The module performs a running crosscorrelation to the sample buffers of left and right microphone according to Eqs. (3.1) and (3.2). If the DDL algorithm is in use, the sample buffers are first Fourier-transformed by one module, and the actual calculation module performs the DDL algorithm on these transformed values. The robot's control architecture is a group, Main Group, composed of several parts, which are Verhalten, DSP–Remote–Part, AbstraktionsSchicht, Edge Translation and of course the added AlsaPart. Abstraktions Schicht and DSP–Remote–Part are below the Verhalten–part and connect the behaviours to the hardware. AlsaPart is also hierarchically below the Verhalten–part and gives sensor input to Verhalten. Edge Translation is hierarchically above the Verhalten–part, and interfaces the Verhalten-part to the in– and outputs of the entire Main Group. The core of the control system is Verhalten–part, which contains several groups. Of these, HigherBehaviours and BasicBehaviours contain the behaviour modules. Other groups, SpeechGenerator and Joystick–Speech–Interface, handle the control of the robot's speech functionality. HigherBehaviours contains modules that react to incoming sensor data and control data from the user interface, and activate other, basic behaviours based on this data. BasicBehaviours–group contains the low–level behaviours that implement the control functionality, like turning the wheels. To use the sound direction information in motion control and environment perception, a new behaviour–based module has been added to group HigherBehaviours. This module translates the sound source direction to control values for the motion control system. Currently the behaviour instructs Marvin to turn to the direction of the sound source. The turning is executed by fusion behaviours TurnLeftFusion and TurnRightFusion, which receive an activation value and an activity value from each behaviour that connects to them. The fusion behaviours fuse these possibly conflicting values with a fusion function to produce one output value, which determines the speed of turning to the corresponding direction. These behaviours are given activity values based on the angle of the detected sound source. The greater the activity value, the faster the turning.

4 Results and discussion

The system was tested in stationary tests, where neither the robot nor the sound source was moving, and in real environment, where the robot was able to move and multiple and moving sound sources were present.

4.1 Stationary tests

In stationary tests, the sound source was positioned in -90..90 degree angle in respect to the microphone system. The angle was changed in 15 degree intervals. The stimulus used was a 30–second sample of electrical noise played out through a PC speaker. The calculation results are plotted in Figure 1 and Figure 2 for correlation algorithm and DDL algorithm respectively.

The system works quite well. The angle values correspond to the real angle, with slight variation in successive results. For larger angle values, the variation is more notable when using the DDL algorithm. Overall, there are no notable differences in the performance of the two algorithms. The angle resolution of both algorithms is greater at the middle of the area of possible angles, since the relationship between the index in the correlation vector, or the index in the delay line, and the corresponding angle value is not linear.

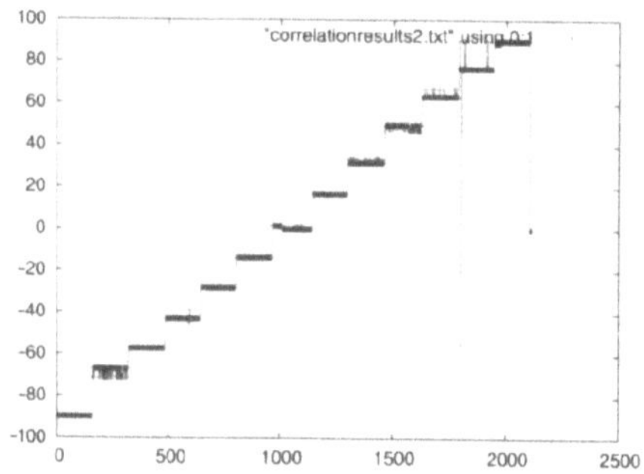

Figure 1. Stationary measurements, correlation algorithm

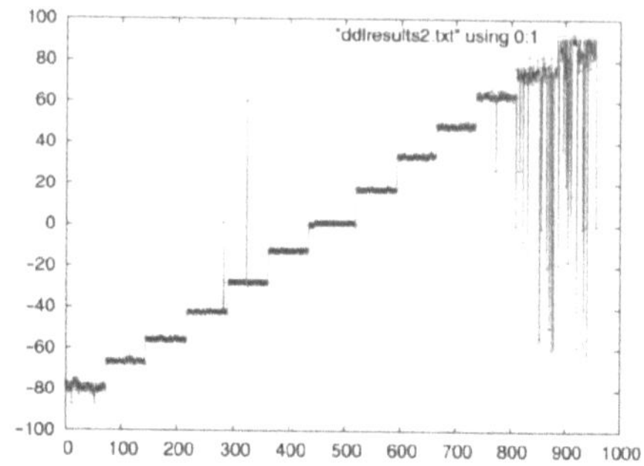

Figure 2. Stationary measurements, DDL algorithm

4.2 Tests in real environment

The microphone system was also tested in a real environment (a large laboratory room), where there was background noise and multiple sound sources present. We wanted to test the detection ability in such an environment, and also the reaction to a moving sound source. The robot's reaction was observed visually. The observations were necessarily subjective, since the tester's angle in respect to the robot changed constantly with the robot's turning. The results were encouraging. If the sound was continuous, with either algorithm the robot was able to smoothly rotate and follow the sound source. The reaction to sounds depended on the duration of the sound. With very short stimuli, the source was often perceived, but additional zero results caused by noise halted the turning. The reaction to speech sounds was tested with good results. If there were people in the room having normal conversation, the robot rotated towards them. An interesting sidenote is how much the robot's ability to react to sounds increased the appearance of interaction between robot and human user. The testing of the robot in real environment inspired passers–by to start calling the robot by name and try to make it turn towards them. Figures 3 and 4 show some estimated angles when testing the robot in a real environment. The estimated values values start out at zero, grow gradually to around 60 degrees, lower to zero, then grow to around -60 degrees and lower again to zero. This corresponds to the robot detecting a sound source and starting to turn towards it. Simultaneously the source starts to move to the direction, where the robot is turning. The sound source stops moving after making a half-turn around the robot, and then robot is able to "catch up" with it. Then the source starts moving back to its original place, with the robot following. Finally, speech sounds were made in different directions around the robot. This shows as non-zero peaks in the results. Since the correlation algorithm executes faster, more estimated values are present in the results of this algorithm. The results of correlation algorithm in a real environment show more variation than the results of the DDL algorithm. The DDL estimates show more defined peaks corresponding to different arrival directions, the correlation results have a greater number of values that are around the same angle value. The reaction to other kinds of sounds, such as clapping hands and clattering tableware, was also tested. Since these sounds are short and impulse–like in nature, the localization results were not as good as with longer and more even sounds. This is most probably due to electrical noise from the recording equipment masking shorter sounds more than longer sounds. It was noted that the correlation algorithm usually produced totally random results with short sounds, but the DDL algorithm was able to produce correct estimates mixed with zero estimates caused by recording noise.

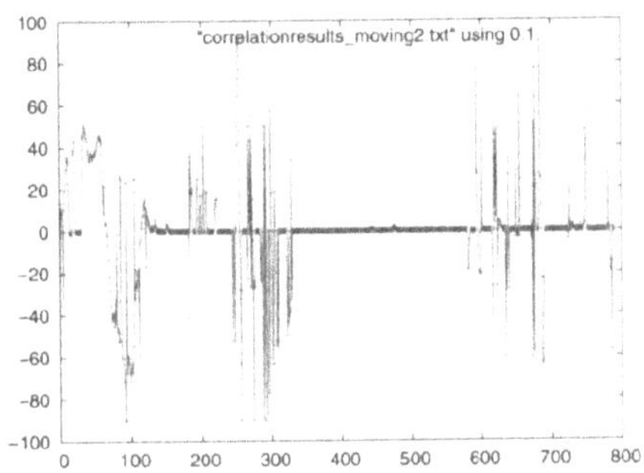

Figure 3. Estimated angles in real environment, correlation algorithm

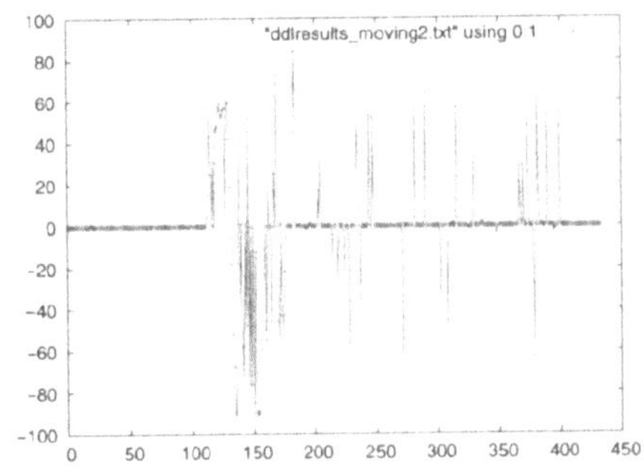

Figure 4. Estimated angles in real environment, DDL algorithm

4.3 Future developments

The current state is a workable platform, on which it is easy to incorporate new parts because of the modular architecture. The most important development would be to eliminate noise from the recording equipment, to avoid false estimates. When this has been done, previous direction estimates can be used in new angle estimation. The possible future developments of the system include the integration of commercial speech recognition software on to the robot, to enable communication between a human user and the robot. Direction estimation combined with speech recognition opens up many possibilities. The audio system could also be used to capture and analyze non–speech sounds. This kind of system is presented in (Hiyane and Lio , 2001). This could be used in environment perception, especially if the system is integrated onto an autonomous outdoor vehicle robot. Sound source detection can be combined with information from robots other sensory systems, like stereo vision and laser scanners. This would allow the robot to detect interesting targets in the environment and distinguish humans among perceived objects based on detected speech sounds.

Bibliography

Arbeitsgruppe Robotersysteme. *AG Robotersysteme: Galerie.* `http://agrosy.informatik.uni-kl.de/galerie/`, 15.8.2005

The ALSA project. *Advanced Linux Sound Architecture - ALSA documentation.* `http://www.alsa-project.org/documentation.php`, 16.8.2005

Laurent Calmes. *A Binaural Sound Source Localization System for a Mobile Robot.* Diploma Thesis, Faculty of Mathematics, Computer Sciences and Natural Sciences, Rheinisch-Westfälische Technische Hochschule Aachen, 2002

FFTW Project.*FFTW Home Page.* `http://www.fftw.org/`, 16.8.2005

Kazuo Hiyane, Jun Lio. *Non-speech sound recognition with microphone array.* Mitsubishi Research Institute,Inc. `http://tosa.mri.co.jp/sounddb/nospeech/system/indexe.htm`, 16.8.2005

Chen Liu, Bruce C. Wheeler, William D. O'Brien, Jr., Robert C. Bilger, Charissa R. Lansing, Albert S. Feng. *Localization of multiple sound sources with two microphones.* J. Acoust. Soc. Am, 108(4):1888-1905, October 2000

Chen Liu, Bruce C. Wheeler, William D. O'Brien, Jr., Charissa R. Lansing, Robert C. Bilger, Douglas L. Jones, Albert S. Feng. *A two-microphone dual delay-line approach for extraction of a speech sound in the presence of multiple interferers.* J. Acoust. Soc. Am, 110(6):3218-3231, December 2001

L. A. Jeffress *A place theory of sound localization.* Journal of Comparative and Physiological Physiology, 41(1):35-39, 1948

Modular Controller Architecture MCA *Modular Controller Architecture: Introduction* `http://mca2.sourceforge.net/`, 15.8.2005

Martin Proetsch. *Behaviour-based Motion Control of an Autonomous Four Wheel Steering Outdoor Vehicle.* Diplomarbeit, Arbeitsgruppe Robotersysteme, Fachbereich Informatik, Technische Universität Kaiserslautern, 2004

Rodney A. Brooks. *A robust layered control system for a mobile robot.* A. I. Memo 864, Massachusetts Institute of Technology, Artificial Intelligence Laboratory, 1985

Development and Control of Wheeled Vehicle that Balances on a Rolling Basketball

Tatsuro Endo[†] and Yosihiko Nakamura[‡]
Department of Mechano-Informatics, University of Tokyo,
E-mail: [†]endo, [‡]nakamura@ynl.t.u-tokyo.ac.jp

Abstract. This paper describes on development and control of a vehicle that moves on a spherical rubber tire. The vehicle obtains 3DOF mobility by rolling the tire and balancing on it. We selected a basketball as a spherical tire, because a rubber tire has three major features, namely durability, ground adaptability and force transferability. Using a basketball as a spherical tire can provide the vehicle with the above-mentioned advantages at low-cost. We discuss mechanical design and statics of the vehicle and its driving system. The statics formulates the force distribution problem among redundant actuators. We also show a preliminary design of controller and the result of experiments.

1 Introduction

Three features of intelligent mobile platforms would be omnidirectionality, ground adaptabilities and maneuverability, whether they are intended for autonomous mobile robots or vehicles of personal mobility.

According to the report of Yamashita et al. (2003), it is difficult for wheeled mobile robots to satisfy both the omnidirectionality and the ground adaptability on rough terrains at the same time. Mechanism of omnidirectional vehicle tends to be complex. Use of omnidirectional wheels for a personal vehicle commonly results in rather small diameter of wheels and therefore reduces the adaptability to the ground unevenness. Although there are many omnidirectional vehicles reported, they haven't solved the problem in full.

Through the history of engineering developments an automobile acquired high ground adaptability, which is due to the suspension system and rubber tires. According to the report of Sakai (1995), rubber tire has three good features. Namely, durability, ground adaptabilities, force transferability. The features are useful not only automobiles but also for omnidirectional vehicles.

"Segway" developed in 2001, is well known for its simple maneuverability. One on Segway operates it by weight shift. The simple and intuitive operations are convenient for everyone. We also seek a simple and intuitive interface for vehicle maneuver.

There are many omnidirectional vehicles and their mechanisms reported."Swedish Wheel" was developed as holonomic omnidirectional wheels by Ilon (1973). A swedish wheel has free wheels along the circular wheel surface, which transmit force only in a certain direction. The idea has been used in various omnidirectional vehicles and

Figure 1. Overall and Downside View of "B. B. Rider"

improved by many researchers. La (1980) and Bradbury (1980) independently developed unique omnidirectional vehicles. Those vehicles move under holonomic constrains. The same functionality was also implemented in different mechanisms like "orthogonal wheel" Killough and Pin (1992). Muir and Neuman (1987) showed a kinematics model of an omni-directional wheeled mobile robot "Uranus". Damoto and Hirose (2001) developed an omnidirectional vehicle "VutonII" using four "Omni-disc". Wada and Asada (2000) designed and developed the variable footprint vehicle with ball wheels.

Nonholonomic design of mobile robots was also atempted using spheres as wheels by several researchers. Ide et al. (2001) suggested that a 2DOF mobile robot could be achieved by using a ball. Tsuruga (1990) actually developed a 2DOF mobile robot driven by an on-board driver, which used a hard spherical tire. More recently, T. Lauwers and Hollis (2005) pososed a similar design and discussed its control issues in depth.

Omnidirectional vehicles are needed for various environments. Our goal is to develop a small-size omnidirectional vehicle for personal mobility and for a platform of intelligent mobile robot. We proposed to use a spherical rubber tire and showed the preliminary design of mobile robot adopting a basketball (Endo and Nakamura (2005)), which is holonomic with three degrees-of-freedom.

2 B. B. Rider

2.1 Specifications of "B. B. Rider"

In this section, we describe a 3DOF omnidirectional mobile platform we have developed (Endo and Nakamura (2005)). We named it "B. B. Rider" (Basketball Rider).

Fig.1 shows overall and downside views of "B. B. Rider". Fig.2 shows a mechanism for driving a ball. Table1 shows the specifications of the robot. In Fig.1, there is a frame to protect the robot from collisions with obstacles.One, who rides on the robot, can rest one's feet on it. The robot has a PC consisting of a CPU board, a back plane board an extended serial interface board, a force torque sensor board and an AD board. In Fig.2

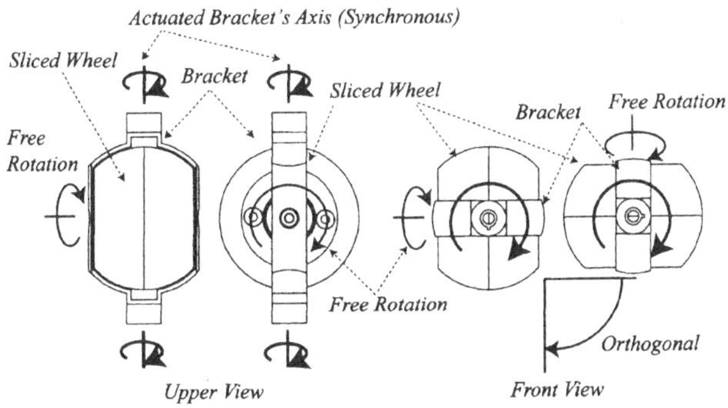

Figure 2. Orthogonal Wheel

Height	0.62 [m]
Width	0.55 [m]
Weight	25 [kg]
Basketball's Diameter	0.245 [m]
Basketball's Inner Pressure	500-600 [Pa]
Basketball's Weight	0.5 [kg]
DC Motor's Output	150 [W]
Gear Ratio	100
Sensor	Gyro,Force Torque Sensor(Under the Seat)

Table 1. Specifications of "B. B. Rider"

shows a force-transfer-mechanism, which were originally developed by Killough and Pin (1992), to drive a basketball by friction force. The mechanism consists of two actuated brackets and two nonactuated wheels, which are mounted in each bracket. The left side of the figure is a upside view of the brackets and wheels. Two brackets are synchronized by a motor with a chain. In the mechanism, one wheel maintains contacts on the ball. The direction of force transfer is orthogonal to the direction of mechanical constraints. The right side of the figure is a front side view of those. The robot has four driving units, each of which has a motor and an encoder respectively. Our robot has two batteries mounted in series. We can select power source from batteries and outer power supplies.

2.2 Why Basketball?

We selected a basketball as a spherical tire for "B. B. Rider". We would like to explain why we selected it.

Basketball is made of rubber. Rubber tire has three good features. Sakai (1995) pointed out durability, ground adaptabilities and force transfer transferability. A bas-

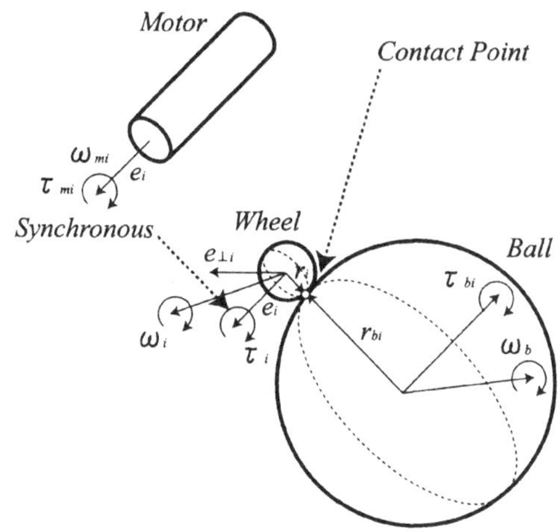

Figure 3. A relationship of a wheel and a basketball

ketball can provide those features to a vehicle at much lower-cost than making a hard rubber-coated ball.It is also significant that a rubber tire naturally possesses a good suspension property, which would result in complex mechanisms if designed for a hard ball.

3 Kinematics and Statics

3.1 Kinematics

We describe that a solution of actuated wheels' kinematics. We need the solution to drive motors in experiments. We explain variables used in this subsection with following Fig.3. ω_i is an (3×1) angular velocity vector of a wheel. Subscript i consists of $1, \cdots, 4$. r_i is a (3×1) radius vector of the wheel from the wheel's center to a contact point. ω_b is a (3×1) angular velocity vector of the basketball. r_{bi} is a (3×1) radius vector of a basketball from the ball's center to a contact point. e_i is a (3×1) base vector, which is orthogonal to r_i, of the wheel's motor. $e_{\perp i}$ is a (3×1) base vector, which is orthogonal to e_i and r_i. ω_{mi} is a angular velocity of the motor. ω_m is a (4×1) matrix of $[\omega_1, \cdots, \omega_4]^T$. E is a (3×4) matrix of $[e_1, \cdots, e_4]$. In design of our robot, we define $|r_w| = |r_1| = |r_2| = |r_3| = |r_4|$. A basketball's radius leads to $|r_b| = |r_{b1}| = |r_{b2}| = |r_{b3}| = |r_{b4}|$. In this solution, there isn't slip among wheels and a basketball.

By a kinematics relationship between a wheel and a basketball, each velocity of a contact point is equal. Then, we obtain an equation as follows:

$$\omega_i \times r_i = \omega_b \times r_{bi} \tag{3.1}$$

Eq.(3.1) includes both actuated and non-actuated velocity by condition of kinematics constraint. By using inner product to $e_{\perp i}$, we obtain an actuated equation as follows:

$$(\omega_i \times r_i)^T e_{\perp i} = (\omega_b \times r_{bi})^T e_{\perp i} \tag{3.2}$$

Scalar triple product in Eq.(3.2) leads to equations as follows:

$$(r_i \times e_{\perp i})^T \omega_i = (r_b \times e_{\perp i})^T \omega_b \tag{3.3}$$

$$|r_i| e_i^T \omega_i = -|r_{bi}| e_i^T \omega_b \tag{3.4}$$

Then, a realtionship of $\omega_{mi} = e_i^T \omega_i$ leads to equations as follows:

$$|r_i| \omega_{mi} = -|r_{bi}| e_i^T \omega_b \tag{3.5}$$

$$\omega_{mi} = -\frac{|r_{bi}|}{|r_i|} e_i^T \omega_b \tag{3.6}$$

Finally, we obtain an actuated kinematics solution as follows:

$$\omega_m = -\frac{|r_b|}{|r_w|} E^T \omega_b \tag{3.7}$$

A coefficient of right-hand side in Eq.(3.7) allows a relationship among wheels and a basketball like gear ratio.

3.2 Statics among redundant actuators

We describe that a solution of actuated wheels' static. We need the solution to simulate dynamics and control.

We explain variables used in this subsection. τ_i is a torque of a wheel on a contact point between the wheel and a basketball. τ_{bi} is a torque of the basketball on the contact point. τ_{mi} is a torque of the wheel's motor. τ_m is a (4×1) matrix of $[\tau_1, \cdots, \tau_4]^T$. τ_b is a (3×1) torque vector of the basketball. Other variables needed in this subsection are defined in foregoing subsection.

In balancing relationship, we obtain an equation as follows:

$$\frac{\tau_i}{|r_i|} e_i = -\frac{\tau_{bi}}{|r_{bi}|} e_i \tag{3.8}$$

A relationship of $\tau_i = \tau_{mi}$ leads to an equation as follows:

$$\frac{\tau_{mi}}{|r_i|} e_i = -\frac{\tau_{bi}}{|r_{bi}|} e_i \tag{3.9}$$

A relationship of $\tau_b = \sum \tau_{bi}$ leads to equations as follows:

$$\sum \frac{\tau_{mi}}{|r_i|} e_i = -\sum \frac{\tau_{bi}}{|r_{bi}|} e_i \tag{3.10}$$

$$E \tau_m = -\frac{|r_w|}{|r_b|} \tau_b \tag{3.11}$$

A pseudo inverse matrix of E leads to an equation as follows:

$$E\tau_m = -\frac{|r_w|}{|r_b|}EE^T(EE^T)^{-1}\tau_b \tag{3.12}$$

Finally, we obtain Eq.(3.13) as a actuated static solution.

$$\tau_m = -\frac{|r_w|}{|r_b|}E^T(EE^T)^{-1}\tau_b \tag{3.13}$$

3.3 Virtual Work between a Ball and a Robot

We explain that a relationship of our robot and a basketball. Eq.(3.7) and Eq.(3.13) lead to an equation as follows:

$$\begin{aligned} \tau_m^T\omega_m &= \left[-\frac{|r_w|}{|r_b|}\{E^T(EE^T)^{-1}\tau_b\}^T\right]\left[-\frac{|r_b|}{|r_w|}E^T\omega_b\right] && (3.14)\\ &= \tau_b^T\{E^T(EE^T)^{-1}\}^TE^T\omega_b && (3.15)\\ &= \tau_b^T\{EE^T(EE^T)^{-1}\}^T\omega_b && (3.16)\\ &= \tau_b^T\omega_b && (3.17)\end{aligned}$$

Eq.(3.17) allows virtual work principle among wheels and the basketball.

4 Control

We describe a method to control "B. B. Rider" in this section. The control method is for stabilizing robot's orientation. The control method is able to stabilize 3 Axes (Roll, Pitch and Yaw) of the robot.

We explain variables used in this subsection. R is a (3×3) matrix of robot's orientation. θ_b is a (3×1) angle vector of the ball. θ_r is a (3×1) angle vector of the robot. ω_b is a (3×1) angular velocity vector of the ball. ω_r is a (3×1) angular velocity vector of the robot. $A(\theta_b, \theta_r)$ is a (6×6) inertia matrix. $B(\theta_b, \omega_b, \theta_r, \omega_r)$ is a (6×1) bias vector matrix. τ is a (3×1) torque vector generated by the robot. err is a (3×1) error vector of the robot's orientation. k_p is a propotional gain of the robot's orientation. k_d is a derivative gain of the robot's angular velocity.

A motion equation of the ball and the robot is given by:

$$A(\theta_b, \theta_r)\begin{pmatrix}\dot{\omega}_b\\ \dot{\omega}_r\end{pmatrix} + B(\theta_b, \omega_b, \theta_r, \omega_r) = \begin{pmatrix}-\tau\\ \tau\end{pmatrix} \tag{4.1}$$

The robot has a gyroscope for sensing the robot orientation and angular velocity in world coordinate. The control method uses these data. Then, we need to know errors of the

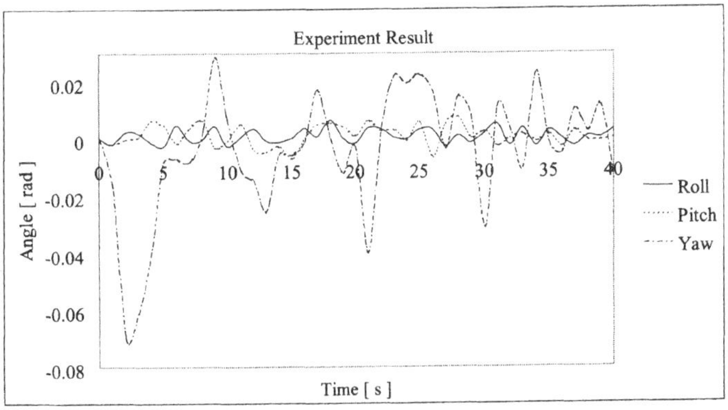

Figure 4. Experiment Result

robot orientation from these data. The error data is given by:

$$err = \frac{1}{2}\left(R - R^T\right) \tag{4.2}$$

$$\cong \frac{1}{2}\begin{pmatrix} 2\theta_{roll} - \theta_{pitch}\theta_{yaw} \\ 2\theta_{pitch} - \theta_{roll}\theta_{yaw} \\ 2\theta_{yaw} - \theta_{roll}\theta_{pitch} \end{pmatrix} \tag{4.3}$$

$$\cong \begin{pmatrix} \theta_{roll} \\ \theta_{pitch} \\ \theta_{yaw} \end{pmatrix} \tag{4.4}$$

We obtain R from the gyroscope's angle data. And we linearlize angles around the equilibrium position of the robot. We propose a feedback controller by using Eq.(4.4). Then, the feedback is given by:

$$\tau = -k_p err - k_d \omega_r \tag{4.5}$$

We obtain the robot's angular velocity from the gyroscope. We control the robot by using Eq.(4.5).

5 Results of Experiments

We describe on the result of experiments of "B. B. Rider". We stabilized the robot by implimenting the method proposed in the previous section. Fig.4 shows the result of the experiments. Roll, Pitch and Yaw angles are stabilized by the control method. We saw vibrations in Fig.4, which turned out that they were due to the contacts between brackets of wheels' and the ball. We are currently modifying the mechanical design of brackets.

6 Conclusions

We developed a 3DOF wheeled robot "B. B. Rider" that balances on a basketball. The robot's driving system has three advantages, which are omnidirectionality, ground adaptabilities and maneuverability. Those advantages were realized by using the basketball as a rubber spherical tire. Mobile robot platforms are expanding their field of activities. We developed a mobile robot platform with high mobility and simple mechanisms.

We are currently designing nonlinear control law for stabilization and positioning. The full dynamics would need to be considered to maximize the potential high mobility of the Basketball Rider. Development of simple and intuitive interface for personal maneuver would be one of the central issues to be discussed.

Acknowledgement

This research was supported by Category S of Grant-in-Aid for Scientific Research (No.15100002), Japan Society for the Promotion of Science.

Bibliography

H. M. Bradbury. Omni-directional transport device, 1980. U. S. Patent No.4223753.

R. Damoto and S. Hirose. Development of vutonii with omni-disc. In *The Japan Society of Mchanical Engineers ROBOMEC 2001 Conference*, 2001.

T. Endo and Y. Nakamura. An omnidirectional vehicle on a basketball. In *IEEE International Conference on Advanced Robotics*, 2005.

K. Ide, T. Matsuyoshi, Y. Adathi, and T. Masuda. Study on mobile robot with ball drive mechanism. In *The 19th Annual Conference of the Robotics Society of Japan*, 2001.

B. E. Ilon. Directionally stable self propelled vehicle, 1973. U. S. Patent No.3746112.

S. M. Killough and F. G. Pin. Design of an omnidirectional and holonomic wheeled platform prototype. In *International Conference on Robotics and Automation*, pages 84–90, 1992.

H. T. La. Omnidirectional vehicle, 1980. U. S. Patent No.4237990.

P. F. Muir and C. P. Neuman. Kinematic modeling for feedback control of an omnidirectional wheeled mobile robot. In *International Conference on Robotics and Automation*, pages 1772–1778, 1987.

H. Sakai. The abc's of wheeled vehicle: Tire mechanics. *Journal of the Robotics Society of Japan*, 13(1):113–118, 1995.

G. Kantor T. Lauwers and R. Hollis. One is enough! In *Preprints of International Symposium of Robotics Research*, 2005.

T. Tsuruga. A vehicle like an inverted pendulum (2) one sphere machine. *Journal of the Robotics Society of Japan*, 8(3):365, 1990.

M. Wada and H. Asada. Design and control of a variable footprint mechanism for holonomic omnidirectional vehicles and its application to wheelchair. *IEEE Transactions on Robotics and Automation*, 16(1):978–989, 2000.

A. Yamashita, H. Asama, T. Arai, J. Ota, and T. Kaneko. A survey on trends of mobile robot mechanisms. *Journal of the Robotics Society of Japan*, 21(3):282–292, 2003.

Durability of Large-Deflective Hinges made of Blend Polypropylene used for Molded Pantograph Mechanisms

Mikio Horie [1], Yudai Okabe[2], Masahiro Yamamoto[2], and Daiki Kamiya[1]

[1] Precision and Intelligence Laboratory, Tokyo Institute of Technology,
4259-R2-14, Nagatsuta-cho, Midori-ku, Yokohama, 226-8503, Japan, e-mail: mahorie@pi.titech.ac.jp
[2] Graduate student of Tokyo Institute of Technology

Abstract. A molded pantograph mechanism consisted of large-deflective hinges and links is used in a new surface mount system proposed by Horie, et al.(2000). To obtain long life, large-deflective hinges, we showed the method to conduct the repetitive tensile-compressive and bending test on a hinge specimen, and manufactured experimental equipment for a fatigue test. We also evaluated the test equipment and confirmed its effectiveness. The materials of the large-deflective hinge were selected as follows: (1) polypropylene (PP), (2) polymerization blend PP, (3) machine blend PP, and (4) another machine blend PP(+MAH-PP). The hinge specimen used for the fatigue test was made with the injection molding machine. From the relationship between the number of cycles leading to the fracture of the hinge and the repetitive tensile-compressive loads in the fatigue test, it has been confirmed that the difference in hinge thickness and the difference in the materials greatly influenced the inner damage of the hinges.

1 Introduction

Portable electronic device, audio-video equipment, and computer peripherals have been getting smaller and smaller these days. Micro-parts even less than a few-millimeters squared, such as 1608 (1.6mm×0.8mm), 0603 (0.6mm×0.3mm) (Mounting Process Manual(1998)), or 0402 (0.4mm×0.2mm), are attached to the electronic circuit cards of these devices. However, most of the surface mounting system is measured in meters, which is huge relative to the micro-parts. M. Horie et al.(2000) have proposed a small surface mount system with a molded pantograph mechanism placed in parallel as shown in Fig. 1. Large-deflective hinges are placed in contraposition to each other in the mechanism. The system is better suited for the size of micro-parts, and enables the making of a large quantity of various parts. It would realize a one-room factory, with which sales territory and manufacture territory are directly connected to each other. The pantograph mechanism has two degrees of freedom, and the input-output characteristics have been determined. A pantograph mechanism with an output node end, whose maximum repeat accuracy is 11μm during pick and place work, has been achieved. During the pick and place work, the output link end describes a trajectory(E_1-E_2-E_3-E_4-E_5) shown in Fig. 1. However, distortion beyond the bounds of its elastic region occurred on the hinges in contraposition to each other, which led to a large angular displacement. Moreover, the hinges underwent displacement by varying combinations of load, such as bending, tension, compression, and shear. This has most commonly caused the pantograph mechanism to break. Therefore, this study aims to clarify the fatigue damage processes in large-deflective hinges through a fatigue test. In the fatigue test, bending, tension, and compression loads will be applied synchronously on hinge specimens, which are comprised of two links and a large-deflective hinge part between the links.

2 Hinge Specimen

2.1 Hinge Specimen Material

Hinge specimens used in this study have two links and a large-deflective hinge part between the links, whose dimensions are shown in Fig. 2. The specimen molded of polypropylene (PP),

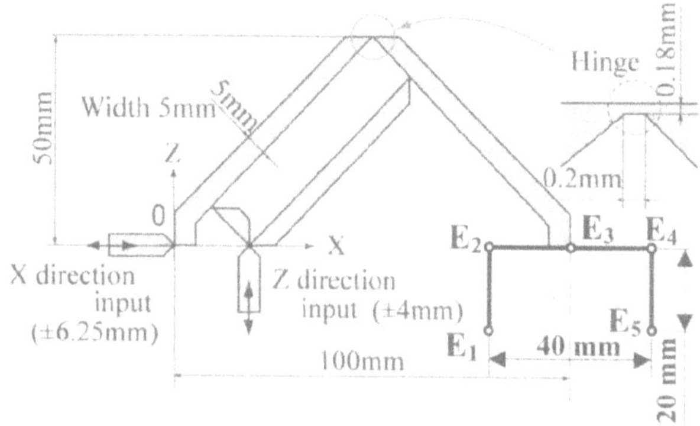

Figure 1. Dimension of the pantograph mechanism

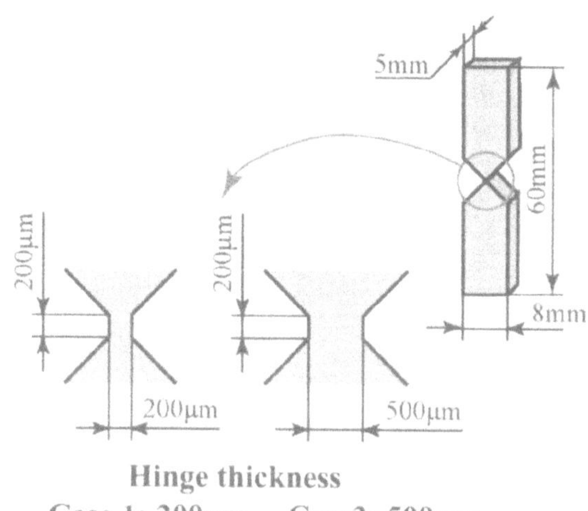

Figure 2. Dimension of the fatigue specimen

PM731M of SunAllomer Ltd., mixed with elastomer. The PP used in this study is said to be more durable than conventional studies (M. Horie et al. (2004), M. Horie, M. Yamamoto et al. (2004)). Moreover, elastomer was blended with the PP so that the material would have rubber-like characteristics, aiming to reduce stress by bending load and fatigue damage. Two types of elastomer were used: one being R-TPO, zealsTM 7023, of Mitsubishi Chemical Corporation, and the other EPT3012P of Mitsui Chemicals. The former is called polymerized polyolefin thermoplastic elastomer. R-TPO is made by multiple-stage polymerizing ethylene. Ethylene is a monomer, which enables PP to have elastomer characteristics when it is polymerized. Hereafter, polypropylene blended with zealsTM will be called Polymerized Blend PP (A1~A4). The latter was, EPDM, Ethylene Propylene-Diene Monomer, which was used in the study by M. Horie, M. Yamamoto et al. (2004). EPDM is blended with PP using a machine. Hereafter, PP blended with EPDM will be called Machine Blended PP (B1~B3). Moreover, Mah-PP, Maleic anhydride modified PP, was added to some Machine Blend PP so that EPDM could be better blended with PP. All names of the specimens used in this study are shown in Table 1.

2.2 Phase Structure of Hinge Specimen

It is said that granules or clusters of EPDM exist in the mother phase of PP when PP and EPDM are blended. In order to confirm this, phase structure of specimens was observed by a chrome acid etching method in this study. This method makes use of different solubility of PP and elastomer against chrome acid. Specimens were soaked in chrome acid, causing the elastomer on the surface to dissolve, which was observed with SEM, Scanning Electron Microscope. The elastomer phase was observed as a concave portion. It was found that there was a large concave

Table 1. Name of specimens and their blending ratio

Name of specimens	Thickness of hinge [μm]	Blending ratio (PP:Elastomer)	Blending ratio of Mah-PP [%]
PP-1	200	100:0	0
A1-1	200	70:30 (R-TPO)	0
A2-1	200	50:50 (R-TPO)	0
A3-1	200	30:70 (R-TPO)	0
A4-1	200	0:100 (R-TPO)	0
B1-1	200	85:15 (EPDM)	3
B2-1	200	70:30 (EPDM))	3
B3-1	200	70:30 (EPDM))	0
PP-2	500	100:0	0
A1-2	500	70:30 (R-TPO)	0
A2-2	500	50:50 (R-TPO)	0
B1-2	500	85:15 (EPDM))	3

portion in the flow direction of resin in hinge part of Machine Blend PP as shown in Fig. 3. The EPDM phase was conducted separating from its mother phase of PP, which was not observed in PP or Polymerized Blend PP. Also, it was found that some parts about 30~50μm below the surface of the specimens of all materials were dissolved differently from the inner part. Hereafter, the part will be called 'Surface Layer'. Observation results showed that the concave portion of

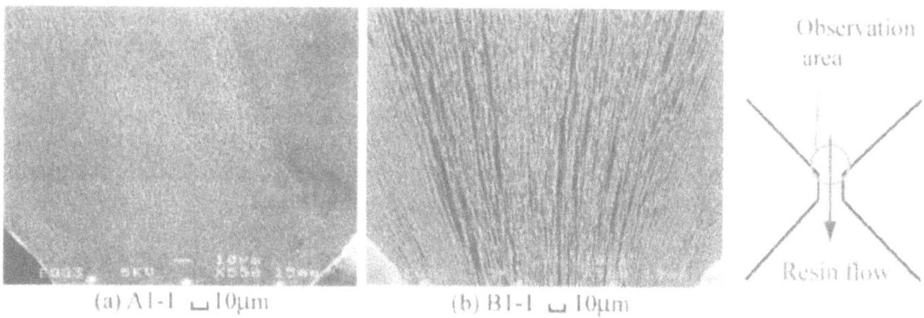

Figure 3. Observation of the fatigue specimen after the chromic acid etching

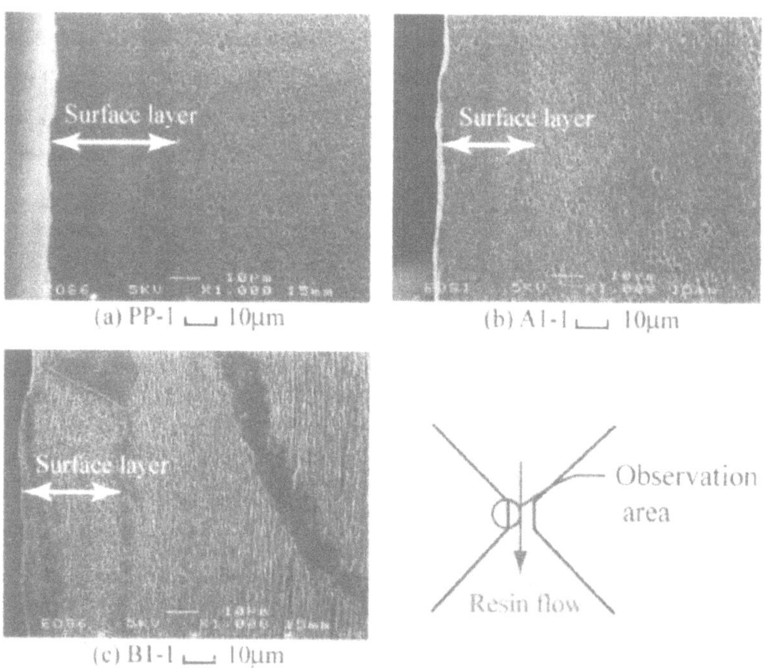

Figure 4. Observation of the fatigue specimen after the chromic acid etching

Surface Layer was smaller or less than the inner part. This is because the surface of specimens is cooled rapidly when specimens are molded. When PP blended with elastomer is molded, resin flows by shear force with the elastomer phase separating. The surface of specimens gets cooled rapidly and becomes hardened with fragmented crystals, while the inner part gets cooled gradually, developing phase separation. Also, it is assumed that crystals in the surface layer of hinges with only PP are smaller than inner part.

3 Fatigue Test Apparatus

Fatigue tests were conducted taking into account the load on hinges in a pantograph mechanism when it is practically used. Two loads: cyclic bending load with angular displacement, and a sine-wave-form cyclic tensile-compressive load, were applied on specimens synchronously. The test apparatus was manufactured by M. Horie, M. Yamamoto et al. (2004), which is shown in Fig.5. The apparatus consists of a cyclic bending unit and a cyclic tensile-compressive unit. The cyclic bending unit is a four-bar link mechanism and provides $\pm\pi/4$ [rad] of bending load under 10~40Hz of impressed voltage when it is actuated by a motor. Conversely, the cyclic tensile-compressive unit provides a cyclic tensile-compressive load when an electric current is sent to the coil in the unit and a magnetic field near the permanent

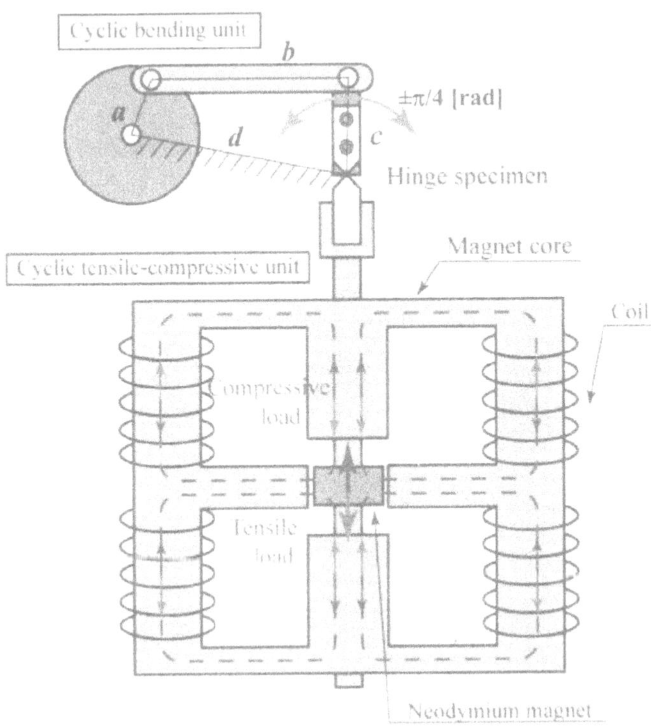

Figure 5. A fatigue test apparatus (*a*=30mm, *b*=3*a*, *c*=SQRT(2)**a*, *d*=SQRT(10)**a*)

magnet. This study provided sine-wave-form tensile-compressive force by passing sine wave currents to the coil as shown in Fig. 6. The maximum cyclic tensile-compressive force was 11.8N when the current was 10Hz while the force was 13.1N when the current was 20Hz. It was also found that cyclic tensile-compressive force was proportional to the impressed voltage as shown in Fig. 7.

4 Fatigue Test

A fatigue test was conducted on hinge specimens shown in Table 1 using the test apparatus in Fig. 5. A cyclic bending load with $\pm\pi/4$ [rad] angular displacement, and a sine-wave-form cyclic tensile-compressive load were applied on the specimens. Cyclic frequency of 10Hz was applied on hinges 200μm thick, while 20Hz was applied on hinges 50μm thick. The results are shown in Fig. 8. The horizontal axis represents the number of cycles when hinges were broken while the vertical axis represents cyclic tensile-compressive load by the fatigue test apparatus. The results show that hinges 500μm thick are more durable than hinges 200μm thick. This is because stress by the tensile-compressive load gets smaller as the cross section of hinge gets larger. However, the number of cycles of tensile-compressive load stayed at about one and a half times as many when the cross section of the hinge was two and a half times as large. It is conceivable that damages by bending load got severer as hinges were thicker. It was also confirmed that PP was the most durable among 200μm-thick hinges, while PP was as durable

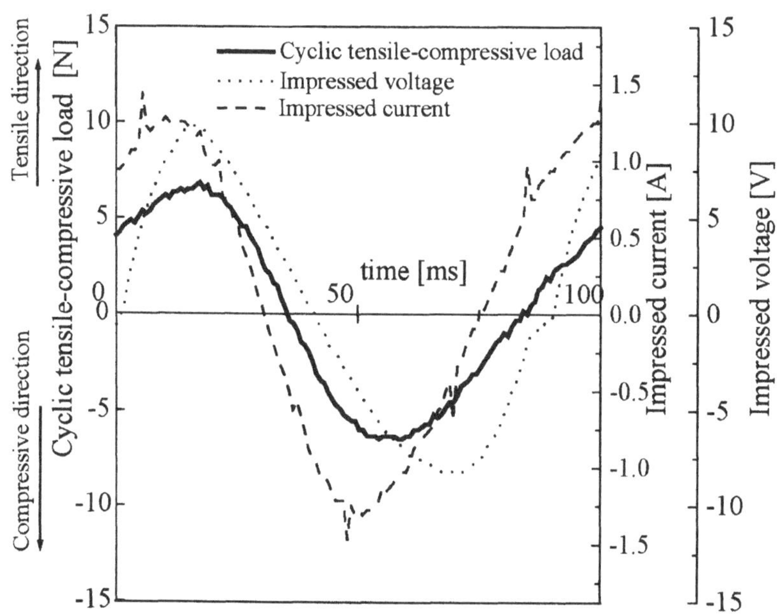

Figure 6. Waveform of the cyclic tensile-compressive load

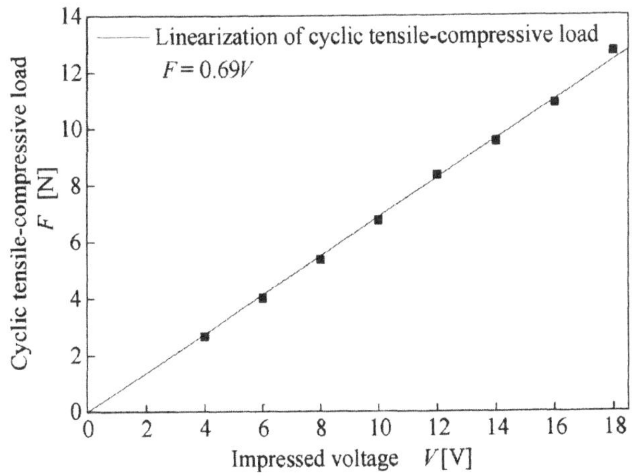

Figure 7. Impressed voltages versus cyclic tensile-compressive loads

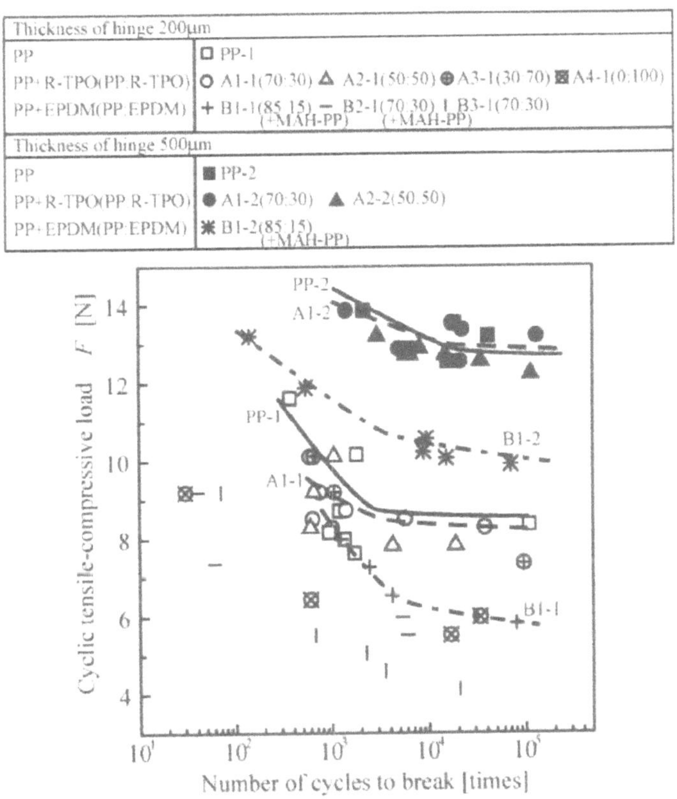

Figure 8. Results of fatigue test

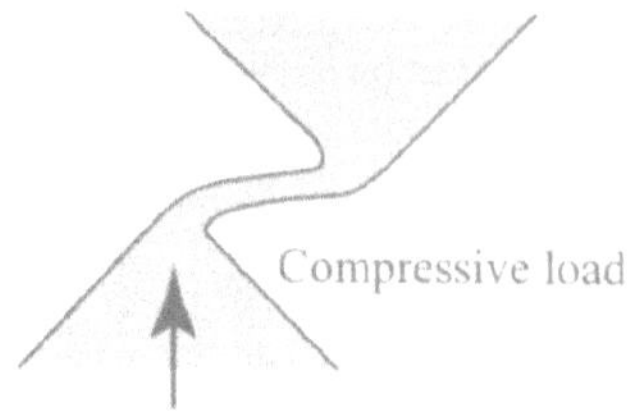

Figure 9. Bucking at the fatigue specimen (Hinge thickness 200μm)

as A1 regarding 500μm-thick hinges. It is because 200μm-thick hinges with elastomer, such as A1, got significantly longer and buckling in the hinge part became prominent during the test as shown in Fig.9. This had an adverse effect on hinge durability. It was also found that the durability of hinges with a thickness of 200μm or 500μm became less durable as the proportion of elastomer blended in the material became larger. But the result could be reversed when hinges are made thicker and the fatigue damage by bending load got severer. Moreover, it was found that Machine Blend PP was far less durable than Polymerized Blend PP.

5 Conclusions

(1) This study decided materials and dimensions of large-deflective hinges and made hinge specimens by injection molding for a fatigue test.
(2) The layer structure of the hinge specimens was observed using chrome acid etching. It was found that there were separated phases in Machine Blend PP, and there were differently dissolved parts in a layer form on the surface of PP, Polymerized Blend PP, and Machine Blend PP.
(3) From the results of a fatigue test on the hinge specimens, it was found that hinges 500μm thick were more durable than 200μm thick specimens, and that Machine Blend PP was less durable compared to Polymerized Blend PP.
(4) Observing the fatigue damage process of the specimens, it was found that damages occurred near Surface Layer, which was mentioned in (2), in PP, Polymerized Blend PP, and Machine Blend PP hinges. Also, damages in resin flow direction, when hinges were molded, were confirmed in Machine Blend PP. These damages were not observed in PP or Polymerized Blend PP.

Bibliography

M. Horie, T. Uchida, and D. Kamiya. A Pantograph Mechanism with Large-Deflective Hinges for Compact Surface-Mount System, *Journal of Japan Society of Mechanical Engineers (Series C)*, Vol. 66, No. 648, 2804-2809, 2000.

M. Horie, D. Kamiya, and M. Tomimatsu. Load Cyclic Characteristics of Large-Deflective Hinges in Moulded Pantograph Mechanisms with Hinges and Links, *Proceeding of the Romansy2004*, CD-ROM, 2004.

Mounting Process Manual. *Japan Robot Association*,1998(in Japanese).

M. Horie, M. Yamamoto, and D. Kamiya. Bending and Tecnsile-Compressive Fatigue Test of Hinges in Pantograph Mechanisms made from Injection Molding Machines, *Procceding of the 6th International Conference on Electronics Materials and Packaging*, 517-522, 2004.

Mobile Robots Based on Magnetizable Elastic Elements and Ferrofluids

Klaus Zimmermann and Igor Zeidis [1]

Technische Universitaet Ilmenau, Faculty for Mechanical Engineering, Ilmenau, Germany
e-mail: klaus.zimmermann@tu-ilmenau.de, igor.zeidis@tu-ilmenau.de

Abstract. In the present paper, we analyze the locomotion using the effects of deformation of magnetizable elastic material and deformation of the surface of a membrane filled with a magnetic fluid under the influence of a magnetic field. Devices based on these principles can be used in medicine and biology. Prototypes implementing these principles have been constructed.

1 Introduction

Non-pedal forms of locomotion in the present time show their advantages in inspection techniques or application in medical technology for diagnostic systems and minimally invasive surgery. In the locomotion of worms one recognizes a conversion of periodic internally driven motions into the change of the position (undulatory locomotion).

The motion of a chain of masses placed on a rough even surface and connected consecutively by identical springs has been studied by Zimmermann et al. (2002a), Zimmermann et al. (2002b), Zimmermann and Zeidis (2003), and Zimmermann et al. (2004a). It is supposed that the system moves along a straight line and Coulomb dry friction force is exerted by the surface upon each mass point opposite to the direction of motion. Thereby, the magnitude of this force is also dependent on the direction of the motion. The excitation is carried out due to internal harmonic forces. Such forces arise from a spring made from a magnetizable elastic material under the influence of an external magnetic field (Turkov, 2002).

In connection with the issue of peristaltic motion, we have considered a symmetric flow of two layers of viscous fluids with different properties separated by a thin membrane (Zeidis and Zimmermann, 2000). The flow of the fluids occurs in a vessel with fixed boundaries. The undulatory perturbations of a membrane are assumed to be given. The physical cause of their appearance is not discussed. For perturbations in the form of a sinusoidal travelling wave we calculate the average flow rate in the layers of the viscous fluids.

One of the possibilities to generate perturbations is a substitution of a normal viscous fluid with a magnetic fluid and application of a travelling magnetic field.

The possibility of creating a viscous magnetic fluid flow for designing autonomous mobile robots has been investigated by Zimmermann et al. (2004b), Zimmermann et al. (2004c),

---

[1] This work is supported by Deutsche Forschungsgemeinschaft (DFG, Zi. 540/6-1, Zi 540/7-1).

Zimmermann et al. (2004d), and Zimmermann et al. (2005). The flow of an incompressible magnetic fluid layer over a rigid substrate, which is generated by a time-varying nonuniform magnetic field, is analyzed.

2 Deformation of a Magnetic-elastic Elements

Turkov (2002) has considered the deformation of an elastic composite body under the influence of a magnetic field. If the Lame coefficients λ and η satisfy the condition $\lambda >> \eta$, then the relative elongation in the direction of the axis OX (for small strains), when the magnetic field is applied in an orthogonal direction, is given by $u = \frac{p}{E_y^{eff}} + u_0$. Here p is the density of the surface force, u_0 is the deformation of the sample, produced by the action of the magnetic field in the absence of the surface force ($p = 0$), and E_y^{eff} is the effective value of Young's modulus.

Due to the action of a magnetic field, the sample of the composite with a drop of a magnetic fluid elongates, and the value modulus E_y^{eff} decreases. For the experimental relationships between the stress and the strain without field ($H = 0$) and in the presence of the field $H = 1.5 \cdot 10^5$ A/m, see Turkov (2002).

2.1 Mathematical Model of the Motion

The motion of a system of two material points x_1 and x_2 of mass m, connected by a spring of stiffness $c = E_y^{eff} \cdot S_0 / l_0$ ($\omega^2 = c/m$), along the axis x is considered (Figure 1). It is assumed that the points are acted upon by small non-symmetric Coulomb dry friction force $\varepsilon m F(\dot{x})$, $\varepsilon << 1$, depending on the velocities $\dot{x} = \dot{x}_i (i = 1,2)$, where $F(\dot{x}) = F_+$ if $\dot{x} > 0$, $F(\dot{x}) = -F_-$ if $\dot{x} < 0$, and $F(\dot{x}) = F_0$ if $\dot{x} = 0$; $-F_- < F_0 < F_+$, $F_- \geq F_+ \geq 0$.

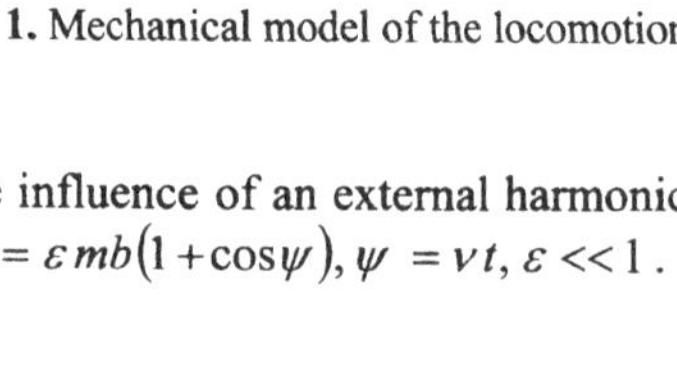

Figure 1. Mechanical model of the locomotion system.

According to Section 1, the influence of an external harmonic magnetic field causes a small harmonic internal force $G(t) = \varepsilon m b (1 + \cos\psi)$, $\psi = \nu t$, $\varepsilon << 1$. The equations of motion have the form

$$\ddot{x}_1 + \omega^2(x_1 - x_2) = -\varepsilon[F(\dot{x}_1) + b(1+\cos\psi)],$$
$$\ddot{x}_2 + \omega^2(x_2 - x_1) = -\varepsilon[F(\dot{x}_2) - b(1+\cos\psi)]. \tag{1}$$

We apply the averaging procedure (Bogoljubow and Mitropolski, 1965) to system (1). To that end, we introduce the new variables, the velocity of the center of mass $V = (\dot{x}_1 + \dot{x}_2)/2$ and the deviation relative to the center of mass $z = (x_2 - x_1)/2$. We utilize the change of variables $z = a\cos\varphi$, $\dot{z} = -a\Omega\sin\varphi$, $\varphi = \Omega t + \vartheta$, $\Omega = \sqrt{2}\,\omega$, where V, a, and ϑ are slow variables, to obtain system (1) in a standard form:

$$\begin{aligned}
\dot{V} &= -\frac{\varepsilon}{2}[F(V + a\Omega\sin\varphi) + F(V - a\Omega\sin\varphi)],\\
\dot{a} &= -\frac{\varepsilon}{2\Omega}\sin\varphi[F(V + a\Omega\sin\varphi) - F(V - a\Omega\sin\varphi) + 2b(1+\cos\psi)],\\
\dot{\varphi} &= \Omega - \frac{\varepsilon}{2a\Omega}\cos\varphi[F(V + a\Omega\sin\varphi) - F(V - a\Omega\sin\varphi) + 2b(1+\cos\psi)],\\
\dot{\psi} &= \nu.
\end{aligned} \tag{2}$$

We will investigate system (2) in a neighborhood of the main resonance $\nu = \Omega + \varepsilon\Delta$, $\Delta \neq 0$. We introduce the new slow variable $\xi = \psi - \varphi$ and eliminate the fast variable ψ. After averaging with respect to the fast variable φ we obtain

$$\begin{aligned}
\dot{V} &= \begin{cases} -\varepsilon\left(\dfrac{F_- + F_+}{\pi}\arcsin\dfrac{V}{a\Omega} - \dfrac{F_- - F_+}{2}\right) & \text{if } 0 \le V < a\Omega,\\ -\varepsilon F_+ \quad \text{if } V \ge a\Omega, \end{cases}\\
\dot{a} &= \begin{cases} -\dfrac{\varepsilon}{\Omega}\left(\dfrac{F_- + F_+}{\pi}\sqrt{1 - \dfrac{V^2}{a^2\Omega^2}} + \dfrac{ab}{2}\sin\xi\right) & \text{if } 0 \le V < a\Omega,\\ -\varepsilon\dfrac{ab}{2\Omega}\sin\xi \quad \text{if } V \ge a\Omega, \end{cases}\\
\dot{\xi} &= \varepsilon\left(-\frac{b}{2\Omega}\cos\xi + \Delta\right).
\end{aligned} \tag{3}$$

We are interested in a steady motion ($\dot{V} = 0$). Then from (3) we have $\dot{a} = \dot{\xi} = 0$ and

$$V=\frac{\sin\Phi}{|\Delta|}\sqrt{\frac{b^2}{4}-E^2\cos^2\Phi},\quad a=\frac{1}{\Omega|\Delta|}\sqrt{\frac{b^2}{4}-E^2\cos^2\Phi},\quad b\ge 2E\cos\Phi,$$
$$\xi=\arccos\left[-\frac{\Delta}{|\Delta|}\cdot\frac{2}{b}\sqrt{\frac{b^2}{4}-E^2\cos^2\Phi}\right],\quad E=\frac{F_-+F_+}{\pi},\quad \Phi=\frac{\pi}{2}\cdot\frac{F_--F_+}{F_-+F_+}. \tag{4}$$

The characteristic polynomial of the linearized system (3) is

$$P(\lambda)=\lambda^3+\lambda^2\cdot\varepsilon\frac{2E}{a\Omega}\cdot\frac{1}{\cos\Phi}+\lambda\cdot\varepsilon^2\left[\frac{E^2}{a^2\Omega^2}\left(1+\sin^2\Phi\right)+\Delta^2\right]+\varepsilon^3\frac{E\Delta^2}{a\Omega\cos\Phi}.$$

The Hurwitz condition holds for all values of the parameters and, hence, the motion with the velocity V of (4) is stable. The numerical integration of the exact system (1) results in an average velocity of $0.3\,m/s$. The values of the parameters for a magnetic-elastic sample described in Section 1 are taken as follows: $m=3\times10^{-3}kg$, $E_y^{eff}=1.6\times10^4\,N/m^2$, $u_0=0.071$, $l_0=29\,mm$, $S_0=5.5mm^2$, $\omega=32\,s^{-1}$, $\varepsilon=\frac{1}{2}u_0$, $b=E_y^{eff}\cdot S_0/m$, $\varepsilon\cdot b=1m/s^2$, $\varepsilon\cdot F_+=0.5\,m/s^2$, $\varepsilon\cdot F_-=1\,m/s^2$, and $\varepsilon\cdot\Delta=0.5\,s^{-1}$. The formula (4) gives the velocity of the center of mass $V=0.29\,m/s$.

2.2 Parametric Resonance

Let the system of two material points connected by a spring be actuated. We consider the case of small change of stiffness under action of a magnetic field (a parametric resonance). According to Section 1, it is possible to approximate the decrease of the stiffness c by $c(t)=m[\omega^2-\varepsilon\, b(1+\cos\psi)]$, $\psi=\nu t$, $\varepsilon\ll 1$. The equations of motion in this case become

$$\ddot{x}_1+[\omega^2-\varepsilon\, b(1+\cos\psi)](x_1-x_2)=-\varepsilon\, F(\dot{x}_1),$$
$$\ddot{x}_2+[\omega^2-\varepsilon\, b(1+\cos\psi)](x_2-x_1)=-\varepsilon\, F(\dot{x}_2). \tag{5}$$

We reduce system (5) to a standard form and consider the behavior of this system in a neighborhood of the main demultiplication resonance $\nu=2\Omega+\varepsilon\Delta$. We introduce, as was the case previously, the new slow variable $\xi=\psi-\varphi$. After averaging we obtain the system that differs from system (3) only in the last equation: $\dot{\xi}=\varepsilon\left(\frac{b}{2\Omega}\cos\xi+\Delta_1\right)$, $\Delta_1=\Delta+\frac{2b}{\Omega}$. The characteristic polynomial of the system is

$$P(\lambda)=\left(\lambda+\varepsilon\,\frac{2E\cos\Phi}{a\Omega}\right)\left(\lambda^2+\lambda\cdot\varepsilon\frac{2\cdot\sin^2\Phi}{a\Omega\cdot\cos\Phi}-\varepsilon^2\frac{E^2}{a^2\Omega^2}\right). \tag{6}$$

One root of the polynomial (6) has a positive real part and, hence, the steady-state motion is unstable.

3 Locomotion using a magnetic fluid in a traveling magnetic field

We consider a plane flow of an incompressible viscous magnetic fluid layer along a horizontal surface in a nonuniform magnetic field (Figure 2). The magnetic permeability of the fluid μ is assumed to be constant. The pressure on the free fluid surface is constant. In the case of a constant magnetic permeability, the body magnetic force is absent and the magnetic field manifests itself in a surface force acting on the free surface (Rosenzweig, 1985). The gravity is not taken into account. In this case, the system of equations consists of the continuity and Navier-Stokes equations:

$$\operatorname{div} \vec{V} = 0, \quad \frac{\partial \vec{V}}{\partial t} + (\vec{V} \cdot \nabla) \vec{V} = -\frac{1}{\rho} \cdot \operatorname{grad} p + \nu \cdot \Delta \vec{V}. \tag{7}$$

Here, $\vec{V} = (u, w)$ and p are the velocity vector and the fluid pressure, $\nu = \eta / \rho$ and η are the kinematic and dynamic fluid viscosity coefficients, ρ is the fluid density, and t is time.

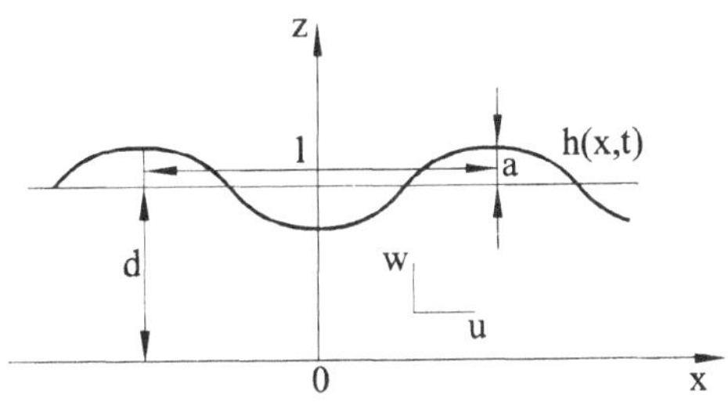

Figure 2. Magnetic fluid layer.

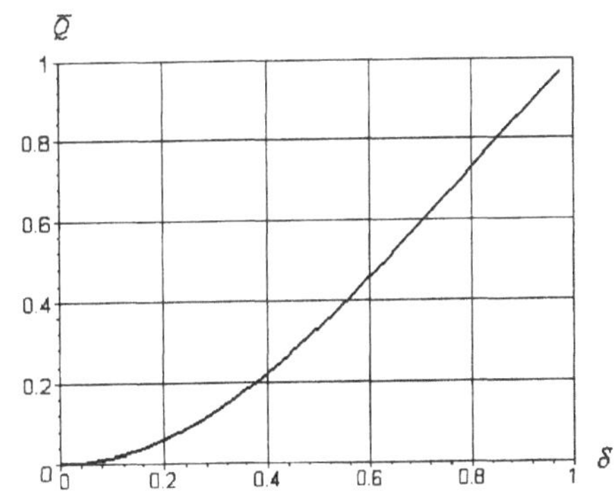

Figure 3. Average volume flow rate $\overline{Q}$ vs. δ.

On the rigid substrate $z = 0$, the no-slip condition is satisfied:

$$\vec{V}(z = 0) = \vec{0}. \tag{8}$$

On the free surface $z = h(x,t)$, conditions of two types, kinematic and dynamic ones, should be satisfied. The kinematic condition is of the form:

$$\frac{dh}{dt} = \frac{\partial h}{\partial t} + u\frac{\partial h}{\partial x} = w. \tag{9}$$

The dynamic conditions of continuity of the normal and tangential stresses on the free surface $z = h(x,t)$ take the form (neglecting the influence of the environment)

$$\left[-p + \gamma\frac{1}{R} + \frac{B_n^2}{8\pi}\left(\frac{1}{\mu} - 1\right) - \frac{H_t^2}{8\pi}(\mu - 1)\right]\vec{n} + \tau_{ij} n^j \vec{e}^i = 0. \tag{10}$$

Here, τ_{ij} are the viscous stress tensor components, R is the radius of curvature of the surface $z = h(x,t)$, $\vec{n}$ is the vector of outward normal to the surface, $\vec{e}^j$ are the basis vectors, γ is the film surface tension coefficient, $B_n = \mu H_n$ is the normal component of the magnetic field strength vector. The magnetic field $\vec{H}$ is assumed to be fixed, since the non-inductive approximation implies $\mu - 1 << 1$. We will assume that the magnetic field creates the periodic travelling wave on the surface of a sufficiently thin layer (with the asterisk * we denote the dimensional variables):

$$h^*(x,t) = d + a\cos\left(\omega t^* - kx^*\right), \quad \varepsilon = d \cdot k << 1.$$

We introduce the dimensionless variables

$$(x,z) = (x^*, z^*)\cdot k, \quad h = h^*/d, \quad \delta = a/d, \quad u = u^*/U_c, \quad w = w^* U_c/\varepsilon, \quad U_c = \omega/k,$$
$$t = t^*\cdot\omega, \quad p = p^*/P, \quad P = \eta\,\omega/\varepsilon^2, \quad H^2 = H^{*2}/P, \quad \mathrm{Re} = \rho U_c d/\eta, \quad W = \gamma\, d k^2/P.$$

For $\varepsilon << 1$, we will seek a solution in the form of a power series in ε ($A \equiv h, u, w, p$):

$$A(x,z,t) = A_0(x,z,t) + \varepsilon \cdot A_1(x,z,t) + \ldots.$$

In the zero approximation in ε, for $\mathrm{Re} < 1$ and $W = O(1)$, using the equations (7) and the condition (8) – (10), we obtain an expression for the velocity components (subscript "0" is omitted):

$$u(x,z,t) = F(x,t)\left(\frac{z^2}{2} - hz\right), \quad w(x,z,t) = F(x,t)\frac{\partial h}{\partial x}\cdot\frac{z^2}{2} + \frac{\partial F}{\partial x}\left(\frac{h}{2}\cdot\frac{z^2}{2} - \frac{z^3}{6}\right), \tag{11}$$

where $F(x,t) = \dfrac{\partial p}{\partial x}$.

The mass conservation law (7), the condition (9), expressions (11), and the assumption that $h(x,t) = h(\xi)$, $\xi = t - x$ imply the following equation for the flow rate Q:

$$\frac{\partial h}{\partial t}+\frac{\partial Q}{\partial x}=0, \quad Q(x,t)=-F\,h^3/3=h+C. \tag{12}$$

If the flow is T-periodic, we can introduce the average flow rate: $\overline{Q}(x)=\frac{1}{T}\int_0^T Q(x,t)dt$. For $h(\xi)=1+\delta\cos(\xi)$, the dimensionless average flow rate is $\overline{Q}=1+C$.

Relation (10), with reference to (12), implies

$$W\frac{\partial^3 h}{\partial \xi^3}+\frac{(\mu-1)}{8\pi}\frac{\partial H^2}{\partial \xi}=-\frac{3}{h^2}-\frac{3C}{h^2}. \tag{13}$$

Using equation (13) it is possible to find the magnetic field creating the prescribed film shape:

$$H^2=H_0^2+\frac{8\pi}{\mu-1}\left(-W\frac{\partial^2 h}{\partial \xi^2}-3\int\frac{1}{h^2}d\xi-3C\int\frac{1}{h^3}d\xi\right), \quad h(\xi)=1+\delta\cos(\xi), \quad \delta<1. \tag{14}$$

The requirement for the magnetic field magnitude to change periodically leads to the expression $C=-2(1-\delta^2)/(2+\delta^2)$.

The dependence of the average volume flow rate $\overline{Q}=1+C$ on the surface oscillation amplitude δ is shown in Figure 3.

The expression for the magnetic field (14) takes the form

$$H^2=H_0^2-D, \quad D=\frac{8\pi}{\mu-1}\left(-W\delta\cos(\xi)+\frac{3\delta\sin(\xi)(2+\delta\cos(\xi))}{(2+\delta^2)(1+\delta\cos(\xi))^2}\right).$$

The constant H_0^2 can be chosen arbitrarily. However, this constant must not be less than the maximum value, $D_{\max}>0$, of the function D in order not to violate the condition $H^2\geq 0$. From now on, we assume $H_0^2=D_{\max}$.

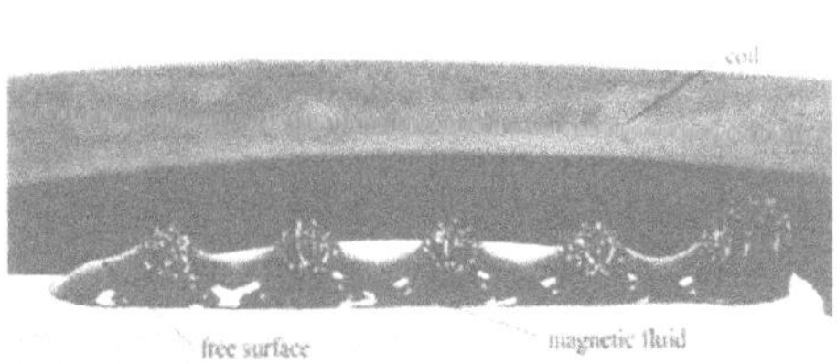

Figure 4. Ferrofluid in a traveling magnetic field.

Figure 5. Inchworm from magnetizable materials.

4 Conclusion

1. The possibility of using a magnetic-elastic element under the action of an external magnetic field as a drive of the microrobot is investigated. It is shown that if non-symmetric Coulomb dry friction is present, a motion of the system with a constant "on the average" velocity is possible.
2. The motion excited by periodic elongation of the element under the action of a magnetic field is stable. The periodic change of the stiffness of the magnetic-elastic element (parametric resonance) results in an unstable motion.
3. The expression for the magnetic field strength creating a sinusoidal wave on the surface of a viscous magnetic fluid as a function of the characteristics of the fluid (viscosity, surface tension, and magnetic permeability) and the parameters of the wave are obtained.
4. Together with experimental investigations of ferrofluids (Figure 4), first prototypes of artificial worms using the principles outlined above have been constructed (Figure 5).

Bibliography

Bogoljubow, N. N., and Mitropolski, Yu. A. (1965). *Asymptotische Methoden in der Theorie der nichtlinearen Schwingungen*, Akademie-Verlag, Berlin.

Rosenzweig, R. E. (1985). *Ferrohydrodynamics*, Cambridge University Press.

Turkov, V. A. (2002). Deformation of an elastic composite involving a magnetic fluid. *Journal of Magnetism and Magnetic Materials*, 252:156-158.

Zeidis, I., and Zimmermann, K. (2000). Ein mathematisches Modell für die peristaltische Bewegung als Grundlage für das Design wurmartiger Mikroroboter. *Technische Mechanik* 1:73-80.

Zimmermann, K., Zeidis, I., and Steigenberger, J. (2002a). Oscillation of a Chain of Pointmasses as a Model of Worm-Like Motion Systems. In *Proceedings of the 4th European Nonlinear Oscillations Conference*, Moscow. 49.

Zimmermann, K., Zeidis, I., and Steigenberger, J. (2002b). Mathematical Model of Worm-like Motion Systems with Finite and Infinite Degree of Freedom. **In** *Proceedings of the 14th CISM-IFToMM Symposium (RoManSy 14)*, Italy. 507-516.

Zimmermann, K. and Zeidis, I. (2003). On artifical Worms as Chain of Mass Points. **In** *Proceedings of the 6th International Conference of Climbing and walking Robots (CLAWAR 2003)*, Italy. 11-18.

Zimmermann, K., Zeidis, I., Steigenberger, J., and Pivovarov, M. (2004a). An Aproach to Worm-Like Motion. **In** *Proceedings of the 21st International Congress of Theoretical and Applied Mechanics*, Poland. 371.

Zimmermann K., Zeidis I., Naletova V. A.,. and Turkov V. A. (2004b). Modelling of worm-like motion systems with magneto-elastic elements. *Physica Status Solidi* (c), 1(12):3706-3709.

Zimmermann K., Zeidis I., Naletova V. A., and Turkov V. A. (2004c). Waves of on the surface of a magnetic fluid layer in a travelling magnetic field. *Journal of Magnetism and Magnetic Materials*, 268:227-231.

Zimmermann K., Zeidis I., Naletova V. A., and Turkov V. A. (2004d). Travelling waves on a free surface of a magnetic fluid layer", *Journal of Magnetism and Magnetic Materials*, 272-276:2343-2344.

Zimmermann K., Zeidis I., Naletova V. A., Turkov V. A., and V.E. Bachurin (2005). Locomotion based on a two-layers flow of magnetizable nanosuspensions", *Journal of Magnetism and Magnetic Materials*, 290-291:808-810.

A Flexure-Based Electromagnetic Linear Actuator for Nano-Positioning

T. J. Teo [1,2], I.-M. Chen [1], G. L. Yang [2] and W. Lin [2]

[1] School of Mechanical and Aerospace Engineering, Nanyang Technological University, Singapore
[2] Singapore Institute of Manufacturing Technology, Singapore[1]

Abstract. In this paper, a novel nano-positioning actuator with large travelling range and large thrust force has been proposed, designed and developed. Known as a Flexure-Based Electromagnetic Linear Actuator (FELA), it consists of two main elements, i.e., an Electromagnetic Driving Mechanism (EDM) and a Flexure-Based Mechanism (FBM). In this FELA, its EDM is developed through a new dual magnet configuration that delivers large force sensitivity and good-to-size ratio. Its actuating body is supported through flexure-based bearings and form a FBM, which ensure its positioning repeatability. A bi-stable mechanism design has been adopted by this FBM to deliver low stiffness in the actuating direction and high stiffness in rest of the non-actuating directions. Lastly, a prototype of the FELA has been developed and preliminary experiments have been conducted to evaluate its performances.

1 Introduction

An actuator with nano-positioning capabilities has long been the key technology in nano-/micro- manufacturing processes such as, nano-imprint lithography, fibre alignment, MEMS assembly, and nano-machining etc. Among various types of nano-positioning actuators, the solid-state Piezoelectric (PZT) actuators have been the most popular choice due to their large actuating force and high stiffness (Dong et al., 2000; Kimball et al., 2000; Sun et al., 2002; Tan et al., 2001). However such PZT actuators have limited displacements that make them unsuitable to drive a high-precision manipulator targeted for large travelling range (Kenji, 1997). Although some existing high-precision positioning actuators are able to eliminate such limitations, the displacement amplification techniques that are used within these actuators inherit other drawbacks. For example, PZT-driven actuators that use high-pitch screw actuating-shaft to achieve millimetres of displacement have poor repeatabilities due to backlash and Coulomb frictions (New-Focus, 2005). Others using PZT or magnetostrictive clamping method to drive an internal shaft for displacement amplification purposes have low payload capacities (EXFO-Burleigh, 2005). Most importantly, these existing actuators have a slow response speed that makes them inefficient in fast speed applications.

An electromagnetic driving scheme has the potential of delivering millimetres of travelling range with nanometers of positioning resolution. Actuators of such frictionless drive, i.e. the voice-coil linear actuators and the electromagnetic suspended (or solenoid type) actuators, have

[1] This work is supported and funded by Singapore Institute of Manufacturing Technology (U03-R-360B)

also been employed in high-precision manipulation. However, a voice-coil linear actuator produces relatively small output forces (or poor force sensitivity), which limit its applications to light-duty tasks such as pick-n-place of miniature parts and micro-EDM etc. Although, a solenoid type of actuators offers good force sensitivity, it is unable to achieve a constant output force throughout its allowable travelling range. Furthermore, magnetization on the ferromagnetic stator introduces magnetic hysteresis, which will affect the positioning accuracy of such actuators at sub-micron level. Lastly, most of the existing electromagnetically driven actuators employ mechanical bearings, e.g., linear guides, to guide the output shaft. These mechanical-bearings introduced dry frictions that will significantly reduce the positioning repeatability of those actuators.

The drawbacks of existing nano-positioning actuators motivate the development of a Flexure-based Electromagnetic Linear Actuator (FELA) (Yang and Teo, 2005). This novel nano-positioning actuator is designed to achieve ±2mm of travelling range, good force sensitivity, e.g., generating at least 50N of continuous thrust force with 1 Amp of input current, fast actuating speed and high stiffness at all non-actuating axes. A FELA consists of an Electromagnetic Driving Mechanism (EDM), which forms the main element in driving the actuating body of this actuator. This actuating body is then support by the flexure-based joints and forms a Flexure-based Bearing Mechanism (FBM). This paper focuses on the design of a FELA with detail considerations on the proposed EDM and the FBM. A prototype of the FELA has been developed and preliminary experiments have been conducted to validate its performances.

2 Concepture Design

2.1 Electromagnetic Driving Mechanism

To drive this proposed FELA, a magnetic linear direct drive (or voice-coil) technique is employed for its linearity between input current and output force. Its competency of achieving infinite positioning resolution and constant output force through the entire range of travel also makes it a promising approach to drive a FELA. From the pre-determined capabilities of a FELA, it must have the ability to generate at least 50N of continuous thrust force with 1Amp of input current. Considering that a certain amount of force is needed to drive the actuating body, a proposed EDM is required to produce more than 50N/Amp of driving force. Based on this requirement, two potential EDMs have been proposed (Table 1).

Table 1. Proposed EDM designs

EDM-1	EDM-2
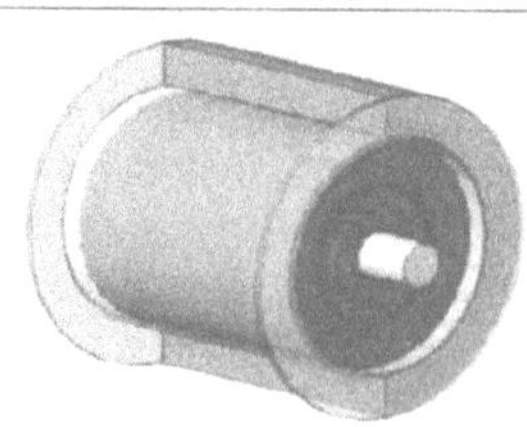	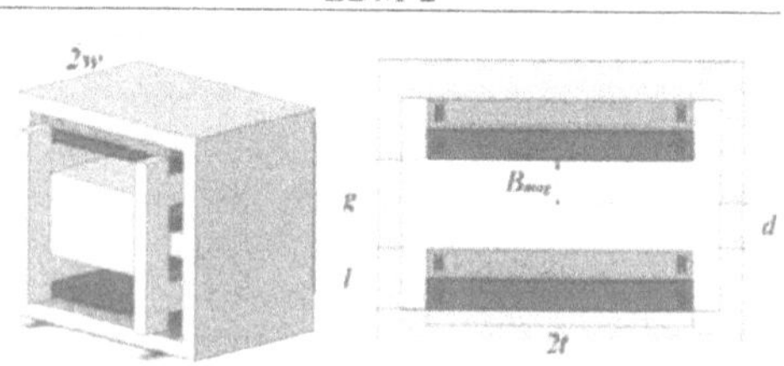

EDM-1 is proposed based on a conventional magnetic circuit that realises a voice-coil type of driving scheme. However, a moving magnet arrangement has been used instead of a conventional moving coil arrangement. In this design, a cylindrical-shaped Permanent Magnet (PM) is employed with the coil wound around it. Placing a steel shelling outside the coil allows the magnetic flux, which emanates from the surface of the PM, to be evenly distributed within the effective coil area. Consequently, every turn of coil is fully utilize to generate the desired output force. Lastly, the moving PM that carry a connecting rod will be attached to the actuating body of the actuator.

In EDM-2, a new type of magnetic circuit design that realizes large force sensitivity and good force-to-size ratio EDM is proposed. Known as a dual magnet configuration, this magnetic circuit is form by two pairs of rectangular PMs and a closed-loop iron core. Each pair of the PMs (backed by a common iron path) is arranged as such that each PM is facing the other in an attracting position. With this configuration, the magnetic flux that emanates from the two PMs amalgamates at the mid-section of the air gap. Consequently, flux leakage from both ends of the PM will be minimized. The actuating body of the actuator, which is wounded with coil, will be inserted at the centre of the iron core. Unlike EDM-1, EDM-2 has a moving coil with fixed PMs arrangement instead.

2.2 Flexure-Based Bearing Mechanism

In this proposed FELA, flexure-based bearings will be employed to support the actuating body of this actuator. Such flexure-based bearings are leaf-spring type of compliant joints, which provides a smooth, rectilinear, frictionless motion and millimetres of travel at low driving force. Consequently, an actuating body with flexure-based bearing forms a FBM that retains the capabilities of an electromagnetic driving scheme, e.g., frictionless and large motions. Most importantly, it provides necessary stiffness to eliminate the stationary positioning noise exhibited by this driving scheme. To facilitate a FELA in achieving its predetermined capabilities, the proposed FBM must meet three main requirements. First, only a small amount of force is require to drive the actuating body in achieving ±2mm of displacement range. Second, the parasitic errors at the non-actuating directions must be minimal, e.g., <0.1μm. Third, the stiffness at non-actuating directions must be a few hundreds times more than the actuating direction. Two promising designs (Table 2) have been proposed in the development stage of a FBM.

Table 2. Proposed FBM designs

FBM-1	FBM-2
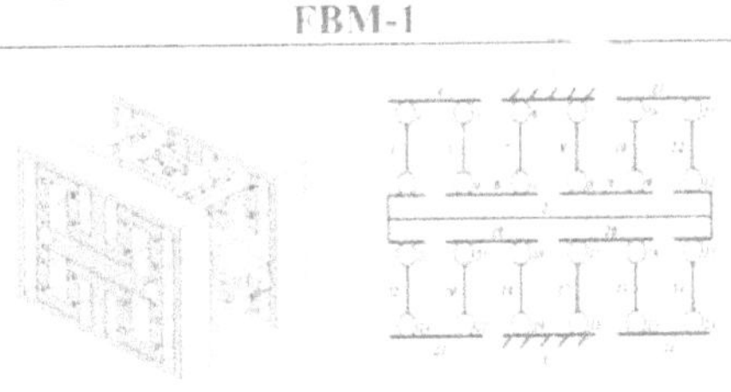	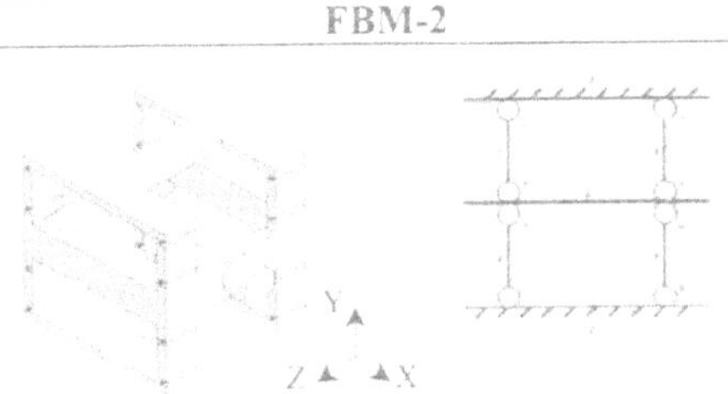

In FBM-1, a double-compound leaf-spring flexure joint is used to join the fixed body and each corner of an actuating body. A symmetry design is adopted to minimise the parasitic errors at non-actuating directions. This configuration allows a large displacement using relatively small driving force. In FBM-2, a pair of leaf-spring flexure joint is used to join the fixed body and an actuating body to form a linear spring mechanism. A symmetry design is adopted to minimise the parasitic errors at non-actuating directions. This result in a bi-stable design that offers small driving force within its permissible travel range.

3 Design Analysis

3.1 Analysis of the Proposed EDM Designs

Lorentz force is commonly known in governing the electromechanical conversion of a magnetic linear direct drive technique. It states that if a current-carrying conductor (or coil) is placed in a magnetic field, a force (*F*) will be act upon it. The magnitude of this force determine by the magnetic flux density (*B*) current (*i*) and orientation of the field and current vector. With a total length (*L*) of coil, being placed in the magnetic field, the output force can be estimated by the following equation:

$$F \approx iL \times B \tag{1}$$

In this principle, the magnetic flux density at the coil area plays an important role in determine the output force. A charge model can be used to estimate the magnetic flux density that emanate from a PM. In this model, only the dimensions and residual induction (*Br*) of the material are required. For a PM magnetized along its length, with thickness (*2t*), width (*2w*) and length (*l*), the flux density at a distance (*d*) from the surface, along its magnetized axis is given by (Furlani, 2001):

$$B_{magj} = \frac{B_{rj}}{\pi_j}\left[\tan^{-1}\left(\frac{D+l_j}{t_j w_j}\sqrt{t_j^2 + w_j^2 + (d+l_j)^2}\right) - \tan^{-1}\left(\frac{D}{t_j w_j}\sqrt{t_j^2 + w_j^2 + d^2}\right)\right] \tag{2}$$

For dual magnet configuration,

$$D = \begin{cases} d, & B_{mag1} \\ g-d, & B_{mag2} \end{cases} \tag{3}$$

$$B_{total} = B_{mag1} + B_{mag2} \tag{4}$$

Each of the proposed EDMs has different size and dimensions of PMs, and length of coil For EDM-1, Br = 13300 Tesla, $2t = 2\pi r$ =157mm, l = 40mm, $2w$ = 54mm and d = 1 – 11mm For EDM-2, Br = 13300 Tesla, $2t$ = 54mm, l = 15mm, $2w$ = 50mm and d = 1 – 5.5mm. The magnetic flux density within the air gap of EDM-1 can be determined through Eq. (2). As for EDM-2, the magnetic flux within the air gap of a dual magnet configuration can be determined using Eqs. (2) to (4). Through ANSYS 8.1, Finite Element Analysis (FEA) simulations (or numerical analyses) are also conducted to determine the output force of each proposed EDM design. The results obtained from both analytical and numerical analyses are summarised in Table 3.

Table 3. Results obtained from the analytical and numerical analyses

Proposed EDM	Analytical, Output force (N),	FEA, Output force (N)
EDM-1	254	274
EDM-2	219	262

Through the analyses, both EDM-1 and EDM-2 generates more than 200N of output force with 1 Amp of input current. However, a dual magnet configuration within EDM-2 offers a constant magnetic flux density (Figure 1a) and evenly distributed magnetic flux throughout the effective air-gap (Figure 1b). This enables EDM-2 to provide a linear output force and high-precision positioning resolution throughout its allowable travelling range. In addition, EDM-2's moving coil arrangement delivers faster speed actuation as compared to EDM-1's moving magnet arrangement. Thus, proposed EDM-2 will be used as the driving mechanism for the proposed FELA.

(a) Magnetic flux density within the air gap (b) Magnetic flux distribution

Figure 1. Numerical analysis of EDM-2 using ANSYS 8.1

3.2 Analysis of the Proposed FBM Designs

Using ANSYS 8.1, all proposed FBM designs underwent four stages of numerical analyses. First, to analyse the force require for driving the actuating body to achieve 2mm at X-direction. Second, to analyse parasitic errors at both Y- and Z- directions during the actuation. Third, to analyse the stiffness at Y-direction when an external force, i.e., the driving force at X-axis, is applied to Y-axis of the actuating body at actuated (+2mm), neutral and retracted (-2mm) position respectively. Forth, to analyse the stiffness at Z-direction, when the same external force is applied to Z axis of the actuating body at the similar actuated, neutral and retracted position respectively. In the analyses, stainless steel has been selected as the material for all proposed FBM designs. This material is assumed to be linearly elastic with Young's modulus of 190GPa, Poisson's ratio of 0.31 and element type chosen as SOLID 92 (10 node).

Based on the results, FBM-2 requires slightly more driving force to achieve the desired displacement as compared to FBM-1. However, FBM-2 offers the much higher stiffness at all non-actuating directions as compared to FBM-1. This is a very much-desired feature for an actuating module. Hence, the proposed FELA's FBM will be developed based on the design of FBM-2. Table 4 summarised the results obtained from the numerical analyses.

Table 4. Results of the numerical analysis for each proposed FBM design

FBM	Driving force at ±2mm (N)	Parasitic error Y-direction (μm)	Parasitic error Z-direction (μm)	X-:Y-direction stiffness ratio (Extend/retract)	X-:Z-direction stiffness ratio (Extend/retract)	X-:Y-direction stiffness ratio (Neutral)	X-:Z-direction stiffness ratio (Neutral)
1	10	0.01	$0.25\text{x}10^{-3}$	$1{:}2.5\text{x}10^{3}$	1:542	1:429	1:499
2	11.5	0	$2.2\text{x}10^{-3}$	1:INF	$1{:}3.1\text{x}10^{3}$	$1{:}1.7\text{x}10^{3}$	$1{:}3\text{x}10^{3}$

4 Prototype of the FELA

A prototype of the FELA is developed based on the selected EDM and FBM designs. This FELA prototype is dimensioned at 100mm x 100mm x 70mm (Figure 2). The leaf-spring flexure joints are form using stainless steel shims and the remaining parts of the FBM are form using aluminium. For the EDM, AWG24 wire (DIA 0.45mm) is selected based on the estimated operating temperature and current of a FELA, which are 120°C and 2 Amp respectively. In addition, rare-earth PM (NdFeB-Type N45M) with residue induction of 1.33 Tesla and maximum operating temperature of 120°C has been selected. The following sections discuss about the preliminary experiments conducted using this prototype.

Figure 2. A prototype of the proposed FELA

4.1 Magnetic Flux Density of the EDM With Dual Magnet Configuration

A LAKESHORE hall-sensor probe and gaussmeter are employed to measure the magnetic flux density within the air gap of the developed EDM (Figure 3a). In this experiment, only half of the air gap volume has been measured as it is assumed that the magnetic flux density distribution will be symmetrical for the other half of the air gap volume. Based on a PM with 54 mm in length and 50 mm in width, the air gap volume measured is 27 mm (X-axis) x 50 mm (Y-axis) x 7mm (Z-axis). The magnetic flux density distribution within the air gap of a dual magnet configuration is illustrated in Figure 3. Based on Figure 3b, where Y-coordinate (along the width of the air gap) is fixed at 25mm, there is a sign of leakage in magnetic flux at the start of the air gap ranging from 0 to 5 mm. However, the magnetic flux density becomes evenly distributed from 5 mm onwards. Such behaviour of the magnetic flux density can be found through the air gap volume. In addition, a constant 0.5 Tesla of magnetic flux density is obtained throughout 70% of the air gap volume.

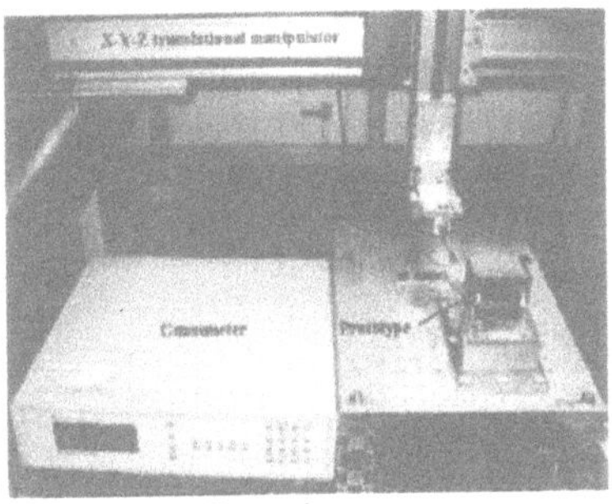

(a) Experiment setup

(b) Magnetic flux density at Y = 25 mm

Figure 3. Magnetic flux density measured within the air gap of the EDM with dual magnet configuration

4.2 Relationship Between Input Current and Output Force of the FELA

The entire setup of this experiment consists of a linear amplifier from Precision Microdynamics, a force/torque sensor from ATI and a FELA prototype driven by an EDM with dual magnet configuration (Figure 4a). In this experiment, a 28VDC linear amplifier has been used to drive the FELA through the proposed EDM. Based on the amount of coil being used in the EDM, its total resistance is measured to be 42.2 Ω. As a result, the highest input current is restricted to ≈ 0.6 Amp.

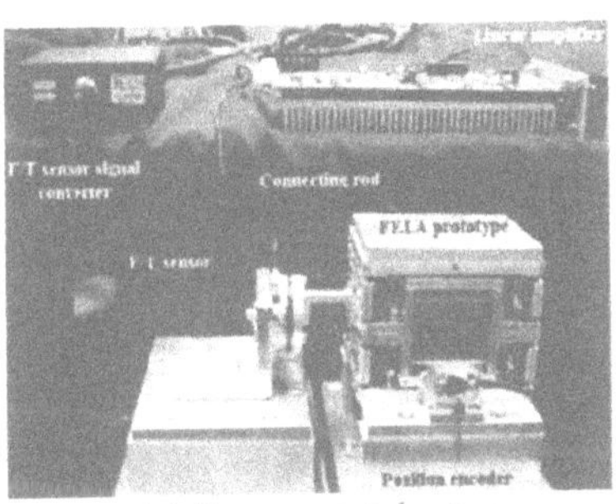

(a) Experimental setup

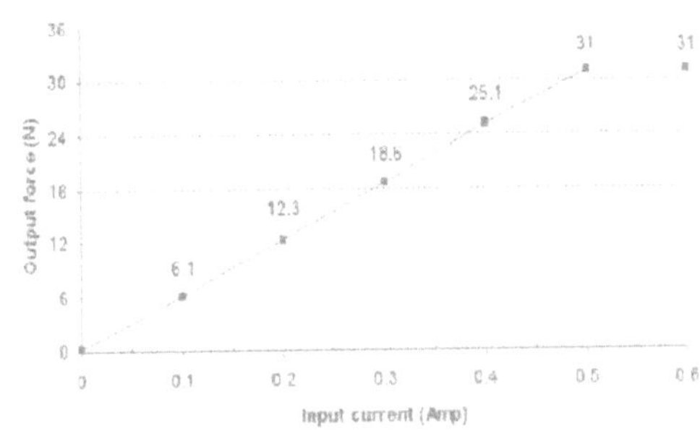

(b) Graph of input current VS output force

Figure 4. Experiment evaluation of the input current and output force of a FELA

The acquired data from this experiment has shown that the relationship between the input current and the output force is almost linear (Figure 4b). The FELA generates a 6 N of continuous thrust force using 0.1 Amp of input current. The maximum input current supposes to be limited to 0.6 Amp, however, the output force of the FELA remain stagnate after 0.5 Amp. With a larger VDC linear amplifier, this FELA will be capable of generating 60 N of output force with 1 Amp of input current judging from the experimental results. There is also a difference between the actual output force of the FELA prototype as compared to the analytical and numerical results. This is because the diameter of the coil chosen for the developed EDM is 0.45mm, while the diameter of the coil used in the analytical and numerical analyses are 0.3mm. As a result, the effective coil length of the actual coiling is approximately 150m instead of 438m that have been used in the analytical and numerical analyses. With the actual effective coil length being reduced by 3 times, it explains why the actual output force is also three times smaller as compared to the output force obtained from the initial analyses.

5 Conclusion

A novel nano-positioning actuator, i.e., FELA, has been proposed, designed and developed. This FELA consists of an EDM that drives an actuating body electromagnetically and a FBM that is formed by this actuating body with flexure-based bearings. A new type of magnetic circuit, i.e., a dual magnet configuration, is proposed to realise a direct electromagnetic driving scheme with large force sensitivity and good force-to-size ratio It is adopted by the EDM to ensure the consistency of the magnetic flux density within the effective air gap volume. A bi-stable mechanism design has been adopted by the FBM to ensure high stiffness at non-actuating directions while retaining low stiffness at the actuating direction. Preliminary experimental results have shown that the magnetic flux is evenly distributed within the effective air gap volume of the EDM. The magnetic flux density is constant throughout 70% of the air gap volume due to the minimal flux leakage. Experiment results have shown that the proposed dual magnet configuration offers a high-force sensitivity EDM, which allows the FELA to achieve large continuous thrust force using small amount of input current. In addition, a compact-sized FELA can be realised because a good force-to-size ratio EDM is achievable using the proposed dual magnet configuration. More experiments will be conducted in near future to further evaluate the performance of this FELA, e.g., its positioning resolutions. Together with a suitable control scheme, this FELA will provide a promising solution in driving high-precision manipulators targeting for large workspaces.

6 Bibliography

Dong, L., Arai, F. and Fukuda, T. (2000). 3D Nanorobotic Manipulation of Nano-order Objects inside SEM. *Proceedings of the 2000 International Symposium on Micromechatronics and Human Science*, 151 - 156.

EXFO-Burleigh (2005). Piezoelectric Inchworm actuator. EXFO Burleigh.

Furlani, E.P. (2001). *Permanent Magnet and Electromechanical Devices*. Academic Press

Kenji, U. (1997). *Piezoelectric Actuator and Ultrasonic Motors*. Kluwer Academic Publishers, USA

Kimball, L.W., Tsai, D.D. and Maloney, J. (2000). Modeling and Batch Fabrication of Spatial Micro-manipulators. *Proceedings of the 2000 ASME Design Engineering Technical Conferences.*

New-Focus (2005). Pico Motor. New Focus.

Sun, Y. et al. (2002). A Novel dual-axis electrostatic microactuation system for macromanipulation. *Proceedings of the 2002 IEEE/RSJ International Conference on Intelligence Robots and Systems*, 1796 - 1801.

Tan, K.K., Lee, T.H. and Zhou, H.X. (2001). Micro-positioning of linear-piezoelectric motor based on a learning nonlinear PID controller. *IEEE/ASME Tr. on Mechatronics*, 6: 428 - 436.

Yang, G. and Teo, T.J. (2005). *A Flexure-Based Electromagnetic Linear Actuator*. Intellectual Property Office of Singapore, IPOS patent application no.: 200508479-3 (filed).

An Infrared Location System for Relative Pose Estimation of Robots

Anssi Kemppainen *‡ , Janne Haverinen *‡ and Juha Röning °‡

* Postgraduate Researcher in Computer Engineering Laboratory, e-mail: pikinen@ee.oulu.fi
* Postdoctorial Researcher in Computer Engineering Laboratory, e-mail: johannes@ee.oulu.fi
° Professor in Computer Engineering Laboratory, e-mail: jjr@ee.oulu.fi
‡ University of Oulu, Faculty of Technology, Department of Electrical and Information Engineering, Oulu, Finland

Abstract. In this work we present an infrared location system for relative pose (position and orientation) estimation in a multi-robot system. Pose estimates are essential for tasks like cooperative simultaneous localization and mapping (C-SLAM), and formation control. In simultaneous localization and mapping (SLAM) relative pose estimates enable more accurate and less time-consuming map building. Respectively, formation control requires accurate pose estimates of other robots to enable robot cooperation in required formation. To address these challenging tasks for small-sized robots, we present a small-sized infrared location system with low current consumption. In the location system, robots use intensity and bearing measurements of received infrared signals to estimate the positions of other robots in polar coordinates. In addition, each robot has a unique modulation frequency from which they are recognized. The location system performs position estimation by rotating a beam collector at constant rotation speed and by measuring the bearing and intensity of the received signal. Infrared signals are received through a small aperture in the beam collector enabling accurate bearing measurements. In order to maximize the measurement range, infrared radiation is reflected sideways into a uniform zone using a conical mirror. Experiments were performed in a group of three robots with a measurement range of up to 3 m while the maximum number of robots was eight. The location system implemented enables relative position estimation among a group of small-sized robots without exchanging position estimates. This is advantageous since the robots are able to maintain formation also in the absence of a radio link.

1 Introduction

Surveillance, rescue and exploration tasks have motivated robot scientists in developing multi-robot systems capable of cooperating in unknown environments. Multi-robot systems compared to single mobile robots can carry out more complex tasks, through their potential to use a greater amount of sensors and actuators. In addition, performing time-critical tasks, for example, finding and rescuing people from a burning house is faster using multiple robots instead of one. Location is essential in multi-robot systems since cooperation, for example, carrying injured people in a group of robots requires each robot to be aware of other's locations.

Location in a multi-robot system consists of self-localization and relative location of other robots. A typical approach to self-localization is simultaneous localization and mapping (SLAM), where the robot's pose (position and orientation) in an unknown environment is defined by measuring position estimates of landmarks. A map is built by fusing landmark perceptions and the robot's pose estimate from dead reckoning. In addition, by measuring the relative locations between robots, formation control and optimal map building, often referred to as exploration (Burgard et al., 2005), can be realized. Cooperative simultaneous localization and mapping (C-SLAM) improves the maps by fusing the relative pose estimates of landmarks among a group of robots. These systems have recently been examined by several authors (Fox et al., 2000), (Howard, 2005),(Mourikis and Roumeliotis, 2005), (Hajjdiab and Laganiere, 2004), (Rodriguez-Losada et al., 2004). By fusing map estimates, the map becomes more accurate and mapping time decreases substantially.

A formation control is task where robots are required to maintain a movement formation as they travel from their starting point to the goal target avoiding collisions with obstacles. Maintenance of a particular formation is required in many applications. For example, in military applications, formation allows each robot to concentrate their sensors across one portion of the environment, while other robots cover the rest(Balch and Arkin, 1998). Secondly, tasks like demining, harvesting (agriculture) and security patrol may require specific formations depending on the surroundings. For example, in narrow corridors or on footpaths, column formations are recommended, whereas in huge hallways or wide areas, diamond or wedge formations are more suitable. Formation control does not require a global location system. Instead, robots need to have position estimates on where other members of the group are located. However, without an absolute positioning system, uncertainty about the absolute position will grow continuously without bound (Roumeliotis and Bekey, 2002).

In this paper, we present an infrared location system for relative pose estimation where each member of the robot group locates and recognizes other robots. The location system is designed to be suitable for both large and small robots. It was developed to be a part of a swarm robotic research programme to be performed with miniature robots (Haverinen et al., 2005). One advantage in the presented infrared location system is that the robots' positions are estimated and they are recognized without data transmission between robots. However, relative orientations can be estimated only by fusing location estimates among a group.

2 Design considerations

In our study, the purpose was to design a relative location system for a group of robots. As robots build and maintain different formations as well as perform exploration tasks in an unknown environment, the system accuracy must be sufficient and the measurement range large. In hallways, as robots carry out exploration tasks in a minimal group, distances between robots may grow to close to 10 metres. Respectively, when moving through narrow corridors or in small disorganized rooms among a large group of robots, the robots may be required to operate very close to each other. Because of these factors, there is the requirement that the accuracy must be sufficient, to be precise, smaller than the lowest distance between robots. Secondly, in order to operate in both small and large groups, as tasks and environments vary, the measurement range must be wide-ranging.

The primary goal of this work was to implement an exact and yet wide-ranging relative location system for small mobile robots. In the related systems, except for one (Grabowski et al., 2000), systems have functioned with larger robots. In order to be suitable for small robots, several restrictions must be introduced. Because our miniature robots (Haverinen et al., 2005) use 1000 mAh lithium batteries, in order to enable a few hours operation time, current consumption was expected to be only a couple hundred milliamperes. In addition, due to the small size of the robots, the sensors have to be light and small enough.

The design considerations focused tightly on selecting appropriate sensors from between laser, vision, ultrasound and IR (infrared) sensors to attain the required accuracy and measurement range, and to achieve small enough size with low current consumption. Related systems have been presented exploiting several techniques including laser range finders (Schneider and Wildermuth, 2004), (Montesano et al., 2004), (Howard et al., 2003),(Moors et al., 2003), ultrasonic TOF measurement (Shoval and Borenstein, 2001), (Grabowski et al., 2000) and vision (Montesano et al., 2005), (Spletzer et al., 2001) for location and recognition of other robots. Laser range finders are accurate and have wide-ranging sensors which measure relative distance and angle to a target. However, range finders cannot distinguish objects from each other, and in addition, sensors are in general large. Vision offers reliable robot recognition for a relative multi-robot location system, provided that differences between robots are visible. Relative locations can be estimated by extracting the robots' features from a captured image. Feature extraction requires, however, a great deal of performance from the location system. Ultrasound sensors are mostly used as sonar sensors in mapping tasks. TOF (time-of-flight) measurements (Grabowski et al., 2000) can be also used to estimate relative distances between robots. Systems of this kind require, however, at least three robots presence since relative positions among a group of robots are estimated by trilateration. In this work, we selected IR sensors since they are small and capable of relative angle measurements between an emitter and receiver. In addition, infrared radiation does not reflect from walls and object surfaces as much as ultrasound. In relative angle measurements, the effects of multipath reflection would be crucial because of ambiguous angle estimates.

3 The architecture

The implemented location system consists of mechanics and control electronics. The mechanics and control electronics were separated in order to minimize size, and to ease the manufacturing of the system (Fig. 2). Mechanics is used to emit infrared radiation and to receive radiation from other robots, whereas control electronics drive the beam collector at a constant angular velocity and estimate the positions of the detected robots. Figure 1 presents a block diagram of the system.

The mechanics consists of an IR emitter and receiver, a DC motor, Hall-effect-sensors, a rotating beam collector and a see-through cylindrical body (Fig. 2). In order to increase

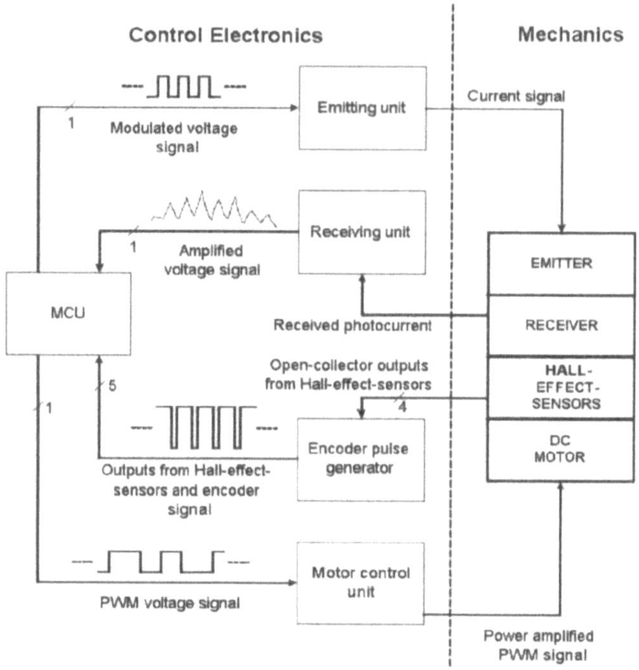

Figure 1. In the location system, the MCU drives beam collector at a constant angular velocity and estimates the positions of the detected robots.

the measurement range, conical mirrors were exploited to reflect the emitted signals sideways into a narrow zone. Signals were received through a small aperture in the beam collector and reflected to the receiver using a mirror. Scanning the surroundings at a constant rotation speed is realized using a DC motor, Hall-effect-sensors and discrete PID controller performed in an MCU.

The control electronics consists of an MCU, an emitting unit, a receiving unit, an encoder pulse generator and a motor control unit (Fig. 1). In order to recognize the emitting robot, the MCU generates in each robot unique modulation frequencies. The emitting unit drives a current-controlled IR emitter by transforming a frequency modulated voltage signal into the current signal. The receiving unit consists of a preamplifier in which the amplification can be changed to enable a wider measurement range. The encoder signal is generated from four Hall-effect-sensor outputs. Using four encoder pulses in a round, the lowest rotation speed in which rotation was sufficient smooth was five revolutions per second. The motor drive was implemented in a motor control unit by amplifying the PWM signal with a half-bridge circuit.

The MCU performed both the rotation speed control and the position estimation. The stable rotation speed of the beam collector was realized by a controlling DC motor

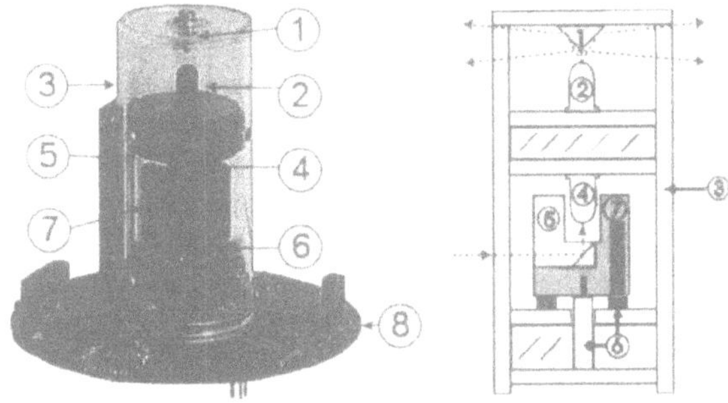

Figure 2. The actual system and the illustration of mechanics: 1) mirror, 2) emitter, 3) see-through body, 4) receiver, 5) aperture, 6) DC motor and Hall-effect-sensors, 7) beam collector, 8) control electronics.

with a discrete PID controller using encoder pulses as speed measurements. The position estimates were calculated by triggering the rising edge of the received signals. As each new signal was detected, the bearing of the beam collector was used to estimate the angular coordinate. Similarly, the radial coordinate was estimated using the intensity of the received signal. In addition, each emitting robot was recognized by the modulation frequency of the received signal.

4 Results

The performance of the implemented location system, presented in figure 2, was measured in a group of three robots. The position error and standard deviation were measured in three different position configurations where one of the robots was locating and the others were targets. In addition, the performance of the location system was measured as one of the targets was moving along a short route.

Figures 3a–c present three test cases where two targets were located in different places around the locating robot. Figure 3d presents a fourth test case where one of the targets was moving along a short path for around one minute. Depending on the distance to the targets, the location estimates were updated at intervals of 1–2 seconds.

The location system located robots in polar coordinates, and position estimates were transformed into cartesian coordinates to be able to define average position errors (see Table 1). The position errors were at their highest over 20 cm, the main reasons for this being faults in the distance measurements (see Table 2).

The standard deviation was calculated for the angular and radial coordinates (see Table 2). The angular coordinate was obtained by using the bearing of the beam collector. In all measurements the standard deviation for the angles was less than 1.5°. Standard deviation for distances suffered from both variations of intensities and different clusters of estimates caused by different amplifications in the preamplifier. Added to this, the standard deviation for the radial coordinate depends on the distance. Since the distance

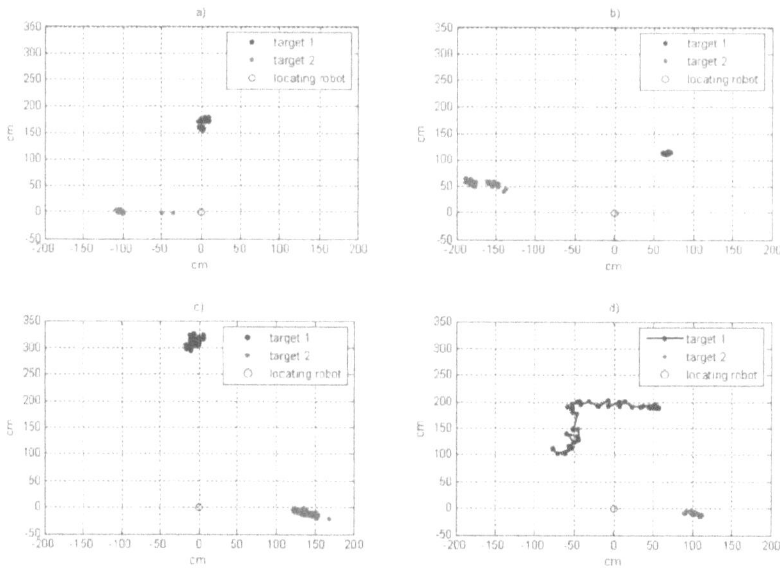

Figure 3. Four test cases: a) Target 1 at (0, 150) and target 2 at (-100, 0), b) Target 1 at (50, 100) and target 2 at (-150, 50), c) Target 1 at (0, 300) and target 2 at (150, 0), d) Target 1 moving from (-50, 100) through (-50, 200) to (50, 200) and target 2 at (-100, 0)

is an inversely exponential function of the intensity of the received signal, the further away the target is located, the sharper the curve is. For this reason, small variations in the received intensity and small noises will significantly increase the standard deviation as the located robot is far away.

In all four test cases target recognition using different identification frequencies for targets was faultless. Target 1 used 1 kHz and target 2 used 2 kHz modulation frequency while permitted frequencies were between 800 Hz and 2.2 kHz at intervals of 200 Hz:s. In addition, the functionality of the target recognition for a group of 8 robots was witnessed by changing the different modulation frequencies and by testing that the recognition was faultless.

5 Conclusions and future work

In this paper, we presented a relative infrared location system for a group of robots. Relative locations among a group of robots are essential for tasks where robots have to cooperate, for example, in carrying injured people. In addition, relative pose estimates can be exploited in simultaneous localization and map building, enabling faster and more

Table 1. Position errors.

Robot	Real positions		Mean value		Position error
	x(cm)	y(cm)	x(cm)	y(cm)	e(cm)
target 1	0	150	3.7	169.8	20.2
target 1	50	100	66.8	111.7	20.5
target 1	0	300	-7.1	310.0	12.2
target 2	-100	0	-101.5	0.8	1.7
target 2	-150	50	-170.1	55.0	20.7
target 2	150	0	139.4	-11.7	15.8

Table 2. Standard deviation.

Robot	Real coordinates		Mean value		Standard deviation	
	r(cm)	θ(deg)	r(cm)	θ(deg)	r(cm)	θ(deg)
target 1	150	90	169.9	88.7	6.3	1.0
target 1	111.8	63.4	130.2	59.1	1.9	0.7
target 1	300	90	310.1	91.3	6.8	1.3
target 2	100	180	101.5	179.6	11.2	1.4
target 2	158.1	161.6	178.8	162.0	13.7	1.2
target 2	150	0	139.9	355.3	8.8	1.3

reliable exploration in unknown environments. In the location system design, infrared sensors were selected since they are suitable for small robots, and since infrared enables reliable bearing measurements. The system uses the bearing and intensity measurements of a received signal to estimate the emitting robot. In addition, the frequency information of the received signal is used for robot recognition. The location system designed is advantageous since IR sensors are small, and since the relative positions of the other robots are estimated without data transmission between robots.

The location measurements testified that the location system was able to recognize targets faultlessly at a distance of 3 m. The bearing estimates were accurate, unlike the distance estimates which in several measurements increased the position error and produced a large standard deviation for the radial coordinate. A three metres measurement range may prove to be inadequate for certain tasks, and yet accurate position estimates must be obtained. For these reasons, among others, the location system has been redesigned and the first measurements have proved an improvement with an increase in the measurement range to close to 10 m.

Bibliography

T. Balch and R.C Arkin. Behavior-based formation control for multirobot teams. *IEEE Transactions on Robotics and Automation*, 14(6):926–939, 1998.

W. Burgard, M. Moors, C. Stachniss, and F. Schneider. Coordinated multi-robot exploration. *IEEE Transactions on Robotics and Automation*, 21(3):376–378, 2005.

D. Fox, W. Burgard, H. Kruppa, and S. Thrun. A probabilistic approach to collaborative multi-robot localization. *Autonomous Robots*, 8(3):325–344, 2000.

R. Grabowski, L.E. Navarro-Serment, C.J.J. Paredis, and P.K. Khosla. Heterogeneous teams of modular robots for mapping and exploration. *Autonomous Robots*, 8(3): 293–308, 2000.

H. Hajjdiab and R. Laganiere. Vision-based multi-robot simultaneous localization and mapping. In *Conference on Computer and Robot Vision*, 2004.

J. Haverinen, M. Parpala, and J. Röning. A miniature mobile robot with a color stereo camera system for swarm robotics research. In *International Conference on Robotics and Automation*, 2005.

A. Howard. Multi-robot simultaneous localization and mapping using particle filters. In *Proceedings of Robotics: Science and Systems*, 2005.

A. Howard, M. Mataric, and G. Sukhatme. Putting the 'i' in team: an ego-centric approach to cooperative localization. In *International Conference on Robotics and Automation*, 2003.

L. Montesano, J. Gaspar, J. Santos-Victor, and L. Montano. Fusing vision-based bearing measurements and motion to localize pairs of robots. In *International Conference on Robotics and Automation*, 2005.

L. Montesano, L. Montano, and W. Burgard. Relative localization for pairs of robots based on unidentifiable moving features. In *International Conference on Intelligent Robots and Systems*, 2004.

M. Moors, F. E. Schneider, and D. Wildermuth. Relative position estimation in a group of robot. In *International Conference on Climbing and Walking Robots*, 2003.

A.I. Mourikis and S.I. Roumeliotis. Performance bounds for cooperative simultaneous localization and mapping (C-SLAM). In *Proceedings of Robotics: Science and Systems*, 2005.

D. Rodriguez-Losada, F. Matia, and A. Jimenez. Local maps fusion for real time multi-robot indoor simultaneous localization and mapping. In *International Conference on Robotics and Automation*, 2004.

S.I. Roumeliotis and G.A. Bekey. Distributed multirobot localization. *IEEE Transactions On Robotics and Automation*, 18(5), 2002.

F.E. Schneider and D. Wildermuth. Using an extended kalman filter for relative localisation in a moving robot formation. In *International Workshop on Robot Motion and Control*, 2004.

S. Shoval and J. Borenstein. Measuring the relative position and orientation between two mobile robots with binaural sonar. In *International Topical Meeting on Robotics and Remote Systems*, 2001.

J. Spletzer, A. Das, R. Fierro, C. Taylor, V. Humar, and J. Ostrowski. Cooperative localization and control for multi-robot manipulation. In *Conference on Intelligent Robots and Systems*, 2001.

Design and Control of the Ball Wheel Drive Mechanism for a Robust Omnidirectional Wheeled Mobile Platform

Young-Chul Lee[1], Danny V. Lee[1], Jae H. Chung[2], Duane A. Bennett[1] and Steven A. Velinsky[1*]

[1] Department of Mechanical and Aeronautical Engineering, University of California-Davis, United States
[2] Department of Mechanical Engineering, Stevens Institute of Technology, Hoboken, NJ, United States

Abstract. This paper discusses design and control of a new ball wheel drive mechanism for a robust omnidirectional wheeled mobile platform. This platform is designed for use in the highway maintenance and construction area, which is a generally an unstructured and congested environment. The proposed ball wheel mechanism can move in all directions on the plane, instantaneously and isotropically. Other novel features include: active regulation of traction drive contact pressure and an optimally controlled reconfigurable drive system.

1 Introduction

Wheeled Mobile Platforms (WMPs) designed for maintenance and construction must often operate in unstructured environments such as worksites, roadway shoulders, and congested factories or warehouses. Generally, conventional vehicles can reach an arbitrary final configuration from some initial configuration. However, the trajectory the vehicle must follow may be relatively complex and dictated by the motion constraints of its conventional wheels. Therefore, much research effort has been devoted to WMPs without motion constraints, herein called Omnidirectional Wheeled Mobile Platforms (OWMPs).

OWMPs have three degrees-of-freedom in planar motion; thus they are also known as holonomic mobile platforms. Two desired attributes are: instantaneous and isotropic omnidirectionality. Instantaneous omnidirectionality exists when a platform can move in any direction without pausing to change its configuration. Isotropic omnidirectionality exists when a platform has an identical capability to climb over obstacles in all directions. OWMPs can be classified by wheel type, which include the Mecanum wheel (Muir and Neuman, 1987), the orthogonal wheel (Byun, Kim and Song, 2002), the spherical wheel (Ferriere, Raucent and Samin, 1997; Tahboub and Asada, 2002), the caster wheel (Holmberg and Khatib, 2000; Wada and Mori, 1996; Yu, Dubowsky and Skwersky, 2000), and the steering wheel (Moore and Flann, 2000).

Steered-wheel platforms do not have true instantaneous omnidirectionality, because their wheels must be reoriented to follow discontinuous paths. A platform with the Mecanum wheel or the orthogonal wheel has instantaneous omnidirectionality, but it does not have isotropic omnidirectionality. A platform with caster wheels has a singular point when it changes its motion direction 180 degrees. A platform with spherical wheels, in general, has both instantaneous and isotropic omnidirectionality, but the existing spherical wheels have only one fixed driving direction and one passive direction. The passively rotated direction cannot be controlled by itself;

* Email: savelinsky@ucdavis.edu

it is controlled by the other spherical wheels through the ground. Therefore, it is not robust when the ground is irregular.

In this paper, a robust ball wheel mechanism is presented where both rotation axes are active and thus the velocity vector of the wheel itself can be fully prescribed. Like other ball wheel designs, this mechanism will have instantaneous and isotropic omnidirectionality. Other novel features include: active regulation of the traction-drive contact pressure and an optimally-controlled reconfigurable drive system. The end-goal is to develop a robust drive mechanism for an OWMP that can operate in unstructured environments with non-ideal terrain.

2 Robust Ball Wheel Mechanism Design

Consider a sphere with its center fixed relative to an inertial reference frame. The axis of rotation of the sphere can be completely defined by setting the tangential velocity of two, non-coaxial points on its surface. This is most conveniently realized with two contact drive wheels, 90 degrees apart, on a great circle of the sphere.

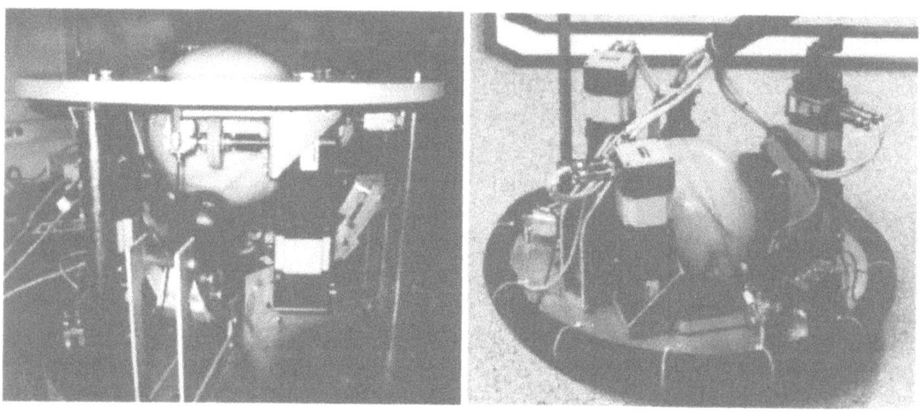

(a) (b)

Figure 1. (a) Photo of bench-top proof of concept. (b) Photo of mobile platform.

The bench-top proof of concept of the robust ball wheel mechanism, shown in Figure 1(a), is based on a 30.5 cm (12 in) diameter, pneumatic sphere composed of highly abrasion resistant polyurethane. Four polyurethane contact wheels are equally spaced along the equator of the sphere. Two adjacent wheels are actuated by servo motors through a precision gear head. Opposite each drive wheel is a passive wheel equipped with a high resolution optical encoder (36000 pulses/rev). The contact pressure between the wheels and the sphere is generated with pneumatic cylinders.

The proof of concept provides empirical data that quantifies the accuracy of the kinematic model, which neglects the inertial and material affects. Moreover, the relationship between the sphere velocity and drive wheel velocities is based on a kinematic model with point contacts in pure rolling. In the experiment, the drive wheels are regulated such that the axis of rotation of the sphere cycles through 360°. The experiments show that, in the absence of a ground load, the kinematic model produces the desired motion, through the majority of the operating range, as shown in Figure 2. However, tracking errors occur whenever heading direction is a multiple of 90 degrees, which corresponds to the sphere rotation axis entering a wheel contact patch.

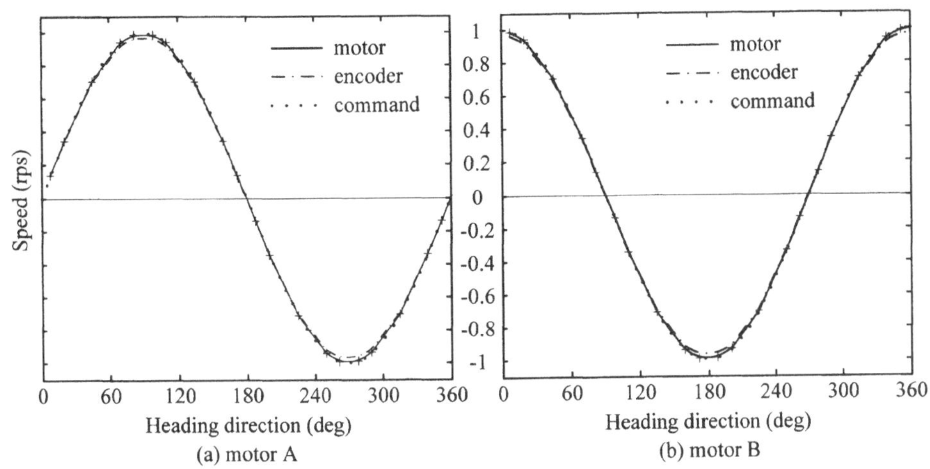

Figure 2. Plots showing ball wheel tracking accuracy.

As the drive wheels press against the sphere, the relatively soft polymers deform, generating a finite contact patch, which is necessary for traction. Due to the geometry difference between the sphere and the drive wheel, slip is always present but is a maximum when the sphere rotation axis enters the contact patch. According to Johnson (1985) this spin motion within the contact patch produces lateral traction and microslip. Gillespie et al. (2002) investigated rolling contact with spin on a spherical continuously variable transmission. Their study shows that extreme transmission ratios, which occur when the rotation axis approaches the drive wheel contact patch, are difficult to regulate.

Practically, spin motion generates shear forces that act as a disturbance moment on the dynamics of the sphere. As a result, the complimentary drive wheel must compensate for the disturbance by generating more torque. The cost of maintaining contact in this configuration is power dissipation and excessive wear on both the sphere and the drive wheel. Active control of the contact pressure allows the drive wheel to be retracted from the sphere in this situation, thereby eliminating the disturbance moment. However, once a drive wheel is retracted, controllability of the sphere in that direction is compromised.

To avoid this configuration, a third drive assembly is added to the mechanism and the contact pressure between the sphere and each drive wheel is independently regulated. This allows the contact pressure to be released when the sphere rotation axis approaches a contact patch.

Moreover, the ability to reconfigure the drive wheels allows the rotation axis to be placed between optimal pairs of drive wheels at all times. The mobile proof of concept is shown in Figure 1(b), while not instrumented for feedback, demonstrates the enhanced maneuverability and robustness of the ball wheel mechanism. Configured for human-in-the-loop operation, the operator provides the feedback and compensation at the joystick to generate the desired trajectory. Robustness is demonstrated by climbing steps, with a net height of 10 cm (4 in), as well as climbing over, and maneuvering around, other obstacles representative of a worksite, e.g. cables, tools, etc. Robust, isotropic omnidirectionality was demonstrated by climbing the steps from an arbitrary orientation in an arbitrary direction.

3 Kinematics

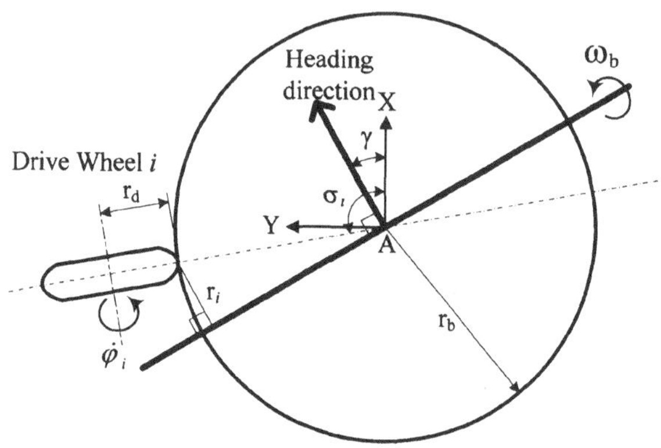

Figure 3. A schematic of ball wheel kinematics.

The schematic model of the ball wheel mechanism is shown in Figure 3. A reference frame is instantaneously fixed on the ground and the origin A of the frame is located at the contact point of the ball wheel and the ground. With the rolling without slip condition, the angular velocity of the ball wheel and that of the drive wheel i are related as follows:

$$\omega_b = \frac{r_d}{r_i} \cdot \dot{\varphi}_i \qquad (1)$$
$$r_i = r_b \cdot \cos(\sigma_i - \gamma)$$

where ω_b denotes the angular velocity of the ball wheel, r_d denotes the radius of the drive wheel, r_i denotes the distance from the ball rotation axis to the contact point between the ball wheel and the drive wheel, $\dot{\varphi}_i$ denotes the angular velocity of drive wheel i, r_b denotes the radius of the ball wheel, the drive wheel angle σ_i denotes the angle between the X axis and the wheel plane, and γ denotes the angle between the X axis and the heading direction of the ball wheel. The velocities $\dot{x}$ and $\dot{y}$ of the center of the ball wheel are

$$\begin{bmatrix} \dot{x} \\ \dot{y} \end{bmatrix} = \begin{bmatrix} r_b \cdot \cos\gamma \cdot \omega_b \\ r_b \cdot \sin\gamma \cdot \omega_b \end{bmatrix}. \tag{2}$$

In the case of three drive wheels on one ball wheel, the drive wheels are equally spaced on the equatorial plane such that $\sigma_1 = 0°$, $\sigma_2 = 120°$ and $\sigma_3 = 240°$. The angular velocities of the drive wheels can be derived from combining Eqns. (1) and (2), which are written as

$$\begin{bmatrix} \dot{\varphi}_1 \\ \dot{\varphi}_2 \\ \dot{\varphi}_3 \end{bmatrix} = \frac{1}{r_d} \begin{bmatrix} \cos(\sigma_1) & \sin(\sigma_1) \\ \cos(\sigma_2) & \sin(\sigma_2) \\ \cos(\sigma_3) & \sin(\sigma_3) \end{bmatrix} \begin{bmatrix} \dot{x} \\ \dot{y} \end{bmatrix}, \text{ or} \tag{3}$$

$$\dot{\mathbf{\Phi}} = \mathbf{C}\,\dot{\mathbf{x}}.$$

Equation (3) shows that any desired planar velocity can be achieved by specifying three angular speeds. Though a single ball wheel has the ability to move in any direction with arbitrary speed, it is not omnidirectional because orientation is neither defined nor controlled. However, a platform with two ball wheels is fully omnidirectional and the velocity vector of each mechanism can be fully prescribed.

4 Control

By appropriate definitions, the dynamic equations of motion can be rewritten as

$$\mathbf{M}\,\ddot{\mathbf{x}} + \mathbf{D}\,\dot{\mathbf{x}} + \mathbf{F}_f\,\mathrm{sgn}(\dot{\mathbf{x}}) = \mathbf{B}\,\mathbf{u}, \tag{4}$$

where **M** denotes the inertia matrix, **D** denotes the damping forces, $\mathbf{F}_f$ denotes the friction forces between the drive wheels and the ball, **B** denotes the input transformation matrix, and $\mathbf{u} = [f_x\ f_y]^T$ denotes the input torques. The control input **u** are chosen as follows:

$$\mathbf{u} = \overline{\mathbf{M}}\left(\ddot{\mathbf{x}}_m + 2\mathbf{\Lambda}\dot{\tilde{\mathbf{x}}} + \mathbf{\Lambda}^2\tilde{\mathbf{x}}\right) + \overline{\mathbf{D}}\,\dot{\mathbf{x}} + \overline{\mathbf{F}}_f\,\mathrm{sgn}(\dot{\mathbf{x}}), \tag{5}$$

where $\overline{\mathbf{M}}$, $\overline{\mathbf{D}}$, and $\overline{\mathbf{F}}_f$ are estimations of $\mathbf{M}$, $\mathbf{D}$, and $\mathbf{F}_f$, respectively. The method of estimation of $\mathbf{F}_f$ is beyond the scope of this paper, so it is assumed known herein. $\mathbf{x}_m$ is a vector of desired positions, $\tilde{\mathbf{x}} = \mathbf{x}_m - \mathbf{x}$, and $\mathbf{\Lambda}$ is a positive definite matrix. The control law, Eqn. (5), derives the operational space forces $\mathbf{F} = [f_x\ f_y]^T$, which should be produced by the three motors connected to the drive wheels.

One method to distribute the operational force **F** to the joint space torque when a system has redundancy in joint space is using the kinematic relation (Holmberg and Khatib, 2000), shown in Equation (3). However, with reconfiguring, the proposed ball wheel system is not redundant at all times. When the heading direction is around ± 90 degrees from the i'th drive wheel angle, the i'th drive wheel should be released. Then, the kinematic relation between the operational space and the joint space is as follows:

$$\begin{bmatrix} \dot{\varphi}_j \\ \dot{\varphi}_k \end{bmatrix} = \frac{1}{r_d} \begin{bmatrix} \cos(\sigma_j) & \sin(\sigma_j) \\ \cos(\sigma_k) & \sin(\sigma_k) \end{bmatrix} \begin{bmatrix} \dot{x} \\ \dot{y} \end{bmatrix}, \text{ or} \qquad (6)$$
$$\dot{\mathbf{\Phi}} = \mathbf{C}_{jk}\, \dot{\mathbf{x}}\,.$$

The joint space torques can be derived by the following equation

$$\mathbf{T}_{jk} = \mathbf{C}_{jk}^{-T}\, \mathbf{F}$$
$$\tau_i = 0 \qquad (7)$$

When the heading direction is away from this region, the operational space forces should be distributed to all three motors optimally, which means that the total torque produced by the motors should be minimized, and the variations in input torque for each motor also should be minimized. To solve this optimization problem, the cost function and the equality constraints are formed as

$$\min_{\mathbf{T}} f(\mathbf{T}) = \|\mathbf{W}\mathbf{T}\|^2$$
$$\mathbf{F} = \mathbf{C}^T\, \mathbf{T} \qquad (8)$$

where $\mathbf{W}$ is a diagonal matrix, which consists of $\varpi_i = \tan(\gamma - \sigma_i)$. The matrix $\mathbf{W}$ applies an infinite weight on motor i, when the heading direction is perpendicular to the drive wheel angle, and a zero weight on motor j, when the ball wheel's rotation axis is parallel to the drive wheel angle σ_j. Therefore, the solution of this optimization guarantees the minimum total torques and the minimum torque variations. Note that this solution is closed-form and not iterative and therefore its applicable to real-time control.

5 Simulation

To show the applicability of the control method, a simulation is developed using SIMULINK. The desired trajectory is a 4 meter (13 ft) diameter circular path. The ball wheel system starts from the center of the circle and follows the trajectory without changing its orientation. The proposed optimal torque distribution method is compared with the method using the pseudo-inverse of the kinematic constraint, which means that instead of solving the optimization problem, Eqn. (8), the following equation is used for deriving the motor torques:

$$\mathbf{T} = \mathbf{C}^{+T}\, \mathbf{F}\,. \qquad (9)$$

Note that $\mathbf{C}^{+T} = (\mathbf{C}^T(\mathbf{C}\mathbf{C}^T)^{-1})^T$ is the pseudo-inverse transpose. The desired and simulated positions of the center of the ball wheel system are shown in Figure 4(a). Figure 4(b) shows the ball wheel position errors; the dashed line is the result using the proposed torque distribution (TD) method, and the solid line is the result using the pseudo-inverse (PI) method. Figure 4(b) shows that errors increased periodically with the PI method while the TD method reaches a steady value. This phenomenon is caused by discontinuous motor torques, as shown in Figure 5. Sudden changes of motor torque can produce slip between the ball wheel and the drive wheels, and can trigger high frequency dynamics of the system. The norms of the motor torques with the two methods are almost the same at each sampling time.

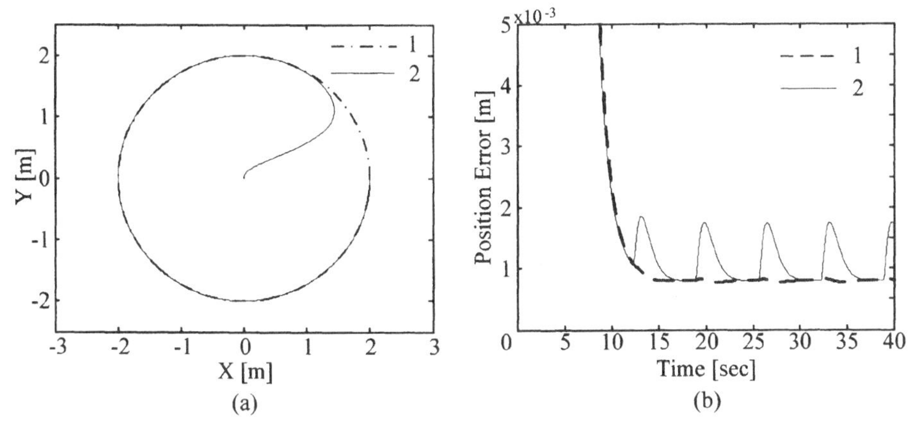

Figure 4. (a) Position-1:Desired position, 2:Ball wheel position (b) Position error-1:Proposed method, 2:Pseudo inverse method.

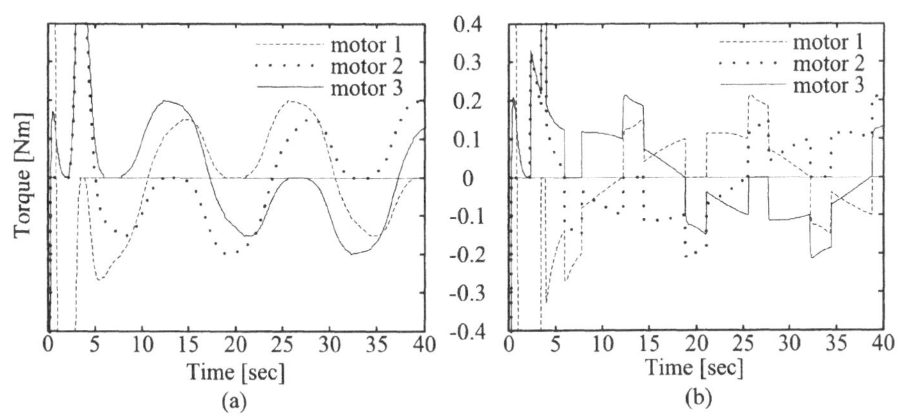

Figure 5. Motor Torque from (a) Proposed method and (b) Pseudo-inverse method.

6 Conclusions

In this paper, a robust ball wheel mechanism for an omnidirectional wheeled mobile platform was presented. The system has dual active drive axes and an optimally-controlled, reconfigurable drive system. It was shown through simulation that the proposed method out performs the conventional pseudo-inverse method in terms of position error. In real applications, the discontinuities in the pseudo-inverse method will lead to slip and increased positioning error.

Acknowledgements

The authors gratefully acknowledge the Division of Research and Innovation of the California Department of Transportation for their support of this work through the Advanced Highway Maintenance and Construction Technology (AHMCT) Research Center at the University of California, Davis.

Bibliography

Byun, K. S., Kim, S. J., and Song, J. B. (2002). Design of a four-wheeled omnidirectional mobile robot with variable wheel arrangement mechanism. In *Proceedings of the IEEE International Conference on Robotics and Automation*, 720-725.

Ferriere, L., Raucent, B., and Samin, J. C. (1997). Rollmobs, a new omnimobile robot. In *Proceedings of the IEEE/RSJ International Conference on Intelligent Robots and Systems*, 913-918.

Gillespie, R. B., Moore, C. A., Peshkin M., and Colgate, J. E. (2002). Kinematic creep in a continuously variable transmission: Traction drive mechanics for cobots. *Journal of Mechanical Design* 124,713-722.

Holmberg, R. and Khatib, O. (2000). Development and control of a holonomic mobile robot for mobile manipulation tasks. *The International Journal of Robotics Research* 19,(11):1066-1074.

Johnson, K. L. (1985). *Contact mechanics*. Cambridge University Press.

Moore, K. L. and Flann, N. S. (2000). A six-wheeled omnidirectional autonomous mobile robot. *Control Systems Magazine, IEEE* 20,(6):53-66.

Muir, P. and Neuman, C. (1987). Kinematic modeling for feedback control of an omnidirectional wheeled mobile robot. In *Proceedings of the Robotics and Automation. Proceedings. 1987 IEEE International Conference on*, 1772-1778.

Tahboub, K. A. and Asada, H. H. (2002). Dynamics analysis and control of a holonomic vehicle with a continuously variable transmission. *Journal of Dynamic Systems, Measurement, and Control* 124,(1):118-126.

Wada, M. and Mori, S. (1996). Holonomic and omnidirectional vehicle with conventional tires. In *Proceedings of the IEEE International Conference on Robotics and Automation*, 3671-3676.

Yu, H., Spenko, M., and Dubowsky, S. (2004). Omni-directional mobility using active split offset castors. *Journals of Mechanical Design*, 126, 822-829.

Chapter VIII

Space Robotics

Results and Verification of Spacecraft Docking Emulation using Hardware-in-the-Loop Simulation

Sébastien Laurier Chapleau[†‡], Éric Martin[†*] and Luc Baron[‡]

[†] Canadian Space Agency, St-Hubert (QC), Canada, e-mail: Eric.Martin@space.gc.ca
[‡] École Polytechnique, Montréal (QC), Canada,
e-mail: Sebastien.Laurier-Chapleau@polymtl.ca e-mail: Luc.Baron@polymtl.ca

Abstract. The interest in new demonstration missions of space servicing is decidedly rising in this beginning of the twenty-first century, as attested by several well-known projects like the Orbital Express, TECSAS, CX-OLEV and others. Such missions require a thorough knowledge of the mechanical behaviour of spacecrafts during docking or berthing. In order to adequately study the dynamics involved, a number of docking emulation test-beds have been elaborated worldwide and tested for their efficiency in providing realistic results, thereby preventing an eventual failure of the docking process. The Canadian Space Agency is developing such a test-bed for spacecraft docking emulation by using hardware-in-the-loop simulation. This paper describes this test-bed and presents the verification procedure used to verify the functionality of this type of simulation.

1 Introduction

As attested by such demonstration missions as the Orbital Express (Potter, 2002), TECSAS (Martin, Dupuis et al., 2005) and CX-OLEV (http://www.orbitalrecovery.com), there is a worldwide interest in space servicing. These missions involve the docking or berthing of spacecrafts and necessitate the anticipation of their mechanical behaviour during impact before performing those operations in space. To this end, analyses can be done on ground by three different docking system emulations: numerical simulation, hardware test-bed and hybrid test-bed (Yokota et al., 1998).

For numerical simulations, mathematical models are used to simulate the behaviour of the spacecrafts during contact. One such model is the Contact Dynamics Toolkit (CDT) developed by MDA (Ma, 2000). This toolkit can perform different types of contact simulation such as bouncing, sliding, rolling, spinning, sticking, and jamming. It is a useful model for the modelling and simulation of intermittent contact/constrained dynamics of mechanical bodies (Martin, Parsa et al., 2005). The Canadian Space Agency (CSA) is also developing a contact module to accurately model complex contact behaviours (Gonthier et al., 2004). The major disadvantage of numerical simulations is that they are very sensitive to the variation of mechanical properties such as the geometry and the surface properties of the model, for example the friction coefficient. Consequently, these properties have to be defined with precision, and they can be difficult to evaluate.

One alternative is the use of a hardware test-bed with free-floating physical mockups to emulate the dynamic movements and determine the contact forces during the docking of spacecrafts. Examples of such models are the Air-Bearing Mobility Simulators (ABMS) of the George C. Mashall Space Flight Centre (MSFC) (Hays et al., 2003), the Experimental Free-Floating Robot Satellite Simulator (EFFORTS) of the Tokyo Institute of Technology (TIT) (Yoshida, 2003) and the MDA Space Missions capture test-bed. When the effect of earth gravity and friction with the exterior is eliminated effectively, this kind of test-bed provides more realistic data than numerical simulations, but is less effective for the validation of a large range of spacecraft configurations and cannot be used with all real spacecraft prototypes (Uchiyama et al., 2003).

Hybrid test-beds use mockups to measure the contact forces, and numerical simulations to compute the trajectory of one or both spacecrafts. Example of such test-beds is the Rendezvous and Docking Operation Test System (RDOTS) of the Japan Aerospace eXploration Agency (JAXA) consisting of a nine-DOF system obtained by a six-DOF parallel platform and a rail-mounted chaser base with two additional degrees of freedom. Several hybrid test-beds using rail and air-table do not permit all DOF motion, and the precision of the results are often affected by friction and inconstant inertia of mockups.

To bypass these problems, the CSA is developing a test-bed for spacecraft docking emulation by using Hardware-in-the-Loop Simulation (HLS) to provide a technology that facilitates the study of satellite docking capabilities. This test-bed, which is described in detail in Section 3, uses an existing test-bed, namely the Special Purpose Dextrous Manipulator (SPDM) Task Verification Facility (STVF) presented in Section 2. Section 4 describes the experimental results that were obtained and the verification processes used.

2 STVF approach

As a partner in the International Space Station (ISS), Canada is responsible for the verification of all tasks involving the SPDM. One component of the verification process is performed using the STVF, which is located at the John H. Chapman space centre. The STVF Manipulator Test-bed (SMT) shown in Fig. 1 is used to refine the analysis of the contact portion of the SPDM tasks before they are conducted in space. Aside from the rigid robot SMT and its controller, the HLS test-bed consists of a simulation of the Space Station Remote Manipulator System (SSRMS) and the SPDM dynamics, as well as a visualisation engine. The motion of the SPDM is controlled by an operator through a simulation engine that generates the endpoint trajectory of the SPDM, which is used as a setpoint for the SMT Robot controller

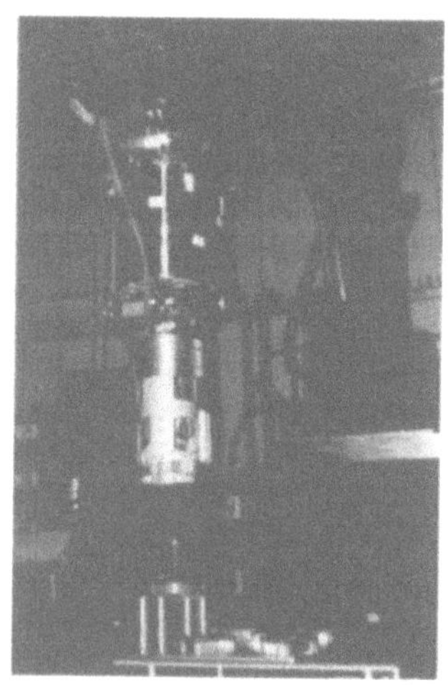

Figure 1. SMT Robot with the MDA end-effector.

to command the SMT end-effector's position.

To allow the dynamics simulation to react to external contact forces, these forces are measured using a force/moment sensor and fed back to the dynamics simulation model. This approach can be used for different types of space robots and can effectively emulate such phenomenons as the vibrations of the space-robot base. The controlled ground robot must be transparent in the frequency band of contact tests, and is made to behave like the simulated space robot by forcing it to track the Cartesian acceleration of the latter. In addition, in order to correct positioning errors, Cartesian position and velocity feedback are used. In the bandwidth of interest for the SPDM, experimentations give good performance and stability (Martin, Doyon et al., 2005).

3 Docking emulation approach

To further develop the HLS approach, the CSA projected to implement this type of simulation to emulate the docking phase of spacecrafts floating in a zero-gravity environment by using STVF as a test environment. This project requires the implementation of a spacecraft simulator in the STVF environment. The major functionalities of the STVF test-bed were kept intact and only the dynamic simulation part was changed. The SPDM simulation was modified to simulate the relative motion of the chaser and target spacecrafts. The contact forces occurring between the chaser's end-effector installed on the SMT Robot, and the target's grapple fixture positioned on the fixed force/moment sensor are measured and applied as external forces on both spacecraft models in the simulator. As shown in Fig. 1, the MDA end-effector is mounted on the SMT Robot as an experimental prototype of docking mechanism.

The dynamics of the chaser and target spacecrafts are modelled with SYMOFROS (L'Archevêque et al., 2000) in the Docking Simulator, within the MATLAB/SIMULINK environment. Contrary to the target model, which is a rigid body, the chaser model has the particularity of having a 6-DOF spring and damper compliant mechanism between its body and its end-effector. This mechanism allows for the docking of two spacecrafts even if the chaser has some misalignment approach. It could also represent the compliance of an uncontrolled robot arm. The simulator also includes the necessary functions to determine the relative trajectory of the chaser frame with respect to the target frame required to perform HLS.

The Docking Simulator has the capabilities of running not only in HLS mode but also with a contact dynamics model based on the CDT of MDA (introduced in Section 1). The operator can use both methods by toggling a switch in the command console. More detailed information is given in Martin, Doyon et al. (2005).

4 Experimental tests and verifications

This section presents the Docking Simulator verification procedures and the results obtained. Subsection 4.1 presents the results obtained with contact dynamics simulation as the active mode and Subsection 4.2 describes the verification procedures and results obtained in HLS mode.

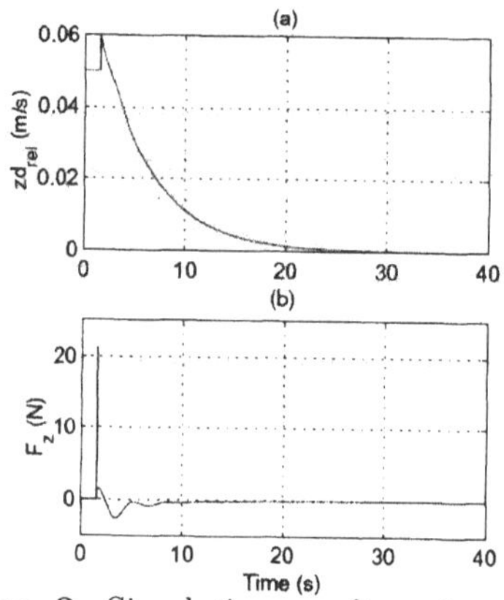

Figure 2. Simulation results using CDT: (a) Relative normal velocity; (b) Normal contact force; under conditions of Table 1.

Table 1. Simulation parameters for Fig. 2.

Parameter	Value
Approach vel.	5 cm/s
Target mass	200 kg
Chaser mass	500 kg
End-effecteor mass	6 kg
Compliance f_n	1,6 Hz
Compl. damp. ratio (ζ)	1,0

4.1 Pure simulation mode

In order to verify the Docking Simulator using the contact dynamics mode, the contact parameters and the geometry of the end-effectors shown in Fig. 1 were provided to CSA by MDA Space Missions. When no contact occurs between the chaser and the target spacecrafts, the verification is performed by using analytic results as a reference.

When contact does occur during a simulation, the analytic results are virtually incalculable and there is no experimental data from an actual docking of spacecrafts available for comparison. Therefore, the accuracy of the simulator cannot be precisely assessed. Based on engineering judgment, however, it is possible to evaluate the validity of the results. For example, Fig. 2 shows the results obtained with the Docking Simulator using the CDT. The parameters used for this simulation are presented in Table 1. As shown in Fig. 2 (a), the relative velocity of the spacecrafts is null after the docking phase. More results will be presented in Subsection 4.2 together with the results obtained in HLS mode.

4.2 HLS mode

This section presents the verification procedure used for the testing of the HLS mode of the Docking Simulator presented in Section 3. After the Docking Simulator was incorporated in the STVF model, tests in HLS mode with the MDA end-effector mounted on the SMT Robot (see Fig. 1) showed instability when contact occurred.

To determine the source of this instability, tests were performed under direct central impact on the sensor force, using a round peg as an end-effector, which has the advantage of having both a more robust structure and less complex geometry. This new configuration served to demonstrate that the system shows more signs of instability with certain inertial properties and approach speed configurations, particularly in the case of small spacecrafts. Fig. 3 shows the time history of the force for the simulation (a) and experimental (b) tests with the properties shown in Table 2. After calculation of the

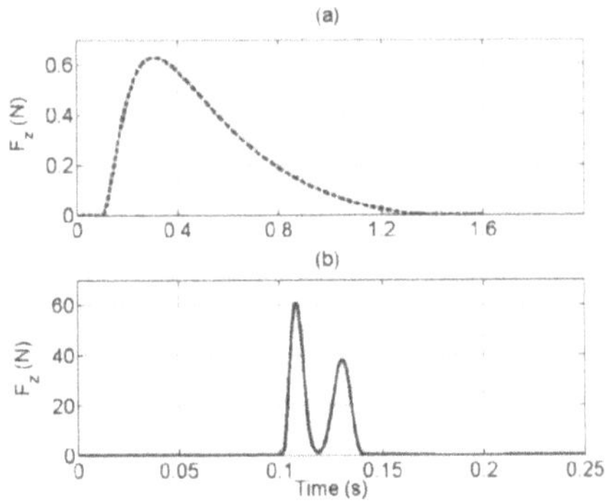

Figure 3. Measured normal contact force: (a) Pure simulation mode; (b) HLS mode; under conditions of Table 2.

Table 2. Spacecrafts configurations for Figs. 3 and 5.

Parameter	Value
Approach vel.	0.2 cm/s
Target mass	90.7 kg
Chaser mass	250 kg
End-effector mass	6 kg
Compliance f_n	1.5 Hz
Compl. damp. ratio (ζ)	1.0

areas under the graphs, which are related to the momentum of each body, it is apparent that, for the experimental test, the linear momentum of each spacecraft had a bigger variation. It is important to note that the force applied to the spacecrafts is different from the measured force because of the fact that a first-order filter is used to eliminate undesired frequency bandwidth, as shown in Fig. 4. In this segment of the verification tests, the cut-off frequency is kept identical to the one used in the STVF model, i.e. 4 Hz.

The conservation of the kinetic energy of the system was a condition used for the verification of the experimental data. The kinetic energy at time t', $K(t')$, for a one dimensional movement is calculated as

$$K(t') = \frac{1}{2} M_{\text{target}} V_{\text{target}}^2(t') + \frac{1}{2} M_{\text{chaser}} V_{\text{chaser}}^2(t'). \tag{4.1}$$

where M and $V(t')$ are respectively the mass and the velocity of the target and the chaser spacecrafts at time t'. Using Eq. (4.1), Fig. 5 shows the kinetic energy of the system using the parameters of Table 2. The solid (blue) line represents the experimental kinetic energy and the dashed (red) line, the simulation. The velocities used for the calculation of the kinetic energy in Fig. 5 are calculated in function of the end-effectors' velocities. Since the chaser is constituted by two rigid bodies, we cannot use the transient part as the real kinetic energy of the system. However after stabilization of the chaser's compliant mechanism at approximately 6 seconds, the velocities of both the end-effector and the spacecraft are the same. As shown in Fig. 5, the final kinetic energy in the experimental test is higher than the initial kinetic energy, which is impossible according to the first law of thermodynamics. The kinetic energy reduction in the simulation test is mainly due to the damping mechanism of the chaser and the restitution coefficient during the impact of the end-effectors.

As the analytic calculation of the real energy loss resulting of the damping mechanism of the chaser is very complex, experimental tests were performed for a direct central impact (without the spring/damper mechanism of the chaser) of the peg probe on the force/moment sensor. Fig. 6 (a) shows the kinetic energy ratio of the system after impact by varying the mass of the spacecrafts for different approach velocities on this series of

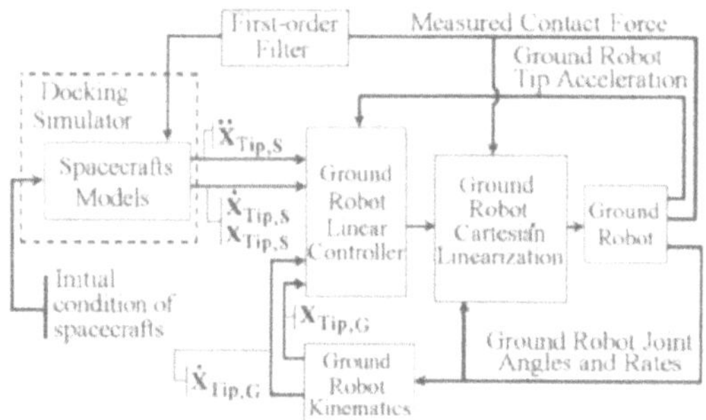

Figure 4. First-order filter impact on measured forces.

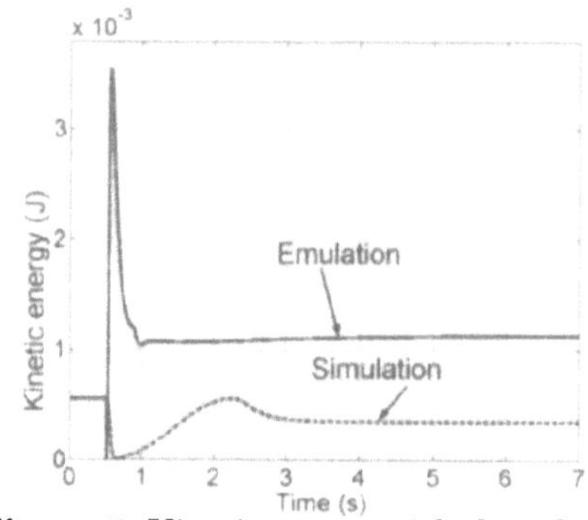

Figure 5. Kinetic energy with the vel. of the end-effectors under conditions of Table 2.

tests. Fig. 6 (b) shows the kinetic energy ratio of the system by varying the spacecraft masses for different cut-off frequencies of the first-order filter, with an approach velocity of 5 mm/s. It is to be noted that an equal mass was used for the two spacecrafts in this series of tests.

The fact that these tests were conducted under very small impact velocity and with high rigidity surfaces allows us to assume that in reality, almost no kinetic energy should be lost during collision. This is equivalent to having a kinetic energy ratio of one. The kinetic energy ratio higher than one as shown in Fig. 6 illustrates that the emulation made with the docking test-bed does not provide conclusive results. One possible reason for this inconsistent data is that the test-bed that was used is made up of a closed-loop emulation (the SMT Robot is fixed to the ground, as is the forces/moments sensor), contrary to free-floating spacecrafts that consist of an open-loop system (no parts retain the two spacecrafts together). Consequently, the STVF configuration has some unknown internal forces resulting from contact that create disturbance in the forces to be applied on spacecrafts. This makes the impedance of the test-bed used different from the impedance of the free-floating spacecraft system to be emulated. The impedance matching of the emulating robot to the free-floating spacecraft system is required to obtain a good level of fidelity (Aghili and Namvar, 2004). However, this process is relatively complex and is limited by the bandwidth of the robot under consideration.

As shown in Fig. 6 (a), when a small mass is used for the spacecrafts the energy ratio is higher than one. One of the reasons for this is the fact that the first force peak obtained in HLS mode, as shown in Fig. 3 (b) between time 0.10s and 0.15s, remains the same, regardless of the inertial configurations and approach velocity. However, as shown in Fig. 3 (a), the expected force is much lower than the applied force of 60N. Applying this non-representative force on the two spacecrafts adds energy to the system, which explains that the energy ratio is higher than one.

Under that assumption, increasing the mass of the spacecrafts should improve the performance so that the non-representative force has less impact. A different sequence of tests done with higher masses demonstrates that once again, an energy ratio higher than one is observed. The explanation in this case is that, due to errors in the controller (wrong impedance), the integrator is building up during contact and needs to discharge

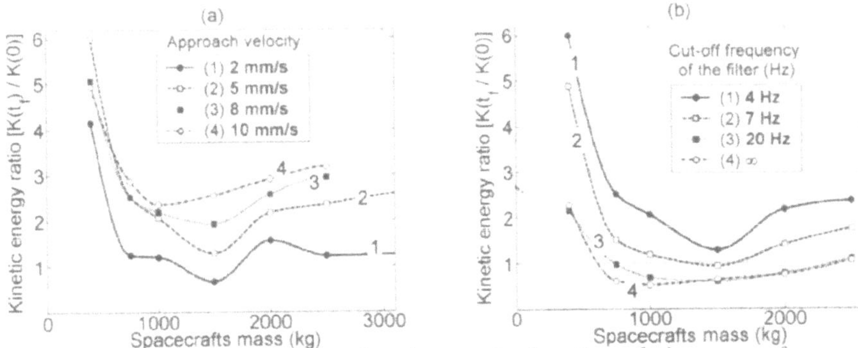

Figure 6. Kinetic energy ratio after impact in function of the mass of spacecrafts: (a) for various approach velocities; (b) for various cut-off frequencies of the first-order filter; $M_{\text{chaser}} = M_{\text{target}}$.

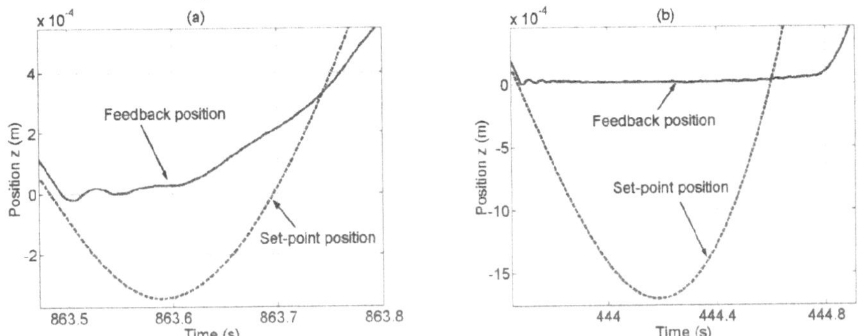

Figure 7. Positioning delay due to the integrator with spacecrafts mass of: (a) 400kg; (b) 8000kg; with an approach velocity of 5 mm/s.

after the separation of the spacecrafts. For this reason the robot continues to apply a force although it should have separated from the force/moment sensor, as shown in Fig. 7 (b), and this results in an accumulation of energy in the systems. As shown in Fig. 6 (b), the kinetic energy ratio can be considerably reduced by decreasing the cut-off frequency of the first-order filter. By this verification process, it is apparent that the SMT Robot is not precise enough for the emulation of spacecraft docking; perhaps a parallel platform with a higher bandwidth could provide better results.

5 Conclusion

Several space missions involving contacts between chaser and target spacecrafts are currently in development. Ground-based validation of the procedures for such demonstrations and the physical properties of the spacecrafts used are important to the success of these missions. Therefore, the CSA developed a Docking Simulator to study the phases of contact during the docking, berthing or capture phases. This simulator can be used in pure simulation mode as well as in HLS mode. The HLS test-bed borrowed the concept of the STVF, the verification facility developed by the CSA to verify the contact portion of the SPDM robot's tasks on the ISS. The simulation mode of the Docking Simulator does provide promising results. However, when tests are performed in HLS mode, inconsistent data show that the simulator cannot be used with the SMT Robot, as the impedance matching is not sufficiently accurate. A faster and more precise robot, such as a parallel platform with higher bandwidth, could enhance results.

Bibliography

F. Aghili and M. Namvar, *A robust impedance matching scheme for emulation of robots*, In Proceedings of 2004 IEE/RSJ, vol.3, pages 2142–2148, Sept. 28 2004.

R. L'Archevêque, M. Doyon, J.-C. Piedboeuf and Y. Gonthier, *SYMOFROS: Software architecture and real time issues*, In DASIA 2000 - Data systems in Aerospace, vol. SP-457, Montréal, Canada: ESA, pages 41–46, May 22-26 2000.

Y. Gonthier, J. McPhee, C. Lange and J.-C. Piedboeuf, *A regularized contact model with asymetric damping and dwell-time dependent friction*, In Multibody System Dynamics, 11(3), pages 209–233, April 2004.

A. B. Hays, P. Tchoryk, Jr., J. C. Pavlich, and G. Wassick, *Dynamic simulation and validation of a satellite docking system*, In Proceedings of SPIE - Space Systems Technology and Operations, vol.5088, pages 77–88, Aug. 2003.

O. Ma, *CDT - a general contact dynamics toolkit*, In Proceedings of the 31st International Symposium on Robotics: ISR 2000, Montréal, Canada, pages 468–473, May 2000.

E. Martin, M. Doyon, Y. Gonthier and C. Lange, *Validation process of the STVF Hardware-in-the-Loop Simulation facility*, In Proc. of the 8th International Symposium on Artificial Intelligence and Robotics & Automation in Space: i-SAIRAS 2005, Munich, Germany, Sept. 5-8 2005.

E. Martin, E. Dupuis, J.-C. Piedboeuf and M. Doyon, *The TECSAS mission from a Canadian perspective*, In Proc. of the 8th International Symposium on Artificial Intelligence and Robotics & Automation in Space: i-SAIRAS 2005, Munich, Germany, Sept. 5-8 2005.

E. Martin, K. Parsa, S. Laurier Chapleau and L. Baron, *Towards spacecraft docking emulation using Hardware-in-the-Loop Simulation*, In Proc. of the 8th International Symposium on Artificial Intelligence and Robotics & Automation in Space: i-SAIRAS 2005, Munich, Germany, Sept. 5-8 2005.

S.D. Potter, *Orbital Express: Leading the way to a new space architecture*, In 2002 Space Core Tech Conf., pages 160–165, Nov. 2002.

M. Uchiyama, S. Tarao and H. Kawabe, *A New Class of Hybrid Motion Simulation Using a Very Fast Parallel Robot*, In Springer Tracts in Advanced Robotics, Robotics Research: The Tenth International Symposium, vol.6, pages 531–544, 2003.

K. Yokota, K. Yamanaka, S. Shirasaka, H. Koyama, M. Inoue, T. Shima, K. Yamada, *Evaluation of contact dynamics simulation fidelity of RDOTS (Rendezvous and Docking Operation Test System)*, In Proceedings of the twenty-first International symposium on Space Technology and Science, vol.1, pages 685–690, Omiya, Japan, 1998.

K. Yoshida, *Engineering Test Satellite VII Flight Experiments for Space Robot Dynamics and Control: Theories on Laboratory Test Beds Ten Years Ago, Now in Orbit*, In The International Journal of Robotics Research, vol.22, no.5, pages 321–335, 2003.

On Dynamic Analysis and Control of a Novel Orbital Debris Disposer

Georgios Rekleitis and Evangelos Papadopoulos[1]

National Technical University of Athens, Department of Mechanical Engineering
15780 Athens Greece
e-mail: {georek | egpapado@central.ntua.gr}

Abstract. The removal of orbital debris is a special on-orbit service that gains interest. In this paper, we present the novel idea of an orbital debris disposer, based on a number of cooperating robotic free-flyers equipped with a flexible surface, such as a net. A uni-dimensional model of the system is studied as a proof of concept, to help in the understanding of system dynamics and physical constraints. This understanding is used in the design of a control strategy aiming at successful debris captures. The response of a simulated system shows that in principle, the constrained control method can achieve the task, and is sufficiently robust in the presence of inaccurate debris parameters.

1 Introduction

During space activities, astronauts have been required to work on extra-vehicular activities (EVA). Most of these operations fall under the general area of on-orbit servicing (OOS) and include tasks like refueling, upgrading, assembling, and repairing of orbital systems. In recent years, OOS is studied as a potential field of operation for orbital robotic systems, (Landzettel et al., 2004), (Dupuis et al., 2004). Systems must be able to perform the EVA needed, relieving astronauts from hostile environment work, or must be able to perform tasks impossible by EVA, such as salvaging uncontrolled satellites. Examples of such systems include NASA's Robonaut, (Bluethmann et al., 2004), or JAXA's ETS-VII, (Yoshida, Hashizume and Abiko, 2001).

Decades of space program activities have burdened the orbital environment with thousands of orbital debris, sized from one square millimeter to one square meter or even larger. These are the products of several collisions as well as of various intentional or unintentional breakups (NASA, 2004). Adding to these the meteorites and other space objects that end up orbiting around the earth, it is easy to understand why the orbital environment includes tens of thousands of orbiting debris (NASA, 2005). These objects pose serious dangers to operating orbital systems and their presence has initiated several studies to estimate the impact results (Lacome, et al. 2002). The actions taken up to now include tracking large debris and imposing avoidance maneuvers on threatened systems, while letting small ones hit a system's armor. Recently, the danger of collision with debris has become a serious concern. Methods for reducing man-made orbital debris, or for debris mitigation have been studied (Smith, D.A. et al, 2003). A special OOS activity, the removal of debris from the orbital environment, has emerged, and methods such as

[1] Support of this work by the EPAN Cooperation Program 4.3.6.1.b (Greece-USA 035) of the Hellenic General Secretariat for Research and Technology is acknowledged.

re-orbiting debris at much higher altitudes, (Todd and Bowling, 2004), or pushing the debris into catastrophic re-entry, (Patera and Ailor, 1998) are being proposed.

In this paper, the novel idea of an orbital debris disposer is proposed. A number of cooperating robotic free-flyers equipped with a flexible surface, such as a net, are controlled aiming at debris capture. A uni-dimensional model of the system is studied, to help in the understanding of system dynamics and of the physical constraints involved, and in designing control algorithms. A control method is developed, that after contact, forces the robotic system and the object to move with the same speed, keeping thus the debris on the net. Simulation results validate the control method, which is robust in the presence of debris parameter uncertainty.

2 Concept Overview

A main problem in capturing orbital debris is that there is no predetermined way of grabbing them, since they include no attachment points. Moreover, their motion can be complex, as they may be tumbling in orbit, making it even more difficult to track and capture, even if their motion itself is predictable. This work studies a method of capturing orbital debris, employing a coordinated swarm of free-flying robots. Using their manipulators, these free-flyers move a rigid ring with a net, and fly before the debris with a relative velocity such that the debris is slowly approaching the net, see Fig. 1. As the debris is approaching, for safety and fuel economy reasons, the robots turn to free-floating mode, i.e. their satellite base is uncontrolled (Papadopoulos and Dubowsky, 1991), and wait for the collision. The collision must be such that the robot system and the debris will continue moving together, with the debris in the net, thus making it much easier to remove it from the orbit. To do this, the robot system initial velocities, as well as the gains of the manipulator joint control algorithms must be chosen appropriately. In the present work, the study of a uni-dimensional model of the interaction is presented as a proof of concept, but also aiming at a better understanding of the behavior of such systems.

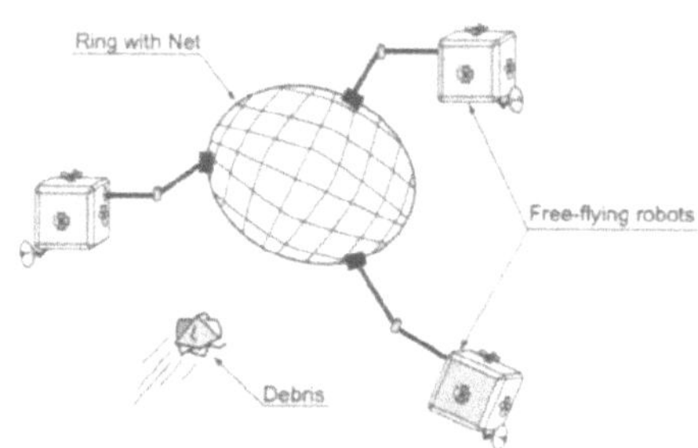

Figure 1. Graphic representation of the Orbital Debris Disposer.

3 Simplified Model Analysis

The system in Fig. 1 is modeled as a uni-dimensional three-body system, see Fig. 2. Body 1 represents the debris, body 2, the spring k, and the damper, b, represent the net, and the body 3, the free-flying satellites. The manipulators apply forces u. The natural spring length is x_0.

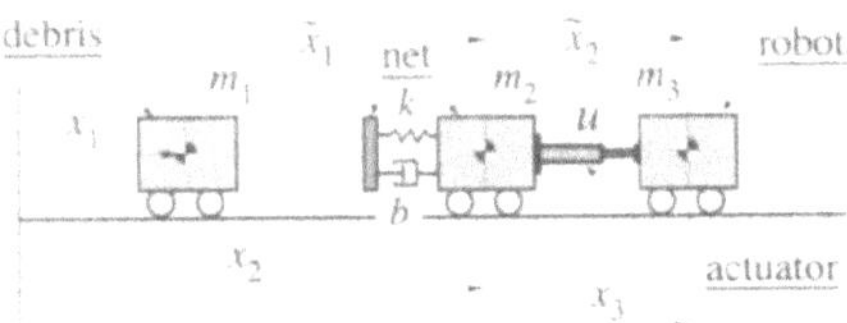

Figure 2. A 1D model of the orbital debris disposer.

The system equations of motion, obtained after the debris / net contact, are:

$$\begin{aligned} &m_1\ddot{x}_1 + b(\dot{x}_1 - \dot{x}_2) + k(x_0 - (x_2 - x_1)) = 0 \\ &m_2\ddot{x}_2 - b(\dot{x}_1 - \dot{x}_2) - k(x_0 - (x_2 - x_1)) = -u \\ &m_3\ddot{x}_3 \qquad\qquad\qquad\qquad\qquad\quad = u \end{aligned} \tag{1}$$

The above system has no external forces acting on it, therefore its linear momentum is conserved. This can be seen by adding all three of them and integrating the result once to obtain,

$$m_1\dot{x}_1 + m_2\dot{x}_2 + m_3\dot{x}_3 = L \tag{2}$$

where L is the system linear momentum. Note that if two of the velocities are driven to their desired values, the third will be given by Eq. (2). This allows us to study the first two equations of motion, in which x_3 is missing. Without loss of generality, we assume that $x_0 = 0$, and obtain,

$$\begin{aligned} &m_1\ddot{x}_1 + b(\dot{x}_1 - \dot{x}_2) + k(x_1 - x_2) = 0 \\ &m_2\ddot{x}_2 - b(\dot{x}_1 - \dot{x}_2) - k(x_1 - x_2) = -u \end{aligned} \tag{3}$$

4 Control Law and Constraints

The problem addressed next is how to actuate the manipulators so that after the collision transient, all bodies move with the *same* velocity. This is important, because otherwise, in the absence of mechanical handles or special latching mechanisms, after the impact, the debris will separate from the net and be lost before it is captured.

From an analytical point of view, this means that two relative speeds must be set with a single control force (underactuation). To achieve this, the control force must be such that the net (spring-damper) returns to its equilibrium point without overshooting, while at the same time the net and the debris acquire the desired speed. Should the spring-damper system overshoot the equilibrium, the debris will acquire a velocity resulting in its loss from the net.

The analysis starts by observing that if all body steady state velocities $\dot{x}_{i,ss}$ are equal, then they must be equal to the system center of mass (CM) velocity given by:

$$\dot{x}_{1,ss} = \dot{x}_{2,ss} = \dot{x}_{3,ss} = L / M \tag{4}$$

where M is the total system mass. This observation directs us to choose the controller force as:

$$u(\dot{x}_1, \dot{x}_2, L) = K_1\left(\dot{x}_2(t) - \dot{x}_1(t)\right) + K_2\left(\dot{x}_2(t) - L / M\right) \tag{5}$$

where K_1 and K_2 are unknown control gains. In Eq. (5), the first term attempts to make the relative velocity of the debris and the net zero, while the second one to make the absolute velocity of the net equal to the CM velocity. If two of the three bodies' velocities attain the velocity given by Eq. (4), then, Eq (2) guarantees that the third will also be the same. Therefore, in principle this controller can achieve the goal of equal velocities.

The general form for the response for Eqs. (3) and (5) is given by:

$$x_1(t) = c_{11}t + c_{12} + c_{13}e^{\lambda_1 t} + c_{14}e^{\lambda_2 t} + c_{15}e^{\lambda_3 t} \tag{6a}$$

$$x_2(t) = c_{21}t + c_{22} + c_{23}e^{\lambda_1 t} + c_{24}e^{\lambda_2 t} + c_{25}e^{\lambda_3 t} \tag{6b}$$

where the c_{ij} (i =1, 2, j = 1,...,5) coefficients depend on system parameters and initial conditions, while the λ_i eigenvalues depend on system parameters, only.

Next, the constraints under which the system will yield the desired response are studied. These constraints are classified as (a) "stability" constraints (SC), so that the velocities of the debris, the net and the robots will converge to certain real values, (b) "contact" constraints (CC), so that the debris will not separate from the net after the impact and, (c) "assumption" constraints (AC), so that the development of the previous types of constraints hold true. For example, in Eqs. (6) real λ_i's are assumed and this is an AC type of constraint. Different ACs result in different groups of constraints to be satisfied. SC, CC, and AC combinations are grouped together and used to compute feasible solutions. In more detail, using a set of initial conditions, namely the observed relative velocities, and the system parameters, such as the body masses and the spring-damper coefficients, the various groups of constraints result in some feasible control pairs K_1 and K_2. If at least one such pair exists, it will ensure the desired response of the system; otherwise, the approach velocity of the free-flyer / net system must be modified till a feasible pair is found. For brevity, here we present the analysis of one such group of constraints, only.

With the AC of real λ_i's, it can be shown that the following condition must be hold,

$$c_{11} = c_{21} = L / M \tag{7}$$

By differentiating Eqs. (6), while incorporating Eq. (7), it can be shown that all steady state velocities can be made equal to L / M if,

$$\lambda_1, \lambda_2, \lambda_3 < 0 \tag{8}$$

Eq. (8) yields the only set of SC constraints, which must be satisfied to attain the desired goal.

Next, the required CCs are derived. These result from the fact that, after the first time at which the net starts deforming, the debris must be in contact with the net. Using Eq. (7), Eqs. (6) yield the initial form of the single CC, (where for simplicity we take $x_0 = 0$):

$$x_2(t) - x_1(t) = (c_{22} - c_{12}) + \sum_{i=3}^{5}\left((c_{2i} - c_{1i})e^{\lambda_{i-2} t}\right) \leq x_0 = 0 \tag{9}$$

Next, Eq. (9) is analyzed aiming at obtaining conditions that are independent of time. To this aim, additional conditions are derived from Eq. (9). Different assumptions will lead to a different development of the contact constraint and, thus, to a different group of constraints.

The steady state of Eqs. (6) is reached at infinite time. Then, the following must hold,

$$\lim_{t\to\infty}\left(x_2(t)-x_1(t)\right)=x_0=0 \Rightarrow c_{22}-c_{12}=0 \tag{10}$$

Thus, the CC given by Eq. (9) becomes:

$$(c_{23}-c_{13})e^{\lambda_1 t}+(c_{24}-c_{14})e^{\lambda_2 t}+(c_{25}-c_{15})e^{\lambda_3 t}\le 0 \tag{11}$$

At $t=0$ (first contact) we have $x_2(0)-x_1(0)=x_0=0$, thus, from Eqs. (6), (7) and (10), we have:

$$(c_{23}-c_{13})+(c_{24}-c_{14})+(c_{25}-c_{15})=0 \tag{12}$$

It is clear that the three terms in Eq. (12) cannot have the same sign. Two possibilities exist here, i.e. we may have (a) two negative terms and one positive or (b) two positive terms and one negative. For brevity, we will present only case (b). Without loss of generality, we assume:

$$c_{23}-c_{13}<0, \quad c_{24}-c_{14},\ c_{25}-c_{15}\ge 0 \tag{13}$$

Inequalities in Eq. (13) yield another set of AC, which must hold in order for the following development of the CC given by Eq. (11) to hold. Dividing Eq. (11) by $c_{23}-c_{13}$, the CC yields:

$$\frac{c_{24}-c_{14}}{c_{23}-c_{13}}e^{(\lambda_2-\lambda_1)t}+\frac{c_{25}-c_{15}}{c_{23}-c_{13}}e^{(\lambda_3-\lambda_1)t}\ge -1 \tag{14}$$

From Eq. (12), we see that constraint (14) holds as equality for $t=0$. Thus, Eq. (14) holds true iff:

$$\lambda_2-\lambda_1<0 \text{ and } \lambda_3-\lambda_1<0 \tag{15}$$

Inequalities (15) are the only needed CC's in this case. SC (8), AC (13) and CC (15) along with the AC that λ_i are real, make the first group of constraints that, if satisfied, the system will have the desired response. For different assumptions, such as Ineqs. (13), different groups of constraints have been derived. As mentioned above, to each such group, a range of control gains may correspond and yield the desired response. To this end, for a set of system parameters and initial conditions, plots of a group of constraints such as of Eqs. (8), (13) and (15) are drawn in the gain space, for identifying the feasible subspaces in which the acceptable sets of control gains lie. If such a subspace does not exist, a different set of constraints is tried, until a feasible set of gains results. Otherwise, since the debris velocity cannot be changed, the approach velocity of the capturing system may have to be adjusted by firing its thrusters.

5 Simulation Results

The system described by Eq. (1) is simulated in MATLAB. The system parameters are m_1 = 20 kg, m_2 = 20 kg, m_3 = 140 kg, k = 500 kg/s^2, b = 40 kg/s, and x_0 = 0. The initial conditions are $x_1(0)$ = 200 m/s, $x_2(0)$ = 199 m/s and $x_3(0)$ = 196.5 m/s. The result of the impact of the three bodies, for a given set of system parameters and initial conditions, with random control gains, are shown in Fig. 3. As seen in Fig. 3 by the relative distance $x_2 - x_1$, the debris bounces off the net at the moment this distance exceeds x_0=0, and then they move with different speeds, parting rapidly.

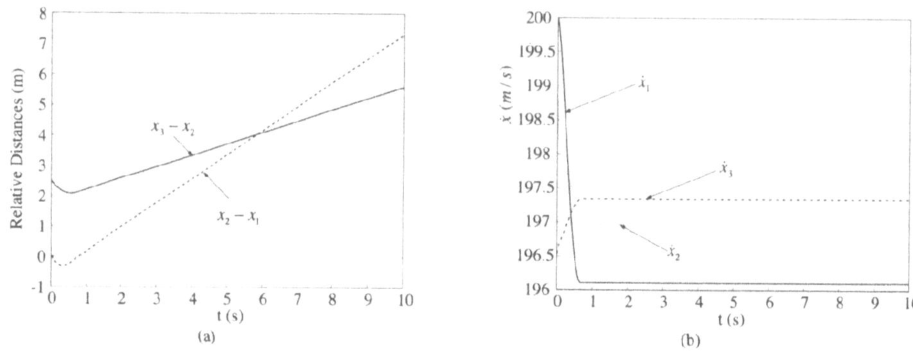

Figure 3. (a) Relative positions response and (b) velocities for control gains $K_1 = 80$ kg/s and $K_2 = 400$ kg/s, that do not satisfy the constraints for successful capture.

Next, the control gains are chosen according to the presented method, while the initial conditions and system parameters remain the same as before. Note that the presented gains are obtained by the same group of constraints as the one presented at the previous section. The velocity response of the three objects for this case is shown in Fig. 4, where we can see that they end up moving with the same speed $\dot{x}_1 = \dot{x}_2 = L / M$.

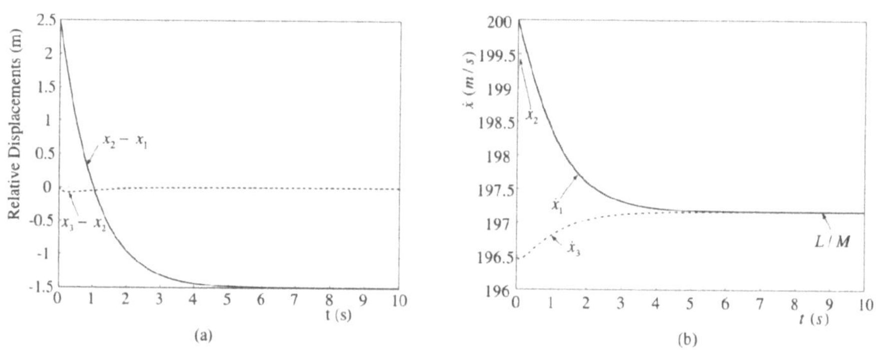

Figure 4. (a) Relative positions response and (b) velocities for control gains $K_1 = 250$ kg/s and $K_2 = 30$ kg/s, that satisfy the constraints for successful capture.

From Fig. 4, we can see that the distance between debris and net, does not exceed $x_0 = 0$ after the first impact, as desired, while the net is displaced for about -1.5*m* with respect to the robot base. Fig. 5 shows three snapshots of the system motion, namely the instant of the first contact, that of the net's maximum deformation and finally the instant in which the steady state is reached.

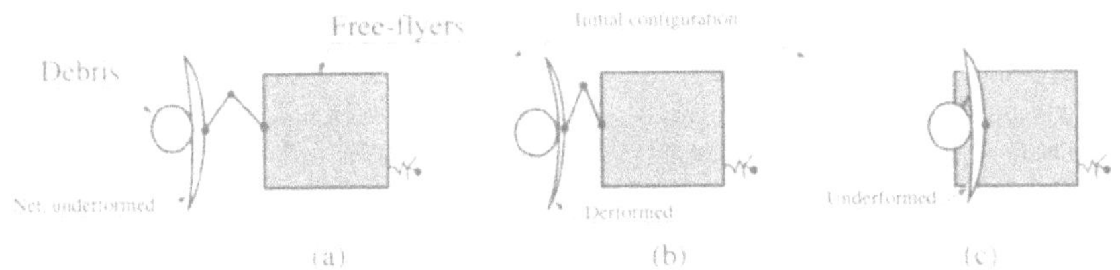

Figure 5. Motion snapshots of the orbital debris disposer during capture. (a) $t = 0.0$ s, first contact, (b) $t = 0.32$ s, net's maximum deformation, (c) $t = 7.8$ s, all bodies move together.

Next, the robustness of the controller strategy is investigated. Typically, the debris mass, its initial velocity, or both, may be inaccurately estimated. Thus, another simulation was run with estimated debris mass 10% less than the actual. As shown in Fig. 6, the velocities of the debris and the net converge on the estimated CM velocity L_{est}/M_{est} and not on the correct L / M. Due to momentum conservation, this results in the robot bases moving at a lower steady state velocity. In this case, at the instance the debris and the net attain their common velocity, a convergence force is applied between the net and the robot bases, so as to decelerate the net and accelerate the robot bases. Then, the system reaches a common velocity, avoiding the possibility of a extending the manipulator to its limits. Also, note that since this force decelerates the net, it will not result in debris loss, but will keep it captured. For a 3 m reach manipulator, we assume that the common velocity must be attained before the net carrying end-effector reaches 2.75 m. In Fig. 6, the transient of the debris-net system is over when the end-effector is at 2.60 m, i.e. less than 2.75 m. At this point, the convergence force starts acting to make all three velocities equal to L / M.

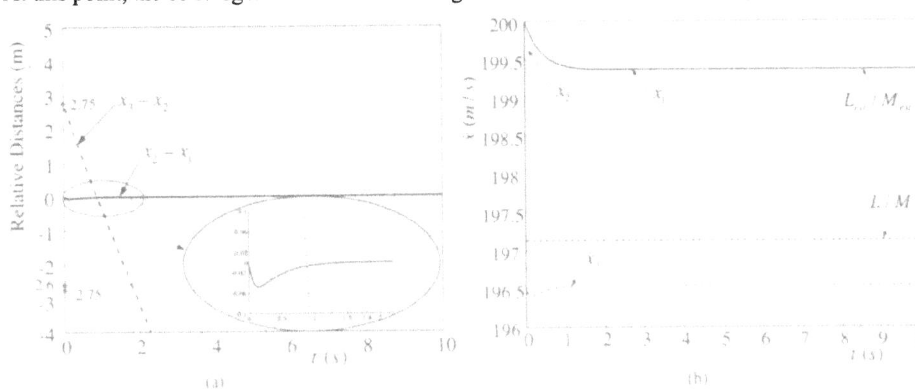

Figure 6. (a) Relative positions response and (b) velocities for control gains $K_1 = 250$ kg/s and $K_2 = 30$ kg/s that satisfy the constraints for successful capture, but with inaccurate estimation of the debris mass.

If the debris mass is overestimated, then the free-flying robots move at steady state with greater velocity than the net-debris system. A way to compensate for this, after the net and the debris reach the same velocity, is to fire the jets of the free-flyers, to slowly decelerate the whole system. This motion will keep the debris on the net, while at the same time will stop the distancing of the robots from the net.

This one-dimensional model analysis demonstrates the concept feasibility. During our current work, we extend the dynamics and control scheme, as well as the accompanying constraints to a more realistic system with higher dimension.

6 Conclusions

A novel concept for an orbital debris capturing system has been presented, including a number of orbital free-flying robots, holding a net, by which the orbital debris is to be captured. In order to better understand the behavior of this complex system, a simplified uni-dimensional system has been derived. The dynamics and control of the system and the desired system response led to the derivation of sets of constraints that must be hold between system parameters and initial conditions. Satisfying the constraints results to a set of control gains that are used in a velocity-based controller. The response of the simulated system has confirmed the constrained control method. The robustness of the method was studied in the case of inaccurate estimation of the debris parameters. It was found that the method shows robustness, keeping the debris in the net, although an additional force may be needed.

Bibliography

Bluethmann, W. et al. (2004). Building an Autonomous Humanoid Tool User. *Proc. IEEE-RAS Int. Conf. on Humanoid Robotics.* Los Angeles, CA, Nov. 2004.

Dupuis, E. et al. (2004). Autonomous Operations for Space Robots. *55th International Astronautical Congress.* Paper #IAC-04-IAA.U.5.03. Vancouver, Canada, Oct. 4-8, 2004.

Lacome, J. L., Espinosa Ch. and Gallet, C. (2002). Simulation of Hypervelocity Spacecrafts and Orbital Debris Collisions Using Smoothered Particle Hydrodynamics in LS-DYNA. *5th Dynamics and Control of Systems and Structures in Space Conference.* King's College, Cambridge, UK, July 14th – 18th.

Landzettel, K. et al. (2004). System Prerequisites and Operational Modes for On Orbit Servicing. *Int. Symposium Space Technology and Science (ISTS)*, Miyazaki, Japan, May 30–June 06 2004.

NASA (2005) *The Orbital Debris Quarterly News*. Vol. 9, Issue 1, pp 3. Available on-line in the address: http://www.orbitaldebris.jsc.nasa.gov/newsletter/newsletter.html

NASA (2004) *History Of On-Orbit Satellite Fragmentations*, 13th Edition. Lyndon B. Johnson Space Center, Houston, TX, May 2004.

Papadopoulos, E. and Dubowsky, S., On the Nature of Control Algorithms for Free-floating Space Manipulators. *IEEE Tr. on Robotics and Automation*, Vol. 7, No. 6, December 1991, pp. 750-758.

Patera, R. P. and Ailor, W. H. (1998) The Realities of Reentry Disposal. *Proceedings of the AAS/AIAA Space Flight Mechanics Meeting*, Monterey, California, February 9–11; also in *Advances in Astronautical Science* Vol. 99, pp 1059–1071, September 1998.

Smith, D.A. et al. (2003) An Assessment of Geostationary Intervention Missions. *Proc. 54th Int. Astronautical Congress*, Vol. 109, pp. 167-182. Sept. 29-Oct. 3, 2003, Bremen, Germany.

Todd, L. and Bowling, T. (2004). Debris Mitigation in Geostationary Earth Orbit. *6th Dynamics and Control of Systems and Structures in Space Conference.* Rionaggiore, Italy, 2004.

Yoshida, K., Hashizume, K. and Abiko, S. (2001) Zero Reaction Maneuver: Flight Validation with ETS-VII Space Robot and Extension to Kinematically Redundant Arm. *Proceedings of the IEEE International Conference on Robotics and Automation*, pp. 441-446, Seoul, Korea, 2001.

Trajectory Generation for Satellite Capture Using a Redundant Manipulator

Guy Rouleau[†‡], Eric Martin[†*] and Inna Sharf[‡]
[†] Canadian Space Agency, St-Hubert, Canada, eric.martin@space.gc.ca
[‡] McGill University, Department of Mechanical Engineering, Montreal, Canada

Abstract. One important area for application of space robotics is autonomous on-orbit servicing of failed or failing spacecraft. In this work, we describe laboratory experiments that verify the feasibility of autonomous capture of a slowly spinning non-cooperative satellite by a manipulator. The developed algorithms have been implemented and tested with the Canadian Space Agency's Automation and Robotics Test-bed, a two-arm, seven degree-of-freedom (dof) each, manipulator system. In the first phase of this work, a redundancy resolution scheme was implemented to maximize the robots' manipulability and increase their functional workspace. In the second phase, an online vision-based trajectory generation algorithm generating a velocity command to safely approach the target satellite and match its motion was developed. An overview of this work and the important results are presented.

1 Introduction

Over the past few decades, robots have played an increasingly important role in the success of space missions. The Canadarm on the Space Shuttle has enabled the on-orbit maintenance of precious space assets such as the Hubble Space Telescope. On the International Space Station (ISS), Canadarm2 has been a crucial element in all construction activities. Its sibling, Dextre, will be essential to the maintenance of the ISS. In light of the missions currently being planned by space agencies around the world, the coming years will likely show an increase in the number and the importance of robots in space missions. Examples include the US Defense Advanced Research Project Agency's (DARPA) Orbital Express mission (Whelan et al., 2000) and the German Space Agency's (DLR) TECSAS mission (Martin et al., 2005).

For TECSAS, autonomy capabilities are required for the capture of the target satellite. In this context, the laboratory experiments performed at the Canadian Space Agency (CSA) to verify the feasibility of the autonomous capture of a tumbling satellite are described in this paper. More specifically, the control of the manipulator and the trajectory generation for the capture are described. Using CSA's Automation and Robotics Test-bed (CART) (Ranjbaran et al., 2001), a mock-up satellite, the target, is mounted on one manipulator while a second manipulator is equipped with a SARAH hand (Laliberté and Gosselin, 2001). The latter, serving as a chaser, approaches and captures the target, guided by a Laser Camera System (LCS) from Neptec (Samson et al., 2004).

2 Redundancy Resolution

In the present section, we present the algorithms implemented for the control of the manipulator. In Fig. 1, the control architecture of CART is shown. The light grey blocks are the new additions implemented in this work. Due to the modularity of the implementation, any block can be replaced by an improved one without influencing the rest of the control architecture.

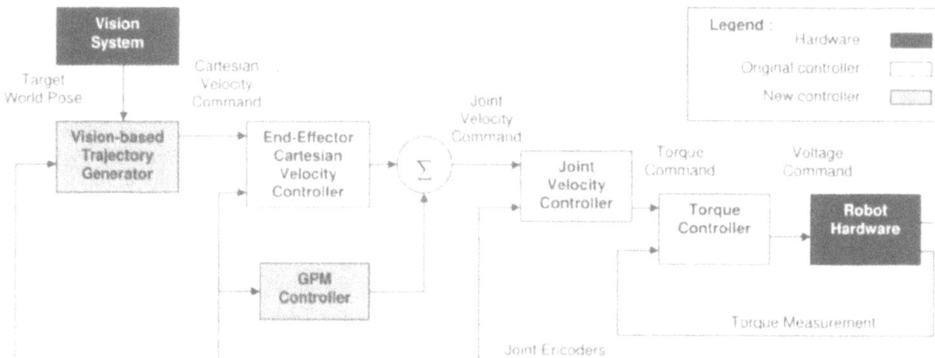

Figure 1. CART control architecture

Both arms of CART are seven-dof redundant serial manipulators. As stated in Chaumette and Marchand (2001) and Nelson and Khosla (1993), it is possible to increase the tracking region of a redundant manipulator by properly controlling its joint motion. Different methods have been proposed to locally optimize the manipulator's configuration and avoid mechanical limits. In Nakamura (1991), the well known Gradient Projection Method (GPM) is described and different numerical applications are shown.

2.1 Gradient Projection Method

For a given manipulator, the n-dimensional joint velocity vector $\dot{\mathbf{q}}$ is related to the m-dimensional end-effector task velocity vector $\dot{\mathbf{x}}$ by the Jacobian matrix $\mathbf{J}$, as

$$\dot{\mathbf{x}} = \mathbf{J}\dot{\mathbf{q}} \tag{2.1}$$

As shown in Nakamura (1991), a scalar function $p(\mathbf{q})$ dependent on the joint configuration can be used to define a solution for $\dot{\mathbf{q}}$ using the GPM:

$$\dot{\mathbf{q}} = \mathbf{J}^{\#}\dot{\mathbf{x}} + \left(\mathbf{1} - \mathbf{J}^{\#}\mathbf{J}\right)\left(-k_{SM}\frac{\partial p\left(\mathbf{q}\right)}{\partial \mathbf{q}}\right) \tag{2.2}$$

where k_{SM} is a control gain that must be tuned, $\mathbf{J}^{\#}$ is the Moore-Penrose inverse of the Jacobian matrix and $\mathbf{1}$ is the $n \times n$ identity matrix. The first term in Eq. (2.2) is the minimum-norm solution and it only affects the end-effector task, while the second term is the gradient projection that controls the self-motion of the manipulator without affecting the end-effector task $\dot{\mathbf{x}}$.

As described in Nelson and Khosla (1993), the self-motion of the manipulator can be used to increase the workspace by avoiding singularities and joint limits. To obtain a measure of the distance from joint limits, the following formula can be used:

$$w(\mathbf{q}) = 1 - e^{-k_{jt} \prod_{i=1}^{n} \frac{(q_i - q_{i\,\min})(q_{i\,\max} - q_i)}{(q_{i\,\max} - q_{i\,\min})^2}} \tag{2.3}$$

where $q_{i\,\min}$ and $q_{i\,\max}$ are the limits for joint i and k_{jt} is a parameter controlling the shape of the cost function. With this equation, the value of $w(\mathbf{q})$ is close to one when far from joint limits and it equals to zero at the joint limits. To measure the distance from a singularity, the most commonly used measure is the manipulability as defined in Yoshikawa (1985). In Marani et al. (2003) it is shown that the manipulability $m(\mathbf{q})$ can be computed as the product of the singular values (σ_i) of the Jacobian $\mathbf{J}$ computed as

$$m(\mathbf{q}) = \prod_{i=1}^{n} |\sigma_i| \tag{2.4}$$

The main advantage of this form is that the gradient of $m(\mathbf{q})$ can be computed easily based on the joint configuration $\mathbf{q}$. Finally, the cost function in Eq. (2.2) is computed by combining Eqs. (2.3) and (2.4), namely

$$p(\mathbf{q}) = m(\mathbf{q})\, w(\mathbf{q}) \tag{2.5}$$

2.2 Experimental Results

To illustrate the effect of the self-motion control, Fig. 2 shows four possible configurations of CART's left arm for the same end-effector pose. In the first picture, the arm is very badly configured and is close to its joint limits. By contrast, the manipulator is very well configured in the last picture. This redundancy resolution scheme allows the manipulator to maintain a good configuration while following a desired trajectory.

3 Trajectory Generation

For capture operations, the solution to the trajectory generation problem is dependant upon the motion of the target. For fast-moving or unpredictable targets, the methods typically used fall in the domain of guidance and navigation theory. These techniques are usually designed to track free-flying targets (e.g., missiles tracking evasive aircraft). In the interception of targets by missiles, the Ideal Proportional Navigation Guidance (IPNG) described in Yuan and Chern (1992) is the most common technique. This method can also be applied to robotic systems as demonstrated by Mehrandezh et al. (2000).

In this paper, we focus on the capture of a slowly tumbling satellite. Since the prediction of orientation for a tumbling satellite is a complex problem, an approach based on the PNG is chosen. A simple method requiring minimal computation and generating a Cartesian tool-frame velocity command for the chasing manipulator is presented.

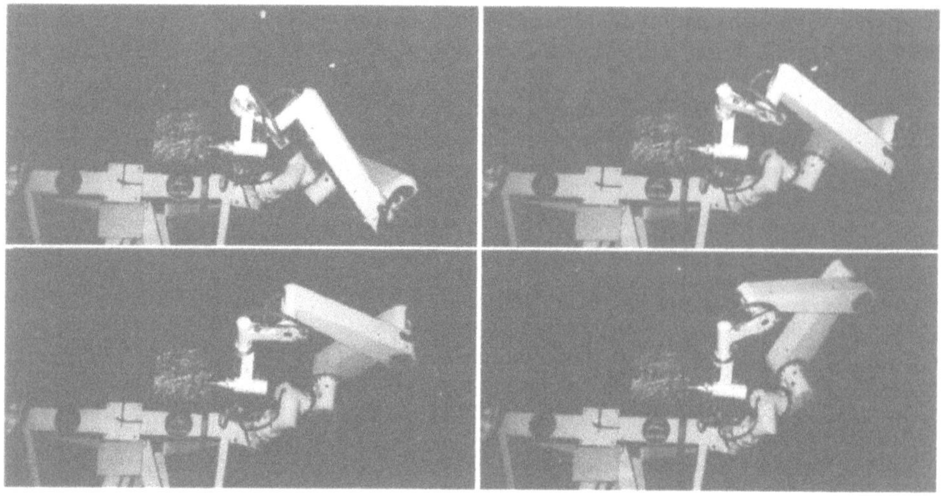

Figure 2. Four possible elbow configurations for the same end-effector pose

3.1 Strategy

In the strategy adopted in our work, a tool frame velocity command is generated based on the 3D pose of the target relative to the chaser. Using a CAD model of the tracked object, the position and orientation data, 6 coordinates, are calculated by the Laser Camera System at a rate of 2 Hz with a 0.5 second delay with an accuracy within 1 cm in translation and 5 deg. in orientation. An extended Kalman filter is used to filter the LCS raw measurements and to provide a smoothed pose of the target satellite at 1 kHz. The Kalman filter also includes a predictor to estimate the pose of the target satellite 0.5 sec forward in time. This predictor is used to eliminate the 0.5 sec delay in obtaining the pose of the target satellite using the LCS sensor.

In this section, the following notation is used: The frame $\mathcal{F}_T$, shown in Fig 3, is the tool frame attached to the SARAH hand mounted on the chaser manipulator and its position and orientation in the inertial reference frame $\mathcal{F}_w$ are respectively denoted by $\left[\mathbf{r}_{T/w}\right]_w$ and $\mathbf{R}_T$. The frame $\mathcal{F}_C$ is the capture frame attached to the handle of the target satellite and its position and orientation are denoted by $\left[\mathbf{r}_{C/w}\right]_w$ and $\mathbf{R}_C$. To express the position of the capture frame $\mathcal{F}_C$ in the tool frame $\mathcal{F}_T$ the notation $\left[\mathbf{r}_{C/T}\right]_T$ is used.

The trajectory is generated in two main phases: approach and capture. To generate the approach trajectory, the six independent degrees-of-freedom of the tool frame $\mathcal{F}_T$ are separated and assigned to independent tasks described below. For each task, a proportional controller to generate a velocity command in the tool frame $\mathcal{F}_T$ has been designed. The control gains have been tuned experimentally. At the completion of these tasks, the chaser should be very close to the target and the algorithm switches to realize the capture, as explained in the Capture subsection.

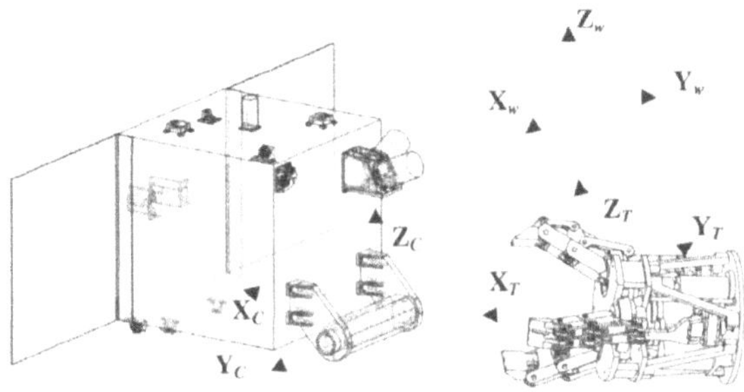

Figure 3. Reference frames used in the trajectory generation algorithm

Tracking The goal of the tracking is to orient the manipulator's end-effector toward the target. This task requires two rotational degrees-of-freedom to generate a pan-tilt motion as described by

$$\begin{aligned} \omega_y &= -k_{Ry} \arctan\left(r_z/r_x\right) \\ \omega_z &= k_{Rz} \arctan\left(r_y/r_x\right) \end{aligned} \tag{3.1}$$

where k_{Ry} and k_{Rz} are control gains for the respective axes, and r_x, r_y, and r_z are the components of $\left[\mathbf{r}_{C/T}\right]_T$.

Approach The goal of the approach task is to control the distance between the chaser and the target. This task requires only one translational degree-of-freedom. With the tracking task activated, the end-effector points toward the target, such that the distance can be controlled by a velocity command forward or backward generated with

$$v_x = k_x \left\|\mathbf{r}_{C/T}\right\| \tag{3.2}$$

where k_x is the control gain and $\left\|r_{C/T}\right\|$ is the distance between the capture frame $\mathcal{F}_C$ and the tool frame $\mathcal{F}_T$.

Translational Alignment For a successful capture, the grasping device must be aligned in front of the target. To do so, we first compute $\left[\mathbf{r}_{T/C}\right]_C$ the position of the tool frame $\mathcal{F}_T$ with respect to the capture frame $\mathcal{F}_C$. In a manner similar to the tracking task, two degrees-of-freedom are employed for the alignment but now it is the two remaining translations that are used, as shown by

$$\begin{bmatrix} 0 \\ v_y \\ v_z \end{bmatrix} = \mathbf{K}_{yz}\mathbf{R}_T^T\mathbf{R}_C \left[\mathbf{r}_{T/C}\right]_C \tag{3.3}$$

where $\mathbf{K}_{yz}$ is a matrix of gains, $\mathbf{K}_{yz} = diag\left\{0, k_y, k_z\right\}$. The first component of $\mathbf{K}_{yz}$ is set to zero because this dof is already assigned to the approach task in Eq. (3.2).

Roll Alignment For the final alignment, the rotational velocity needed to be completely aligned with the target is generated as

$$\omega_x = k_{Rx}\theta_{x_{C/T}} \tag{3.4}$$

where k_{Rx} is the control gain, and $\theta_{x_{C/T}}$ represents the orientation of the capture frame about the x-axis of the tool frame.

Capture Once the chaser achieves perfect tracking of the target and the distance separating them is small, the trajectory generator is changed to achieve the final approach and to match the motion of the target during the activation of the grasping device. The resulting close-range control law is defined as

$$\begin{aligned} \mathbf{v} &= \mathbf{R}_T^T \mathbf{K}_T \left[\mathbf{r}_{C/T}\right]_T \\ \boldsymbol{\omega} &= \mathbf{R}_T^T \mathbf{K}_R \boldsymbol{\lambda}_{C/T} \end{aligned} \tag{3.5}$$

where $\mathbf{K}_T$ and $\mathbf{K}_R$ are diagonal matrices of control gains for translation and orientation. The symbol $\boldsymbol{\lambda}_{C/T}$ is the skew-vector representation of the rotation matrix giving the orientation of the capture frame relative to the tool frame $\mathbf{R}_{C/T} = \mathbf{R}_T \mathbf{R}_C^T$. From Eq. (3.5), it becomes clear that the ultimate goal of the capture operation is to overlay the tool frame $\mathcal{F}_T$ exactly onto the capture frame $\mathcal{F}_C$.

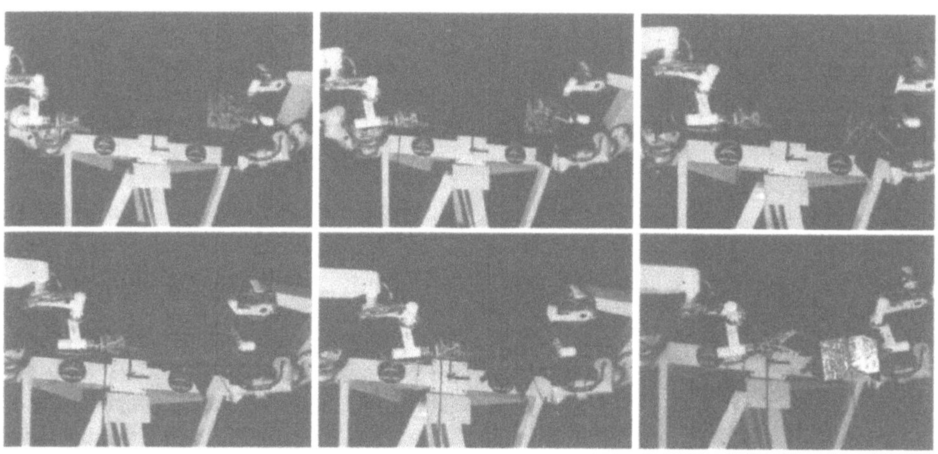

Figure 4. The chaser satellite approaching the target satellite to match its motion.

3.2 Experimental Results

A variety of experiments were performed in the laboratory. Different trajectories were tested for the target satellite. In all the experiments, the chaser manipulator first tracks the target and approaches it to within a safe distance without taking care of

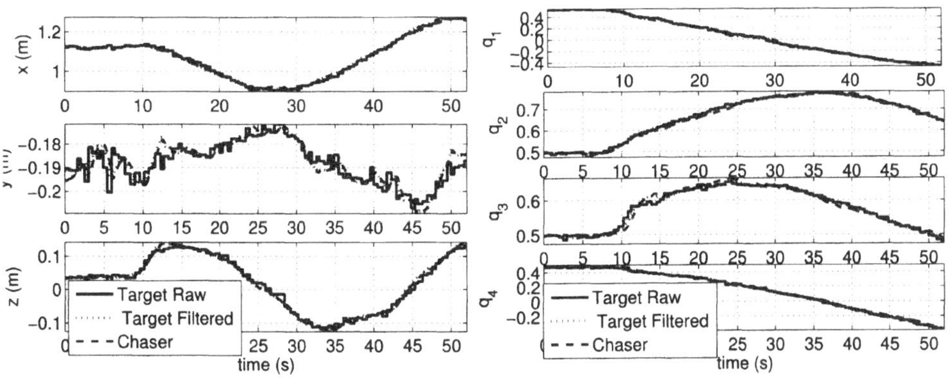

Figure 5. Cartesian position and orientation (quaternion) of the grasping device matching the motion of the target

its orientation. When sufficiently close, the chaser aligns with the target satellite and maintains the same safe constant distance. Finally, the command is given for the arm to perform the final approach and to grasp the handle mounted on the mock-up satellite using the SARAH hand. Fig. 4 presents a sequence of images from the autonomous capture scenario performed.

Selected experimental data are presented in Fig. 5. On the left, the tracking of the x, y, and z coordinates of the origin of a frame located 15 cm in front of the centre of the capture handle and alligned with the capture frame is illustrated. On the right, the four components of the quaternion representing the orientation of the target are shown. The solid (black) lines show the desired tracking position calculated from the raw data provided by the LCS. The dotted (blue) lines are the outputs of the Kalman filter, namely, the desired tracking position. Finally, the dashed (red) lines show the actual coordinates of the end-effector of the chaser, which is commanded to track the desired tracking motion. As observed in Fig. 5, the position tracking is within 0.01 m and the orientation is tracked within a few degrees.

4 Conclusion

In this paper, we presented some aspects of the experimental implementation of the autonomous tracking and capture of a tumbling satellite. The control of the redundant manipulator and a trajectory generation algorithm have been presented. The redundancy resolution scheme improves the behaviour of the manipulator and should always be used to control this type of redundant manipulator. The trajectory generation algorithm allows to safely track, approach and capture a tumbling satellite with a minimum of computation. The algorithm is easy to implement and can be used for other applications where a chaser needs to capture a moving target. In the future, the developed control architecture will be used to test and validate other more advanced capture algorithms and new technologies for autonomy and remote-operation of space robots.

Bibliography

F. Chaumette and T. Marchand. A redundancy-based iterative approach for avoiding joint limits: application to visual servoing. In *IEEE Transactions on Robotics and Automation*, volume 17, pages 719–730, Octobre 2001.

T. Laliberté and C. M. Gosselin. Underactuation in space robotic hands. In *Proceeding of the 6th International Symposium on Artificial Intelligence and Robotics & Automation in Space: i-SAIRAS 2001*, St-Hubert, Quebec, Canada, 2001.

G. Marani, J. Kim, J. Yuh, and W. K. Chung. Algorithmic singularities avoidance in task-priority based controller for redundant manipulator. In *IEEE/RSJ International Conference on Intelligent Robots and Systems*, pages 3570–3574, 2003.

É. Martin, É. Dupuis, J.-C. Piedboeuf, and M. Doyon. The tecsas mission from a canadian perspective. In *Proceeding of the 8th International Symposium on Artificial Intelligence, Robotics and Automation in Space: i-SAIRAS 2005*, Munich, Germany, 2005.

M. Mehrandezh, N. M. Sela, R. G. Fenton, and B. Benhabib. Robotic interception of moving objects using an augmented ideal proportional navigation guidance technique. *IEEE Transaction on Systems, Man, and Cybernetics - Part A: Systems and Humans*, 30, 2000.

Y. Nakamura. *Advanced Robotics: Redundancy and Optimization.* Addison-Wesley, Reading, MA, USA, 1991. ISBN 0-201-15198-7.

B.J. Nelson and P.K. Khosla. Increasing the tracking region of an eye-in-hand system by singularity and joint limit avoidance. In *IEEE International Conference on Robotics and Automation*, volume 3, pages 418 – 423, May 1993.

F. Ranjbaran, F. Shadpey, K. Khial, J.-C. Piedboeuf, S. Kalaycioglu, and M. Dub. Csa automation and robotics testbed - current capabilities. In *Proceeding of the 6th International Symposium on Artificial Intelligence and Robotics & Automation in Space: i-SAIRAS 2001*, St-Hubert, Quebec, Canada, 2001.

C. Samson, C. English, A. Deslauriers, I. Christie, F. Blais, and F. Ferrie. The neptec three-dimensional laser camera system: From space mission sts-105 to terrestrial applications. In *Canadian Aeronautics and Space Journal*, volume 50, pages 115–123, June 2004.

D.A. Whelan, E.A. Adler, S.B. Wilson, and G.M Roesler. Darpa orbital express program: effecting a revolution in space-based systems. In *Small Payloads in Space*, volume 136, pages 48–56, November 2000.

T. Yoshikawa. Manipulability of robotic mechanisms. In *International Journal of Robotic Research*, volume 4, pages 3–9, 1985.

P.J. Yuan and J.S. Chern. Ideal proportional navigation. In *Journal of Guidance, Control and Dynamics*, volume 15, pages 1161–1165, 1992.

Innovative methods of evaluation in space robotics and surgical robotics design

Prof. Alberto Rovetta[1] , Ph.d. Stud. Elena Cristina Paul[2], Ph.d. Stud. Chiara Zocchi[3]
[1,2,3] *Politecnico di Milano, Dipartimento di Meccanica*
Piazza Leonardo da Vinci, 32 – 20133 Milano (Italy)
e-mail: alberto.rovetta@polimi.it, cristina.paul@polimi.it , chiara.zocchi@polimi.it

ABSTRACT This paper deals with innovative methods of parameters evaluation in projects related to a locomotion system for Mars. The motion subsystem configurations are developed at a detailed level compatible with the assessment of project feasibility and with the identification of required technological development. Also the analysis of rover performances in overcoming obstacles and of robot movement are here reported. The design and verification of the analytical models, needed for robot action planning and describing how the robot interacts with its environment, are essential for design evaluation. This paper starts from the descriptions of the robotic characteristics of the systems, after the evaluation of the other robotic systems up to now applied. Then two innovative methods are presented: the Preferential Compatibility and the Prospect of INdipendent Decisions ARea (Pindar) techniques.

1 Introduction

In this paper the application of the preferential compatibility and the Prospect of INdipendent Decisions ARea (Pindar) techniques are presented in order to show the methodology of analysis and evaluation parameters in robotics design. These techniques are used in the Robotics Laboratory, as a first step in all our projects: the application in space is here reported.

In the last years space robotics is assuming a more and more important role thanks to a greater interest for the exploration of the celestial bodies. The high project and planning costs induce researchers to study robots that, beyond to being suitable for the exploration of the spaces, are applicable also in other fields. Future rovers will need to travel several kilometres, over periods of months and manipulate rock and soil samples in an autonomous way: rocker-bogie based rovers are likely candidates for these missions, but their physics is quite complex.

In this paper initially an analysis of the requirements of the motion subsystems, with reference to design concepts, sensors and motion control has been effectuated. Then the Preferential Compatibility and the Prospect of INdipendent Decisions ARea (Pindar) techniques are applied in order to find the best solutions to our projects.

2 Principles of design

It is necessary and comfortable for the designers to present and to interpret the main concepts of design for the robotic systems. The design is the main phase in the creative ideas generation and development and the presentation of design methodology and the validation of the concepts are a step of honesty and clearness.

3 Basic elements of robotics and parameters of evaluation and choice

The basic elements considered are: robotics and reliability, duration, dexterity, sensors, safety, stability. After a punctual analysis of the single topics, reported in the following sections, the criteria of structures evaluation have been elaborated. It is concluded with a ROBOTICS PROPOSAL articulated in a logical of ideation, creativity, relative evaluation and comparison: it reaches innovative concepts, with a scientific and technical validation under robotic aspect. Final summary of the proposal makes usable the job of formulation here carried out. Two methods, up to now, have been used. They are called: the method of Preferential Compatibility and the Pindar one (Prospect of Independent Decision Area).

4 Motivations of roboticS

A. ROBOTICS AND RELIABILITY: mechanic structure, resistance, no attack from environmental variations, dexterity of the motion, small sizes, light weight. In front of unexpected events, solution with the maintenance of at least one of the functions. Objectives in development: To maintain the state of the robot; To maintain the connection with human, without damage itself; Not damage other robots; Not loose the acquired knowledge; Robot morphology; Motion in narrow areas; Connection with the navigator system; Vision system.
B. ROBOTICS AND DURATION: Materials and reactions to the environment; elements of control and mechanical properties, elements in motion and static and dynamic duration.
C. ROBOTICS AND STABILITY: Position of the gravity centre under internal and external actions during the motion, the transitory phases, the start and the stop.
D. ROBOTICS AND DEXTERITY: motion in the area of work, vision system connected to the navigator system.
E. ROBOTICS AND SENSOR for static and dynamics duration, balancing, orientation, intervention in front of unexpected events. Forces and positions, accelerations and shock hits.
F. ROBOTICS AND SAFETY: safe motion, interface of stabilization, action, milling. Security towards the robot, the communication with the navigation system, the environment of the planet to be explored, the human communication.

5 Methods of preferential compatibility

The method uses all the possible logical combinations between input of the project and output demands, for a crossed evaluation. To be able to begin this type of analysis we must have good acquaintance of the state of the art.

The first phase is about the description of the currently existing locomotion robots and the projects carried until the phase of feasibility. They must be estimated on the bases of the project requirements which are put into effect and they must be submitted to the comparison with the demands for the objective: robots for the action on Mars. It's clear that the proposition phase must exceed the phase already got from Spirit and Opportunity, in January 2004, in order not to repeat results already acquired, with an automatic obsolescence of the project.

In this phase the functional robotic characteristics emerged are: Environmental conditions: Temperature, Wind, milling's path, chemical and physical properties. Composition ground Sand, Rock, and Mud. Shape of the land: Plane, Sinking, Slope Functional characteristics:

Weight, Dimension, Speed Stability, Motion Control, Auto stand-up, Vehicle Aerodynamics, Motors / Actuators, Suspensions, Energy Consumption, Dexterity, End effector, Materials, Sensors and Velocity. Possibility of unexpected events: failure, inadequate calibration, loss of energy or communication with navigation system.

6 Decisional table of design evaluation

The Decisional Table contains the definition of the Groups of characterization of the project. They are: the functional characteristics; the typologies of robot; the possibility of unexpected events; the robot position and the environmental conditions. It is evident that the Project of the Laboratory of Robotics, under the robotic aspect, must consider all the possibilities of events, expectable and those cannot be foreseen, as it is rule also in industrial environment.

With the aim of clarification, robot colony means a number of advanced robots equal or greater than three that cooperate; swarm means a very high number of mini robots that act in the same way, but which offer data and various information. Information is elaborated from the same swarm or from an external entity; society means a robotic organized structure with tasks and various priorities.

Flat Decisional table, of first approximation The Evaluation index can be developed at more levels and can be carried out in several phases. The first phase is the exploratory one.

In the Decisional Table a column with the figure of merit FM of every robotic system is presented. This figure varies from 0 to 1, where 1 represents the maximum compatibility and suitability towards the space robotics. It is a statistical concept used for multivariate analysis. The personal value PV for every crossing varies also from 0 to 1, where 1 represents the maximum compatibility, thought from the author of the Evaluation, for the robotics in space, and 0 means no compatibility. The product of the figure of merit for the personal value defines the "value of comparison" CV.

The successive phase consists in using the third dimension: it must be performed that everyone executes its personal job and then at the end the final values converge in a single big table, with the only values of comparison. At last the final average value is calculated and the solution / solutions which today seems to us the most compatible and suitable will emerge.

In vertical of Decisional Table, there are:

First group: Size, weight, stability, motion control, auto stand-up, energy consumption, materials, motors / actuators , speed, sensors, suspension, vehicle aerodynamics.

Second group: wheels, tracks, legs, wheels + legs, skates, ball, propeller, legs, snake, jump systems, beetle, ski, track + ski, egg, autogyro, swarm, airship.

Third group: society, swarm, colony, single.

Environmental conditions: ground, wind, temperature, duration of the day, atmospheric composition, gravity.

The results of Preferential Compatibility are reported in Fig.1

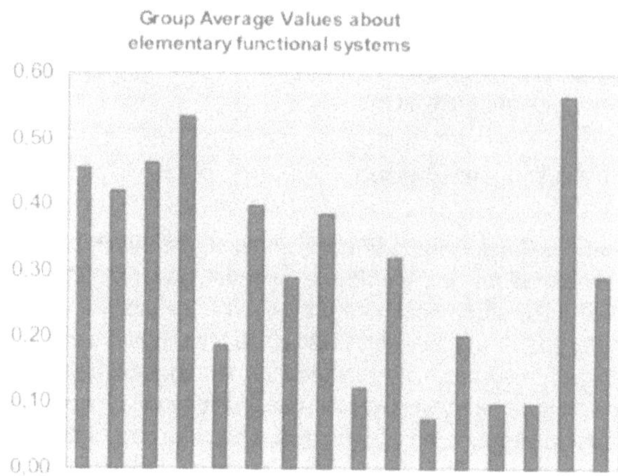

Fig. 1 Results of the Preferential Compatibility analysis, where the solution "wheels + legs" looks as the most preferable due to high group average values.
On the abscissa: wheels, tracks, legs, wheels + legs, skates, ball, propeller, legs, snake, jump systems, beetle, ski, track + ski, egg, swarm, airship.

7 The application of the technique pindar - prospect of independent decisions area

The Pindar technique is constituted by a sequence of steps during which the Evaluations of the project proposals are executed in series, eliminating gradually those with less suitability to the objective of the project.

This technique represents one combination between the morphologic and the criterion analysis: it eliminates some imperfection of the 2 types of analysis and it is very easy to apply, because it can be realized also without the aid of the computer. It is a fast and multi objective analysis, because it uses voting from 1 to 10. The rapidity turns out from the process of 'aim', because the possibilities of solution become smaller and smaller.

The Pindar technique presents 5 steps. Step 1 generally defines the objectives and the general criteria which are the basis for the choice. They make preferential one solution with regards to the others. The input variables are introduced as possible topics of design and planning. Step 2 defines the evaluation criteria. Step 3 introduces the weights for the evaluation. Step 4 begins to quantify the results, in a weighted evaluation and executes the first selective choices of draft work. Step 5 begins the crossed evaluations in order to determine numerically the indices of suitability in function of the objectives of the project.

STEP 1: Objectives: robotic small size system for bone milling during knee replacement / rover for the exploration of the Mars planet.

The criteria for the evaluation and the variable connected to, are here defined:

Space Robotics: *System of land locomotion*: wheels, tracks, sledge snake, legs, skates, wheels and legs, tracks and wheels, linkages, wheels and jump. *Aerial system locomotion:* wings; propeller; ball and airship. *Functional Characteristics:* actuators, sensors, motion control,

breakdown auto stand-up, vehicle aerodynamics, materials and energy storing. *Unexpected events*: losses of energy and upsetting. *Motion:* speed, stability and overcoming of obstacles. *Typology of rover*: single, colony and swarms.
STEP 2: Evaluation criteria for the choices: agreement with the scope, overcoming of several tests (hypothesis – simulations), temperature, wind, sand, stones, mud, slopes, variation of the land (for space),
STEP 3: Importance of the evaluation criteria in space robotics: for each parameter considered a weight is defined.
STEP 4: It analyzes the criteria with successive eliminations each designer gives a value, between 1 and 10, to the variables in relation to the evaluation criteria; then these values are multiplied for the criteria weights and their mean value is calculated and reported in the last column.
A. FUNCTIONAL CHARACTERISTICS: The same for functional characteristics and the results are: the aerodynamics of the vehicle (space) having the lowest average value of 5, 32 16 are eliminated. By holding account of the fact that there are systems of locomotion of aerial type, it is maintained in the calculation.
B. UNEXPECTED EVENTS: The same for unexpected events and the results are: the failure (space) 6, 30 are eliminated.
C. MOTION (SPACE): The same for Motion and the result is that between the three variables, stability will be the less important.
FINAL STEPS. The final steps compare the results and get the final table, which is a planning of choices and methodologies.
The first classification and the score after interpolation of the first characteristics present the following results: the highest value has been reached by the following type of robot: propeller + sensors + actuators + motion control + vehicle aerodynamics+ auto stand up+ materials + energy storing ((6.74+ 6,69+ 6,92+ 7,76+ 6,16+ 5,32+6,45+6,16)/8= 6,52)

8 Description of the decisional tables and final project propoSAL

Decisional tables have been developed, tested and come true in order to facilitate the project formulation and the design of the locomotion systems. The numerous options have been put in evidence and this represents an excellence of the project, which is innovative and full of traditional procedures, interpreted in a more modern support of knowledge.
Every reference designer has proposed his / her **figure of merit**_for each elementary system: they vary from 0 to 1, where 1 represents the maximum compatibility and suitability towards space robotics. In this design content, concepts of multivariate analysis have been strongly and directly used. Each reference person has used just an evaluation criterion, which is expressed with **a personal value,** for each relation. At the end the final values are reported: they converge in a single table, with the values of comparison. After the final average, the solutions, that seem more compatible and suitable to new proposed projects for Mars exploration with robot, emerge in evaluative shape.

9 Application in robotic surgery

Biorobotics has significant applications in all fields of modern medicine concerning

biomechanics, control and electronics, repeatability, execution rapidity, elimination of infection and decrease of invasiveness. The problems to be faced to make this surgical system possible are mainly the development of technical solutions that guarantee suitable performances, the reduction of the high costs that these solutions involve and above all the safety study in order to guarantee the absence of risks, since it's a system that will operate in contact with man. Even in this direction the research is moving to offer to the medical personnel high technology products capable to restore a state of health and wellness in pathological subject close to normality. The parameters considered are *System for bone milling*: end effector and its velocity. *System for motion in surgical area:* actuators, dexterity, size, motion control, robot velocity, vision system. *Functional characteristics:* weight, materials, sensors and energy consumption. *Unexpected events*: failure, inadequate calibration, loss of energy or communication with navigation system, damage of the robot or of the soft tissues/ligaments. *Robot typology*: serial, parallel and anthropomorphic arm. *Robot position*: lateral with respect to the surgical table, bridge-shape, on leg or on ceiling.
Applying both techniques of Preferential Compatibility (Table 1) and Pindar, we detect that the better robot for knee surgery was a serial + lateral + bridge-shape + on bone robot (fig.2)

			Robot Typology						
			arm		parall el		Serial		
		FM	0,95	**PV**	0,8	**PV**	0,8	**PV**	**CV**
Functional characteristics	Dimension	1,000	0,50	0,500	0,50	0,400	0,70	0,560	**0,49**
	weight	1,000	0,50	0,500	0,90	0,720	0,70	0,560	**0,59**
	dexterity	0,950	0,90	0,855	0,80	0,608	0,90	0,684	**0,72**
	control	0,950	0,90	0,855	0,70	0,532	0,90	0,684	**0,69**
	vision	0,800	0,00	0,000	0,00	0,000	0,00	0,000	**0,00**
	energy consumption	0,800	0,40	0,320	0,50	0,320	0,60	0,384	**0,34**
	Materials	0,800	0,50	0,400	0,50	0,320	0,50	0,320	**0,35**
	actuators	0,800	0,40	0,320	0,70	0,448	0,60	0,384	**0,38**
	Velocity	0,700	0,90	0,630	0,50	0,280	0,80	0,448	**0,45**
	Sensors	0,900	0,90	0,810	0,60	0,432	0,80	0,576	**0,61**
	End effector	0,700	0,90	0,630	0,60	0,336	0,90	0,504	**0,49**

Tab. 1 Application of Preferential Compatibility to the choice of syrgical robotics, where the figure of merit FM of every robotic system is presented. This figure varies from 0 to 1, where 1 represents the maximum compatibility and suitability towards the surgical robotics. The personal value PV for every crossing varies also from 0 to 1, where 1 represents the maximum compatibility, thought from the author of the Evaluation, for the surgical robotics,

and 0 means no compatibility. The product of the figure of merit for the personal value defines the "value of comparison" CV.

10 General conclusion

From the results obtained with the interpolations and also with the table of the unexpected events, it is conclusive that it is consistent with our objective to have a robot with a mixed system of locomotion, and at the first place a “wheels + legs” robot (fig.2 right). Like typology, a colony or a swarm of robots, instead of a single robot, is presented as the best system. In robotic surgery field the conclusion is to have a serial small size bridge-shape robot (fig. 2 left). The technique applied shows the better design choice in front of project requirements and specifics, giving a more objective approach to the design phase.

Fig.2 The robots for knee surgery (left) and Ladyfly (right) designed after the application of the Preferential Compatibility and Pindar techniques

Acknowledgements

The Authors thank ASI (Italian Space Agency) for the research development, ESA (European Space Agency) for the partecipation at Project Aurora Exomars and Eng. Ignazio Barraco, Alcatel Alenia Space, for their engagement and support for this work. Thanks to Prof. Norberto Confalonieri, Head Surgeon on Hospital CTO in Milan, and Prof. Giovanni Peretti, Head Surgeon on G. Pini Hospital in Milan, for helpful dicussions.
Thanks also to the researchers, doctorate students and students who have cooperated.

Bibliography

Science, vol. 305 no 5685- 6.8.2005,vol 306 no 5702- 3.12.2005, vol. 306 no 5704- 17.12.2005

Rodney A., Brooks A, robust layered control system for a mobil robot .IEEE Journal of Robotics and Automation April 1986

Ostergaard E., Mataric J and Gaurav S,Distributed multi-robot task allocation for emergency handling, 2001 IEEE/ RSJ, IROS

Schenker P.S., et al, Reconfigurable robots for all terrain exploration, Proc. of SPIE. November 2000

Kucherenko , Bogatchev, Chassis Concepts for the ExoMares Rover ESA Workshop on Advanced Space Technology for Robotics and Automation ASTRA 2004 Estec, Norwijk

Schulte , Hofmann, Naumann, A Sample Preparation and Handling System for ESA's Pasteur Exobiology Paylod for Mars ESA Workshop on Advanced Space Technology for Robotics and Automation, ASTRA 2004 Estec, Norwijk

Attila , Matti Concept study for the subsurface sampling system for the Pasteur paylod of the ExoMars Mission ESA Workshop on Advanced Space Technology for Robotics and Automation, ASTRA 2004 Estec, Norwijk

Rovetta A., Cosmi F., Romano V. F., Kinematics and dynamics of single and multiple arm robotic system, a practical application, Robotics in Alpe Adria Region, June 21-23,20-27,1992

Rovetta A., Robotica , Hoepli, 2005

Rovetta A, FRIEND: space telerobot for rescue and recovery of astronauts in space stations, IEEE/ RSJ International Worksop on Intelligent Robots and Systems, IROS'91,3-5 November,1991

Rovetta A.,Romano V.F, Cosmi F.,Nenchev D., Friend :an extra vehicular activity retriever telerobotic project, Space Technology , Vol .15, No.3,1995, pp123-131

Ellery, A., An Introduction to Space Robotics, Springer , 2000

Cosmi F., Rovetta A, Development of Dynamics for a Space Telerobot, Iros '93 , Yokohama, July 1993

L.P. Nolte, L.J. Zamarano, Z. Jiang, Q. Wang, F. Langlotz, E. Arm, H. Visarius, A Novel Approach to Computer Assisted Spine Surgery, Proceedings of the First International Symposium on Medical Robotics and Computer Assisted Surgery, vol. 2, pp. 323-328, 1994.

N. Confalonieri, D. Saragaglia, F. Picard, P.Cerea, K.Motavalli, Computerised OrthoPilot system in arthroprosthesis surgery of the knee. J Orthopaed Traumatol (2000) 2:91-98

M.Shoam, M.Burman, E.Zehavi, L.Joskowicz, E.Batkilin, Y.Kunicher, Bone-Mounted Miniature Robot for Surgical Procedures: Concept and Clinical Applications IEEE Trans. Robotics Automation, vol. 19, October 2003.

Shin, Yoon: Force prediction model for robotic surgery in orthopedics, Proceedings . CAOS 2004

Chapter IX

Vision and navigation

Visual Target Detection in Unstructured Environments – A Novel Technique for Robotic Navigation

Andrzej Śluzek[1,2] and Md Saiful Islam[1]

[1] NTU, School of Computer Engineering, Singapore, e-mail: assluzek@ntu.edu.sg
[2] SWPS, Warsaw, Poland

Abstract. We report a novel method of visual modeling and detecting 3D rigid objects randomly located in complex cluttered environments. Models are built (in *reference scale*) using visual saliencies (interest points) in template images presenting the object of interest from various viewpoints. By matching interest point detected in camera-captured images (*relative scale* is used there) to visual saliencies from the database, the target objects are detected and verified. If a stereovision system is available, the procedure is separately performed for both cameras. Subsequently, the corresponding matches from both images can be used to solve the fundamental correspondence problem. The paper briefly discusses methodology and presents exemplary experimental results in vision-based robotic aplications.

1 Introduction

In advanced applications of robotics, there is often a requirement to visually identify target objects (or objects of interest) and to determine their locations. Typical examples are: detection of objects to be subsequently manipulated by a robot, identification of landmark objects for autonomous navigation, surveillance and monitoring problems, etc. Generally, it should be possible to recognize and localize such objects in cluttered environments and/or in partial occlusion conditions. Additionally, objects in captured images can be visually distorted (compared to the model objects) by various geometric and photometric transformations. The recognition and localization methods should be robust to all such transformations and conditions.

In this paper, we report development of a novel method of modeling and detecting (by model matching) 3D rigid objects randomly located in complex cluttered environments. The objects may be partially occluded and (visually) distorted, i.e. the above mentioned requirements are satisfied. The method is superior to existing equivalent alternatives in terms of computational complexity (real-time performance is achievable) without sacrificing efficiency of object detection and localization. Additionally, the method significantly simplifies the major issue of stereo vision, i.e. the image correspondence problem.

In the proposed method, objects of interest are both modeled and recognized using easily perceivable local-shape features (visual saliencies) which is a commonly accepted solution for partial occlusion conditions, e.g. Hausler and Ritter (1999), Wong et al. (1998). However, most of the methods represent objects by features from the reference images associated with 3D information with respect to a given reference frame while human vision only memorizes 2D features (e.g. shape) of the object and localizes objects using stereovision-like mechanisms. Only in a few works (e.g. Joseph, 1999) this approach in investigated. We also believe that human vision (which is the most successful vision system so far to localize objects in cluttered environments) should be mimicked in the robotic navigation systems. In the proposed method, human vision-like approach is emulated using the following procedures applied first to object modeling, and subsequently to object detection/localization:

1. Detection and invariant description of visual saliencies (interest points) in template images presenting the object of interest from various viewpoints. So-called *reference scale* of images is used at this phase. Local-shape models of the object are subsequently produced based on geometric relations between the saliencies. This procedure is performed offline during the database building for the intended applications.
2. Detection and description of visual saliencies (interest points) in camera-capture images. Determining matches between detected interest point and visual saliencies from the database. So-called *relative scale* is applied to camera-captured images. If a stereovision system is available, the procedure is separately performed for both cameras.
3. Building and verification of hypotheses about presence of target objects in the captured image (including determination of the viewpoint).
4. For stereovision only, 3D localization of target objects using the correspondence between visual saliencies identified in both images.

Steps 2 – 4 are online procedures to be performed by computational resources of a navigating robot. Therefore, their computational complexity is the fundamental issue.

Section 2 of the paper briefly discusses theories and tools used in Steps 1 and 2. The problems of hypothesis building and verification (including stereovision, i.e. Steps 3 and 4)) are presented in Section 3. Application aspects of the method are discussed in Section 4.

2 Fundamentals of the Reference/Relative Scale Method

Local-shape features are generally accepted tools for modeling/detection objects in cluttered environments (e.g. Hausler and Ritter, 1999, Jurie and Schmid, 2004). This is mainly because no segmentation of the object is required and the methods are robust to partial occlusions and background changes. Other practical issues to be considered for visual object detection are illumination changes and pose changes (geometric transformations of the object in the image). Particularly, the change of scale of the object is one of the most concerning issues for the local-shape methods. In order to cope with the scale changes, multi-scale methods (pyramids or scale-space) have evolved. In multi-space methods, model representation and recognition is carried out in multiple scales so that object modeling requires huge memory, and the recognition process becomes computationally expensive. Nevertheless, the method has been successfully applied to various image analysis tasks (e.g. Lowe, 2004 or Mikolajczyk and Schmid, 2004). In many applications, local-shape methods are based on detection of Harris interest points (e.g. Schmid et al., 2000). Interest points are easily perceivable discontinuities of image intensity (usually associated with corners or corner-like features). As such discontinuities are generally not affected by partial occlusions, they are more suitable for identification of partially visible objects than region-based or contour-based local features. Although alternative interest point detectors exist (e.g. SIFT detector, Lowe 2004) we found them computationally too complex for our needs.

2.1 Detection and Invariant Description of Interest Points

We propose the reference/relative scale approach where interest points are extracted using the scale normalized auto-correlation matrix of Harris detector given by

$$N(X, \sigma_I) = \sigma_D^2 g(\sigma_I) \otimes \begin{bmatrix} I_x^2(X,\sigma_D) & I_x I_y(X,\sigma_D) \\ I_x I_y(X,\sigma_D) & I_y^2(X,\sigma_D) \end{bmatrix} \tag{2.1}$$

where X is a point of an image, and I is the image intensity function. The scale is represented by σ_I, i.e. $g(\sigma_I)$ is a circular Gaussian integration window with σ_D variance (proportional to σ_I).

Subsequently, interest points are defined as points where the value of

$$C(X, \sigma_I) = \det(N(X, \sigma_I))\text{-}\lambda \mathrm{tr}^2(N(X, \sigma_I)) \tag{2.2}$$

exceeds a predefined threshold.

Although the number of interest points detected in a given object may vary depending on viewing aspects (scale, orientation) and illumination conditions, we have proposed a method (details to be found in Islam, 2005, Chapter 3) that assures a high level of repeatability in interest point detection, significantly superior to results reported by other authors (e.g. e.g. Schmid et al., 2000). Figure 1 gives two examples illustrating the level of repeatability in our method.

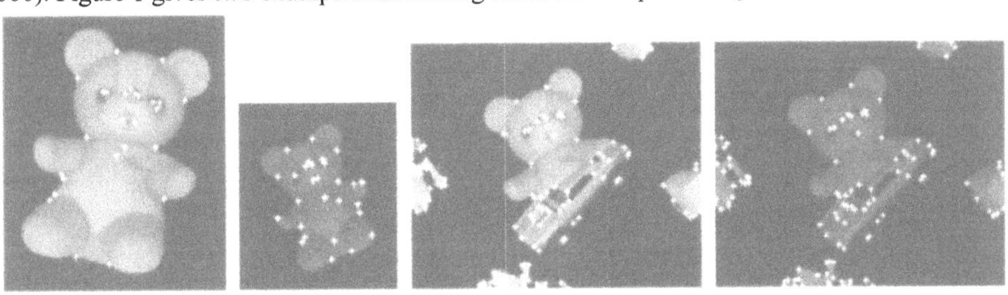

Figure 1. Examples of repeatability in interest point detection under changing geometry (including occlusions) and varying illumination conditions.

In order to characterize detected interest points (and to match them to the corresponding interest points from other images) we have proposed a simple yet very powerful local descriptor (LID) that is invariant under 2D similarity transformations (rotation, scaling) and contract changes. LID is computed for n concentric circular patches around the interest point as shown in Figure 2 so that a higher dimensionality (i.e. a better discriminatory quality) can be achieved.

The proposed LID can be any of Maitra invariants (Maitra, 1979) that are derived from Hu moment invariants (Hu, 1962) to achieve invariance under contrast changes. The first Hu invariants are:

$$\varphi_1 = \eta_{20} + \eta_{02} \qquad \varphi_2 = (\eta_{20} - \eta_{02})^2 + 4(\eta_{11})^2 \qquad \varphi_3 = (\eta_{30} - 3\eta_{12})^2 + (3\eta_{21} - \eta_{03})^2 \tag{2.3}$$

where η_{pq} are normalized central moments, i.e. $\eta_{pq} = \mu_{pq} / (\mu_{00})^{0.5(p+q)+1}$.

The corresponding Maitra invariants are:

$$\beta_1 = \frac{\sqrt{\varphi_2}}{\varphi_1} \qquad \beta_2 = \frac{\varphi_3}{\varphi_2} \cdot \frac{\mu_{00}}{\varphi_1} \tag{2.4}$$

Our experiments have indicated that the simplest Maitra invariant β_1 has the best information content (see Islam, 2005) combined with the lowest noise sensitivity so that the implemented LID for n concentric circles is an n-dimensional vector:

$$\text{LID} = \{\beta_1^{(1)}, \beta_1^{(2)}, \ldots, \beta_1^{(n)}\} \tag{2.5}$$

For a typical reference scale, we use 5 circles of the corresponding radii of 8, 12, 16, 18, 20 pixels. For the relative scale, the radii of the circles are proportionally rescaled.

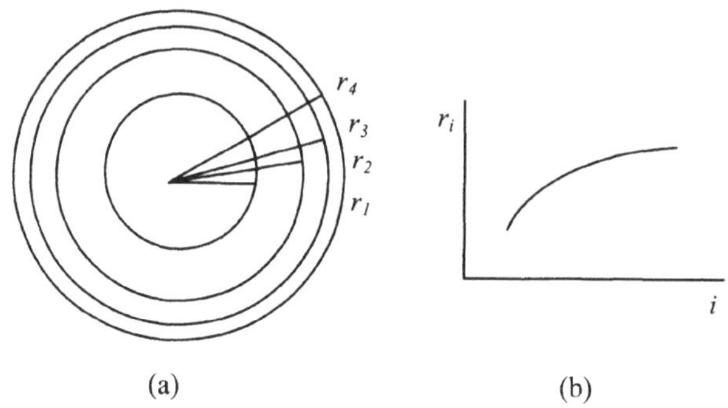

Figure 2. Concentric circles centered at the interest point (a). The radii of the circles increase slowly (b).

2.2 Matching Interest Points

Using LID proposed above, visual saliencies (interest points) detected in various images can be matched regarding their visual content. In particular, camera-captured images can be matched to a template image of a target object. Since the proposed detection and description method is invariant to geometric and photometric transformations, we expected a robust performance under realistic conditions.

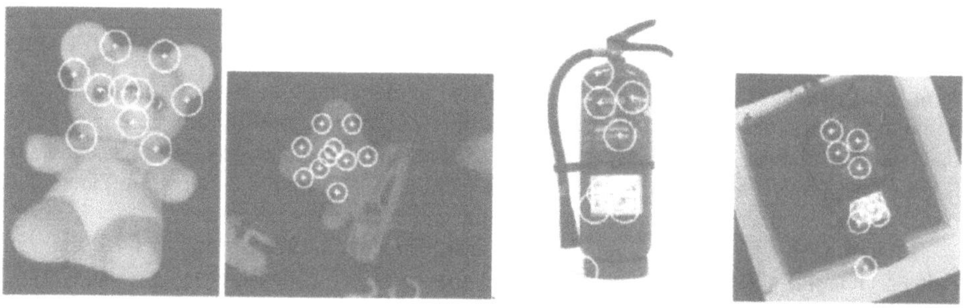

Figure 3. Interest points matched in template images and test images.

The fundamental assumption in our matching method is a priori knowledge of the relative scale. In typical robotic applications, this is not a restrictive assumption since we can either approximate the scale from the geometry of the application or we expect to identify the target object when it appears at a certain distance form a navigating robot. In the worst case scenario, the range of reference scale can be assessed, and matching interest points is simultaneously performed in a few different scales. Section 3 discusses how to resolve possible ambiguities in interest point matching and the relative scale selection.

Figure 3 presents exemplary matching results for two objects (one taken from a popular database and the other on is camera-captured). A significant number of matches found indicate that objects can be reliably identified (see Section 3) in spite of partial occlusions and other degrading effects. The examples do not cover the problem of 3D matching (even though the objects are solids) but this aspect will be presented in the next section.

3 Modeling and Detection of 3D Objects

3D objects are modeled using local-shape features (interest points) computed from multiple views of the object. Obviously, several images of the object are captured from significantly different viewpoints, as a representative set of *reference images* for the whole object. The number of required views depends on the object itself. For simplicity, we consider views from a constant distance with uniform intervals (e.g. 15°) of orientation angles. Arbitrary reference scale is assigned to all reference images. The proposed local invariant descriptor (LID) for each of detected interest points is subsequently computed and memorized with the database of modeled objects.

Unfortunately, even for simple objects the overall number of interest points in multiple reference images taken from different viewpoints may be very large. Therefore, the matching process for interest points extracted from camera-captured scenes could be prohibitively time-consuming (it should be noted that in unstructured environments the majority of interest points would belong to the background objects rather than to the targets). Therefore, multidimensional LID descriptors should be properly indexed to optimize search during interest point matching. As an illustration, Figure 4 shows the total number of interest points in just one reference image, and the number of interest points in a camera-captured image.

We have implemented an efficient hashing technique. In general, for hashing multidimensional data, values of few dimensions having highest variance are used as the hash-key (e.g. Stein and Medioni, 1992). This method seldom reorganizes data space to make it uniform over the indexing space. Our hashing technique, called *PCA-hashing*, uses principal component analysis method to obtain most uniform distribution of data. During the off-line process of object modeling we accumulate all the LIDs extracted form the reference images and select proper hash-key to create a lookup table.

Therefore, matching between interest points extracted from a camera-captured image (assuming that the relative scale is known) and the database interest points can be performed very efficiently.

Additionally, for each reference image we compute the spatial relations between all interest points. These spatial relations are represented by a connected graph called *shape-graph*, which allows partial matching. If the viewpoint change is small, the distances between two interest points are proportional to the relative scale of the object, and remain invariant to other transformations of the image.

Figure 4. Interest points detected in a single reference image, and in an exemplary test image.

3.1 Detection of Objects: Hypotheses Generation and Verification

When interest points are extracted in a scene image, they can be matched to the interest points of all reference images of the target objects, but those matches are usually highly ambiguous. Therefore, from matches of individual interest points, a consistent hypothesis regarding presence of the target object should be proposed and subsequently verified.

The matched interest points from the camera image are used to generate hypotheses about the target object and its orientation using generalized Hough transformation (GHT), e.g. Ballard (1981). GHT is an extension of classical Hough transformation used to detect arbitrary shapes. It is also used for object recognition (e.g. Ulrich et al., 2003).

Although the target object interest points should match only with the similar points of the reference images, there could be false matches so that a large number of hypotheses can be expected. We make use the spatial constraints represented by *shape-graph* for a robust verification for each of the hypotheses produced. Only a small number of correctly matched pairs (so-called *seed*) is sufficient to ensure the identity of the object in the scene, i.e to verify the hypothesis.

Figure 5. Matched pairs of interest points (after hypothesis verification) for Fig.4 scene.

It should be noticed that a target object in the scene could be matched with more that one reference image (of similar viewpoints). In such cases we take all these matches as correct matches. In other words, an estimation of exact orientation of the target could be obtained by interpolation technique using all matched orientation. Figure 5 presents the matched pairs for the hypothesis derived from the situation shown in Figure 4.

3.2 Localization of Objects UsingStereovision

Stereovision in usually often available in camera-equipped robotic systems. Target-detection methodology presented above can greatly improve efficiency of 3D reconstruction by reducing the number of correspondences and precise localization of points the are mutually corresponding.

The same procedure of interest point matching, and subsequent hypothesis verification are applied to both stereo-images. Once the interest points from both stereo-images are matched to the same model interest point, they can be immediately used as correspondences to determine the scene depth. Figure 6 illustrates this mechanism. It can be noticed that at least some template interest points are matched in both stereo images. Thus, the scene depth (distance to the target object) can be computed.

Figure 6. Matched pairs of interest points in a pair of stereo images.

Our experiments have shown that depth estimates errors in this method do not exceed 2% of the actual distance.

4 Discussion

The above target detection has been tested in exploration of hallway environment, where the robot has to find various objects (e.g. a fire extinguisher) (or other objects). A typical scenario is presented in Figure 7. Although no real robot was employed (instead, we used a trolley with a mounted computer and stereovision installed) the system was controlled according to the commands produced by a vision-guided algorithm.

The proposed methods and algorithms have been coded in Matlab, which is not intended for real-time applications. Nevertheless, the overall execution time for a single round (i.e. image capturing, interest point detection and matching, hypothesis verification, and stereovision depth reconstruction) typically does not exceed 10 seconds. Even this time is significantly lower than results reported for systems intended for similar tasks, For example, the vision system for mobile robot Antigone (Kim and Nevatia, 1999) needs about 2.5 minutes to process one frame of stereo images. We expect that with careful coding using a time-efficient language (and, possibly, hardware-supported implementation of low-level algorithms) real-time performance is achievable.

Figure 7. Examples of hallway exploration scenes.

The paper has briefly discussed the crucial elements of the proposed method of detecting and localizing 3D objects in cluttered environments. However, there are numerous problems that are also important for improving robustness of the method (e.g. minor deformations of circular patches to compensate perspective projection effects, etc.). Details can be found in Islam (2005).

Bibliography

Ballard, D.H. (1981). Generalizing the Hough transform to detect arbitrary shapes. *Pattern Recognition* 13(2): 111-122.

Hausler, G., and Ritter, D. (1999). Feature-based object recognition and localization in 3D-space using a single video image. *Computer Vision and Image Understanding* 73(1): 64-81.

Hu, M.K. (1962). Visual pattern recognition by moment invariants. *IRE Transactions on Information Theory* 8: 179-187.

Islam, M.S. (2005). Relative scale method to recognize and locate objects for robotic applications. *PhD thesis*, Nanyang Technological University, Singapore

Joseph, S.H. (1999). Optimal pose estimation in two and three dimension. *Computer Vision and Image Understanding* 73(2): 215-231.

Jurie, F. and Schmid, C. (2004). Scale-invariant shape features for recognition of object categories. *Proceedings of IEEE Conf. on Computer Vision and Pattern Recognition* 2: II.90 – II.96.

Kim, D. and Nevatia, R. (1999). Symbolic navigation with a generic map. *Autonomous Robots* 6(1): 69-88.

Lowe, D. G. (2004). Distinctive image features from scale-invariant keypoints. *International Journal of Computer Vision* 60(2): 91-110.

Maitra, S. (1979). Moment invariants. *Proceedings of IEEE* 67(4): 697-699.

Mikolajczyk, K. and Schmid, C. (2004). Scale & affine invariant interest point detectors. *International Journal of Computer Vision* 60(1): 63-86.

Schmid, C., Mohr, R., and Bauckhage, C. (2000). Evaluation of interest point detectors. *International Journal of Computer Vision* 37(2): 151-172.

Stein, F. and Medioni, G. (1992). Structural indexing: efficient 3-D object recognition. *IEEE Transactions on Pattern Analysis and Machine Intelligence* 14(2): 125-145.

Ulrich, M., Steger, C., and Baumgartner, A. (2003). Real-time object recognition using a modified generalized Hough transformation. *Pattern Recognition* 36: 2557-2570.

Wong, A.K.C., Rong, L., and Liang, X. (1998). Robotic vision: 3D object recognition and pose determination. *Proceedings of IEEE/RSJ Int.Conf. on Intelligent Robots and Systems*, 1202-1209.

Hand image interpretation based on double active contour tracking

Włodzimierz Kasprzak [†*] and Piotr Skrzynski [†]
[†] Warsaw University of Technology, Institute of Control and Computation Eng.,
ul. Nowowiejska 15/19, PL-00-665 Warsaw, Poland
e-mail: W.Kasprzak@ia.pw.edu.pl

Abstract. We propose an approach to hand sign interpretation in image that is based on active contour tracking. We can decompose our approach into 5 steps: a color-based skin pixel detection, a double hand contour detection, the localization of fingers and palm (the hand description generation), the detection of a final position (with respect to considered signs) and finally, the interpretation of a single position or a sequence of positions in terms of a hand sign. We employ a double active contour-based finger and palm localization in the image and a subsequent interpretation in terms of signs. As a final result, 21 different signs are recognized, that correspond to hand positions (i.e. the visibility of palm, fingers and thumb).

1 Introduction

Hand image recognition is mainly considered in the context of man-machine communication (Rehg and Kanade, 1993) and person identification (biometrics) (Sanchez-Reillo et al., 2000). In this paper we are interested in man-machine communication aspects, assuming that a single view is available, the hand can be freely oriented in space, while finger positions can also be different. Our goal is to interpret the recognized hand in terms of commands given to the machine by a human operator. The creation of such image recognition system requires solutions in several image analysis areas: color-based region-of-interest detection (Yining et al., 1999; Fu et al., 2000), hand contour detection, 3-D hand instance generation and interpretation.

A straightforward approach to object detection is to extract its shape (or contour) information from the image. A popular approach to closed contour detection for free-form and deformable objects uses "active contours" (Kass et al., 1988) or "snakes" (Berger, 1990). Snakes (active contours) (Kass et al., 1988; Berger, 1990; Osher and Paragios, 2003) are curves defined in image plane that can change their shape and can move under the influence of forces. Typically we consider parametric curves and allow them to move toward desired features, usually edges, under the influence of *external* and *internal* forces.

*The authors gratefully acknowledge the support from the Faculty for Electronics and Information Technology, Warsaw University of Technology.

There are two key problems with parametric active contours. First, if the initial contour is not sufficiently close to the true boundary it will likely converge to a wrong result. The basic idea is to increase the capture range of the external force fields and to guide the contour toward the desired boundary (Cohen and Cohen, 1993). The second problem is that active contours only partially progress into boundary concavities (Abrantes and Marques, 1996). In our work we adapt the approach of GVF snakes (Xu and Prince, 1998), in which external forces for active contour models take the form of gradient vector flow (GVF) fields. Particular advantage of the GVF snake over a traditional snake is its ability to move into boundary concavities.

The final contour should be interpreted in terms of a 3-D object. In this paper we compute two different active contours and analyze their differences in terms of finger and palm detection, i.e. we analyze the waveform of a 1-D difference function, that expresses contour point distances from the contour's center of mass (Kauppinen et al., 1995).

2 Color-based ROI detection

The sensor data is initially given in a 24-bit RGB color scheme. It is evident that this color space is not well designed for color-based object recognition. Among different alternative color spaces, like YUV, HSV and YCbCr, we found the YCbCr space to be most suitable for our task. In this scheme the Y component represents intensity, the Cb component - the "chromatic distance" of given color from pure blue color, and the Cr component - such a distance from pure red color.

We applied a normalization of the Cb and Cr components driven by the intensity Y normalization to a predefined value (let us fix it to 128). Due to this procedure we can achieve very stable and narrow intervals around 100 (for Cb) and 150 (for Cr) that represent the skin color.

$$\begin{bmatrix} Y_{norm} \\ Cb_{norm} \\ Cr_{norm} \end{bmatrix} = \begin{bmatrix} 128 \\ (Cb - 128) * 128/Y + 128 \\ (Cr - 128) * 128/Y + 128 \end{bmatrix} \tag{2.1}$$

Without the intensity-based normalization a color-based skin detection may sometimes fail, especially in difficult lighting conditions. If shadows or strong lighting fall onto the shin surface then the colors significantly change. Also in case of a very structured background containing colors similar to the skin color, detection problems may appear. Obviously other color correction solutions may be tried, like for example density or luminance regularization (Chai and Nagan, 1999).

3 Active contour detection

Snakes (active contours) (Kass et al., 1988; Berger, 1990; Osher and Paragios, 2003) are curves defined in image plane that can change their shape and can move under the influence of forces. These are decomposed into *internal* forces, as a result of expected curve's *stiffness*, and *external* forces computed from the image data and distributed over the image. After the snake is initialized inside-of or outside-of an object boundary it is expected to evolve to this boundary while being controlled by these forces. The internal forces are designed to hold the curve together (*elasticity* forces) and to avoid a large bending (*bending* forces).

We consider parametric curves and allow them to move toward desired features, usually edges, under the influence of *potential* forces, which are defined to be the negative gradient of a potential function. The active contour is defined in the image plane as a sequence of control points $p_i = (x_i, y_i)$, $(i = 0, ..., n-1)$ and connecting line segments $L_i = [p_i, p_{i+1}]$. During contour evolution the total energy of points is minimized.

3.1 The GVF snake

The external energy E_{ext} is traditionally estimated as a potential energy (the sum of image function $I(x, y)$) measured in control points:

$$E_{\text{ext}} = K_0 * \sum_{i=0}^{n-1} I(x_i, y_i), \tag{3.1}$$

where K_0 is a scaling coefficient. The appropriate external force is computed as the negative gradient of this potential energy function ($F_{\text{ext}} = -\nabla E_{\text{ext}}$). It pulls the contour toward the desired image edges.

We have applied another approach for the computation of external forces, called *gradient vector field* (GVF) (Xu and Prince, 1998), that allows us better than before to move the snake into boundary concavities. The GVF is a dense vector field, $\mathbf{v}(x, y) = [u(x, y), v(x, y)]$, derived from the image by minimizing a certain energy functional in a variational framework (Figure 1). The energy functional is similar to the one formulated for the computation of optical flow (Horn and Schunck, 1981):

$$\mathcal{E} = \int\int m(u_x^2 + u_y^2 + v_x^2 + v_y^2) + |\nabla f|^2 \cdot |\mathbf{v} - \nabla f|^2 dxdy \tag{3.2}$$

where $f(x, y) = \nabla I(x, y)$ is the gradient image and m is some weight parameter.

The minimization of $\mathcal{E}$ is achieved by solving a pair of decoupled linear partial differential equations that diffuses the gradient vectors of a gray-level or binary edge map computed from the image.

The internal energy E_{int} consists of two components:

$$E_{\text{int}} = E_{\text{elastic}} + E_{\text{stiffness}} \tag{3.3}$$

The *elasticity* energy is proportional to the squared contour's length:

$$E_{\text{elastic}} = K_1 * \sum_{i=0}^{n-1} |p_i - p_{i-1}|^2, \tag{3.4}$$

where K_1 is an elasticity coefficient and the p_i-s are the contour's control points. The *elasticity* force tends to shorten the line segments, i.e. to move the control points closer to their neighbors. The *stiffness* energy is proportional to the squared curvature measured in control points:

$$E_{\text{stiffness}} = K_2 * \sum_{i=0}^{n-1} |p_{i-1} - 2p_i + p_{i+1}|^2, \tag{3.5}$$

where K_2 is an stiffness coefficient. The *stiffness* force tends to lower the curvature, i.e. it avoids bending the contour without shortening it.

The appropriate elasticity and stiffness forces can be derived by using the energy gradient (i.e. $\mathbf{F} = -\nabla E$). In the final stable state of the contour the balance of forces for given contour in given image is achieved:

$$\mathbf{F}_{\text{int}} + \mathbf{F}_{\text{ext}} = 0 \tag{3.6}$$

We repeatedly apply for every control point p_i the pair of equations:

$$x_i = x_i + a \cdot F_{\text{elastic}}(X, i) + b \cdot F_{\text{stiff}}(X, i) + g \cdot F_{\text{ext}}(X, i) \tag{3.7}$$

$$y_i = y_i + a \cdot F_{\text{elastic}}(Y, i) + b \cdot F_{\text{stiff}}(Y, i) + g \cdot F_{\text{ext}}(Y, i) \tag{3.8}$$

where a, b, g are some weight parameters and $F(X, i)$ or $F(Y, i)$ denote the force component along the X axis or Y axis, respectively, at control point p_i.

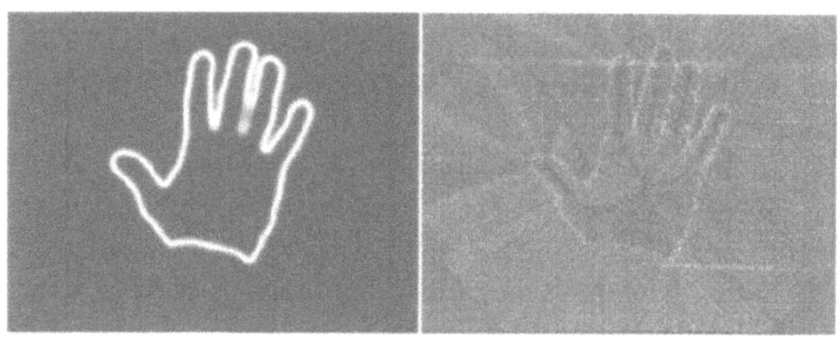

Figure 1. The gradient image (on the left) and the gradient vector field (with $m = 0.2$) (on the right) computed from a hand image with a uniform background.

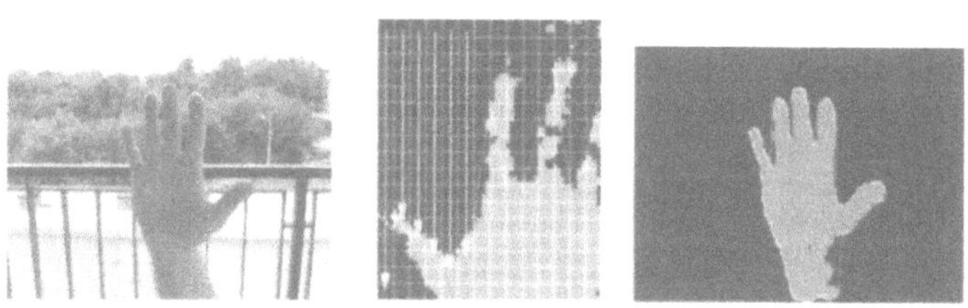

Figure 2. The detection of initial active contour in the segmented image: (a) sample image, (b) the detection of consecutive initial contour points, (c) examples of a well-detected initial contour.

3.2 Initialization of the active contour

The input data to this step is the binary image containing skin-color pixels. The binary image is divided into 3x3 or 4x4 blocks (the block size is a parameter). At first the blocks are examined from left to right and from top to down. If a block contains a sufficient amount of skin pixels (let us call it the "block filling" threshold) then the next

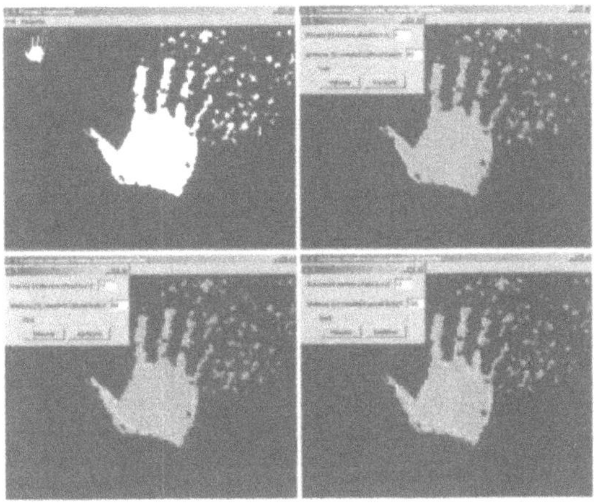

Figure 3. The detection of initial active contour in a deteriorated segmentation image - its dependence from the block size and the "block filling" threshold.

active contour point is located in the left upper corner of this block. The search moves to the next column on the right side. This cycle is iterated until the end of image or a no-skin column is found (see Figure 2). Next, the blocks are examined from right to left and from bottom to up. An eventual control point of the active contour is now located in the bottom right corner of the examined block.

There is no guarantee that the initial contour matches sufficiently exact the hand shape. In a deteriorated segmentation image (Figure 3) different block sizes and "block filling" parameters may lead to correct or quite incorrect initial contours.

3.3 Outer (i.e. hand) and inner (i.e. palm) contour detection

Two active contours are computed out from the same initialization. In the image of size 640×480 in our experiments the hand covered the image part of 7 - 20% . For the first (outer hand) contour the *snake* parameters were set to: $a = 0.8, b = 0.1, g = 0.6$. The second contour should detect the palm area of the hand, i.e. without covering the fingers. In this case we changed the parameter g to 0.15. Under these conditions the *elasticity* force is more dominant than in first case and this leads to a shorter contour than the initial one (Figure 4).

4 Contour difference description

At first a new relocation of contour's control points is established due to an approximation of existing points. The new control points are uniformly distributed and its number is fixed. Next the center of mass of the inner contour is determined. This center point together with the direction of the longer boundary box of the outer contour allows for a first contour post-processing step - the elimination of a forearm part (Figure 5).

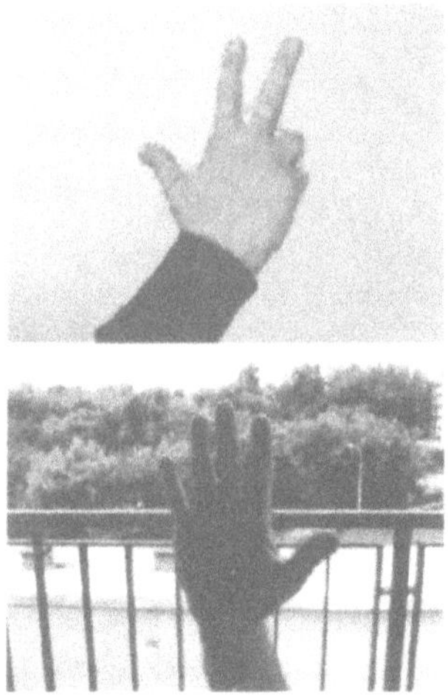

Figure 4. Typical results of double active contour detection.

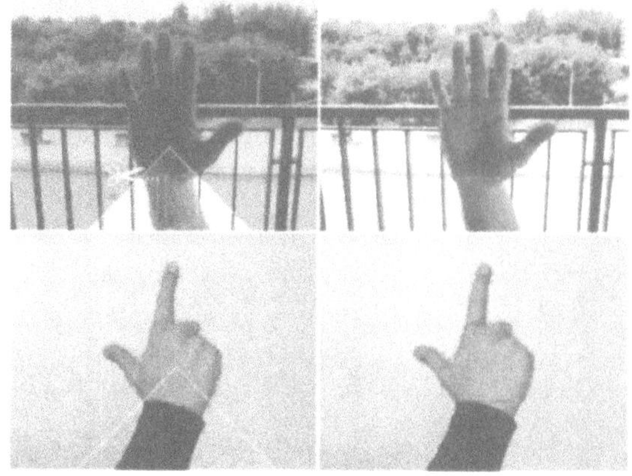

Figure 5. The elimination of a forearm part from both contours: (left column) the part contained in the triangle area is eliminated, (right column) the post-processed contours.

The distances of all contour points from this mass are computed, starting from the point that is located lowest in the image and nearest to the image column, in which the center of mass is located. We obtain two 1-D distance distributions. These distributions are subtracted one from the other, length-normalized and all negative values are reset to zero. In the resulting function the visible tips of fingers correspond to local maximum points (Figure 6).

5 Hand instance detection and sign interpretation

Finally a hand instance detection on base of two contour descriptions can be performed and this hand instance can then be interpreted in terms of a specific sign from a set of 21 static signs. The implicit hand model consists of a box representing the palm and 5 cylinders that represent the 5 fingers.

Current hand instance is generated out from the hand contour description in three steps. The center of mass and other specific boundary points of the inner contour determine the location and size of the palm (Figure 7(a)). The fingers are detected on base of the locations of finger tips with respect to the projection point of the center of mass onto the palm rectangle side, opposite to the tips (Figure 7(b)). The thumb is detected among the points if its tip projection onto the palm rectangle is located below the mass center point (Figure 7(c)).

The relation of inner contour's height-to-width induces the front or side-view, while the height-to-height relation for both contours induces the fraction of top-view.

Figure 6. Example of the difference between distance-to-mass center values for both active contours. We have here 5 easy-detectable extremum points along the horizontal axis.

Figure 7. The detected hand instance and its corresponding sign: (a) palm detection, (b) finger detection, (c) thumb detection, (d) the GUI of the application showing the final sign and hand description.

Table 1. The set of 21 signs and their codes corresponding to hand instances.

Sign	Thumb position	Detected fingers	Hand view
A	No	closed (0)	Front/Back
B	No	5in1 (0)	Front/Back
C	No	1	Front/Back
D	No	2	Front/Back
E	No	3	Front/Back
F	No	4	Front/Back
G	No	5in1 (0, 1)	Side
H	Left	thumb (1)	Front/Back
I	Left	4in1 + thumb	Front/Back
J	Left	2	Front/Back
K	Left	3	Front/Back
L	Left	4	Front/Back
M	Left	5	Front/Back
N	Left	4in1 + thumb	Side
O	Right	thumb (1)	Front/Back
P	Right	4in1 + thumb	Front/Back
Q	Right	2	Front/Back
R	Right	3	Front/Back
S	Right	4	Front/Back
T	Right	5	Front/Back
U	Right	4in1 + thumb	Side

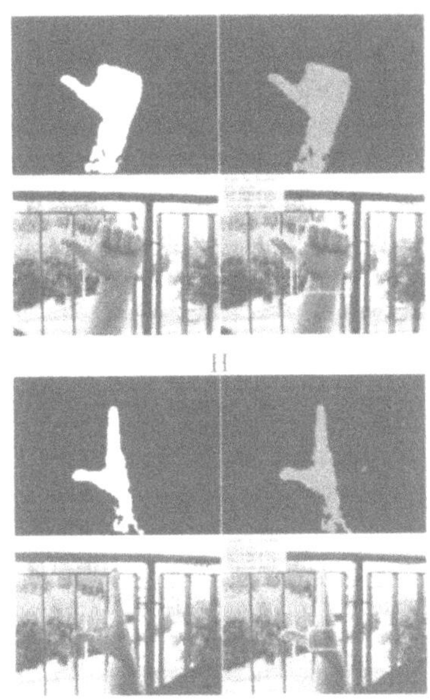

Figure 8. Some steps of sign detection of letters H and N.

6 Test examples

We have experimentally verified that our hand description can in an unique way be transformed into a set of 21 signs (Table 1). In Figure 8 examples of intermediate steps of our hand sign detection approach are shown. If the fingers are well separated one from the other then the first active contour is progressing in-between them. The second active contour is then matching the boundary palm area of the hand. Otherwise, both contours are similar, except of the visible thumb position.

Bibliography

J. M. Rehg, and T. Kanade. Digit Eyes: Vision-Based Human Hand Tracking. *Report CMU-CS-93-220*, School of Computer Science, Carnegie Mellon University, 1993.

D. Yining, B. S Manjunath, and H. Shin. Colour image segmentation. *Computer Vision and Pattern Recognition, IEEE Computer Society Conference*, 2:2446–2451, 1999.

C.-S. Fu, W. Cho, and S. K. Essig. Hierarchical colour image region segmentation for content-based image retrieval system. *IEEE Transactions on Image Processing*, 9:156–162, 2000, No.1.

M. Kass, A. Witkin, and Terzopoulos D.. Snakes: Active Contour Models. *International Journal of Computer Vision*, 18:321–331, 1988, No. 1.

M. O. Berger. Snake growing. *Lecture Notes in ComputerScience*, 427:387–396, 1990.

S. Osher, and N. Paragios, Editors. *Geometric Level Set Methods in Imaging, Vision, and Graphics*. Springer Verlag, New York, 2003.

H. Kauppinen, T. Seppanen, and M. Pietikainen. An Experimental Comparison of Autoregressive and Fourier-Based Descriptors in 2D Shape Classification. *IEEE Trans. Patt. Anal. Mach. Intell.*, 17:201–206, 1995, No. 2.

R. Sanchez-Reillo, C. Sanchez-Avila, and A. Gonzalez-Marcos. Biometric Identification through Hand Geometry Measurements, *Biometrics*, 22, 2000, No. 10.

C-Y. Xu, and J. L. Prince. Snakes, Shapes and Gradient Vector Flow. *IEEE Trans. on Image Processing*, 7:359–369, 1998, No. 3.

L. D. Cohen, and I. Cohen. Finite-element methods for active contour models and balloons for 2-D and 3-D images. *IEEE Trans. Pattern Anal. Machine Intell.*, 15:1131–1147, 1993, No. 11.

A. J. Abrantes, and J. S. Marques. A class of constrained clustering algorithms for object boundary extraction. *IEEE Trans. on Image Processing*, 5:1507–1521, 1996, No. 11.

D. Chai, and K. N. Nagan. Face Segmentation Using Skin Color Map In Videophone Applications. *IEEE Trans. on Circuits and Systems for Video Technology*, 1999. (www.soem.ecu.edu.au/ dchai/public/papers/ csvt99.ps).

B. K. P. Horn, and B. G. Schunck. Determining optical flow. *Artificial Intelligence*, 17:185–203, 1981.

Mobile robot localization using laser range scanner and omnicamera

Mariusz Olszewski [†‡] Barbara Siemiatkowska, [*‡] Rafal Chojecki [†‡] Piotr Marcinkiewicz [†‡] Piotr Trojanek [†2‡] Marek Majchrowski [†2‡]

[*] Institute of Fundamental Technological Research,
Polish Academy of Sciences,
Swietokrzyska 21, 00-049 Warsaw, Poland
e-mail: bsiem@.ippt.gov.pl

[†] Warsaw University of Technology
Institute of Automatic Control and Robotics
sw. Andrzeja Boboli 8, 02-525 Warsaw, Poland

[†] Warsaw University of Technology
Institute of Control and Computation Engineering
Nowowiejska, 15/19, 00-665 Warsaw, Poland

Abstract. In this paper a method of mobile robot localization in an unknown indoor environment is presented. The robot is equipped with a single omni-directional camera and a laser range finder. The data of the 2D scanner is used to detect walls in the robot environment. The images taken from omnicamera allow to detect texture on the walls. The robot's position can easily be estimated by combining information taken from the camera and from the laser scanner and this data can be used to follow the robot's path. The method has been tested with the use of mobile robot *E*LEKTRON R1 in a real office environment.

1 Introduction

Autonomous mobile robot has to be able to determine its position and orientation while moving and the most fundamental part of navigation systems is self-localization. The pose of a robot is denoted by a triple (x, y, θ), where (x,y) is the position of the robot and θ is its orientation. Until now numerous localization methods have been proposed (Thrun et al., 1998), (Jones and Crowley, 1999), (Betke and Gurvits, 1997), (Borges and Aldon, 2003), (Davison and Murray, 1998). The most widely used is odometry but error in determining the position of the robot increases proportionally to the distance traveled by the vehicle. In order to improve localization additional methods should be used. The methods belong to one of two classes:

- *R*elative - the current pose p_t of the robot is determined relative to the previous one p_{t-1}.
- *G*lobal - the position and orientation of the robot is computed within a priori given exact map or a known scene .

In this paper relative localization technique is considered but proposed techniques can be also used for global localization.

Usually, an on-board laser range finder is used in order to trace the position of a robot. If the robot moves along a corridor the orientation of the robot and its displacement in the direction perpendicular to the wall can be computed very accurately but it is difficult to determine the displacement of the vehicle paralelly to the walls. In order to improve the localization technique additional sensor should be used.

In our experiments the mobile platform is equipped with the omni-directional camera with hyperbolic mirror. The main advantage of omni-directional sensors is that they give a 360-degrees view of the robot's environment and they simplify the interpretation of a scene. Based on images taken from an omnicamera the angular information about position of the object relatively to the robot is available immediately but the distance can be estimated after calibrating of the sensor. In our approach the calibration is performed using the method described in (Siemiatkowska and Chojecki, 2004). Figures 1a), 1b) present the readouts of the laser range finder taken in the corridor in two different positions. The displacement of the robot equals 40cm. Figure 2a), 2b) present the images taken from the omnicamera.

Analysis of the images allows to determine the displacement of the robot pararelly to the walls.

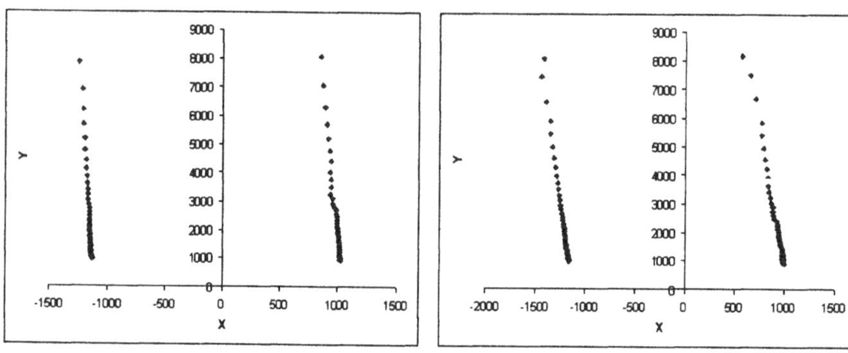

Figure 1. Laser readings

The method presented in this paper is a development of the algorithm proposed in (Siemiatkowska and Chojecki, 2004) and consists of following steps:

1. The omnicamera sensor is calibrated. The method of calibration is described in (Siemiatkowska and Chojecki, 2004);
2. Readings from laser range finder and omni-camera images are collected;
3. 2D map of an environment based on laser readings is build. The map is represented as a set of segments.
4. The images taken from omni-camera sensor are analysed. The images of the walls are transformed.

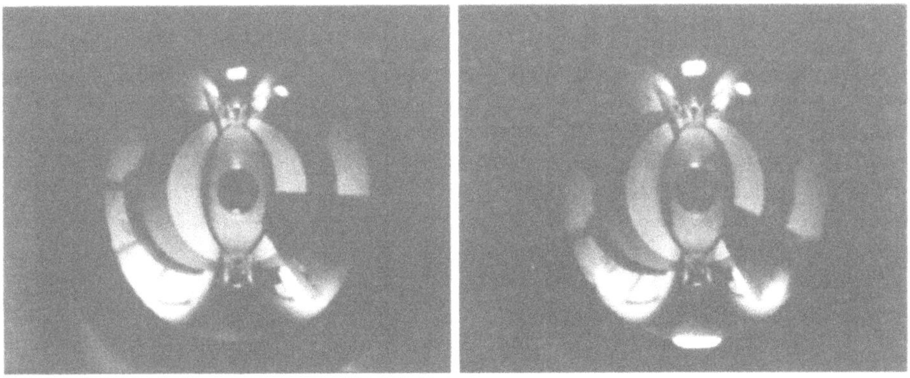

Figure 2. Image of the wall taken from omnicamera

5. The displacement along the wall is computed.

2 Wall detection

The method of building a 2-D map of environment is described in (Thrun et al., 1998), (Siemiatkowska and Dubrawski, 1998).

The method of walls detection based on laser scanner readings consists of two stages:

- Firstly the modification of Hough transform is used in order to built sets of collinear points. A line segment is described using normal notation:

$$x \cdot cos\alpha + y \cdot sin\alpha = c \tag{2.1}$$

  where c is the distance from the origin to this line computed along a normal, and α is the orientation of the normal with respect to the X axis. For points (x_i, y_i), which belong to the same line, the parameters (α, c) are the same.
- Secondly in order to improve the precision of computed values of parameters we look for c and α which minimize the error:

$$\delta = \sum_{i=0}^{N} (R_i \cdot cos(\alpha - \phi_i) - c)^2 \tag{2.2}$$

  where N is the number of collinear readings, R_i is the distance to the nearest obstacles in direction ψ_i. The straight line is indicated by the pair of parameters (c^*, α^*) for which the value of δ is minimal. The method described above allows to determine the orientation of the robot and its displacement in direction perpendicular to the wall.

When the robot acts in the corridor the error of determinig the robots orientation does not exceed 0.3^o and the error of determining the displacement in the direction perpendicular to the wall 0.2cm. The error does not increase in time.

3 Cosine transform

The discrete cosine transform (DCT) is a technique for converting a signal into elementary components. It is widely used in image compression. It was developed by Ahmed, Natarajan and Rao (Ahmed et al., 1974). Chen and Pratt (Chen and Pratt, 1984) started to use DCT for image compression. 1D sequence of length N is defined by equation:

$$C(u) = \alpha(u) \sum_{x=0}^{N-1} f(x) cos(\frac{\pi(2x+1)u}{2N}) \tag{3.1}$$

for u = 0, 1,2,...,N-1. $\alpha(u)$ is defined as:

$$\alpha(u) = \begin{cases} \sqrt{\frac{1}{N}} & for \quad u = 0 \\ \sqrt{\frac{2}{N}} & for \quad u \neq 0 \end{cases} \tag{3.2}$$

The inverted cosine transform is defined by:

$$f(x) = \sum_{n=0}^{N-1} \alpha(u) C(n) cos(\frac{\pi(2n+1)x}{2N}) \tag{3.3}$$

When cosine transform is applied to an input signal, it yields a vector of weighted values corresponding to the amount to which each basis function is present in the original signal. For most cases, much of the signal energy lies at low frequencies.

4 Displacement detection along the wall

The positions of observed walls are acquired by the laser scanner. Inverse perspective transform (Adorni et al., 2003) can be used to find texture of wall based on omnidirectional image. The angular image coordinates (ψ, β) are transformed to wall coordinates (w, h). where w - horizontal position, h - vertical position.

Transformation is given by

$$w = \frac{d}{tg(\psi)} \tag{4.1}$$

$$h = \sqrt{\frac{tg(\beta)}{d^2 + w^2}} \tag{4.2}$$

where d is distance to wall measured by laser scanner. Figure 4a) presents the fragment of the image taken from the omnicamera. Figure 4b) presents transformed image of the wall.

The features from the walls are projected onto horizontal edges. For each vertical line of the wall the 1D cosine transform is performed. The weighted values of the cosine transform describe a texture of a line and they are the features which are projected. The lines of similar texture correspond to similar weighted values. In our experiment 8 values are used.

Considering transformed images of the walls for two different positions of the robot we can determine the robot displacement along the wall by comparing the position of the areas of similar texture.

To figure out best fit movement along the wall, one should find minimum of the error given by following formulae:

$$E(x,y) = \sum_{\hat{x}=0}^{M-1} [i_1(\hat{x}) - i_2(x+\hat{x})]^2 \tag{4.3}$$

i - is a vector of feature of a line $\hat{x}$. M is the number of vertical lines in the image of the wall.

When omnicamera sensor is used the resolution depends on the distance between the center of the camera and the observed region. In order to reduce an impact of errors we are searching for minimum of a function:

$$E(x) = \sum_{\hat{x}=0}^{M-1} a_i [i_1(\hat{x}) - i_2(x+\hat{x})]^2 \tag{4.4}$$

where coefficient a_i qualify reliability of camera readings. The values are computed during calibration process. In our approach the value of a_i is described by formulae 4.5.

$$a_i = \frac{1}{1+R_i^2} \tag{4.5}$$

Where R_i is the distance between corresponding point on the wall and the robot position.

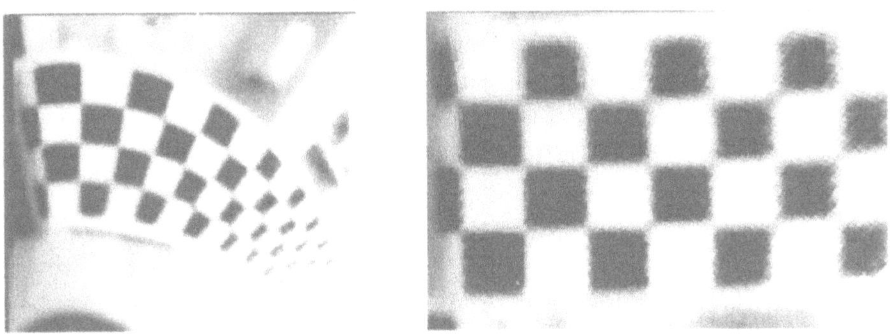

a) b)

Figure 3. Transformation of the image

5 Experimental results

In performed experiments the robot ELECTRON 1 is used 5. The robot was developed in the Institute of Automatic Control and Robotics and Institute of Control and Computation Engineering of Warsaw University of Technology. The six wheels mobile platform is equipped with on-board Biscuit PC computer (Celeron 650 MHz). Communication between the robot and the host computer is done via wireless Ethernet. The robot is powered by two reversible DC motors. It can move with an approximate maximum speed of 0,2m/s. The vehicle is equipped with odometry sensors, 18 optical distance sensors, SICK LMS 200 laser scanner and a panoramic camera. The omni-directional sensor is composed of C-MOS color camera pointed upwards at the vertex of the hyperbolical mirror. The optical axis of the camera and the optical axis of the mirror are aligned.

Figure 4. The mobile robot ELECTRON 1

1) catadioptric sensor, 2) camera, 3) laser scanner, 4) robot

Experiments were performed in the corridor. The error of determining the robot displacement along the wall depends not only on the distance between the robot and the wall but also on the texture of the wall. The method is useless when there are not any characteristic area on the wall.

The result depends also on the distance between the robot and characteristic places on the wall. Figures 5-6 present the result of our experiment. The task for the robot is to determined its distance to the center of the door. Error of determining robots position was computed for three different paths of the robot. In the first situation approximated distance between the robot and the wall equals 0.5m, in the second one the distance equals 1m and in the last one 1.5m.

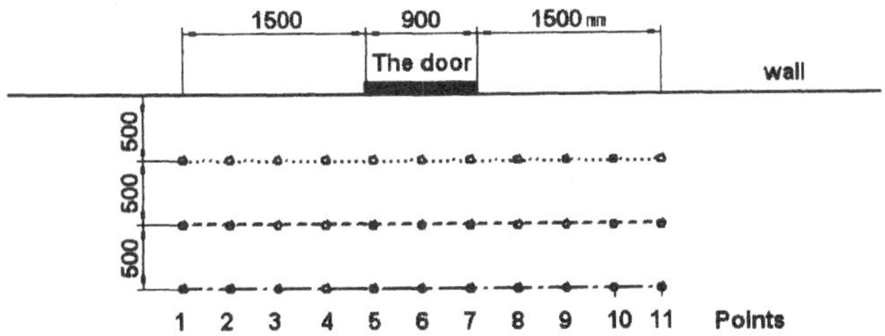

Figure 5. The path of the robot

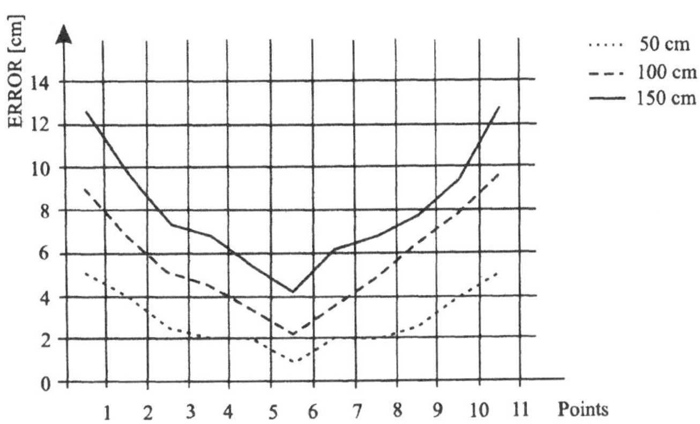

Figure 6. Error of determining the position of the robot

6 Conclusion

The main motivation of the work presented in this article was to built the system for mobile robot localization. The environment is represented as 2D map of walls. Each wall is described using linear cosine transform. Values of coefficients of cosine transform are the features which are traced. The performed experiments proved the efficiency of the proposed method as a tool for local navigation. Presented method can be used not only as a localization technique but also as a tool for 3D map building.

Experiments were supported by grants MEiN 3 11C 38 29 and 1525/T11/2005/29.

Bibliography

G. Adorni, S. Cagnoni, M. Mordonini, and A. Sgorbissa. Omnidirectional stereo systems for robot navigation. In *Proceedings of IEEE Workshop on Omnidirectional Vision*, pages 79–89, 2003.

N. Ahmed, T. Natarajan, and K. R. Rao. On image processing and a discrete cosine transform. *IEEE Transaction on Computers*, pages 90–93, 1974.

M. Betke and L. Gurvits. Mobile robot localization using landmarks. *IEEE Transactions on Robotics and Automation*, pages 135–142, 1997.

G.A. Borges and M. J. Aldon. Robustified estimation algorithms for mobile robot localization based on geometrical environment maps. *Robotics and Autonomous Systems*, 45:131–159, 2003.

W. H. Chen and W. K. Pratt. Scene adaptive coder. *Transaction on Communication*, pages 225–232, 1984.

A.J. Davison and D. W. Murray. Mobile robot localization using active vision. In *Proceedings of European Conference on Computer Vision*, pages 809–825, 1998.

C.S. Andersen S. Jones and J.L. Crowley. Appearance based processes for visual navigation. In *Proceedings of Symposium on Intelligent Robotics Systems*, pages 227–236, 1999.

B. Siemiatkowska and R. Chojecki. Mobile robot localization based on omincamera. In *Proceedings of IAV04*, pages 123–126, 2004.

B. Siemiatkowska and A. Dubrawski. Cellular neural network for navigation of a mobile robot. In *Proceedings of RSTC'98*, pages 201–211, 1998.

S. Thrun, D. Fox, and W. Burgard. Probabilistic mapping of an environment by a mobile robot. In *Proceedings of ICRA*, pages 1546–1551, 1998.

Navigation of an Autonomous Ground Vehicle - Gate Recognition and Crossing

Yi Lu[1], Vladimir Polotski[2] and Jurek Sasiadek[1]

[1] Department of Mechanical and Aerospace Engineering, Carleton University, Ottawa, Ontario, K1S 5B6, Canada, e-mail: jsas@ccs.carleton.ca
[2] Frontline Robotics, Ltd. of Ottawa, Ontario, Canada

Abstract: Guiding an autonomous ground vehicle through the gate based on the range data from the laser scanner is considered. This kind of operation is important for security mobile robots. Information about the gate shape is supposed to be more accurate than the geographical location of the gate. This necessitates the transition from GPS-based guidance in the open field to range-based guidance in the proximity of the gate. Recognition procedure based on the proposed concept of the gate signature is followed by Map-matching and Localization. Navigation controller makes use of the polar coordinate representation and discontinuous feedback adapted for real time applications.

1 Introduction

Autonomous navigation in outdoor environment can be effectively performed by small robots but remains a challenging task as soon as real scale ground vehicles are concerned. The problem of gate recognition and crossing addressed in this paper is a part of a larger outdoor navigation project mainly oriented to security applications such as patrolling critical areas, namely: airports, gas plants, etc. Sensing equipment for outdoor navigation usually consists of (1) basic devices providing absolute geographical position such as compass, wheel encoders, GPS receiver, and (2) additional sensors perceiving the vehicle surroundings. Video cameras and/or range measuring laser scanners are most often used for this purpose. A typical mission for an autonomous security vehicle consists of patrolling a predefined area. During the patrol task, GPS/INS/odometry provides basic measurements for global scale navigation along a predefined path. Range-sensor is primarily used for detecting unknown objects and triggering avoidance maneuvers. A few meters precision of basic localization is not sufficient for guiding through the gate or narrow passage between two buildings, where typical width is of 4-6 meters and required precision is of 0.3-05 m. Considering the above reasons, combining GPS with other navigation technologies would be very effective. In fact this integration of sensors was found to produce very reliable autonomous operations in some proven mobile robotic navigation systems [See Ye and Borenstein (2003), Vale, et al. (2004), Sasiadek and Wang (2003)]. The gate crossing problem addressed here consists of detecting the entrance using a proximity sensor and then reactively navigating through this entrance. The gate-crossing problem differs from range-based wall following, since it requires the transition from open field essentially GPS or fusion-based navigation to range-based navigation. This transition itself requires a feature extraction and recognition phase usually not necessary in the environments as underground mines, where range-based wall following is sufficient [See Bakambu, et al. (2000), Roberts, et al. (2000)].

2 Problem Description

The problem to be addressed consists of guiding a vehicle through the area delimited by two objects of known shape called *posts* located at the known distance from each other (*entrance width*). The geographical location of the whole structure (called *gate*) is supposed to be only roughly known. This means that the longitude/latitude of the *entrance* center point can be transformed to its position on the area map with few meters accuracy, but the geometry of the gate can be known with few centimeters accuracy. The typical task to be performed consists of navigating along the path that leads to the gate area and passing through the gate. Typical entrance width is *6 m*, typical size of the post is about *1 m*, experimental vehicle used for a proof of concept stage is about *2 m* wide and *3.5 m* long. We are currently using an ARGO conquest 6x6 platform manufactured by Ontario Drive & Gear Ltd. (ODG) retrofitted, for computer control and equipped with NovAtel GPS, MicroStrain inertial sensor, built-in wheel odometry, and LMS-221 laser scanner (manufactured by SICK). The angular resolution of the scanner was set to set to $0.5°$. The maximum range is *80* meters, although we have limited it to *30 m* in order to get reliable and meaningful data. The scanner provides the sequence of range values (indexed according to a scanning angle). The data is transferred to the controlling computer by RS422 interface.

3 Zone and Signature Concept

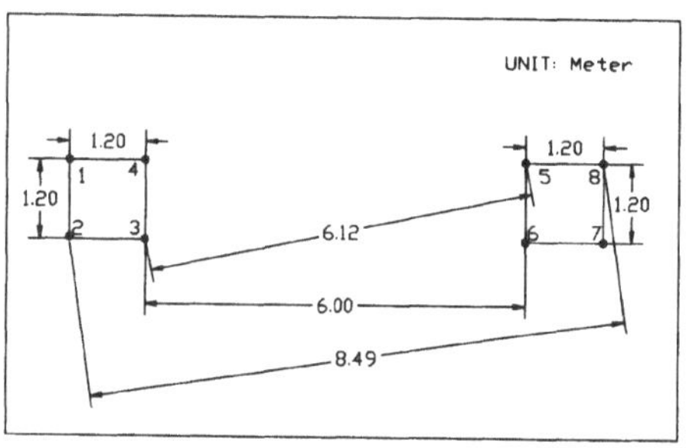

Figure 1. Gate dimension and vertex points

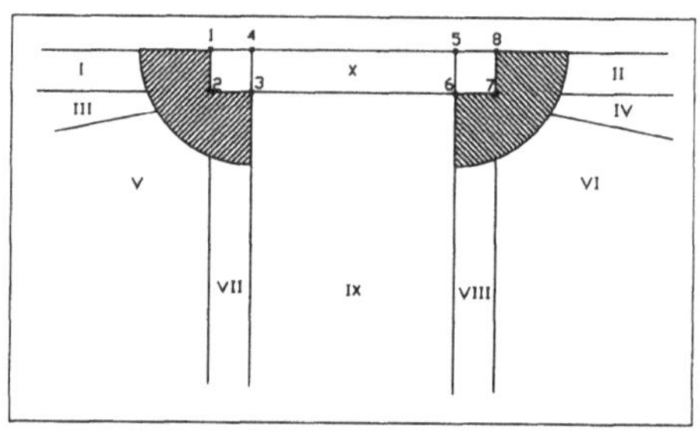

Figure 2. Proximate gate zones for navigation.

In order to recognize the gate using information provided by the LMS scanner, we have to address first the perception of the gate geometry by the sensor. We formalize our description dividing the neighborhood of the gate into proximity zones and introducing the concept of the gate signature. For simplicity reasons our approach description refers to a particular setup used in our field experiments for which wooden boxes (cubes of *1.20 m*) have been used to construct the gate. The distance between the boxes – the gate entrance - is set to *6.00 m*. The vertices are indexed for reference as shown in Figure 1 (the edge is referenced by delimiting vertices). Depending of the position of the sensor the perception of the gate obtained using LMS scanner is different. To handle this difference, the gate area is partitioned into ten zones as illustrated in Figure 2. Scanning from some zones (I-IV, X in Figure 2) gives poor representations of the gate as illustrated by icons in the table 1. For Zone I and II there is only one visible edge, in zone III and IV the reflection degrades since the part of the surface is illuminated at a very small angle. Our control algorithm will try to avoid guiding through those zones I-IV, using zones V-IX instead. Zone X also gives a poor representation of the gate, but this zone is inevitably visited by the vehicle after it has almost crossed the gate area. Recognition is not an issue by that time – the gate is almost crossed but the special procedure is needed to estimate the relative pose of the vehicle in this zone.

In order to compute the vehicle's location relative to the gate, we have first to distinguish the zones from which the scan is obtained. The concept of the "gate signature" has been proposed. Signature vector is computed as follows: we consider the length of each observed segment, the distance between two continuous portions of the scan (gap), and the distance between the scan start point and the scan end point (size). An example of signature computation is given in Figure 3: there are four segments, ab, bc, de, and ef; cd is a gap between two posts and af is the total width of the scan. A vector becomes $[ef, de, -cd, bc, ab, af]$. The negative sign of the third component emphasizes the fact that it corresponds to the gap and not to the edge.

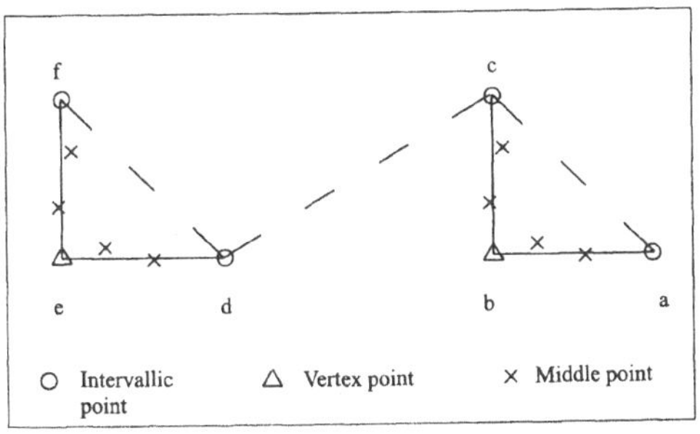

Figure 3. Signature of gate segment.

Table 1 shows all possible canonical signature vectors for the scans obtained from the zones VI, VII. One can see that even the dimension of the signature varies from zone to zone. Canonical signatures are computed for every zone. By comparing this on-line signature against the set of canonical zone signatures we may roughly determine the robot location with respect to the gate (corresponding zone) and then proceed to the gate recognition, and finally to the refinement of the localization result and guidance.

Table 1. Signatures of the gate scans.

Zone	Image	Vector	Dim
VI	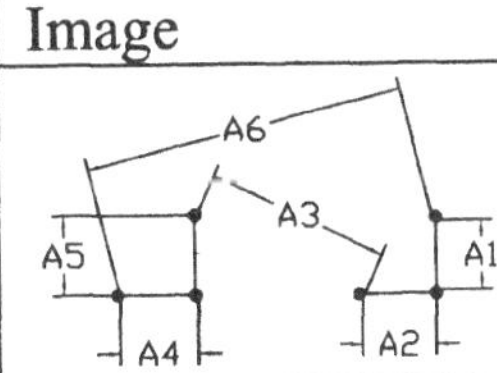	[A1,A2,-A3, A4,A5,A6]	1x 6

VII	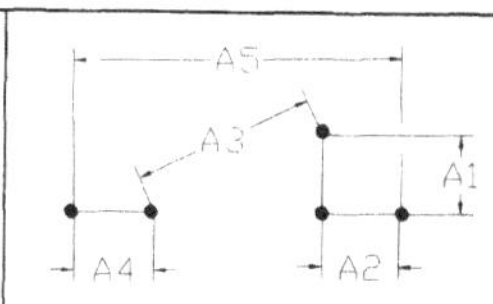	[A1,A2,-A3, A4,A5]	1x 5

4 Gate Recognition

This section describe an algorithm of gate recognition containing several modules: the GPS/INS/LMS data collection module, the map building module, the map filtering module, the map fitting module, the gate signature calculation module and the unexpected error check module. Data acquisition from GPS/INS/LMS is the first step of the algorithm. The map building module computes the map of the current environment using row data from LMS. Some of these objects are false positives, they do not correspond to real objects but to the noisy LMS measurements (physically attributed to multi-path reflections, dust, etc.). The encircled object is a gate. The map filtering module removes all the noisy points and non-gate objects from the preliminary map. It evaluates all the detected objects by their dimension and relative position keeping only objects corresponding to the gate posts. Only approximations for post size and entrance width are used for this computation (50% difference in size and width is tolerated). Figure 4 shows the filtered map. After the above steps, if the filtered map still contains more than two objects, the tolerance conditions for the dimension and distance are tightened and the filtering algorithm is re-applied. Each edge is delimited by vertex points and can be identified by those points. The vertex points are located on the border of the point cloud (have neighbors only on the right or only on the left hand side) and thus correspond to the corner of the object. When all the vertexes/corners are identified, the current signature for the scanned gate is constructed as explained in section III. Comparing this current signature with canonical signatures we assume that the gate is recognized if a canonical signature matches the current signature. For matching decision we use the squared norm of the difference between the current and canonical signature.

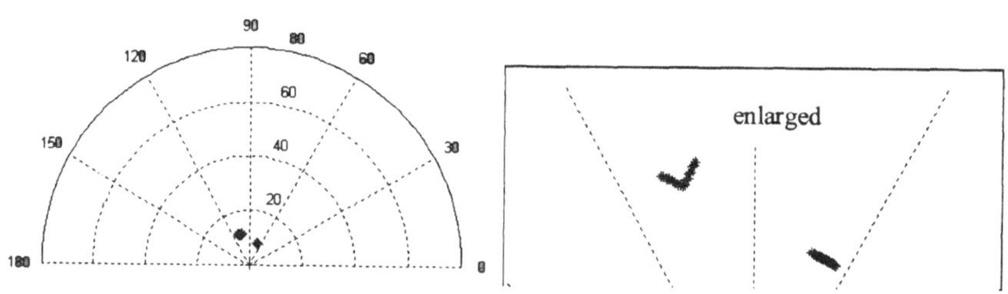

Figure 4. Filtered map and the enlarged gate object.

Let the canonical signature be:

$A_0 = [a_{01}, a_{02}, -a_{03}, a_{04}, a_{05}, a_{06}]$ and the current signature be:

$A = [a_1, a_2, -a_3, a_4, a_5, a_6]$. The criterion norm of $A - A_0$ is:

$$e = [(a_1 - a_{01})^2 + (a_2 - a_{02})^2 + (-a_3 + a_{03})^2 + (a_4 - a_{04})^2 + (a_5 - a_{05})^2 + (a_6 - a_{06})^2]^{\frac{1}{2}}$$

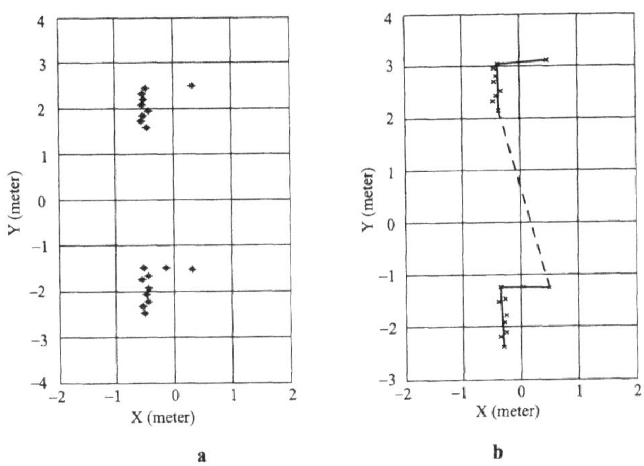

Figure 5. The image of gate.

5 LOCALIZATION AND CONTROL

The localization module outputs the data to the gate crossing control module. Here the final step of processing localization data is described. Let us define a coordinate frame $\{OXY\}$ attached to a vehicle with the original point O – being a point of the platform where LMS is located, OY – being a longitudinal axe of the vehicle aligned with 90° LMS beam. Figure 6 illustrates the layout. The gate recognition procedure is based on the extraction of points *{1, 2, 3, 4, 5, 6, 7, 8}*. Using this information the middle point C of the entrance and the direction of the straight line containing *{2, 3, 6, 7}* (from the left post to right post) can be estimated. More precisely, we are estimating the direction of the beam pointing from the left to the right post – let us call it entrance-direction P_{ENTER}. An angle (ψ) of this beam with OX axe is the first parameter to be output to the control low being designed. Let us defined a target point T somewhere shifted from C along the direction orthogonal to P_{ENTER} called gate direction P_{GTAE}. Second output parameter is an angle, α, between OT and OY. As can be easily seen for the vehicle perfectly aligned with the gate both angles are zero. In order to guide the vehicle through the gate one may apply either planning/correction approach or nonlinear control approach. We have chosen the latter, exploiting the similarity of the problem of guiding a vehicle towards and through the gate with the parking problem. In both cases we have to drive the vehicle to a desired pose (position/orientation). For the "gate through" problem the desired position is a middle point of the entrance segment and orientation – orthogonal to this segment. In contrast with original settings of [Bakambu, et al. (2000)], in our case the target point is not known in advance but has to be continuously estimated on-line. This constitutes a remarkable difference between gating and parking problems. Another difference is that we do not constrain the vehicle speed by zero at the target point, but rather set this speed to the desired value (usually lower than in the field). In order to address these differences, we (1) estimate the entrance middle point *C*, (2) we define a goal point T, (it could be any point located in the middle of the gate and little away from the rear side of the gate post,) as a point moving from the entrance middle point along the gate direction (see definition above). This procedure allows us to keep the vehicle offset from the goal point (singular point [Bakambu, et al. (2000)), thus ensuring a stable motion across the gate. An intermediate control output is defined as follows:

$u=K_1\alpha+K_2\psi$

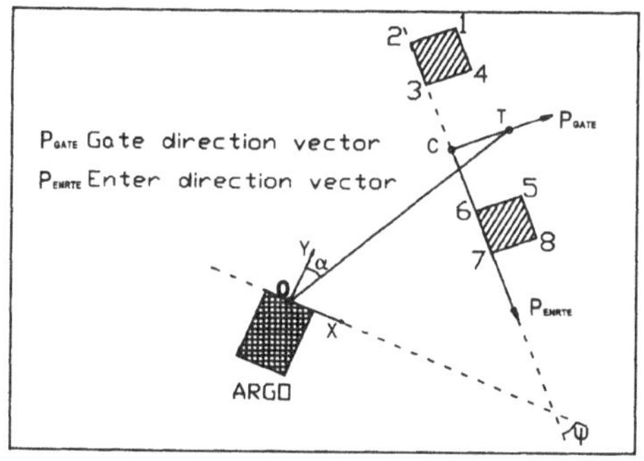

Figure 6. Localization and control method

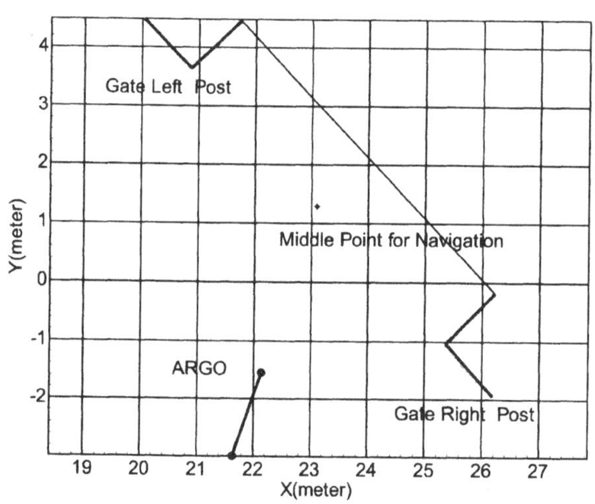

Figure 7. Gate image and the guide mid-point

where K_1 and K_2 are chosen using a pole placement technique. For our field experiments K_1 =0.9 and K_2 =-0.4. This corresponds to both poles set to *-1* on the Z-plane. In order to compute the final control output, a particular steering design of the ARGO vehicle (skid-steering) is taken into account. In order to avoid unnecessary reactions on the small offsets, which could lead to unstable behavior, a dead zone ($-u_{min}$, u_{min}) for u is introduced and final control output is computed as follows:

$$U = N(sign(u - u_{\min}) + sign(u + u_{\min}))/2$$

Here N stands for a value to be applied for initiating a vehicle turn. Figure 7 shows extraction of the middle point and entrance direction from the real LMS image.

6 RESULTS AND DISCUSSION

All the developed gate-recognition and guiding-through the gate modules have been first tested in simulation with synthetic data. In addition, the LMS data collected along the manual driving through the gate have been used to verify filtering and signature-based recognition modules, and finally, the algorithm has been fully integrated on-board and tested in the field. The map filtering module continuously monitors the environment based on the available LMS data. If the gate is actually farther than a soft-upper-limit range the filtered map is empty. Our experiments show that this period is usually about 1 second and never exceeds 3 seconds. As soon as the gate becomes consistently visible, the recognition module computes the gate signature, searches for a closest canonical signature, and estimates the vehicle pose relative to the gate. The localization module computes the gate middle point *C*, and gate direction P_{GTAE} and sends them to the motion control module, which calculates the control outputs *U* and send it to the hardware (low level controller). Another vehicle was parked in the field to prove the ability of filtering out additional (non-gate) objects. In spite of this uncertainty the LMS-based position of the vehicle relative to the gate is precise enough (0.1m, 3 deg) and a vehicle successfully navigates through the gate using our algorithm described above. Stable gate recognition is crucial for subsequent steps of the algorithm. The offset (2-norm of the difference) is between the current signature and a canonical signature (taken from the data base according to the zone where the vehicle is located). It may be seen that the error is reduced significantly in accordance with the distance to the approaching gate. The soft upper-limit range has been set based on theoretical and experimental analysis. If the posts are located too far apart although below the hard-range limit, the recognition results deteriorate. This can be roughly explained as follows: observing a small object (a singular point in the limiting case) we do not get enough information for positioning – lateral and orientation offsets are coupled. In practice the computation remains possible but measurement errors are amplified and results become unusable.

7 CONCLUSION

The performed experiments have demonstrated validity and usefulness of presented concepts for a single gate case. More experiments are needed to verify the conditions for the multi-gate case. The possible improvements of the recognition module would let the navigation algorithm recognize more than one gate and then make its decision according the requirement from the upper level control. The requirement may consist of an entry into a particular gate or the sequence of gates in any given sequence. Conditions for object recognition are difficult in some particular situations. Creating an environment map using multi-sensor information should be considered. In particular, multi-laser systems or a laser scanner linked with radar sensors, sonar, or video cameras ought to be considered. The system improvement with the multi-sensor and sensor fusion procedure would make the recognizing procedure more effective and the overall system more robust.

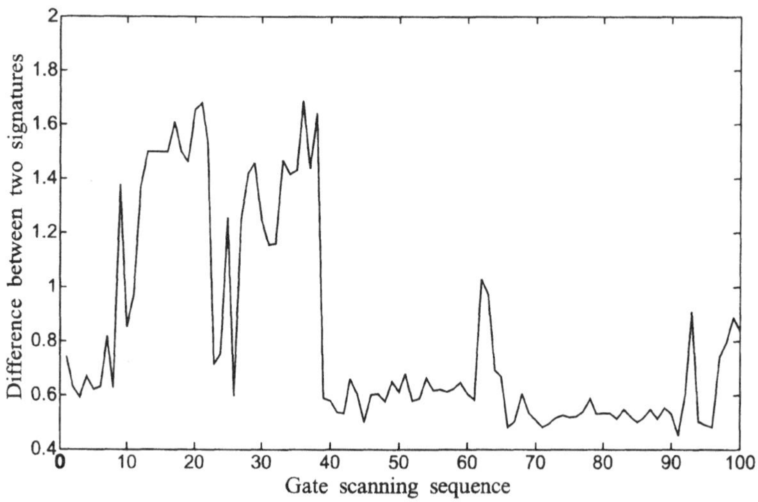

Figure 8. Recognition error.

Bibliography

C. Ye, J. Borenstein (2002), *Characterization of a 2-D Laser Scanner for Mobile Robot Obstacle Negotiation*, Proceedings of the 2002 IEEE ICRA, Washington DC, USA, 10-17 May, pp. 2512-2518.

A. Vale, J. Miguel Lucas, Maria Isabel Ribeiro (2004), *Feature Extraction and Selection for Mobile Robot Navigation in Unstructured Environments*, Proc/ of the 5th IFAC/EURON IAV, Lisbon, 5 - 7 July.

J. Z. Sasiadek and Q. Wang (2003) *Low cost automation using INS/GPS data fusion for accurate positioning*, Robotica, Volume 21, pp. 255-260.

J. Nsasi Bakambu, et al (2000), *Heading-Aided Odometry and Range data Integration for Positioning of Autonomous Mining Vehicles*, Proc. of IEEE Int. Conf. on Control Applications, Anchorage AK, Sept.

J. Roberts et al (2000), *Autonomous Control of Underground Mining Vehicles using Reactive Navigation*, Proceedings of the IEEE Int. Conference on Robotics and Automation, San Francisco, April 2000

K. Astolfi (1999), *Exponential Stabilization of a Wheeled Mobile Robot via Discontinuous Control, ASME* Journal of Dynamical Systems Measurements and Control, v.121, pp.121-125.

P. Goel, S.I. Roumeliotis, and G.S. Sukhatme (1999), *Robot Localization using Relative and Absolute Position Estimates* in Proc IEEE Int. Conference on Robots and Systems, October, Kyongju, Korea.

S.T. Pfister, et al. (2002), *Weighted Range Sensor Matching Algorithms for Mobile Robot Displacement Estimation*. In Proc. 2002 IEEE Int. Conference on Robotics and Automation, May, Washington DC

FSC
www.fsc.org
MIX
Papier | Fördert gute Waldnutzung
FSC® C083411

Kolibri 360 GmbH
Ferdinand-Jühlke-Straße 7
99095 Erfurt, Deutschland
produktsicherheit@kolibri360.de